Handbook of Experimental Pharmacology

Volume 209

For further volumes:
http://www.springer.com/series/164

Hans-Georg Joost
Editor

Appetite Control

Springer

Editor
Hans-Georg Joost
German Institute of Human Nutrition
Department of Pharmacology
Arthur-Scheunert-Allee 114-116
14558 Nuthetal
Germany
joost@dife.de

ISSN 0171-2004 e-ISSN 1865-0325
ISBN 978-3-642-24715-6 e-ISBN 978-3-642-24716-3
DOI 10.1007/978-3-642-24716-3
Springer Heidelberg Dordrecht London New York

Library of Congress Control Number: 2011945416

Printed on acid-free paper

Springer is part of Springer Science+Business Media (www.springer.com)

Preface

The marked and continuing increase in the prevalence of obesity during few decades is one of the most remarkable characteristics of Western civilization. It profoundly affects the society not only with its consequences for public health but also with its economical and sociological/cultural aspects. As has been demonstrated by artifacts such as the Venus of Willendorf as well as by historical reports, obesity occurred in almost all societies as a rare condition among the rich and privileged. However, in the last decades of the twentieth century, a marked increase in its prevalence commenced in the USA, in Europe, and subsequently, also in parts of Asia, reaching epidemic proportions. There is little doubt that this increase reflects the influence of an "adipogenic" environment: the unlimited availability of food in combination with the development of particularly palatable, less satiating foods (the so-called fast food), together with a mechanization of all tasks, causing an inactive, sedentary lifestyle. So far, all efforts to halt this epidemic by changes in diet and lifestyle had little success: Triglycerides lost during an intervention are rapidly regained once the individual returns to the previous food choice and eating behavior, and few individuals manage to continue their intervention indefinitely. This observation appears to reflect the influence of a powerful biological system that controls not only energy balance, i.e., caloric intake, but also food choice and nutritional preferences.

It is obvious that such a biological system controlling energy balance can comprise targets for a pharmacological intervention. Consequently, considerable efforts have been made in the past to elucidate the molecular basis of this system, to identify potential targets for intervention, and to discover agents inhibiting food intake by a reduction of appetite. As this volume will summarize, these efforts have led to a broad understanding of the mechanisms controlling food intake, to the identification of numerous targets for intervention, and also to several agents with appetite-inhibiting potency. However, so far, none of these agents have met the high safety standards required for the chronic treatment of obesity. Agents such as sibutramine and rimonabant that were in use for some time have been withdrawn because of unacceptable side effects. Thus, this volume cannot describe established and accepted therapeutics. Rather, its aim is to summarize the state of knowledge of potential target proteins and mechanism in order to point out promising directions of future research. Thereby, the volume addresses pharmacologists, physiologists, neuroendocrinologists, and all other scientists who are interested in the field of obesity research.

The first part of the volume presents a comprehensive description of the central mechanisms controlling food intake, and thereby energy balance, in the mammalian organism. During the last two decades, research in this area has produced a remarkable wealth of information and has characterized the function of numerous peptides and their receptors in a neuroendocrine network of appetite control. Chapters covering leptin, neuropeptide Y, MSH, AGRP, NMU, the orexins and neuromedins, and their respective receptors highlight this plethora of information. In addition, the central effects of insulin in modulating this network are covered in the first section. An often overlooked aspect of the obesity problem is its analogy with addictive behavior. Thus, the concluding chapter of the first section covers the role of reward systems in appetite control. It has been believed that dysfunction of the neuroendocrine circuitry leads to obesity, e.g., by alterations such as "leptin resistance," but so far, no fully convincing, comprehensive molecular mechanism has been proposed. Thus, the question as to why individuals are sensitive or resistant to the "adipogenic" environment remains to be answered.

The circuits controlling food intake depend on peripheral signals from sensors that monitor the availability of nutrients. Solid evidence has been accumulated, indicating that the gastrointestinal tract plays a major role in generating these signals. Therefore, the second section of the volume includes chapters covering the peptides ghrelin, GLP-1, CCK, PYY, PP, and amylin, as well as their receptors. More recently, it has become apparent that the intestinal tract plays an even more important role because its microbiota appears to modulate energy balance and, consequently, adiposity. Although the underlying mechanisms are largely unknown, it was appropriate to add a chapter covering these novel aspects to the second section of the volume.

Three separate chapters describing the sensing of nutrients are included in the third section of the volume. The availability of glucose is assessed by a complex system of glucosensors in the CNS; at low glucose levels, these neurons trigger the sensation of hunger. The consumption of fat is also monitored, but the underlying molecular mechanisms are less well known. The current status of knowledge is described in the second chapter of the third section, with particular focus on the fatty acid transporter CD36. The molecular mechanisms of chemosensation of glucose, fatty acids, and amino acids in the intestinal mucosa are described in the third chapter of the section. These processes are controlled by nutrient-binding receptors, transporters, or ion channels that trigger the secretion of intestinal hormones. Interestingly, receptors recognizing sweet taste have been found recently in the intestinal mucosa, as is also described in the final chapter of the section.

The current status of antiobesity drugs is described in the fourth section of the volume. In contrast to the plethora of mechanistic information described in the first three sections of the volume, at present, there are no antiobesity drugs available for the therapy of morbid obesity. This situation is not due to a lack of effective agents but to the presence of serious side effects which cannot be tolerated under conditions of a chronic therapy. The section describes the agents that have been in use and later withdrawn (reuptake inhibitors of catecholamines, e.g., sibutramine; cannabinoid antagonists, e.g., rimonabant) or have been investigated in clinical trials (HT-receptor antagonists, e.g., lorcaserine; MCH-receptor antagonists; antiepileptics, e.g., topiramate, zonisamide). Efforts are still being made to develop strategies minimizing the side effects of existing agents, for instance, by reduced dosage in drug combinations. An interesting strategy to prevent the psychiatric side effects of cannabinoid antagonists is the design of peripherally restricted derivatives. In addition, bombesin-3-receptor agonists, a more recently identified, promising group of appetite-inhibiting agents, are covered in a separate chapter.

I am deeply grateful to the authors of the volume who have contributed excellent, informative, and comprehensive chapters to this volume. All authors are distinguished experts in their fields and have contributed important original data to our understanding of appetite control. They have quite different scientific backgrounds, and, together, they represent all relevant disciplines. Thereby, the topics are presented from different points of view, not exclusively from that of pharmacology. Thus, I believe that they have written a timely and unique overview on one of the most important areas of current research. I do hope that this volume will contribute to generate novel ideas, findings, and solutions to the pressing problem of obesity.

Potsdam-Rehbrücke, Germany H.-G. Joost

Contents

Part I
The Neuroendocrine, Hypothalamic Circuitry Regulating Hunger, Satiety, and Energy Expenditure

Leptin Receptors

Elizabeth C. Cottrell and Julian G. Mercer

Contents

Abstract The hormone leptin, secreted predominantly from adipose tissue, plays a crucial role in the regulation of numerous neuroendocrine functions, from energy homeostasis to reproduction. Genetic deficiency as a consequence of leptin or leptin receptor mutations, although rare in humans, leads to early onset of chronic hyperphagia and massive obesity. In most human obesity, however, leptin levels are chronically elevated. Under these conditions of persistent hyperleptinaemia, and particularly when obesity is associated with a high-fat diet, leptin resistance develops, and signalling through the leptin receptor is curtailed, fuelling further weight gain. Here, we review the role of leptin receptors in the regulation of feeding and obesity development. Leptin receptors are found in each of the major components of the CNS "feeding" circuitry—the brainstem, hypothalamus and

E.C. Cottrell (✉)
Centre for Cardiovascular Science, Queen's Medical Research Institute, University of Edinburgh, Edinburgh, Scotland EH16 4TH, UK

J.G. Mercer
Division of Obesity and Metabolic Health, Rowett Institute of Nutrition and Health, University of Aberdeen, Bucksburn, Aberdeen, Scotland AB21 9SB, UK

H.-G. Joost (ed.), *Appetite Control*, Handbook of Experimental Pharmacology 209,
DOI 10.1007/978-3-642-24716-3_1, © Springer-Verlag Berlin Heidelberg 2012

distributed reward centres. Through these receptors, leptin exerts influences on signalling and integration within these circuits to alter feeding behaviours. Although some progress is now being made with peptide analogues, the leptin receptor has not proved to be amenable to small molecule pharmacological intervention to date. Where clinical benefit from recombinant leptin administration has been achieved, this has been under circumstances of complete endogenous leptin deficiency or relative hypoleptinaemia such as in lipodystrophy.

Keywords Leptin • Leptin receptors • Obesity

Abbreviations

AgRP	Agouti-related protein
ARC	Arcuate nucleus of the hypothalamus
CART	Cocaine- and amphetamine-regulated transcript
CNS	Central nervous system
CSF	Cerebrospinal fluid
DIO	Diet-induced obesity
DMH	Dorsal medial nucleus
ERK	Extra-cellular signal-regulated kinase
JAK	Janus kinase
LepR	Leptin receptor
LHA	Lateral hypothalamic area
NPY	Neuropeptide-Y
NTS	Nucleus tractus solitarius
PI3K	Phosphoinositide 3-kinase
POMC	Pro-opiomelanocortin
PVN	Paraventricular nucleus
SH2	Src homology 2
SOCS	Suppressors of cytokine signalling
STAT	Signal transducer and activator of transcription
VMH	Ventromedial nucleus
VTA	Ventral tegmental area

1 Introduction

The regulation of energy balance involves a matching of energy intake with energy expenditure, in order to maintain a stable body weight over time. The relative stability of body weight in most adults over prolonged periods is testament to the accuracy with which this regulatory system functions. Positive energy balance occurs when intake exceeds expenditure and leads to a progressive increase in body weight and adiposity if the appropriate negative regulation does not occur.

In this situation, prolonged positive energy balance will invariably lead to the development of obesity.

It is apparent within human populations that some individuals are better able to maintain an appropriate body weight than others in the face of an "obesogenic" environment. Studies in animal models have begun to shed some light on key pathways that may, at least in part, determine differential susceptibility to diet-induced obesity (DIO). The central nervous system (CNS) plays an essential role in the regulation of feeding and energy expenditure, integrating signals from gastro-intestinal afferents and circulating nutrient-related factors to alter behaviour and neuroendocrine function (Schwartz et al. 2000). Although the discovery of leptin initially highlighted its critical role in the regulation of body weight, leptin receptors (LepRs) have since been identified in virtually every tissue, and diverse roles for this hormone have been shown in reproduction, development, immunity, cardiovascular function, cognition and some forms of cancers (Ahima and Flier 2000; La Cava and Matarese 2004; Harvey et al. 2005; Tune and Considine 2007). In this chapter, we will focus specifically on the role of leptin and LepRs in the regulation of feeding and energy homeostasis.

2 Historical Perspectives on the Discovery of Leptin

The genetically obese *ob/ob* mice were first described in 1950 by Ingalls and colleagues (Ingalls et al. 1950) who reported that in these animals, affected off-spring began to display significant hyperphagia and increased body weight at around 4 weeks of age, and by 3 months, they were approximately twice the weight of non-obese littermates. Subsequent parabiosis experiments in which the circulatory systems of an obese animal were fused with a lean control mouse indicated that it was the lack of a blood-borne signal that caused the massive obesity in *ob/ob* animals, as this surgery resulted in a correction of hyperphagia and significant reduction in body weight of the obese animals (Coleman 1973). In contrast, the surgical union of an obese diabetic (*db/db*) mouse to a lean control was ineffective in rescuing the phenotype of these mice, suggesting a failure of these animals to respond to some circulating factor (Coleman 1978).

Almost 20 years later, the mutations responsible for these forms of genetic obesity were identified. In *ob/ob* mice, Friedman and colleagues identified a non-sense mutation in a gene that is expressed in white adipose tissue and normally produces a peptide hormone of 16 kDa expressed in and secreted by white adipose tissue (Zhang et al. 1994). The mutation causes the production of a truncated protein that is not secreted, and as a consequence of this defective secretion, *ob/ob* mice are massively obese. In addition, the expression of the *ob* gene is markedly elevated (approximately 20-fold) compared with wild-type mice (Zhang et al. 1994).

Following this discovery, a number of groups simultaneously investigated the effects of administering recombinant *ob* protein to *ob/ob*, wild-type and *db/db* animals. These studies showed that the treatment reduced food intake and increased oxygen consumption in *ob/ob* mice (Campfield et al. 1995; Halaas et al. 1995; Pelleymounter et al. 1995) but was without effect in *db/db* animals (Campfield et al. 1995; Halaas et al. 1995). In *db/db* mice, a loss-of-function mutation in the receptor for *ob* was found to be responsible for the obesity in these mice (Tartaglia et al. 1995). Thereafter, a variant of the receptor was identified which is partially responsible for the phenotype of the polygenic New Zealand mouse model of obesity (Igel et al. 1997; Kluge et al. 2000). The product of the *ob* gene was subsequently named leptin, and its receptor, leptin receptor (LepR), derived from the Greek word for thin, "leptos" (Halaas et al. 1995). Both *ob/ob* and *db/db* mice are also infertile and exhibit a reduced longitudinal growth. Thus, leptin clearly has widespread functions in the neuroendocrine regulation of growth and reproduction as well as body weight, the mechanisms of which are discussed in more detail below.

3 Homeostatic Regulation of Feeding in the Central Nervous System

In order to communicate the size of peripheral energy stores to the CNS, a circulating factor needs to be produced in response to changes in energy availability, to be circulated in amounts proportional to energy stores and to be sensed by the brain. Insulin was the first such satiety signal to be identified. Following feeding, circulating concentrations are increased, and insulin enters the brain via a regulated transport mechanism (Baura et al. 1993) where it acts at insulin receptors to bring about changes in feeding-related neuropeptides. Central administration of insulin acts to reduce energy intake and body weight (Woods et al. 1979), indicating a direct effect of this hormone at the level of the brain.

The second key hormone shown to act as a signal of plentiful energy stores was leptin. This peptide hormone is secreted predominantly by adipocytes, and in the adult, circulating levels are related to body mass (Maffei et al. 1995) but more closely correlated with the degree of adiposity (Liuzzi et al. 1999). Leptin, like insulin, is transported across the blood–brain barrier (BBB) by a saturable system (Banks et al. 1996) and also acts within brain regions that regulate feeding behaviour and thermogenic responses.

Numerous early ablation studies indicated a key role for the hypothalamus in the regulation of energy balance. Leptin receptors are widely expressed within the mediobasal hypothalamus (Mercer et al. 1996; Elmquist et al. 1998b). Within the hypothalamus, the arcuate nucleus (ARC) is considered to be a key primary nucleus with respect to reception and relay of nutrient-related signals from the circulation.

Two important populations of ARC cells are considered "first-order" neurons: those co-expressing the anorexigenic pro-opiomelanocortin (POMC) precursor and cocaine- and amphetamine-regulated transcript (CART) neuropeptide, and those co-expressing the orexigenic neuropeptide-Y (NPY) and agouti-related protein

(AgRP). Both of these neuronal populations express the receptors for, and respond to, alterations in concentrations of leptin and insulin (Baskin et al. 1999; Niswender and Schwartz 2003). Raised levels of these hormones, indicative of a fed state and sufficient energy stores, act centrally to increase POMC/CART expression and concomitantly reduce NPY/AgRP levels. Conversely, negative energy balance induced by food shortage has the opposite effects, as depicted in Fig. 1.

These neurons then project to further downstream nuclei within the hypothalamus, where additional information from brainstem and higher cortical centres is integrated. Key hypothalamic targets of ARC projections include the neurons of the paraventricular nucleus (PVN), dorsal medial nucleus (DMH), lateral hypothalamic area (LHA) and the ventromedial nucleus (VMN). Disruption of connections between the ARC and these nuclei is associated with an inability to successfully regulate energy balance (Dawson et al. 1997; Bell et al. 2000; Bouret et al. 2004). However, as discussed in more detail below, although the ARC has received much attention as a primary target in mediating the effects of leptin on energy balance, the role of other sites within the mediobasal hypothalamus and other regions of the brain is increasingly being recognised.

The critical role of neuronal leptin action in body weight regulation was definitively shown through the generation of mice with selective deletion of leptin receptors throughout the CNS (Cohen et al. 2001). These mice exhibit obesity and a peripheral hormone profile very similar to that of *db/db* mice. Conversely, adenoviral expression of leptin receptors within the ARC in Zucker *fa/fa* rats, which lack functional leptin receptors, was reported to ameliorate the obesity in these animals (Morton et al. 2003). However, other studies in mice report only a partial

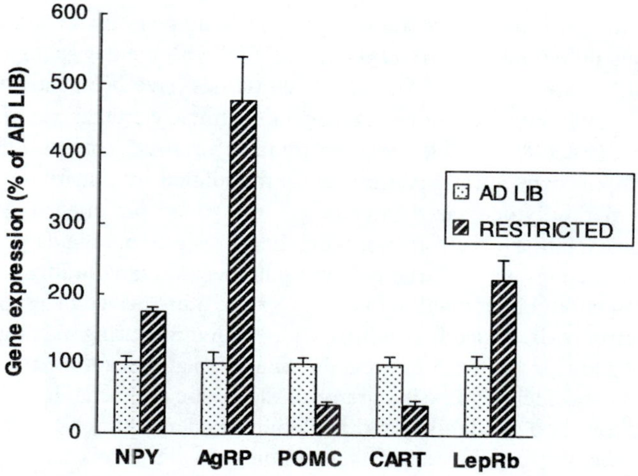

Fig. 1 Arcuate gene expression of the orexigenic neuropeptides NPY and AgRP, the anorexigenic neuropeptides POMC and CART, and long-form leptin receptor (LepRb) following a 96-h period of food restriction in mice (restricted animals fed approximately 50% of ad libitum mice; Mercer et al., unpublished data)

reversal of the obesity in *db/db* mice having transgenic replacement of LepRb in the ARC (Coppari et al. 2005), indicating that other leptin-responsive regions also play a role.

Thus, aside from the well-recognised role of the ARC, it is becoming increasingly apparent that other brain regions are of importance in the regulation of energy balance. Numerous studies have now demonstrated roles for alternative sites of leptin action. Neurons within both the VMH and LHA have been shown to respond directly to leptin administration (Elmquist et al. 1998a; Dhillon et al. 2006; Leinninger et al. 2009). Extra-hypothalamic sites, particularly the brainstem, also contain LepRs (Elmquist et al. 1998b; Mercer et al. 1998), and LepR-containing neurons within the nucleus tractus solitarii (NTS) and area postrema have been demonstrated to be leptin responsive (Bjorbaek and Kahn 2004; Hayes et al. 2011). In addition to the reception of circulating factors, NTS neurons also receive neural inputs from gastrointestinal afferents (e.g., stomach distension that occurs with feeding), which play a role in the termination of feeding (Grill et al. 2002). It has also been shown recently that there are interactions between leptin and signals of gastrointestinal distension, for example, a potentiation of afferent neural signals by hindbrain leptin delivery (Huo et al. 2007). These findings indicate that leptin is involved both directly and indirectly in the regulation of appetite and feeding behaviour, and that the sites at which leptin acts are widely dispersed throughout the brain.

4 Regulation of Leptin Production and Receptor Levels

As mentioned above, leptin is mainly expressed in adipose tissue, and circulating concentrations in the fed state are highly correlated with the degree of adiposity, at least in adults (Maffei et al. 1995). Additional tissues have been shown to express leptin mRNA, including the stomach, placenta, mammary epithelium, pituitary and hypothalamus (Bado et al. 1998; Smith-Kirwin et al. 1998; Hoggard et al. 2001). Circulating leptin concentrations are acutely modulated by nutritional state, with levels being markedly decreased in response to fasting (Ahren et al. 1997a, b) or cold exposure (Hardie et al. 1996). This fall in leptin mediates a number of physiological adaptations to disrupted energy homeostasis, including a stimulation of feeding behaviour and reduction in energy expenditure, as well as suppression of the reproductive axis. Indeed, administration of leptin during fasting is able to ameliorate a number of these neuroendocrine adaptations (Ahima et al. 1996). Thus, leptin is considered a key integrator of metabolic and behavioural adaptation to a reduction in food availability and depletion of energy stores.

As with many ligand–receptor interactions, LepR levels are regulated in response to changes in circulating hormone concentrations. In the case of *ob/ob* mice, these animals exhibit a marked upregulation of hypothalamic LepR in response to the complete absence of hormone, which can be countered with administration of exogenous leptin (Mercer et al. 1997) (Fig. 2, below). Similarly,

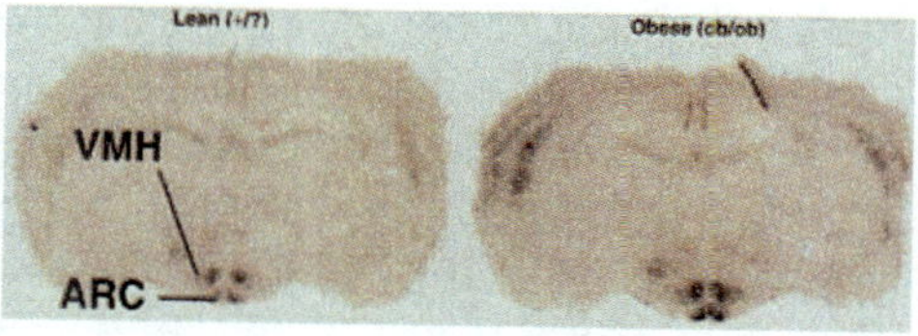

Fig. 2 Leptin receptor gene expression in *ob/ob* mouse and a lean (+/?) littermate control, as demonstrated by in situ hybridisation

cold exposure or food restriction, and the accompanying rapid reduction in circulating leptin that occurs in these states, also result in significant upregulation of LepR within the hypothalamus (Lin and Huang 1997; Mercer et al. 1997) (Fig. 1). These changes in LepR gene expression would presumably enhance leptin signalling in the face of low circulating levels of ligand.

In situations of hyperleptinaemia (as in the *db/db* mouse model or during obesity progression), reduced LepR expression and impaired downstream signalling are observed (Malik and Young 1996; Wilsey and Scarpace 2004). These are discussed in more detail below, in the context of leptin resistance occurring in the face of chronically elevated leptin levels.

5 Leptin Receptor Signalling

The leptin receptor is a member of the class I cytokine-receptor family, having an extra-cellular ligand-binding domain, a transmembrane domain and a cytoplasmic signalling domain. To date, six LepR isoforms have been identified (LepRa-f; reviewed in (Fruhbeck 2006)), resulting from alternative mRNA splicing and/or proteolytic processing. Each of these isoforms shares common extra-cellular and transmembrane domains but differs in their intra-cellular C-terminal sequences. LepRa, c, d and f possess relatively short cytoplasmic domains (and are referred to as "short" isoforms), and LepRe is a secreted form of the leptin receptor. A degree of intra-cellular signalling has been demonstrated for the short isoforms (Bjorbaek et al. 1997); however, only LepRb, with an extended intra-cellular C-terminal region of ~300 amino acids, is capable of complete intra-cellular signal transduction (Baumann et al. 1996). Indeed, it is a mutation in the cytoplasmic region of LepRb that confers the leptin insensitivity in *db/db* mice, as well as in the Zucker fatty rat (Chua et al. 1996). Furthermore, LepRb has been shown to be critical in mediating leptin's actions in terms of body weight regulation, as there is no discernable phenotypic difference between mice which lack all leptin receptor isoforms (db^{3J}/db^{3J}), *db/db* mice (which lack only LepRb function) and leptin-deficient *ob/ob* animals (Chua et al. 1996; Tartaglia 1997).

The main intra-cellular events following the binding of leptin to LepRb involve receptor dimerisation and activation of the janus kinase/signal transducer and activator of transcription (JAK/STAT) pathway (Banks et al. 2000). Heterodimers of LepRa and LepRb appear to be incapable of signalling since the short LepRa isoform lacks the residues Leu896 and Phe897 that are critical for dimerisation

(Bahrenberg et al. 2002). Like many cytokine receptors, LepRs do not possess any intrinsic kinase activity, and, as such, signalling requires interaction with and activation of cytoplasmic non-receptor tyrosine kinases (JAKs). The JAK proteins are non-covalently attached through a proline-rich "box1" motif located in the juxtamembrane region of the intra-cellular domain. LepRb is associated preferentially with JAK2 (Kloek et al. 2002), and upon leptin binding and receptor dimerisation, JAK2 is autophosphorylated on several tyrosine residues and at the same time phosphorylates a number of tyrosine residues on the LepRb intra-cellular domain. These phosphorylated tyrosines provide sites for recruitment of intra-cellular proteins that contain phosphotyrosine-binding domains, such as Src homology 2 (SH2) domains. The transcription factor STAT3, containing a specific SH2 domain, is one of the key intra-cellular signalling pathways activated by leptin signalling through LepRb. Recruited STAT3 proteins are phosphorylated and dimerised, and these active pSTAT-3 dimers translocate to the nucleus where they bind target genes to bring about changes in gene transcription (Vaisse et al. 1996). In addition to JAK/STAT signalling, activation of LepRb results in the activation of extra-cellular signal-regulated kinase (ERK) and phosphoinositide 3-kinase (PI3K) pathways (Fruhbeck 2006). Leptin signalling by LepRb is also under negative feedback regulation, through suppressors of cytokine signalling (SOCS) proteins, specifically SOCS3, which functions to inhibit tyrosine phosphorylation of LepR (Munzberg et al. 2003), and thus attenuates further signalling (Fig. 3).

At least three important tyrosine residues have been identified within the LepRb cytoplasmic domain (Tyr^{985}, Tyr^{1077} and Tyr^{1138}). The Tyr^{985} and Tyr^{1077} residues are considered critical sites for SOCS3 recruitment, and therefore negative feedback of leptin signalling (Eyckerman et al. 2000). Tyr^{1138} has been shown to be crucial in mediating the activation of STAT3 pathways, as mice in which this tyrosine was replaced with a serine residue fail to activate STAT3 and exhibit hyperphagia and early-onset obesity, similar to the *db/db* mice (Bates et al. 2003). Furthermore, through the use of these mice (termed "*s/s*" mice), it has been elegantly demonstrated that although Tyr^{1138}-STAT3 signalling is critical for the actions of leptin on energy balance regulation, disruption does not result in infertility or linear growth impairments. *s/s* mice are reproductively functional without the need for exogenous leptin and can attain a normal body length. Within the ARC, *s/s* mice have reduced POMC and increased AgRP gene expression, similar to *db/db* mice, but levels of NPY mRNA are relatively normal. Thus, these studies show that downstream of the leptin receptor, intra-cellular actions of leptin are mediated by distinct components of the leptin signalling pathway. The normalised levels of NPY gene expression in the *s/s* model and resultant reversal of the hypothalamic–gonadal axis are consistent with earlier reports of *ob/ob/NPY*$^{-/-}$ mice, in which obesity is attenuated and animals have improved fertility and growth profiles (Erickson et al. 1996). More recently, attention has turned to the role of other leptin signalling pathways (PI3K and ERK), as it is clear that not all leptin actions require altered gene expression through STAT3 transcriptional alterations. Leptin can bring about rapid changes in membrane potential, within minutes of application, indicating non-genomic (and therefore pSTAT3-independent) actions (Cowley

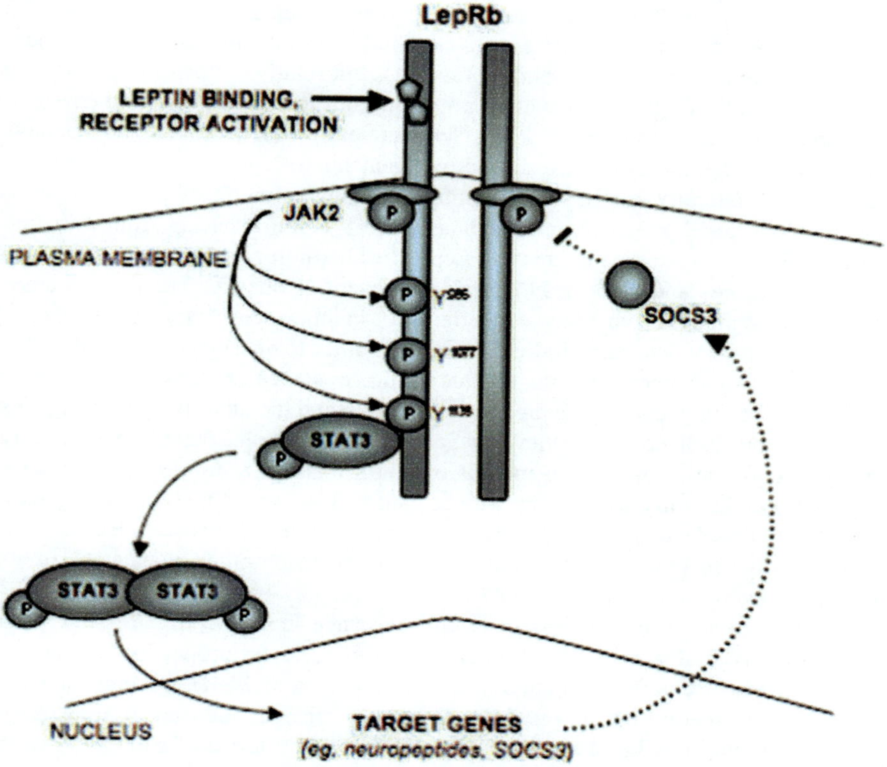

Fig. 3 Simplified diagrammatic overview of main intra-cellular pathway activated following leptin binding to LepRb. Upon leptin binding, JAK2 undergoes autophosphorylation and phosphorylates key tyrosine residues within the cytoplasmic LepRb region (Y^{985}, Y^{1077}, Y^{1138}), providing docking sites for binding of intra-cellular signalling components. In particular, STAT3 transcription factors bind to these activated SH2 domains at Y^{1138} and are phosphorylated, dimerised, and translocated to the nucleus where they alter transcription of target genes. Suppressor of cytokine signalling (SOCS3) is induced by pSTAT3 and acts to negatively regulate leptin signalling by inhibiting phosphorylation of the cytoplasmic tail of LepRb by JAK2

et al. 2001; Solovyova et al. 2009). The three tyrosine residues appear to define the pleiotropy of LepR signalling: Tyr^{985} is necessary and sufficient for activation of ERK, Tyr^{1077} can induce tyrosine phosphorylation of STAT5, whereas Tyr^{1138} is capable of phosphorylating STAT1, 3 and 5 (Hekerman et al. 2005).

6 Obesity and Leptin Resistance

The hope that leptin might be a "miracle cure" in the treatment of obesity was short lived. Rather than being leptin deficient, most obese humans and animals have high levels of circulating leptin (Maffei et al. 1995; Considine et al. 1996). The failure of elevated leptin levels to control or reverse obesity suggests the existence of a resistant state (Scarpace and Zhang 2009). Studies in humans showed that obese individuals exhibit high peripheral leptin levels but relatively lower cerebrospinal fluid (CSF) concentrations, indicating impaired transport of leptin from the peripheral to central sites (Caro et al. 1996; Schwartz et al. 1996). Rodent studies have subsequently demonstrated two distinct components to leptin resistance—a resistance to peripherally administered leptin suggesting a failure of the hormone to access CNS target sites and/or resistance to CNS leptin resulting from impaired responses in CNS neurons expressing LepR. Furthermore, leptin resistance is also considered a key and potentially causal component of obesity. There is evidence from outbred rat strains that inherent differences in leptin sensitivity prior to dietary manipulation may determine individual susceptibility to obesity (Levin et al. 2003, 2004). These and other data suggest that leptin resistance predisposes towards or promotes obesity, although the precise relationship and the underlying mechanisms remain to be resolved (see review by Scarpace and Zhang 2009). A particularly relevant observation is that all models of leptin resistance develop obesity on a high-fat diet, but only some show weight gain and increased body adiposity on a stock diet (Scarpace and Zhang 2009). This suggests a link between leptin resistance and the reward system-driven caloric over-consumption of palatable diets (as addressed in more detail, below).

In order to study the process of leptin resistance in obesity progression, DIO rodents are typically fed a high-energy diet, with a greater proportion of calories being derived from fat compared with chow-fed animals. These models have provided information on the rate of body weight changes induced by transfer to such a diet and in addition have revealed differences in terms of susceptibility to DIO between different mouse strains (Prpic et al. 2003) and within outbred rat strains (Levin et al. 1989; Archer et al. 2003; Enriori et al. 2007). The body weight responses, along with the progressive loss of leptin sensitivity during DIO progression, also resemble the human situation. Therefore, these animal models are of great use in the identification of pathways that are affected during DIO development. Experimental evidence has suggested that during DIO, there is a progressive loss of leptin responsiveness, involving an initial gain in adiposity whilst retaining responses to peripherally administered leptin, followed by a loss of peripheral leptin responsiveness, and finally resistance to both peripherally and centrally administered leptin (El-Haschimi et al. 2000).

Within the leptin signalling pathway, suppressor of cytokine signalling 3 (SOCS3) has emerged as a likely candidate in mediating hormone resistance. Following leptin receptor activation, SOCS3 is induced and acts to negatively regulate leptin signalling, as discussed above (Howard and Flier 2006). Elevated SOCS3 expression in the ARC is observed in DIO mice, and this is associated with a reduced sensitivity to the effect of leptin on food intake, body weight and neuropeptide secretion (Munzberg et al. 2004; Enriori et al. 2007). Neuron-specific removal (Mori et al. 2004) or heterozygous loss of SOCS3 (Howard et al. 2004) confers an enhanced response to leptin administration and a greater resistance to DIO.

The composition and origin of fat in the diet may influence not only the progression towards obesity but also leptin signalling. The direct effect of high-fat diet on leptin resistance may reflect the direct action of fatty acids, triglycerides and lipid molecules at leptin-sensitive neurons, a possible component of hypothalamic nutrient sensing (Jordan et al. 2010) or an effect on leptin transport across the blood–brain barrier (Banks et al. 2004). The picture emerging in high-fat diet-induced obesity is of leptin resistance both causing and resulting from obesity. Thus, high-fat diets can induce leptin resistance either directly in the absence of obesity or indirectly, secondary to obesity and resultant high leptin levels.

Finally, it was recently demonstrated that although ARC leptin responsiveness is reduced in DIO mice, animals are hyper-responsive to the effects of melanocortin agonists, acting on downstream pathways (Enriori et al. 2007). The presence of an intact melanocortin pathway in situations of leptin resistance is considered an attractive target for pharmacological intervention in obesity (Zhang and Scarpace 2006). However, administration of melanocortin agonists to over-weight humans failed to yield successful results (Hallschmid et al. 2006).

7 Leptin Signalling and Reward Pathways in the Control of Feeding

It is frequently observed that laboratory rodents that have predictable growth curves on a stock pellet diet lose this tight regulatory capability and gain weight and body fat when transferred to a more energy-dense, palatable diet. The development of DIO on diets that are high in fat and simple carbohydrates is often taken as evidence that homeostatic (energy demand–driven) systems are being overridden by non-homeostatic (palatability-, pleasure- or reward-based), hedonic systems, i.e. the interaction between these systems may underlie the tendency to "ignore" feedback repletion signals such as leptin and insulin and over-consume calories when rewarding energy-dense and sweetened foods are available. Some of the historical indicators of interaction between homeostatic and non-homeostatic food intake are reviewed by Figlewicz and Benoit (2009). The mechanisms underlying this interaction remain to be fully resolved although it is clear that there is an extensive motivational circuitry involved in the regulation of food intake within the limbic system of the CNS, in addition to the brainstem and hypothalamic circuits referred to earlier. The motivational circuits encompass parts of the cerebral cortex, hippocampus, and amygdala and the anatomical areas and pathways constituting and running between the ventral tegmental area (VTA) and substantia nigra in the midbrain and the striatum and nucleus accumbens in the forebrain. The dopamine and opioid pathways are major players within this widespread circuitry, and leptin is an external hormonal signal that can influence non-homeostatic behaviours, in addition to its role in energy homeostasis in the hypothalamus (Figlewicz 2003b). Leptin receptors are expressed throughout the limbic forebrain, for example, in dopaminergic neurons in the VTA and substantia nigra, suggesting that these neurons are directly targeted by the leptin hormone (Figlewicz 2003a). What is the effect of leptin on these neurons and on the behaviours they influence? Food restriction enhances the rewarding properties of a variety of stimuli including food, and such states of negative energy balance are accompanied by reduced levels of leptin. Consistent with the idea that low levels of leptin in food restriction increase reward sensitivity (Figlewicz and Benoit 2009), fMRI imaging of individuals with congenital leptin deficiency before and after leptin therapy revealed that the leptin-deficient state was accompanied by activation of striatal areas (Farooqi et al. 2007). The relevance of these observations for more common forms of obesity is provided by the observation that motivation for food rewards is elevated in both food restriction and in obesity accompanied by leptin resistance, where circulating leptin levels are high but signal transduction through the leptin receptor is restricted. Thus, in leptin-resistant obesity, there may be enhanced motivation for rewards, including food reward, thereby stoking the potential to over-consume palatable diets (Pandit et al. 2011).

8 Pharmacological Modification of Leptin Signalling

Shortly after leptin was discovered, the recombinant protein was evaluated for systemic effects in *ob/ob*, wild-type and *db/db* mice (Campfield et al. 1995; Halaas et al. 1995; Pelleymounter et al. 1995). The outcomes of these mouse studies established the precedent for the clinical trials that would subsequently be undertaken in humans; recombinant leptin was very active in (naïve) *ob/ob* mice, reduced food intake and body weight, had a more limited effect in wild-type mice where the recombinant protein presumably incrementally augmented endogenous leptin levels, and was ineffective in the receptor-deficient *db/db* mouse. In humans with congenital leptin deficiency, recombinant leptin is extremely effective (Farooqi et al. 1999), decreases food intake and body weight, and normalises a number of other physiological functions including pubertal development. In contrast, clinical trials in common human obesity have produced disappointing results so far (Heymsfield et al. 1999). Leptin is currently available through an Amylin-sponsored compassionate use programme for treatment of congenital leptin deficiency. More recently, the therapeutic potential of leptin has become recognised in a number of states characterised by a relative hypoleptinaemia, as opposed to complete deficiency, and in particular in lipodystrophy (or lipoatrophy).

Lipodystrophy in humans can be congenital or acquired, for example, through antiretroviral pharmacotherapy of HIV infection. It is characterised by localised or generalised loss of adipose tissue but is accompanied by many of the secondary metabolic symptoms that accompany obesity, such as fatty liver, dyslipidaemia, insulin resistance and type 2 diabetes (Savage and O'Rahilly 2010). The extent of fat loss determines the severity of the metabolic consequences and the degree of leptin deficiency, which in turn drives excessive food consumption. Congenital lipodystrophy is a rare autosomal recessive condition, whereas the acquired condition is common amongst HIV-infected patients (Jacobson et al. 2005). Leptin treatment of congenital lipodystrophy markedly improves metabolic indicators (Oral and Chan 2010), and therapy is effective, albeit less so, in partial lipodystrophy, including that seen with HIV infection (Lee et al. 2006; Mulligan et al. 2009). Leptin is currently available (Amylin expanded access programme) for treating congenital lipodystrophy that presents with associated metabolic symptoms. The sustained improvement observed with leptin therapy in all forms of lipodystrophy—acquired, inherited, generalised or partial (Chong et al. 2010)—has led to calls for wider application to be considered in partial and prediabetic severe lipodystrophy (Savage and O'Rahilly 2010).

Although peripheral leptin monotherapy with leptin or pegylated leptin (Heymsfield et al. 1999; Zelissen et al. 2005) in over-weight or obese hyperleptinaemic individuals may not be a realistic intervention for the treatment of obesity due to high endogenous hormone levels and deficiencies in transport into the brain (as discussed above), recent innovation has focused on combination therapy and on developing methods to deliver doses of leptin directly into the CNS, thereby circumventing the issue of BBB passage. A promising alternative to the limited efficacy of monotherapy in hyperleptinaemic obese subjects is combination therapy. One example of this is the co-administration of recombinant leptin (metreleptin) and the amylin analogue, pramlintide. The rationale behind the pramlintide/metreleptin therapy in over-weight and obese patients is the combination of a long-term adiposity signal and a short-term satiety signal. Clinical trials of this combination therapy resulted in greater weight loss (as body fat) than either agent in isolation, a broadly additive effect (Ravussin et al. 2009). Alternative delivery routes to peripheral injection also have potential. In particular, the intra-nasal application of leptin has been an attractive approach (Schulz et al. 2004; Kastin and Pan 2006). Several studies have described in rodents successful uptake of leptin into the brain following intra-nasal delivery, leading to reductions in food intake (Fliedner et al. 2006; Bermudez-Humaran et al. 2007). One previous study reporting on the efficacy of delivering neuropeptides via intra-nasal administration in humans found that alpha-MSH and insulin were both able to induce weight loss in healthy human subjects but not in over-weight individuals (Hallschmid et al. 2004). Thus, it remains to be seen whether these methods might be adapted to present a realistic therapeutic strategy for the treatment of obesity in humans.

Targeting of leptin receptors for the treatment of cancer and cachexia (persistent and potentially harmful weight loss in the presence of a chronic disease, such as cancer) is another expanding field in leptin biology. Although not the focus of this chapter, antagonism of leptin receptors is favourable in the context of some forms of disease, including autoimmune disease, hypertension and tumorigenesis, where leptin plays a stimulatory role in disease progression (Gertler 2006). Severe weight loss, due to reduced appetite and increased sympathetic nervous system activity, is often also experienced in cancer, and in this case, antagonism of LepR is beneficial, bringing about increased appetite and reduced energy expenditure. Several recent reports have developed peptide analogues of leptin that function as selective inhibitors at LepR, and these antagonists have been shown to successfully inhibit cancer cell growth both in vitro and in vivo (Otvos et al. 2010), as well as to increase food intake and weight gain in rodents (Otvos et al. 2010; Shpilman et al. 2011).

9 Conclusions and Future Directions

Since the discovery of leptin, our understanding of the neuroendocrine regulation of feeding and body weight homeostasis has expanded rapidly. Genetic leptin deficiency or mutations impairing leptin receptor signalling within the CNS result in the development of severe obesity. The majority of human obesity, however, is thought to result largely from over-consumption of calories and reduced energy expenditure, lifestyle factors which over time result in progressive increases in body weight. Underlying differences in leptin sensitivity may potentially predispose some individuals to increased susceptibility to obesity in our modern "obesogenic" society. Leptin resistance may also contribute to the over-consumption of highly palatable foods, as reward pathways appear to be enhanced in the absence of leptin signalling.

Pharmacological modulation of leptin receptor signalling has not yet resulted in realistic pharmacological treatments for human obesity, where in most cases hyperleptinaemia and leptin resistance are present. Therapies that are able to circumvent resistance to peripheral leptin treatment, therefore targeting central leptin receptors and overcoming impairments in blood–brain barrier transport, continue to be developed. However, manipulation of leptin receptors may prove useful in the treatment of a range of other diseases, including lipodystrophy and some forms of cancer.

References

Ahima RS, Flier JS (2000) Leptin. Annu Rev Physiol 62:413–437

Ahima RS, Prabakaran D, Mantzoros C, Qu D, Lowell B, Maratos-Flier E, Flier JS (1996) Role of leptin in the neuroendocrine response to fasting. Nature 382:250–252

Ahren B, Larsson H, Wilhelmsson C, Nasman B, Olsson T (1997a) Regulation of circulating leptin in humans. Endocrine 7:1–8

Ahren B, Mansson S, Gingerich RL, Havel PJ (1997b) Regulation of plasma leptin in mice: influence of age, high-fat diet, and fasting. Am J Physiol 273:R113–120

Archer ZA, Rayner DV, Rozman J, Klingenspor M, Mercer JG (2003) Normal distribution of body weight gain in male Sprague-Dawley rats fed a high-energy diet. Obes Res 11:1376–1383

Bado A, Levasseur S, Attoub S, Kermorgant S, Laigneau JP, Bortoluzzi MN, Moizo L, Lehy T, Guerre-Millo M, Le Marchand-Brustel Y, Lewin MJ (1998) The stomach is a source of leptin. Nature 394:790–793

Bahrenberg G, Behrmann I, Barthel A, Hekerman P, Heinrich PC, Joost HG, Becker W (2002) Identification of the critical sequence elements in the cytoplasmic domain of leptin receptor isoforms required for Janus kinase/signal transducer and activator of transcription activation by receptor heterodimers. Mol Endocrinol 16:859–872

Banks AS, Davis SM, Bates SH, Myers MG Jr (2000) Activation of downstream signals by the long form of the leptin receptor. J Biol Chem 275:14563–14572

Banks WA, Coon AB, Robinson SM, Moinuddin A, Shultz JM, Nakaoke R, Morley JE (2004) Triglycerides induce leptin resistance at the blood-brain barrier. Diabetes 53:1253–1260

Banks WA, Kastin AJ, Huang W, Jaspan JB, Maness LM (1996) Leptin enters the brain by a saturable system independent of insulin. Peptides 17:305–311

Baskin DG, Figlewicz Lattemann D, Seeley RJ, Woods SC, Porte D Jr, Schwartz MW (1999) Insulin and leptin: dual adiposity signals to the brain for the regulation of food intake and body weight. Brain Res 848:114–123

Bates SH, Stearns WH, Dundon TA, Schubert M, Tso AW, Wang Y, Banks AS, Lavery HJ, Haq AK, Maratos-Flier E, Neel BG, Schwartz MW, Myers MG Jr (2003) STAT3 signalling is required for leptin regulation of energy balance but not reproduction. Nature 421:856–859

Baumann H, Morella KK, White DW, Dembski M, Bailon PS, Kim H, Lai CF, Tartaglia LA (1996) The full-length leptin receptor has signaling capabilities of interleukin 6-type cytokine receptors. Proc Natl Acad Sci USA 93:8374–8378

Baura GD, Foster DM, Porte D Jr, Kahn SE, Bergman RN, Cobelli C, Schwartz MW (1993) Saturable transport of insulin from plasma into the central nervous system of dogs in vivo. A mechanism for regulated insulin delivery to the brain. J Clin Invest 92:1824–1830

Bell ME, Bhatnagar S, Akana SF, Choi S, Dallman MF (2000) Disruption of arcuate/paraventricular nucleus connections changes body energy balance and response to acute stress. J Neurosci 20:6707–6713

Bermudez-Humaran LG, Nouaille S, Zilberfarb V, Corthier G, Gruss A, Langella P, Issad T (2007) Effects of intranasal administration of a leptin-secreting Lactococcus lactis recombinant on food intake, body weight, and immune response of mice. Appl Environ Microbiol 73:5300–5307

Bjorbaek C, Kahn BB (2004) Leptin signaling in the central nervous system and the periphery. Recent Prog Horm Res 59:305–331

Bjorbaek C, Uotani S, da Silva B, Flier JS (1997) Divergent signaling capacities of the long and short isoforms of the leptin receptor. J Biol Chem 272:32686–32695

Bouret SG, Draper SJ, Simerly RB (2004) Formation of projection pathways from the arcuate nucleus of the hypothalamus to hypothalamic regions implicated in the neural control of feeding behavior in mice. J Neurosci 24:2797–2805

Campfield LA, Smith FJ, Guisez Y, Devos R, Burn P (1995) Recombinant mouse OB protein: evidence for a peripheral signal linking adiposity and central neural networks. Science 269:546–549

Caro JF, Kolaczynski JW, Nyce MR, Ohannesian JP, Opentanova I, Goldman WH, Lynn RB, Zhang PL, Sinha MK, Considine RV (1996) Decreased cerebrospinal-fluid/serum leptin ratio in obesity: a possible mechanism for leptin resistance. Lancet 348:159–161

Chong AY, Lupsa BC, Cochran EK, Gorden P (2010) Efficacy of leptin therapy in the different forms of human lipodystrophy. Diabetologia 53:27–35

Chua SC Jr, Chung WK, Wu-Peng XS, Zhang Y, Liu SM, Tartaglia L, Leibel RL (1996) Phenotypes of mouse diabetes and rat fatty due to mutations in the OB (leptin) receptor. Science 271:994–996

Cohen P, Zhao C, Cai X, Montez JM, Rohani SC, Feinstein P, Mombaerts P, Friedman JM (2001) Selective deletion of leptin receptor in neurons leads to obesity. J Clin Invest 108:1113–1121

Coleman DL (1973) Effects of parabiosis of obese with diabetes and normal mice. Diabetologia 9:294–298

Coleman DL (1978) Obese and diabetes: two mutant genes causing diabetes-obesity syndromes in mice. Diabetologia 14:141–148

Considine RV, Sinha MK, Heiman ML, Kriauciunas A, Stephens TW, Nyce MR, Ohannesian JP, Marco CC, McKee LJ, Bauer TL et al (1996) Serum immunoreactive-leptin concentrations in normal-weight and obese humans. N Engl J Med 334:292–295

Coppari R, Ichinose M, Lee CE, Pullen AE, Kenny CD, McGovern RA, Tang V, Liu SM, Ludwig T, Chua SC Jr, Lowell BB, Elmquist JK (2005) The hypothalamic arcuate nucleus: a key site for mediating leptin's effects on glucose homeostasis and locomotor activity. Cell Metab 1:63–72

Cowley MA, Smart JL, Rubinstein M, Cerdan MG, Diano S, Horvath TL, Cone RD, Low MJ (2001) Leptin activates anorexigenic POMC neurons through a neural network in the arcuate nucleus. Nature 411:480–484

Dawson R, Pelleymounter MA, Millard WJ, Liu S, Eppler B (1997) Attenuation of leptin-mediated effects by monosodium glutamate-induced arcuate nucleus damage. Am J Physiol 273:E202–206

Dhillon H, Zigman JM, Ye C, Lee CE, McGovern RA, Tang V, Kenny CD, Christiansen LM, White RD, Edelstein EA, Coppari R, Balthasar N, Cowley MA, Chua S Jr, Elmquist JK, Lowell BB (2006) Leptin directly activates SF1 neurons in the VMH, and this action by leptin is required for normal body-weight homeostasis. Neuron 49:191–203

El-Haschimi K, Pierroz DD, Hileman SM, Bjorbaek C, Flier JS (2000) Two defects contribute to hypothalamic leptin resistance in mice with diet-induced obesity. J Clin Invest 105:1827–1832

Elmquist JK, Ahima RS, Elias CF, Flier JS, Saper CB (1998a) Leptin activates distinct projections from the dorsomedial and ventromedial hypothalamic nuclei. Proc Natl Acad Sci USA 95:741–746

Elmquist JK, Bjorbaek C, Ahima RS, Flier JS, Saper CB (1998b) Distributions of leptin receptor mRNA isoforms in the rat brain. J Comp Neurol 395:535–547

Enriori PJ, Evans AE, Sinnayah P, Jobst EE, Tonelli-Lemos L, Billes SK, Glavas MM, Grayson BE, Perello M, Nillni EA, Grove KL, Cowley MA (2007) Diet-induced obesity causes severe but reversible leptin resistance in arcuate melanocortin neurons. Cell Metab 5:181–194

Erickson JC, Hollopeter G, Palmiter RD (1996) Attenuation of the obesity syndrome of ob/ob mice by the loss of neuropeptide Y. Science 274:1704–1707

Eyckerman S, Broekaert D, Verhee A, Vandekerckhove J, Tavernier J (2000) Identification of the Y985 and Y1077 motifs as SOCS3 recruitment sites in the murine leptin receptor. FEBS Lett 486:33–37

Farooqi IS, Bullmore E, Keogh J, Gillard J, O'Rahilly S, Fletcher PC (2007) Leptin regulates striatal regions and human eating behavior. Science 317:1355

Farooqi IS, Jebb SA, Langmack G, Lawrence E, Cheetham CH, Prentice AM, Hughes IA, McCamish MA, O'Rahilly S (1999) Effects of recombinant leptin therapy in a child with congenital leptin deficiency. N Engl J Med 341:879–884

Figlewicz DP (2003a) Adiposity signals and food reward: expanding the CNS roles of insulin and leptin. Am J Physiol 284:R882–892

Figlewicz DP (2003b) Insulin, food intake, and reward. Semin Clin Neuropsychiatry 8:82–93

Figlewicz DP, Benoit SC (2009) Insulin, leptin, and food reward: update 2008. Am J Physiol 296: R9–R19

Fliedner S, Schulz C, Lehnert H (2006) Brain uptake of intranasally applied radioiodinated leptin in Wistar rats. Endocrinology 147:2088–2094

Fruhbeck G (2006) Intracellular signalling pathways activated by leptin. Biochem J 393:7–20

Gertler A (2006) Development of leptin antagonists and their potential use in experimental biology and medicine. Trends Endocrinol Metabol: TEM 17:372–378

Grill HJ, Schwartz MW, Kaplan JM, Foxhall JS, Breininger J, Baskin DG (2002) Evidence that the caudal brainstem is a target for the inhibitory effect of leptin on food intake. Endocrinology 143:239–246

Halaas JL, Gajiwala KS, Maffei M, Cohen SL, Chait BT, Rabinowitz D, Lallone RL, Burley SK, Friedman JM (1995) Weight-reducing effects of the plasma protein encoded by the obese gene. Science 269:543–546

Hallschmid M, Benedict C, Born J, Fehm HL, Kern W (2004) Manipulating central nervous mechanisms of food intake and body weight regulation by intranasal administration of neuropeptides in man. Physiol Behav 83:55–64

Hallschmid M, Smolnik R, McGregor G, Born J, Fehm HL (2006) Overweight humans are resistant to the weight-reducing effects of melanocortin4-10. J Clin Endocrinol Metab 91:522–525

Hardie LJ, Rayner DV, Holmes S, Trayhurn P (1996) Circulating leptin levels are modulated by fasting, cold exposure and insulin administration in lean but not Zucker (fa/fa) rats as measured by ELISA. Biochem Biophys Res Commun 223:660–665

Harvey J, Shanley LJ, O'Malley D, Irving AJ (2005) Leptin: a potential cognitive enhancer? Biochem Soc Trans 33:1029–1032

Hayes MR, Skibicka KP, Leichner TM, Guarnieri DJ, DiLeone RJ, Bence KK, Grill HJ (2011) Endogenous leptin signaling in the caudal nucleus tractus solitarius and area postrema is required for energy balance regulation. Cell Metab 11:77–83

Hekerman P, Zeidler J, Bamberg-Lemper S, Knobelspies H, Lavens D, Tavernier J, Joost HG, Becker W (2005) Pleiotropy of leptin receptor signalling is defined by distinct roles of the intracellular tyrosines. FEBS J 272:109–119

Heymsfield SB, Greenberg AS, Fujioka K, Dixon RM, Kushner R, Hunt T, Lubina JA, Patane J, Self B, Hunt P, McCamish M (1999) Recombinant leptin for weight loss in obese and lean adults: a randomized, controlled, dose-escalation trial. JAMA 282:1568–1575

Hoggard N, Crabtree J, Allstaff S, Abramovich DR, Haggarty P (2001) Leptin secretion to both the maternal and fetal circulation in the ex vivo perfused human term placenta. Placenta 22:347–352

Howard JK, Cave BJ, Oksanen LJ, Tzameli I, Bjorbaek C, Flier JS (2004) Enhanced leptin sensitivity and attenuation of diet-induced obesity in mice with haploinsufficiency of Socs3. Nat Med 10:734–738

Howard JK, Flier JS (2006) Attenuation of leptin and insulin signaling by SOCS proteins. Trends Endocrinol Metab: TEM 17:365–371

Huo L, Maeng L, Bjorbaek C, Grill HJ (2007) Leptin and the control of food intake: neurons in the nucleus of the solitary tract are activated by both gastric distension and leptin. Endocrinology 148:2189–2197

Igel M, Becker W, Herberg L, Joost HG (1997) Hyperleptinemia, leptin resistance, and polymorphic leptin receptor in the New Zealand obese mouse. Endocrinology 138:4234–4239

Ingalls AM, Dickie MM, Snell GD (1950) Obese, a new mutation in the house mouse. J Hered 41:317–318

Jacobson DL, Knox T, Spiegelman D, Skinner S, Gorbach S, Wanke C (2005) Prevalence of, evolution of, and risk factors for fat atrophy and fat deposition in a cohort of HIV-infected men and women. Clin Infect Dis 40:1837–1845

Jordan SD, Konner AC, Bruning JC (2010) Sensing the fuels: glucose and lipid signaling in the CNS controlling energy homeostasis. Cell Mol Life Sci 67:3255–3273

Kastin AJ, Pan W (2006) Intranasal leptin: blood-brain barrier bypass (BBBB) for obesity? Endocrinology 147:2086–2087

Kloek C, Haq AK, Dunn SL, Lavery HJ, Banks AS, Myers MG Jr (2002) Regulation of Jak kinases by intracellular leptin receptor sequences. J Biol Chem 277:41547–41555

Kluge R, Giesen K, Bahrenberg G, Plum L, Ortlepp JR, Joost HG (2000) Quantitative trait loci for obesity and insulin resistance (Nob1, Nob2) and their interaction with the leptin receptor allele (LeprA720T/T1044I) in New Zealand obese mice. Diabetologia 43:1565–1572

La Cava A, Matarese G (2004) The weight of leptin in immunity. Nat Rev Immunol 4:371–379

Lee JH, Chan JL, Sourlas E, Raptopoulos V, Mantzoros CS (2006) Recombinant methionyl human leptin therapy in replacement doses improves insulin resistance and metabolic profile in patients with lipoatrophy and metabolic syndrome induced by the highly active antiretroviral therapy. J Clin Endocrinol Metab 91:2605–2611

Leinninger GM, Jo YH, Leshan RL, Louis GW, Yang H, Barrera JG, Wilson H, Opland DM, Faouzi MA, Gong Y, Jones JC, Rhodes CJ, Chua S Jr, Diano S, Horvath TL, Seeley RJ, Becker JB, Munzberg H, Myers MG Jr (2009) Leptin acts via leptin receptor-expressing lateral hypothalamic neurons to modulate the mesolimbic dopamine system and suppress feeding. Cell Metab 10:89–98

Levin BE, Dunn-Meynell AA, Banks WA (2004) Obesity-prone rats have normal blood-brain barrier transport but defective central leptin signaling before obesity onset. Am J Physiol 286: R143–150

Levin BE, Dunn-Meynell AA, Ricci MR, Cummings DE (2003) Abnormalities of leptin and ghrelin regulation in obesity-prone juvenile rats. Am J Physiol Endocrinol Metab 285: E949–957

Levin BE, Hogan S, Sullivan AC (1989) Initiation and perpetuation of obesity and obesity resistance in rats. Am J Physiol 256:R766–771

Lin S, Huang XF (1997) Fasting increases leptin receptor mRNA expression in lean but not obese (ob/ob) mouse brain. Neuroreport 8:3625–3629

Liuzzi A, Savia G, Tagliaferri M, Lucantoni R, Berselli ME, Petroni ML, De Medici C, Viberti GC (1999) Serum leptin concentration in moderate and severe obesity: relationship with clinical, anthropometric and metabolic factors. Int J Obes Relat Metab Disord 23:1066–1073

Maffei M, Halaas J, Ravussin E, Pratley RE, Lee GH, Zhang Y, Fei H, Kim S, Lallone R, Ranganathan S et al (1995) Leptin levels in human and rodent: measurement of plasma leptin and ob RNA in obese and weight-reduced subjects. Nat Med 1:1155–1161

Malik KF, Young WS 3rd (1996) Localization of binding sites in the central nervous system for leptin (OB protein) in normal, obese (ob/ob), and diabetic (db/db) C57BL/6J mice. Endocrinology 137:1497–1500

Mercer JG, Hoggard N, Williams LM, Lawrence CB, Hannah LT, Trayhurn P (1996) Localization of leptin receptor mRNA and the long form splice variant (Ob-Rb) in mouse hypothalamus and adjacent brain regions by in situ hybridization. FEBS Lett 387:113–116

Mercer JG, Moar KM, Hoggard N (1998) Localization of leptin receptor (Ob-R) messenger ribonucleic acid in the rodent hindbrain. Endocrinology 139:29–34

Mercer JG, Moar KM, Rayner DV, Trayhurn P, Hoggard N (1997) Regulation of leptin receptor and NPY gene expression in hypothalamus of leptin-treated obese (ob/ob) and cold-exposed lean mice. FEBS Lett 402:185–188

Mori H, Hanada R, Hanada T, Aki D, Mashima R, Nishinakamura H, Torisu T, Chien KR, Yasukawa H, Yoshimura A (2004) Socs3 deficiency in the brain elevates leptin sensitivity and confers resistance to diet-induced obesity. Nat Med 10:739–743

Morton GJ, Niswender KD, Rhodes CJ, Myers MG Jr, Blevins JE, Baskin DG, Schwartz MW (2003) Arcuate nucleus-specific leptin receptor gene therapy attenuates the obesity phenotype of Koletsky (fa(k)/fa(k)) rats. Endocrinology 144:2016–2024

Mulligan K, Khatami H, Schwarz JM, Sakkas GK, DePaoli AM, Tai VW, Wen MJ, Lee GA, Grunfeld C, Schambelan M (2009) The effects of recombinant human leptin on visceral fat, dyslipidemia, and insulin resistance in patients with human immunodeficiency virus-associated lipoatrophy and hypoleptinemia. J Clin Endocrinol Metab 94:1137–1144

Munzberg H, Flier JS, Bjorbaek C (2004) Region-specific leptin resistance within the hypothalamus of diet-induced obese mice. Endocrinology 145:4880–4889

Munzberg H, Huo L, Nillni EA, Hollenberg AN, Bjorbaek C (2003) Role of signal transducer and activator of transcription 3 in regulation of hypothalamic proopiomelanocortin gene expression by leptin. Endocrinology 144:2121–2131

Niswender KD, Schwartz MW (2003) Insulin and leptin revisited: adiposity signals with overlapping physiological and intracellular signaling capabilities. Front Neuroendocrinol 24:1–10

Oral EA, Chan JL (2010) Rationale for leptin-replacement therapy for severe lipodystrophy. Endocr Pract 16:324–333

Otvos L Jr, Kovalszky I, Scolaro L, Sztodola A, Olah J, Cassone M, Knappe D, Hoffmann R, Lovas S, Hatfield MP, Beko G, Zhang S, Wade JD, Surmacz E (2010) Peptide-based leptin receptor antagonists for cancer treatment and appetite regulation. Biopolymers 96:117–125

Pandit R, de Jong JW, Vanderschuren LJ, Adan RA (2011) Neurobiology of overeating and obesity: the role of melanocortins and beyond. Eur J Pharmacol 660(1):28–42
Pelleymounter MA, Cullen MJ, Baker MB, Hecht R, Winters D, Boone T, Collins F (1995) Effects of the obese gene product on body weight regulation in ob/ob mice. Science 269:540–543
Prpic V, Watson PM, Frampton IC, Sabol MA, Jezek GE, Gettys TW (2003) Differential mechanisms and development of leptin resistance in A/J versus C57BL/6J mice during diet-induced obesity. Endocrinology 144:1155–1163
Ravussin E, Smith SR, Mitchell JA, Shringarpure R, Shan K, Maier H, Koda JE, Weyer C (2009) Enhanced weight loss with pramlintide/metreleptin: an integrated neurohormonal approach to obesity pharmacotherapy. Obesity (Silver Spring) 17:1736–1743
Savage DB, O'Rahilly S (2010) Leptin therapy in lipodystrophy. Diabetologia 53:7–9
Scarpace PJ, Zhang Y (2009) Leptin resistance: a prediposing factor for diet-induced obesity. Am J Physiol 296:R493–500
Schulz C, Paulus K, Lehnert H (2004) Central nervous and metabolic effects of intranasally applied leptin. Endocrinology 145:2696–2701
Schwartz MW, Peskind E, Raskind M, Boyko EJ, Porte D Jr (1996) Cerebrospinal fluid leptin levels: relationship to plasma levels and to adiposity in humans. Nat Med 2:589–593
Schwartz MW, Woods SC, Porte D Jr, Seeley RJ, Baskin DG (2000) Central nervous system control of food intake. Nature 404:661–671
Shpilman M, Niv-Spector L, Katz M, Varol C, Solomon G, Ayalon-Soffer M, Boder E, Halpern Z, Elinav E, Gertler A (2011) Development and characterization of high affinity leptins and leptin antagonists. J Biol Chem 286:4429–4442
Smith-Kirwin SM, O'Connor DM, De Johnston J, Lancey ED, Hassink SG, Funanage VL (1998) Leptin expression in human mammary epithelial cells and breast milk. J Clin Endocrinol Metab 83:1810–1813
Solovyova N, Moult PR, Milojkovic B, Lambert JJ, Harvey J (2009) Bi-directional modulation of fast inhibitory synaptic transmission by leptin. J Neurochem 108:190–201
Tartaglia LA (1997) The leptin receptor. J Biol Chem 272:6093–6096
Tartaglia LA, Dembski M, Weng X, Deng N, Culpepper J, Devos R, Richards GJ, Campfield LA, Clark FT, Deeds J, Muir C, Sanker S, Moriarty A, Moore KJ, Smutko JS, Mays GG, Wool EA, Monroe CA, Tepper RI (1995) Identification and expression cloning of a leptin receptor, OB-R. Cell 83:1263–1271
Tune JD, Considine RV (2007) Effects of leptin on cardiovascular physiology. J Am Soc Hypertens 1:231–241
Vaisse C, Halaas JL, Horvath CM, Darnell JE Jr, Stoffel M, Friedman JM (1996) Leptin activation of Stat3 in the hypothalamus of wild-type and ob/ob mice but not db/db mice. Nat Genet 14:95–97
Wilsey J, Scarpace PJ (2004) Caloric restriction reverses the deficits in leptin receptor protein and leptin signaling capacity associated with diet-induced obesity: role of leptin in the regulation of hypothalamic long-form leptin receptor expression. J Endocrinol 181:297–306
Woods SC, Lotter EC, McKay LD, Porte D Jr (1979) Chronic intracerebroventricular infusion of insulin reduces food intake and body weight of baboons. Nature 282:503–505
Zelissen PM, Stenlof K, Lean ME, Fogteloo J, Keulen ET, Wilding J, Finer N, Rossner S, Lawrence E, Fletcher C, McCamish M (2005) Effect of three treatment schedules of recombinant methionyl human leptin on body weight in obese adults: a randomized, placebo-controlled trial. Diabetes Obes Metab 7:755–761
Zhang Y, Proenca R, Maffei M, Barone M, Leopold L, Friedman JM (1994) Positional cloning of the mouse obese gene and its human homologue. Nature 372:425–432
Zhang Y, Scarpace PJ (2006) Circumventing central leptin resistance: lessons from central leptin and POMC gene delivery. Peptides 27:350–364

The Role of Neuropeptide Y in Energy Homeostasis

Adam P. Chambers and Stephen C. Woods

Contents

Abstract When administered into the brain, NPY acts at Y1 and Y5 receptors to increase food intake. The response occurs with a short latency and is quite robust, such that exogenous NPY is generally considered to be the most potent of a growing list of orexigenic compounds that act in the brain. The role of endogenous NPY is not so straightforward, however. Evidence from diverse types of experiments suggests that rather than initiating behavioral eating per se, endogenous NPY elicits autonomic responses that prepare the individual to better cope with consuming a calorically large meal.

Keywords AgRP • Cephalic responses • Insulin • Leptin • Obesity

---

Supported by NIH DK17844.

A.P. Chambers
Departments of Medicine, University of Cincinnati, 2170 East Galbraith Road, Cincinnati, OH 45237, USA

S.C. Woods (✉)
Department of Psychiatry and Behavioral Neuroscience, University of Cincinnati, 2170 East Galbraith Road, Cincinnati, OH 45237, USA
e-mail: Steve.woods@uc.edu

H.-G. Joost (ed.), *Appetite Control*, Handbook of Experimental Pharmacology 209,
DOI 10.1007/978-3-642-24716-3_2, © Springer-Verlag Berlin Heidelberg 2012

Neuropeptide Y (NPY) is a 36-amino acid member of a family of closely related extracellular signaling peptides that includes peptide YY (PYY) and pancreatic polypeptide (PP) as cousins. Collectively, this group of peptides binds with the Y family of G-protein-coupled receptors (GPCRs) on cell membranes, and NPY in particular binds to Y1, Y5, and perhaps others to influence energy homeostasis. Soon after its discovery in 1982 (Tatemoto et al. 1982), NPY was found to be one of the most ubiquitous peptides in the mammalian brain (Allen et al. 1983) as well as in the peripheral nervous system (Gray and Morley 1986; Lundberg et al. 1982). In 1984, three labs reported that when NPY is administered directly into the brain of rats, food intake increases (Clark et al. 1984; Levine and Morley 1984; Stanley and Leibowitz 1984). This was a highly important observation, for it was the first demonstration of an endogenous peptide that is made in the brain and that has potent orexigenic activity, predating the later discoveries of galanin, ghrelin, the orexins, agouti-related peptide (AgRP) and melanin-concentrating hormone (MCH) and their actions, and being more efficacious than several orexigenic endogenous opioids. The role of NPY in the control of food intake and energy homeostasis is the focus of this review.

1 NPY and Y Receptors in the Brain

Although NPY is expressed throughout much of the brain (Gray and Morley 1986), research on its influence over food intake has focused mainly on likely targets in the hypothalamus since this area is known to be important in the regulation of energy homeostasis (Schwartz et al. 2000a) and since NPY and Y1, Y2, and Y5 receptors are expressed there (Chee and Colmers 2008; Lecklin et al. 2003; Leibowitz and Alexander 1991; Mashiko et al. 2003; Wolak et al. 2003). The most-studied circuits are diagrammed in Fig. 1. NPY-expressing neurons are concentrated in the arcuate (ARC) and dorsomedial (DMN) hypothalamic nuclei, and major projections are to the paraventricular (PVN), ventromedial (VMN), and lateral hypothalamic (LH) areas (Kamiji and Inui 2007).

There is compelling evidence that NPY acts at both Y1 and Y5 receptors to stimulate food intake. Both receptors are widely distributed throughout the brain, including in the cortex, hippocampus, amygdala, and hypothalamus, and they are often co-expressed in the same cells (Wolak et al. 2003). Within the hypothalamus, the two are co-expressed in the ARC, the parvocellular PVN, and the LH (Wolak et al. 2003). While both Y1 and Y5 receptors are also expressed in areas of the hindbrain important in energy regulation such as the nucleus of the solitary tract (NTS) and area postrema (Mahaut et al. 2010), the density of staining is relatively low there (Ishizaki et al. 2003; Wolak et al. 2003) but increases during fasting (Ishizaki et al. 2003; Mahaut et al. 2010). Nonetheless, administration of NPY into the fourth cerebral ventricle, where it is most likely to stimulate hindbrain as opposed to hypothalamic Y receptors, stimulates eating (Corp et al. 1990; Steinman et al. 1994; Taylor et al. 2007; Xu et al. 1995) and elicits c-fos expression in the

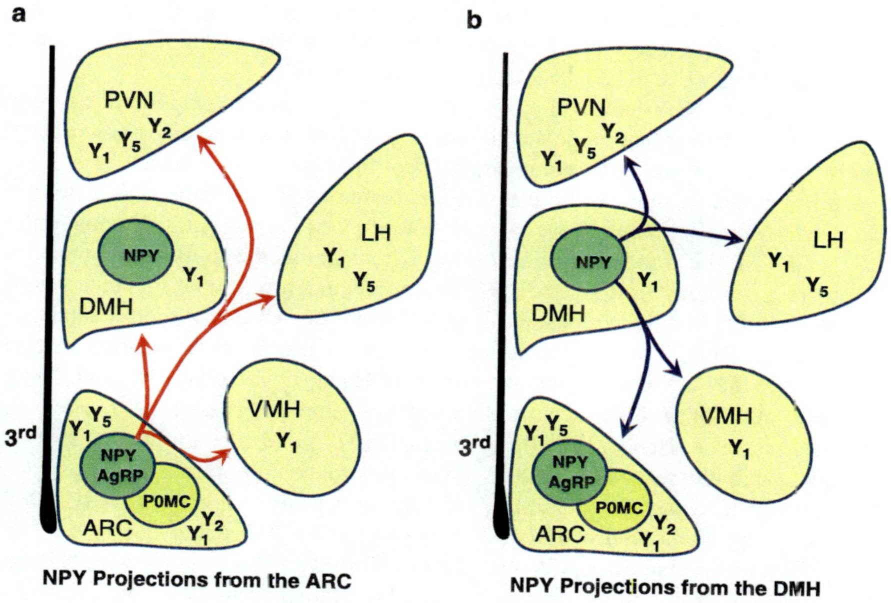

Fig. 1 A unilateral view of the hypothalamus at the level of the arcuate nucleus (ARC). (**a**) NPY-containing neurons (*green*) project from the ARC (*red*) and (**b**) dorsal medial hypothalamus (DMH) (*blue*) to many hypothalamic nuclei including the ventral medial hypothalamus (VMH), lateral hypothalamus (LH), and paraventricular nucleus (PVN). These projections may be excitatory or inhibitory, depending on the area and the receptor subtype expressed on the neurons to which they project. NPY can act on Y_1, Y_2, Y_4, and Y_5 receptors; however, only Y_1, Y_2, and Y_5 receptor subtypes have been implicated in the control of food intake. Within the ARC, reciprocal connections between anorectic POMC and orexigenic NPY-containing neurons innervate one another in a negative feedback manner. Likewise, neurons in the VMH, which provide excitatory input to POMC neurons in the ARC, are inhibited by NPY-expressing neurons in the ARC and DMH. These circuits provide a cellular basis for the in vivo effects of NPY on food intake. In other areas, such as the lateral hypothalamus (LH), the electrophysiological actions of NPY have yet to be defined

PVN (Xu et al. 1995). Thus, the hypothalamic and hindbrain NPY systems appear to be linked and to have similar actions in terms of eliciting food intake.

Intracerebroventricular (icv) administration of Y1 agonists increases food intake in rats (Mullins et al. 2001), whereas administration of Y1 antagonists (Daniels et al. 2001; Kanatani et al. 1996; Wieland et al. 1998) or Y1 antisense oligonucleotides (Lopez-Valpuesta et al. 1996a; Lopez-Valpuesta et al. 1996b; Schaffhauser et al. 1998) attenuates both fasting- and NPY-stimulated food intake. Analogously, icv administration of Y5 agonists (Haynes et al. 1998; Lecklin et al. 2003; McCrea et al. 2000; Parker et al. 2000; Wyss et al. 1998) increases food intake, whereas administration of Y5 antagonists (Daniels et al. 2002; Kask et al. 2001; Polidori et al. 2000; Yokosuka et al. 2001) or antisense oligonucleotides (Flynn et al. 1999; Schaffhauser et al. 1997; Tang-Christensen et al. 1998) decreases food intake. There are many reviews of NPY and the specific Y receptors related to energy homeostasis (Balasubramaniam 1997; Beck 2006; Blomqvist and Herzog 1997; Chee et al. 2010; Kamiji and Inui 2007).

In addition to NPY-expressing cells, the ARC also has a population of neurons whose secretions have an opposite action on food intake and energy homeostasis. These neurons express proopiomelanocorticotropin or POMC, which is a large precursor molecule that can be cleaved into a number of possible active smaller peptides (Cone et al. 2001; Cowley et al. 2003). POMC-expressing neurons in the ARC and in the hindbrain cleave the parent molecule into α-melanocyte-stimulating hormone (αMSH). αMSH acts on melanocortin-3 and melanocortin-4 receptors (MC3 and MC4) located in the PVN and elsewhere to elicit a catabolic response, i.e., animals administered MC3/4 agonists into the third ventricle or else directly into the PVN eat less and lose weight (Giraudo et al. 1998; Kim et al. 2000; Thiele et al. 1998a). NPY acts on some of the same target neurons to elicit an anabolic response. Hence, it is the balance of NPY-induced anabolic and αMSH-induced catabolic responses that is considered to be important, and it is also important to note that POMC/αMSH neurons also project directly to ARC NPY neurons, suppressing their activity (Cowley et al. 1999; Kalra et al. 1999) (Fig. 1).

AgRP is an orexigenic peptide that is co-expressed and co-released with NPY in many ARC neurons (Broberger et al. 1998a). Rather than being ubiquitously expressed throughout the brain like NPY, however, AgRP is made only in the ARC. AgRP is an inverse agonist and at physiological concentrations functions as an antagonist at MC3/4 receptors, i.e., it antagonizes the action of αMSH (Quillan et al. 1998). Thus, when these NPY/AgRP neurons are activated, target neurons receive a double-anabolic push as it were. NPY-stimulated Y receptors activate circuits that ultimately lead to taking in more food and accumulating fat, while at the same time AgRP antagonizes MC3/4 receptors and consequently blunts circuits that reduce eating and fat storage (Chee et al. 2010; Cowley et al. 2003; Kalra et al. 1999; Schwartz et al. 2000b; Seeley et al. 2005).

Several points are pertinent. Most (95%) ARC NPY neurons express AgRP, and all AgRP neurons express NPY (Broberger et al. 1998b). Second, other populations of anabolic NPY neurons do not co-express AgRP, such as those in the DMN and NTS. Thus, different populations of NPY-expressing cells should be anticipated to have differential effects on energy homeostasis. Consistent with this, whereas both ARC and DMN NPY neurons densely innervate the PVN and LH, only the DMN NPY neurons, which do not contain AgRP, directly project to hindbrain areas important in controlling energy homeostasis such as the NTS and the dorsal motor nucleus of the vagus (Yang et al. 2009). Overexpressing NPY in the DMN causes increased food intake and body weight, especially on a high-fat diet, and knockdown of NPY uniquely in the DMN attenuates the development of hyperphagia, obesity, and glucose tolerance in rats fed a high-fat diet as well as in OLETF rats (Yang et al. 2009). OLETF rats are an animal model of obesity with compromised cholecystokinin (CCK) receptors and chronically increased meal size thought to be due to increased DMN NPY activity (Moran and Bi 2006).

2 NPY and Food Intake

While perhaps overly simplistic, a common view of one action of NPY is that it serves as an accelerator of food intake in the hypothalamus, whereas αMSH serves as a brake. Neurons in the PVN, VMN, and LH that express Y1, Y5, MC3, and MC4 receptors are among a number of targets for NPY, AgRP, and αMSH (Fig. 1). ARC neurons in particular are considered to receive and integrate diverse signals related to energy needs, readily available energy, and the status of energy stores throughout the body. Consistent with this view, both ARC NPY/AgRP and POMC neurons have receptors for signals related to fat stores (e.g., adiposity signals such as leptin and insulin, and also ghrelin), for signals related to the caloric content of ongoing meals (e.g., satiation signals such as cholecystokinin [CCK] and glucagon-like peptide-1 [GLP-1]), and for local levels of energy-rich nutrients (e.g., glucose and some fatty acids) (Fig. 2). For example, signals that herald low body fat, such as increased ghrelin or low insulin and leptin, stimulate the ARC NPY system

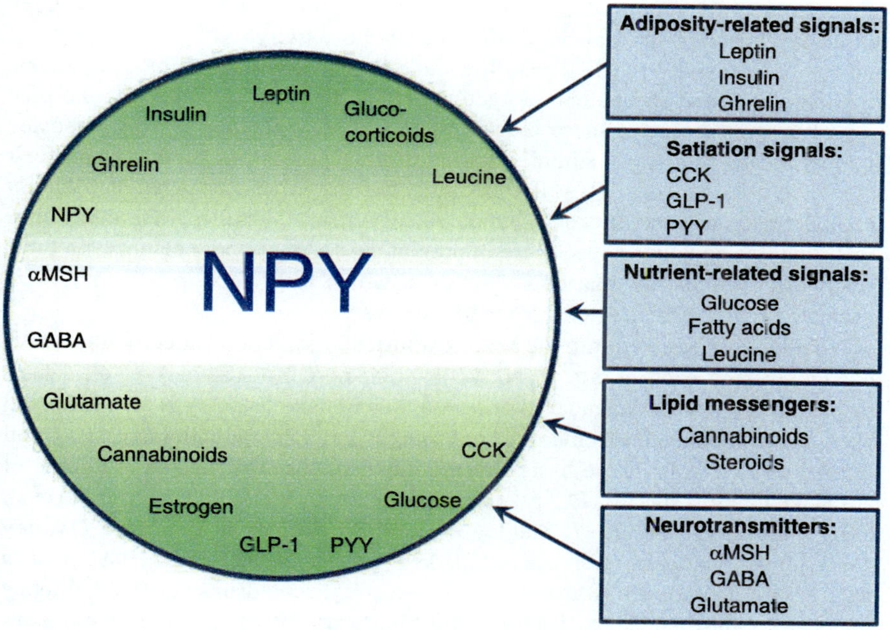

Fig. 2 In addition to neurotransmitters, such as GABA, glutamate, and αMSH, NPY-expressing neurons are responsive to diverse metabolic signals including short-term satiation factors such as cholecystokinin (CCK), glucagon-like peptide-1 (GLP-1), and peptide YY (PYY). NPY-expressing cells also respond to changes in circulating levels of metabolic fuels, including glucose and leucine, and signals of adiposity such as leptin and insulin. As a result, NPY-expressing neurons are able to sense and respond to changes in both short-term, meal-related and long-term, adiposity-related signals important for the control of energy homeostasis, supporting the view the NPY is an integral player in the neural regulation of energy balance

(Baskin et al. 1999; Cone et al. 2001; Hahn et al. 1998; Marks et al. 1992; Schwartz et al. 1993b; Traebert et al. 2002), whereas signals that indicate body fat is elevated, such as increased insulin and leptin, decrease ARC NPY (Cota et al. 2006; Sandoval et al. 2007; Schwartz et al. 1993a, 1991, 1996b, 1992, 2000b; Sipols et al. 1995). Increased levels of satiation signals related to calories being ingested, such as PYY (Riediger et al. 2004) and CCK (Becskei et al. 2009), reduce activity of ARC NPY neurons. Agonists for cannabinoid (Gamber et al. 2005) and opioid receptors that elicit increased food intake (Israel et al. 2005) also increase ARC NPY activity. Administration of drugs that reduce glucose utilization by brain cells increases ARC NPY (Akabayashi et al. 1993; Giraudo et al. 1999; Minami et al. 1995) as well as hindbrain NPY (Li and Ritter 2004), and increasing glucose availability reduces indices of activation in the ARC (Becskei et al. 2009). Corticosterone administered into the brain increases food intake (Green et al. 1992), and corticosterone acts on its receptors on ARC NPY neurons (Ceccatelli et al. 1989) to increase NPY expression (Corder et al. 1988; White et al. 1994a, b), whereas reducing corticosterone levels reduces ARC NPY (Akabayashi et al. 1994; Watanabe et al. 1995).

The important point is that the ARC NPY system is a key component of an important neural circuit where diverse metabolic information is combined and the output is reflected as the balance between anabolic activity (NPY and AgRP tone) and catabolic activity (αMSH) that flows to other circuits controlling food intake, circulating glucose and autonomic activity. It is important to note that the dual control system aspect of these circuits, with different peptides having opposing actions, means that when stimuli such as insulin, ghrelin, or leptin act on their respective receptors on ARC NPY neurons, the consequent changes of behavior (e.g., altered food intake) cannot unequivocally be attributed to NPY alone since AgRP may be equally or even more important depending upon melanocortin tone. Nonetheless, conditions that evoke states in which more food will be eaten are associated with elevated ARC NPY and vice versa.

Whether administered into the cerebroventricular system or directly into sensitive brain areas such as the PVN, VMH, LH, or DMN (Stanley et al. 1985), exogenous NPY dose-dependently increases food intake (Stanley et al. 1986), and when it is administered chronically, body weight and body fat both increase as well (Stanley et al. 1986). These basic observations have been replicated in dozens of labs and numerous species (Henry et al. 2005; Mashiko et al. 2003; Raposinho et al. 2001; Sainsbury et al. 1997; Stanley et al. 1986) including rats, mice (Morley 1987), golden hamsters (Kulkosky et al. 1989), sheep (Miner et al. 1989), snakes (Morris and Crews 1990), white-crowned sparrows (Richardson et al. 2000), rabbits (Pau et al. 1988), chickens (Kuenzel and McMurtry 1988), and rhesus monkeys (Larsen et al. 1999). The major effect of NPY on food intake and increased adiposity is mediated through actions in the central nervous system. However, in the periphery, NPY is released from sympathetic nerve terminals and contributes to increasing vascular tone and blood pressure (Allen et al. 1985; Franco-Cereceda et al. 1985; Petty et al. 1984). Nonetheless, and consistent with its central anabolic function, systemic NPY also reportedly stimulates fat angiogenesis, macrophage infiltration, and proliferation of adipocytes through actions at Y2 receptors (Kuo et al. 2008; Kuo et al. 2007).

Several features of the orexigenic response to NPY are important. It is dose responsive, with higher doses eliciting a greater increment of intake over control injections (Clark et al. 1984; Levine and Morley 1984; Stanley and Leibowitz 1984). The response occurs with a very short latency, typically seconds to minutes. Exogenous NPY also prolongs eating once it has begun, with consequent larger meals being consumed whether NPY acts in the hypothalamus (Leibowitz and Alexander 1991; Lynch et al. 1994; Stricker-Krongrad et al. 1994) or in the hindbrain (Corp et al. 1990; Taylor et al. 2007; Yang et al. 2009). Thus, NPY both facilitates the onset of meals and delays satiation. Finally, the orexigenic response to NPY is long-lasting, often persisting for several hours or more (Clark et al. 1984; Levine and Morley 1984; Stanley and Leibowitz 1984).

When rats are given a choice of diets and administered NPY icv, the bulk of evidence suggests that NPY elicits a relative preference for increasing carbohydrate as opposed to fat intake. This has been observed when rats have a choice of the three individual macronutrients (i.e., access to pure sources of carbohydrate, protein, and fat) (Morley et al. 1987; Stanley et al. 1989, 1985) as well as when they can choose between two complete diets differing in the relative amounts of carbohydrate and fat (Welch et al. 1994). However, it is not entirely clear whether NPY elicits a carbohydrate preference per se as opposed to increasing the consumption of the nutrient that is already preferred by the animals, and there clearly seems to be a strong influence of past experience in the choice situation (Welch et al. 1994). As discussed below, other effects of NPY on energy homeostasis also vary with experience.

3 Endogenous Versus Exogenous NPY

An important question is whether endogenous NPY shares the same properties as exogenous NPY in terms of eliciting eating. There is no doubt that exogenous NPY elicits food intake. In initial attempts to address whether endogenous NPY also elicits food intake, antibodies to NPY were administered intraventricularly or else directly into sensitive areas such as the VMN or PVN, and food intake was decreased relative to controls (Dube et al. 1994; Shibasaki et al. 1993; Walter et al. 1994). Consistent with this, administration of NPY antisense oligonucleotides or else an adeno-associated virus to knockdown endogenous NPY also suppresses food intake (Gardiner et al. 2005; Hulsey et al. 1995). In this regard, it appears that exogenous and endogenous NPY act analogously to increase food intake. In apparent contrast, however, a whole-body genetic knockout of the NPY gene has no obvious phenotype in terms of daily food intake or body weight (Erickson et al. 1996a), but this cannot be easily interpreted since NPY is found in so many brain circuits and influences so many behaviors in addition to those directly related to energy homeostasis, e.g., anxiety (Heilig and Thorsell 2002; Thorsell et al. 2002), reproduction (Kalra and Kalra 2004), and addictions (Thiele et al. 2004). Intake of ethanol, which has both caloric and sedative properties, is actually reduced by NPY, and mice lacking NPY have increased ethanol consumption (Thiele et al. 1998b).

Another way to address the issue is to determine whether the hypothalamic NPY system is regulated by nutritional state or adiposity. Fasting causes both an increased tendency to eat and an upregulation of NPY mRNA in the ARC and NPY protein in the ARC and PVN (Chua et al. 1991; Sahu et al. 1988; Sanacora et al. 1990; Schwartz et al. 1993b), and these increases are reversed by refeeding (Beck et al. 1990b; Sahu et al. 1988). Prolonged food restriction is also associated with increased NPY mRNA in the ARC (McShane et al. 1993; White et al. 1994a, b) and increased NPY protein in the PVN and other areas of the hypothalamus (Beck et al. 1990b; Calza et al. 1989; Sahu et al. 1988). Consistent with these observations, consuming food causes an attenuation of NPY release in the PVN (Sahu et al. 1992a; Yoshihara et al. 1996), whereas food restriction increases PVN NPY release (Kalra et al. 1991; Stricker-Krongrad et al. 1993). Thus, NPY synthesis and secretion are under the influence of nutritional state, with fasting turning the system up and eating turning it down. Certainly, NPY in the hypothalamus is correlated with the likelihood of eating.

4 NPY and Adiposity

Another way to consider the role of endogenous NPY is to see how it varies with body fat. As might be expected, animals with a genetic tendency to become obese have chronically elevated hypothalamic NPY. This has been particularly well documented in animal models of deficient central leptin signaling. Leptin is a hormone secreted from white adipose tissue in proportion to the amount of stored fat; fatter individuals secrete more leptin and leaner individuals secrete less. Leptin is transported through the blood–brain barrier and gains access to leptin receptors throughout the brain (Banks et al. 1999, 1996). In the ARC, NPY-expressing neurons express leptin receptors, and leptin downregulates NPY mRNA. Animals with genetically compromised or absent leptin receptor signaling in the brain have enhanced ARC NPY expression (Schwartz et al. 1996b). This includes the fatty Zucker (*fa/fa*) rat (Beck et al. 1990a; McKibbin et al. 1991; Sanacora et al. 1990), diabetic (*db/db*) (Chua et al. 1991) and leptin-deficient mice (*ob/ob*) (Schwartz et al. 1996a; Wilding et al. 1993), the corpulent (*cp/cp*) rat (Williams et al. 1992), and the Koletsky rat (Keen Rhinehart et al. 2004). When functional leptin receptors were experimentally inserted into the ARC of Koletsky rats, NPY mRNA was reduced (Keen-Rhinehart et al. 2005; Morton et al. 2003). Likewise, reducing leptin signaling in normal mice upregulates ARC NPY and leads to obesity (Cohen et al. 2001).

However, the precise phenotype of leptin-deficient animals is not quite so straightforward as the above discussion might imply. For example, even though fatty Zucker rats have elevated expression of NPY mRNA in the ARC and elevated NPY protein in the PVN and other areas, the expression of Y1 and Y5 receptors in the hypothalamus is downregulated (Beck et al. 2001; Widdowson 1997), and there is less actual binding of NPY to Y5 receptors (Widdowson 1997). Consequently, these animals are less sensitive to the orexigenic action of exogenous NPY (Brief et al. 1992; McCarthy et al. 1991; Stricker-Krongrad et al. 1994). The downregulation of NPY receptors is presumably a compensatory mechanism to counter the increased release of NPY in the PVN in these Zucker fatty rats (Dryden et al. 1995; Stricker-Krongrad et al. 1997). The compensation is not sufficient, however, to keep them from becoming obese.

Like leptin, insulin is a circulating adiposity signal that influences the brain. Both peptide hormones are secreted from peripheral organs in direct proportion to body fat and considered to be adiposity hormones, and both are transported through the blood–brain barrier (Banks et al. 1997, 1996; Banks and Kastin 1998; Levin et al. 2003). ARC neurons express both leptin and insulin receptors, and administration of either leptin or insulin locally in the vicinity of the ARC reduces NPY message and protein and causes a reduction of food intake and body weight, in large part by activating the melanocortin system (Cowley et al. 1999, 2001; Elias et al. 1999; King et al. 2000; Mizuno et al. 1998; Morton and Schwartz 2001). Thus, the data indicate a strong correlation between the actions of endogenous and exogenous NPY.

5 NPY and Energy Expenditure

The above discussion paints a relatively cohesive picture of some of NPY's effects, at least with regard to the actions of exogenous NPY. NPY is generally regarded as a potent signal which elicits an overall anabolic action in the hypothalamus. It is not clear, however, whether NPY works on both sides of the energy equation, i.e., whether NPY influences energy expenditure in addition to energy intake in its anabolic effort. Experiments designed to assess the potential effects of NPY on energy expenditure have provided little evidence to suggest that NPY alters metabolic rate in a way that would contribute to increased adiposity. In one report, NPY had no effect on energy expenditure when administered directly into the PVN, although it did increase the respiratory quotient consistent with an increased utilization of carbohydrates (Menendez et al. 1990). In another report, NPY had differing, sometimes opposing, effects on body temperature when administered into discreet hypothalamic nuclei (Jolicoeur et al. 1995). Specifically, body temperature was decreased when NPY was injected into the ARC and preoptic areas of the hypothalamus, and it was increased when NPY was injected into the PVN. Injections made directly into the LH and VMH had no effect on body temperature, whereas food intake was significantly and dose-dependently increased regardless of the site of injection (Bouali et al. 1995).

Thus, the major effect of NPY on energy balance appears to be on the energy intake side of the equation. That said, the actions of endogenous NPY on energy intake are likely far more complex than the homeostatic model discussed so far. In this regard, it is important to note, and as discussed above, mice that lack NPY, although they have normal body weight and daily food intake (Erickson et al. 1996b), nonetheless eat significantly less food than wild-type mice in the first few hours of the dark (Sindelar et al. 2005), implying that they are not capable of handling the large caloric load in a short period of time as typically occurs early in the dark.

6 Problems with the Model

A number of physiological changes occur prior to a meal that prepare the body to deal with the influx of nutrients. Examples include the cephalic release of insulin (Teff 2000) and other gastrointestinal hormones (e.g., ghrelin (Drazen et al. 2006) and GLP-1 (Vahl et al. 2009)). These changes can be stimulated by the sight and smell of food as well as by the time of day and other meal-related cues. An important question is whether endogenous NPY actually elicits eating as opposed to preparing the body to deal with a pending meal. We believe that the evidence favors the latter possibility for several reasons. For one, NPY message/protein in the ARC is normally increased an hour or so before lights go out, and lights-out is the time that small rodents eat their largest meals of the day (Akabayashi et al. 1994; Jhanwar-Uniyal et al. 1990). While this correlation might be interpreted to indicate that NPY is under the circadian schedule imposed by the light cycle, and that the NPY consequently causes the large meal to be eaten at dark onset, causality is actually more likely in the other direction. This was demonstrated in a study in which rats had only one short window of time to eat each day. When the eating time was situated in the light period, at a time when rats normally eat very little, the NPY pattern shifted, following the food availability, i.e., the ARC-PVN system was turned up an hour or so before the scheduled feeding time (Sahu et al. 1988, 1992b; Yoshihara et al. 1996). This important observation has two major implications. The first is that rather than being under the control of the light cycle per se, increases in NPY signaling are tied to the time of day that food is made available. The second is that the NPY system is increased in anticipation of expected food, i.e., it is under the influence of learning. While it might be argued that the rats that have food available only once a day have increased NPY because they are maximally food restricted at that time, this argument cannot be applied to the increase of NPY prior to lights-out in ad-lib-fed animals (Akabayashi et al. 1994; Jhanwar-Uniyal et al. 1990). A more reasonable conclusion would be that the elevated NPY is preparing the animal to deal with the anticipated caloric load, essentially enabling it to consume a high number of calories in a short period of time (Woods 1991, 2009), i.e., it is as if a pending meal elicits NPY activity rather than NPY activity eliciting a pending meal. Mice that lack NPY, although they have normal body weight and daily food intake (Erickson et al. 1996b), nonetheless eat significantly less food than wild-type mice in the first few hours of the dark (Sindelar et al. 2005), implying that they are not capable of handling the large caloric load typical of wild-type mice.

When NPY is administered into the brain, and especially into the cerebrospinal fluid, it presumably reaches diverse populations of receptors with quite different and perhaps even opposing actions. For example, the increased eating elicited by exogenous NPY might be construed as a hedonically positive behavior. However, the same dose of NPY, administered in the same manner to an animal consuming a novel-flavored food or drink, causes an aversion to be formed to that flavor (Sipols et al. 1992). In one experiment, rats just administered icv NPY were presented with a novel food. They ate more of the food than saline-injected controls, but also formed an aversion to the flavor after a single trial (Sipols 1991). This also occurred in rats whose initial intake of the flavored food was matched to that of controls. Exogenous NPY at doses that increase food intake also cause other symptoms of malaise in rats, including increased geophagia and decreased need-induced sodium intake (Woods et al. 1998). The point is that different populations of NPY receptors have different actions with regard to influencing ingestion. It is reminiscent of some of the central actions of GLP-1 and food intake, with activation of GLP-1 receptors in the hypothalamus contributing to satiation and activation of GLP-1 receptors in the amygdala causing an aversion (Kinzig et al. 2002). The differential actions can only be dissociated by manipulations that impact local receptor populations in discrete brain areas.

Exogenous NPY does not always cause increased food intake. This was originally suggested in rats with intraoral catheters which can deliver small amounts of a liquid directly into the back of the mouth (Seeley et al. 1995b). When a liquid food is slowly infused into their mouths this way, rats swallow the food until they become satiated, at which time they stop swallowing and the infused food simply drains out of the mouth. Such rats typically consume the same amount of food when swallowed in this manner as when they lap the liquid food from a drinking spout, i.e., total meal size is the same whether the rats actively approach the drinking spout, lap up the liquid, and then swallow it (i.e., when they display both the appetitive and the consummatory phases of eating) or else merely swallow food that becomes freely available in their mouth without having to go and obtain it (i.e., when they display only the consummatory phase) (Flynn and Grill 1988). We originally found that when rats are administered icv NPY, they increase their meal size when they have to approach the spout and drink the liquid food, but do not increase their intake when they passively swallow the intraorally infused food, implying that NPY stimulates the appetitive as opposed to the consummatory phases of eating (Seeley et al. 1995b). This finding was replicated by some (Ammar et al. 2000) but not all investigators (Day et al. 2005; Taylor et al. 2007). We subsequently determined that the amount of its prior experience with the testing apparatus is a key determinant of whether or not a rat administered icv NPY increases consumption relative to a control administration (Benoit et al. 2005). There are two important conclusions from these experiments. The first is that the ability of NPY to increase food intake in some situations is modifiable by experience, and the second is that NPY more reliably elicits the appetitive than the consummatory phases of eating. The latter point was dramatically demonstrated by

Ammar and colleagues (Ammar et al. 2000) when they observed that NPY infusion could actually decrease food intake in some social situations.

Using a different approach (Drazen et al. 2005), we found that rats trained to anticipate receiving NPY at a specific time of day learn to eat more at that time even when no NPY is administered. We interpreted these findings to indicate that NPY plays a role in mediating autonomic anticipatory responses that help the individual cope with the effects of large caloric loads, i.e., there is compelling evidence that rather than initiating eating per se, NPY acts upon circuits that prepare and thus enable the animal to consume large meals. The inability of mice lacking NPY to consume large individual meals (Sindelar et al. 2005; Sindelar et al. 2004) is consistent with this interpretation. The observation that animals eat more food when administered exogenous NPY may simply be a manifestation of having been conditioned to eat when NPY receptors in the PVN and elsewhere are stimulated (Woods 2009).

If a primary role of endogenous NPY is to get the individual ready to eat, as opposed to eliciting behavioral eating, administering NPY icv should elicit responses an animal typically makes *before* it starts eating an expected meal, i.e., NPY should elicit the cephalic responses that allow consuming large caloric loads in short periods of time (Strubbe and Woods 2004; Woods 1991, 2009). Consistent with this, icv NPY arouses animals, making them more alert (Szentirmai and Krueger 2006), it causes pancreatic insulin secretion before any food has been ingested (Sainsbury et al. 1997; Yavropoulou et al. 2008), and it increases circulating levels of the gastrointestinal hormones glucose-dependent insulinotropic peptide (GIP) (Yavropoulou et al. 2008) and vasoactive intestinal peptide (Anastasiou et al. 2009).

The distinction we are making is subtle. Does endogenous NPY cause animals to eat or to get ready to eat? If the appropriate cephalic responses have been made, and food is presented, more food can and likely will be eaten (Woods 1991), and if an association has been formed between prior NPY activity and subsequent food availability, exogenous NPY would be anticipated to elicit eating (Woods 2009). The role of conditioning is clearly important since Drazen et al. (Drazen et al. 2005) found that rats presented with stimuli previously associated with receiving NPY ate more food than controls. This is strong evidence that there is a large learning component to NPY and its effects on ingestion.

In a clever series of experiments, Davidson (Seeley et al. 1995a) and colleagues looked at the distinction between getting prepared to eat vs. eating per se in a different way. Rats were trained to associate receiving icv NPY with a mild footshock, whereas no shock was given when the same rats were treated with saline. Rats soon learned to "freeze" when given NPY whether a shock was administered or not, but not when administered saline, indicating that icv NPY activates interoceptive neural circuits that can be used as discriminative stimuli in a conditioning situation. The researchers then asked whether the interoceptive stimuli associated with NPY administration resembled those associated with being food deprived by assessing behavioral immobility or "freezing" inside the chamber during fed and fasted states. The nutritional state of the animals failed to affect freezing behavior indicating that the interoceptive cues-produced NPY were not the same as those produced by caloric deprivation (Seeley et al. 1995a). Interoceptive cues associated with NPY also failed to generalize to cues associated with the glucoprivic drug, 5-thioglucose, which also elicits robust feeding (Altizer and Davidson 1999). These experiments imply that whatever sensations are elicited by NPY, they are not the same as those associated with other states in which rats eat more food, i.e., food deprivation or lowered glucose availability in the brain.

A different point of view as to the effects of NPY is that it increases the motivation to eat (again, as opposed to increasing eating per se). This can be demonstrated by making animals work harder to gain access to food or by making the food less pleasant by either adding a bitter taste to it or applying a mild shock along with the food (Flood and Morley 1991; Jewett et al. 1992, 1995). In all of these instances, icv NPY increases the effort to get food.

To sum up, NPY is certainly not necessary for animals to eat normal amounts of food and maintain a normal amount of body weight (Erickson et al. 1996b). NPY is necessary, however, for animals to be able to eat the larger meals generally consumed early in the dark period (Sindelar et al. 2005) or when they are given drugs that acutely reduce glucose availability to the brain (Sindelar et al. 2004), and all of these increased feeding responses rely upon increased activity of NPY in the PVN (He et al. 1998). Likewise, the hyperphagia typical of insulin-deficient diabetes does not occur in animals lacking NPY (Sindelar et al. 2002). All of these observations are consistent with the concept that endogenous NPY prepares and thus enables the body to be able to cope with the large caloric intake that occurs in these situations.

7 Potential Clinical Applications of NPY

Since the discovery of the orexigenic effects of NPY in rodents, the development of antagonists that block the actions of NPY have been a logical target for the development of pharmacological agents to treat obesity (MacNeil 2007). However, the potential for such agents is limited by the involvement of NPY signaling in so many other behaviors and possibly by the complexity and makeup of the NPY system itself. In part, this relates to the role that NPY plays in stress, anxiety, depression, addiction, cognition, and other behaviors (Eaton et al. 2007). Thus, pharmacological agents that target the NPY system must cross the blood–brain barrier and gain access to relevant sites of action, without causing adverse or off-target effects. In a major clinical trial carried out by Merck and Co., a highly selective Y5 receptor antagonist was well tolerated. However, the overall amount of weight loss was disappointing, with the average participant losing only 3 kg of body weight (Erondu et al. 2006). Additional studies that combined the NPY antagonist with other phamacotherapies also failed to produce greater than expected weight loss when compared to either treatment in isolation (Erondu et al. 2007). This could be because the role of NPY in humans is not entirely analogous to that in other species or because antagonism at the Y5 receptor alone is not sufficient to produce significant weight loss in humans. At present, evidence for the clinical efficacy of NPY receptor antagonists is lacking. Whether or not a selective Y1 receptor antagonist, or pharmacological agents that target the Y1 and Y5 receptor simultaneously, can produce clinically meaningful weight loss is not known, and to this extent, NPY remains a viable target for the treatment of obesity.

8 Conclusions

When administered into the brain, NPY acts at Y1 and Y5 receptors to increase food intake. The response occurs with a short latency and is quite robust, such that exogenous NPY is generally considered to be the most potent of a growing list of orexigenic compounds that act in the brain. The role of endogenous NPY is not so straightforward, however. Evidence from diverse types of experiments suggests that rather than initiating behavioral eating per se, endogenous NPY elicits autonomic responses that prepare the individual to better cope with consuming a calorically large meal.

References

Akabayashi A, Levin N, Paez X, Alexander JT, Leibowitz SF (1994) Hypothalamic neuropeptide Y and its gene expression: relation to light/dark cycle and circulating corticosterone. Mol Cell Neurosci 5:210–218

Akabayashi A, Zaia CT, Silva I, Chae HJ, Leibowitz SF (1993) Neuropeptide Y in the arcuate nucleus is modulated by alterations in glucose utilization. Brain Res 621:343–348

Allen JM, Polak JM, Rodrigo J, Darcy K, Bloom SR (1985) Localisation of neuropeptide Y in nerves of the rat cardiovascular system and the effect of 6-hydroxydopamine. Cardiovasc Res 19:570–577

Allen YS, Adrian TE, Tatemoto K, Crow TJ, Bloom SR, Polak JM (1983) Neuropeptide Y distribution in the rat brain. Science 221:877–879

Altizer AM, Davidson TL (1999) The effects of NPY and 5-TG on responding to cues for fats and carbohydrates. Physiol Behav 65:685–690

Ammar AA, Sederholm F, Saito TR, Scheurink AJ, Johnson AE, Sodersten P (2000) NPY-leptin: opposing effects on appetitive and consummatory ingestive behavior and sexual behavior. Am J Physiol Regul Integr Comp Physiol 278:R1627–1633

Anastasiou OE, Yavropoulou MP, Kesisoglou I, Kotsa K, Yovos JG (2009) Intracerebroventricular infusion of neuropeptide Y modulates VIP secretion in the fasting conscious dog. Neuropeptides 43:41–46

Balasubramaniam AA (1997) Neuropeptide Y family of hormones: receptor subtypes and antagonists. Peptides 18:445–457

Banks WA, DiPalma CR, Farrell CL (1999) Impaired transport of leptin across the blood-brain barrier in obesity. Peptides 20:1341–1345

Banks WA, Jaspan JB, Kastin AJ (1997) Selective, physiological transport of insulin across the blood-brain barrier: novel demonstration by species-specific radioimmunoassays. Peptides 28:1257–1262

Banks WA, Kastin AJ (1998) Differential permeability of the blood-brain barrier to two pancreatic peptides: insulin and amylin. Peptides 19:883–889
Banks WA, Kastin AJ, Huang W (1996) Leptin enters the brain by a saturable system independent of insulin. Peptides 17:305–311
Baskin DG, Breininger JF, Schwartz MW (1999) Leptin receptor mRNA identifies a subpopulation of neuropeptide Y neurons activated by fasting in rat hypothalamus. Diabetes 48:828–833
Beck B (2006) Neuropeptide Y in normal eating and in genetic and dietary-induced obesity. Philos Trans R Soc Lond B Biol Sci 361:1159–1185
Beck B, Burlet A, Nicolas JP, Burlet C (1990a) Hypothalamic neuropeptide Y (NPY) in obese Zucker rats: implications in feeding and sexual behaviors. Physiol Behav 47:449–453
Beck B, Jhanwar-Uniyal M, Burlet A, Chapleur-Chateau M, Leibowitz SF, Burlet C (1990b) Rapid and localized alterations of neuropeptide Y in discrete hypothalamic nuclei with feeding status. Brain Res 528:245–249
Beck B, Richy S, Dimitrov T, Stricker-Krongrad A (2001) Opposite regulation of hypothalamic orexin and neuropeptide Y receptors and peptide expressions in obese Zucker rats. Biochem Biophys Res Commun 286:518–523
Becskei C, Lutz TA, Riediger T (2009) Diet-derived nutrients mediate the inhibition of hypothalamic NPY neurons in the arcuate nucleus of mice during refeeding. Am J Physiol Regul Integr Comp Physiol 297:R100–110
Benoit SC, Clegg DJ, Woods SC, Seeley RJ (2005) The role of previous exposure in the appetitive and consummatory effects of orexigenic neuropeptides. Peptides 26:751–757
Blomqvist AG, Herzog H (1997) Y-receptor subtypes – how many more? Trends Neurosci 20:294–298
Bouali SM, Wimalawansa SJ, Jolicoeur FB (1995) In vivo central actions of rat amylin. Regul Pept 56:167–174
Brief DJ, Sipols AJ, Woods SC (1992) Intraventricular neuropeptide Y injections stimulate food intake in lean but not obese Zucker rats. Physiol Behav 51:1105–1110
Broberger C, De Lecea L, Sutcliffe JG, Hokfelt T (1998a) Hypocretin/orexin- and melanin-concentrating hormone-expressing cells form distinct populations in the rodent lateral hypothalamus: relationship to the neuropeptide Y and agouti gene-related protein systems. J Comp Neurol 402:460–474
Broberger C, Johansen J, Johansson C, Schalling M, Hokfelt T (1998b) The neuropeptide Y/agouti gene-related protein (AGRP) brain circuitry in normal, anorectic, and monosodium glutamate-treated mice. Proc Natl Acad Sci USA 95:15043–15048
Calza L, Giardino L, Battistini N, Zanni M, Galetti S, Protopapa F, Velardo A (1989) Increase of neuropeptide Y-like immunoreactivity in the paraventricular nucleus of fasting rats. Neurosci Lett 104:99–104
Ceccatelli S, Cintra A, Hokfelt T, Fuxe K, Wikstrom AC, Gustafsson JA (1989) Coexistence of glucocorticoid receptor-like immunoreactivity with neuropeptides in the hypothalamic paraventricular nucleus. Exp Brain Res 78:33–42
Chee MJ, Colmers WF (2008) Y eat? Nutrition 24:869–877
Chee MJ, Myers MG Jr, Price CJ, Colmers WF (2010) Neuropeptide Y suppresses anorexigenic output from the ventromedial nucleus of the hypothalamus. J Neurosci 30:3380–3390
Chua SC, Brown AW, Kim J, Hennessey KL, Leibel RL, Hirsch J (1991) Food deprivation and hypothalamic neuropeptide gene expression: effects of strain background and the *diabetes* mutation. Mol Brain Res 11:291–299
Clark JT, Kalra PS, Crowley WR, Kalra SP (1984) Neuropeptide Y and human pancreatic polypeptide stimulate feeding behavior in rats. Endocrinology 115:427–429
Cohen P, Zhao C, Cai X, Montez JM, Rohani SC, Feinstein P, Mombaerts P, Friedman JM (2001) Selective deletion of leptin receptor in neurons leads to obesity. J Clin Invest 108:1113–1121
Cone RD, Cowley MA, Butler AA, Fan W, Marks DL, Low MJ (2001) The arcuate nucleus as a conduit for diverse signals relevant to energy homeostasis. Int J Obes Relat Metab Disord 25 (Suppl 5):S63–67

Corder R, Pralong F, Turnhill D, Saudan P, Muller AF, Gaillard RC (1988) Dexamethasone treatment increases neuropeptide Y levels in rat hypothalamic neurons. Life Sci 43:1879–1886
Corp ES, Melville LD, Greenberg D, Gibbs J, Smith GP (1990) Effect of fourth ventricular neuropeptide Y and peptide YY on ingestive and other behaviors. Am J Physiol 259: R317–R323
Cota D, Proulx K, Smith KA, Kozma SC, Thomas G, Woods SC, Seeley RJ (2006) Hypothalamic mTOR signaling regulates food intake. Science 312:927–930
Cowley MA, Cone RD, Enriori P, Louiselle I, Williams SM, Evans AE (2003) Electrophysiological actions of peripheral hormones on melanocortin neurons. Ann N Y Acad Sci 994:175–186
Cowley MA, Pronchuk N, Fan W, Dinulescu DM, Colmers WF, Cone RD (1999) Integration of NPY, AGRP, and melanocortin signals in the hypothalamic paraventricular nucleus: evidence of a cellular basis for the adipostat. Neuron 24:155–163
Cowley MA, Smart JL, Rubinstein M, Cerdan MG, Diano S, Horvath TL, Cone RD, Low MJ (2001) Leptin activates anorexigenic POMC neurons through a neural network in the arcuate nucleus. Nature 411:480–484
Daniels AJ, Chance WT, Grizzle MK, Heyer D, Matthews JE (2001) Food intake inhibition and reduction in body weight gain in rats treated with GI264879A, a non-selective NPY-Y1 receptor antagonist. Peptides 22:483–491
Daniels AJ, Grizzle MK, Wiard RP, Matthews JE, Heyer D (2002) Food intake inhibition and reduction in body weight gain in lean and obese rodents treated with GW438014A, a potent and selective NPY-Y5 receptor antagonist. Regul Pept 106:47–54
Day DE, Keen-Rhinehart E, Bartness TJ (2005) Role of NPY and its receptor subtypes in foraging, food hoarding, and food intake by Siberian hamsters. Am J Physiol Regul Integr Comp Physiol 289:R29–36
Drazen DL, Vahl TP, D'Alessio DA, Seeley RJ, Woods SC (2006) Effects of a fixed meal pattern on ghrelin secretion: evidence for a learned response independent of nutrient status. Endocrinology 147:23–30
Drazen DL, Wortman MD, Seeley RJ, Woods SC (2005) Neuropeptide Y prepares rats for scheduled feeding. Am J Physiol Regul Integr Comp Physiol 288:R1606–1611
Dryden S, Pickavance L, Frankish HM, Williams G (1995) Increased neuropeptide Y secretion in the hypothalamic paraventricular nucleus of obese (fa/fa) Zucker rats. Brain Res 690:185–188
Dube MG, Xu B, Crowley WR, Kalra PS, Kalra SP (1994) Evidence that neuropeptide Y is a physiological signal for normal food intake. Brain Res 646:341–344
Eaton K, Sallee FR, Sah R (2007) Relevance of neuropeptide Y (NPY) in psychiatry. Curr Top Med Chem 7:1645–1659
Elias CF, Aschkenasi C, Lee C, Kelly J, Ahima RS, Bjorbaek C, Flier JS, Saper CB, Elmquist JK (1999) Leptin differentially regulates NPY and POMC neurons projecting to the lateral hypothalamic area. Neuron 23:775–786
Erickson JC, Clegg KE, Palmiter RD (1996a) Sensitivity to leptin and susceptibility to seizures of mice lacking neuropeptide Y. Nature 381:415–418
Erickson JC, Hollopeter G, Palmiter RD (1996b) Attenuation of the obesity syndrome of ob/ob mice by the loss of neuropeptide Y. Science 274:1704–1707
Erondu N, Addy C, Lu K, Mallick M, Musser B, Gantz I, Proietto J, Astrup A, Toubro S, Rissannen AM, Tonstad S, Haynes WG, Gottesdiener KM, Kaufman KD, Amatruda JM, Heymsfield SB (2007) NPY5R antagonism does not augment the weight loss efficacy of orlistat or sibutramine. Obesity (Silver Spring) 15:2027–2042
Erondu N, Gantz I, Musser B, Suryawanshi S, Mallick M, Addy C, Cote J, Bray G, Fujioka K, Bays H, Hollander P, Sanabria-Bohorquez SM, Eng W, Langstrom B, Hargreaves RJ, Burns HD, Kanatani A, Fukami T, MacNeil DJ, Gottesdiener KM, Amatruda JM, Kaufman KD, Heymsfield SB (2006) Neuropeptide Y5 receptor antagonism does not induce clinically meaningful weight loss in overweight and obese adults. Cell Metab 4:275–282
Flood JF, Morley JE (1991) Increased food intake by neuropeptide Y is due to an increased motivation to eat. Peptides 12:1329–1332

Flynn FW, Grill HJ (1988) Intraoral intake and taste reactivity responses elicited by sucrose and sodium chloride in chronic decerebrate rats. Behav Neurosci 102:934–941
Flynn MC, Turrin NP, Plata-Salaman CR, Ffrench-Mullen JM (1999) Feeding response to neuropeptide Y-related compounds in rats treated with Y5 receptor antisense or sense phosphothio-oligodeoxynucleotide. Physiol Behav 66:881–884
Franco-Cereceda A, Lundberg JM, Dahlof C (1985) Neuropeptide Y and sympathetic control of heart contractility and coronary vascular tone. Acta Physiol Scand 124:361–369
Gamber KM, Macarthur H, Westfall TC (2005) Cannabinoids augment the release of neuropeptide Y in the rat hypothalamus. Neuropharmacology 49:646–652
Gardiner JV, Kong WM, Ward H, Murphy KG, Dhillo WS, Bloom SR (2005) AAV mediated expression of anti-sense neuropeptide Y cRNA in the arcuate nucleus of rats results in decreased weight gain and food intake. Biochem Biophys Res Commun 327:1088–1093
Giraudo SQ, Billington CJ, Levine AS (1998) Feeding effects of hypothalamic injection of melanocortin 4 receptor ligands. Brain Res 809:302–306
Giraudo SQ, Grace MK, Billington CJ, Levine AS (1999) Differential effects of neuropeptide Y and the mu-agonist DAMGO on 'palatability' vs. 'energy'. Brain Res 834:160–163
Gray TS, Morley JE (1986) Neuropeptide Y: anatomical distribution and possible function in mammalian nervous system. Life Sci 38:389–401
Green PK, Wilkinson CW, Woods SC (1992) Intraventricular corticosterone increases the rate of body weight gain in underweight adrenalectomized rats. Endocrinology 130:269–275
Hahn TM, Breininger JF, Baskin DG, Schwartz MW (1998) Coexpression of Agrp and NPY in fasting-activated hypothalamic neurons. Nat Neurosci 1:271–272
Haynes AC, Arch JR, Wilson S, McClue S, Buckingham RE (1998) Characterisation of the neuropeptide Y receptor that mediates feeding in the rat: a role for the Y5 receptor? Regul Pept 75–76:355–361
He B, White BD, Edwards GL, Martin RJ (1998) Neuropeptide Y antibody attenuates 2-deoxy-D-glucose induced feeding in rats. Brain Res 781:348–350
Heilig M, Thorsell A (2002) Brain neuropeptide Y (NPY) in stress and alcohol dependence. Rev Neurosci 13:85–94
Henry M, Ghibaudi L, Gao J, Hwa JJ (2005) Energy metabolic profile of mice after chronic activation of central NPY Y1, Y2, or Y5 receptors. Obes Res 13:36–47
Hulsey MG, Pless CM, White BD, Martin RJ (1995) ICV administration of anti-NPY antisense oligonucleotide: effects on feeding behavior, body weight, peptide content and peptide release. Regul Pept 59:207–214
Ishizaki K, Honma S, Katsuno Y, Abe H, Masubuchi S, Namihira M, Honma K (2003) Gene expression of neuropeptide Y in the nucleus of the solitary tract is activated in rats under restricted daily feeding but not under 48-h food deprivation. Eur J Neurosci 17:2097–2105
Israel Y, Kandov Y, Khaimova E, Kest A, Lewis SR, Pasternak GW, Pan YX, Rossi GC, Bodnar RJ (2005) NPY-induced feeding: pharmacological characterization using selective opioid antagonists and antisense probes in rats. Peptides 26:1167–1175
Jewett D, Cleary J, Levine A, Schaal D, Thompson T (1992) Effects of neuropeptide Y on food-reinforced behavior in satiated rats. Pharmacol Biochem Behav 42:207–212
Jewett D, Cleary J, Levine A, Schaal D, Thompson T (1995) Effects of neuropeptide Y, insulin, 2-deoxyglucose, and food deprivation on food-motivated behavior. Psychpharmacology 120:267–271
Jhanwar-Uniyal M, Beck B, Burlet C, Leibowitz SF (1990) Diurnal rhythm of neuropeptide Y-like immunoreactivity in the suprachiasmatic, arcuate and paraventricular nuclei and other hypothalamic sites. Brain Res 536:331–334
Jolicoeur FB, Bouali SM, Fournier A, St-Pierre S (1995) Mapping of hypothalamic sites involved in the effects of NPY on body temperature and food intake. Brain Res Bull 36:125–129
Kalra SP, Dube MG, Pu S, Xu B, Horvath TL, Kalra PS (1999) Interacting appetite-regulating pathways in the hypothalamic regulation of body weight. Endocr Rev 20:68–100
Kalra SP, Dube MG, Sahu A, Phelps CP, Kalra P (1991) Neuropeptide Y secretion increases in the paraventricular nucleus in association with increased appetite for food. Proc Natl Acad Sci USA 88:10931–10935

Kalra SP, Kalra PS (2004) NPY–an endearing journey in search of a neurochemical on/off switch for appetite, sex and reproduction. Peptides 25:465–471

Kamiji MM, Inui A (2007) Neuropeptide y receptor selective ligands in the treatment of obesity. Endocr Rev 28:664–684

Kanatani A, Ishihara A, Asahi S, Tanaka T, Ozaki S, Ihara M (1996) Potent neuropeptide Y Y1 receptor antagonist, 1229U91: blockade of neuropeptide Y-induced and physiological food intake. Endocrinology 137:3177–3182

Kask A, Vasar E, Heidmets LT, Allikmets L, Wikberg JE (2001) Neuropeptide Y Y(5) receptor antagonist CGP71683A: the effects on food intake and anxiety-related behavior in the rat. Eur J Pharmacol 414:215–224

Keen-Rhinehart E, Kalra SP, Kalra PS (2005) AAV-mediated leptin receptor installation improves energy balance and the reproductive status of obese female Koletsky rats. Peptides 26:2567–2578

Keen Rhinehart E, Kalra SP, Kalra PS (2004) Neuropeptidergic characterization of the leptin receptor mutated obese Koletsky rat. Regul Pept 119:3–10

Kim MS, Rossi M, Abusnana S, Sunter D, Morgan DG, Small CJ, Edwards CM, Heath MM, Stanley SA, Seal LJ, Bhatti JR, Smith DM, Ghatei MA, Bloom SR (2000) Hypothalamic localization of the feeding effect of agouti-related peptide and alpha-melanocyte-stimulating hormone. Diabetes 49:177–182

King PJ, Widdowson PS, Doods H, Williams G (2000) Regulation of neuropeptide Y release from hypothalamic slices by melanocortin-4 agonists and leptin. Peptides 21:45–48

Kinzig KP, D'Alessio DA, Seeley RJ (2002) The diverse roles of specific GLP-1 receptors in the control of food intake and the response to visceral illness. J Neurosci 22:10470–10476

Kuenzel WJ, McMurtry J (1988) Neuropeptide Y: brain localization and central effects on plasma insulin levels in chicks. Physiol Behav 44:669–678

Kulkosky PJ, Glazner GW, Moore HD, Low CA, Woods SC (1989) Neuropeptide Y: behavioral effects in the golden hamster. Peptides 9:1389–1393

Kuo LE, Czarnecka M, Kitlinska JB, Tilan JU, Kvetnansky R, Zukowska Z (2008) Chronic stress, combined with a high-fat/high-sugar diet, shifts sympathetic signaling toward neuropeptide Y and leads to obesity and the metabolic syndrome. Ann N Y Acad Sci 1148:232–237

Kuo LE, Kitlinska JB, Tilan JU, Li L, Baker SB, Johnson MD, Lee EW, Burnett MS, Fricke ST, Kvetnansky R, Herzog H, Zukowska Z (2007) Neuropeptide Y acts directly in the periphery on fat tissue and mediates stress-induced obesity and metabolic syndrome. Nat Med 13:803–811

Larsen PJ, Tang-Christensen M, Stidsen CE, Madsen K, Smith MS, Cameron JL (1999) Activation of central neuropeptide Y Y1 receptors potently stimulates food intake in male rhesus monkeys. J Clin Endocrinol Metab 84:3781–3791

Lecklin A, Lundell I, Salmela S, Mannisto PT, Beck-Sickinger AG, Larhammar D (2003) Agonists for neuropeptide Y receptors Y1 and Y5 stimulate different phases of feeding in guinea pigs. Br J Pharmacol 139:1433–1440

Leibowitz SF, Alexander JT (1991) Analysis of neuropeptide Y-induced feeding: dissociation of Y1 and Y2 receptor effects on natural meal patterns. Peptides 12:1251–1260

Levin BE, Dunn-Meynell AA, Banks WA (2003) Obesity-prone rats have normal blood-brain barrier transport but defective central leptin signaling prior to obesity onset. Am J Physiol Regul Integr Comp Physiol 286:143–150

Levine AS, Morley JE (1984) Neuropeptide Y: a potent inducer of consummatory behavior in rats. Peptides 5:1025–1029

Li AJ, Ritter S (2004) Glucoprivation increases expression of neuropeptide Y mRNA in hindbrain neurons that innervate the hypothalamus. Eur J Neurosci 19:2147–2154

Lopez-Valpuesta FJ, Nyce JW, Griffin-Biggs TA, Ice JC, Myers RD (1996a) Antisense to NPY-Y1 demonstrates that Y1 receptors in the hypothalamus underlie NPY hypothermia and feeding in rats. Proc Biol Sci 263:881–886

Lopez-Valpuesta FJ, Nyce JW, Myers RD (1996b) NPY-Y1 receptor antisense injected centrally in rats causes hyperthermia and feeding. Neuroreport 7:2781–2784

Lundberg JM, Terenius L, Hokfelt T, Martling CR, Tatemoto K, Mutt V, Polak J, Bloom S, Goldstein M (1982) Neuropeptide Y (NPY)-like immunoreactivity in peripheral noradrenergic neurons and effects of NPY on sympathetic function. Acta Physiol Scand 116:477–480

Lynch WC, Hart P, Babcock AM (1994) Neuropeptide Y attenuates satiety: evidence from a detailed analysis of patterns ingestion. Brain Res 636:28–34

MacNeil DJ (2007) NPY Y1 and Y5 receptor selective antagonists as anti-obesity drugs. Curr Top Med Chem 7:1721–1733

Mahaut S, Dumont Y, Fournier A, Quirion R, Moyse E (2010) Neuropeptide Y receptor subtypes in the dorsal vagal complex under acute feeding adaptation in the adult rat. Neuropeptides 44:77–86

Marks JL, Mu L, Schwartz MW, Porte D Jr, Baskin DG (1992) Effect of fasting on regional levels of neuropeptide Y mRNA and insulin receptors in the rat hypothalamus: An autoradiographic study. Mol Cell Neurosci 3:199–205

Mashiko S, Ishihara A, Iwaasa H, Sano H, Oda Z, Ito J, Yumoto M, Okawa M, Suzuki J, Fukuroda T, Jitsuoka M, Morin NR, MacNeil DJ, Van der Ploeg LH, Ihara M, Fukami T, Kanatani A (2003) Characterization of neuropeptide Y (NPY) Y5 receptor-mediated obesity in mice: chronic intracerebroventricular infusion of D-Trp(34)NPY. Endocrinology 144:1793–1801

McCarthy HD, McKibbin PE, Holloway B, Mayers R, Williams G (1991) Hypothalamic neuropeptide Y receptor characteristics and NPY-induced feeding responses in lean and obese Zucker rats. Life Sci 49:1491–1497

McCrea K, Wisialowski T, Cabrele C, Church B, Beck-Sickinger A, Kraegen E, Herzog H (2000) 2–36[K4, RYYSA(19–23)]PP a novel Y5-receptor preferring ligand with strong stimulatory effect on food intake. Regul Pept 87:47–58

McKibbin PE, Cotton SJ, McMillan S, Holloway B, Mayers R, McCarthy HD, Williams G (1991) Altered neuropeptide Y concentrations in specific hypothalamic regions of obese (fa/fa) Zucker rats. Possible relationship to obesity and neuroendocrine disturbances. Diabetes 40:1423–1429

McShane TM, Petersen SL, McCrone S, Keisler DH (1993) Influence of food restriction on neuropeptide-Y, proopiomelanocortin, and luteinizing hormone-releasing hormone gene expression in sheep hypothalami. Biol Reprod 49:831–839

Menendez JA, McGregor IS, Healey PA, Atrens DM, Leibowitz SF (1990) Metabolic effects of neuropeptide Y injections into the paraventricular nucleus of the hypothalamus. Brain Res 516:8–14

Minami S, Kamegai J, Sugihara H, Suzuki N, Higuchi H, Wakabayashi I (1995) Central glucoprivation evoked by administration of 2-deoxy-D-glucose induces expression of the c-fos gene in a subpopulation of neuropeptide Y neurons in the rat hypothalamus. Brain Res Mol Brain Res 33:305–310

Miner JL, Della-Fera MA, Paterson JA, Baile CA (1989) Lateral cerebroventricular injection of neuropeptide Y stimulates feeding in sheep. Am J Physiol 257:R383–387

Mizuno T, Kleopoulos S, Bergen H, Roberts J, Priest C, Mobbs C (1998) Hypothalamic proopiomelanocortin mRNA is reduced by fasting and in ob/ob and db/db mice, but is stimulated by leptin. Diabetes 47:294–297

Moran TH, Bi S (2006) Hyperphagia and obesity in OLETF rats lacking CCK-1 receptors. Philos Trans R Soc Lond B Biol Sci 361:1211–1218

Morley JE (1987) Neuropeptide regulation of appetite and weight. Endocrine Rev 8:256–287

Morley JE, Levine AS, Gosnell BA, Kneip J, Grace M (1987) Effect of neuropeptide Y on ingestive behaviors in the rat. Am J Physiol 252:R599–R609

Morris YA, Crews D (1990) The effects of exogenous neuropeptide Y on feeding and sexual behavior in the red-sided garter snake (Thamnophis sirtalis parietalis). Brain Res 530:339–341

Morton GJ, Niswender KD, Rhodes CJ, Myers MG, Blevins JE, Baskin DG, Schwartz MW (2003) Arcuate nucleus-specific leptin receptor gene therapy attenuates the obesity phenotype of Koletsky (fa(k)/fa(k)) rats. Endocrinology 144:2016–2024

Morton GJ, Schwartz MW (2001) The NPY/AgRP neuron and energy homeostasis. Int J Obes Relat Metab Disord 25(Suppl 5):S56–62

Mullins D, Kirby D, Hwa J, Guzzi M, Rivier J, Parker E (2001) Identification of potent and selective neuropeptide Y Y(1) receptor agonists with orexigenic activity in vivo. Mol Pharmacol 60:534–540
Parker EM, Balasubramaniam A, Guzzi M, Mullins DE, Salisbury BG, Sheriff S, Witten MB, Hwa JJ (2000) [D-Trp(34)] neuropeptide Y is a potent and selective neuropeptide Y Y(5) receptor agonist with dramatic effects on food intake. Peptides 21:393–399
Pau MY, Pau KY, Spies HG (1988) Characterization of central actions of neuropeptide Y on food and water intake in rabbits. Physiol Behav 44:797–802
Petty MA, Dietrich R, Lang RE (1984) The cardiovascular effects of neuropeptide Y (NPY). Clin Exp Hypertens A 6:1889–1892
Polidori C, Ciccocioppo R, Regoli D, Massi M (2000) Neuropeptide Y receptor(s) mediating feeding in the rat: characterization with antagonists. Peptides 21:29–35
Quillan JM, Sadee W, Wei ET, Jimenez C, Ji L, Chang JK (1998) A synthetic human Agouti-related protein-(83–132)-NH_2 fragment is a potent inhibitor of melanocortin receptor function. FEBS Lett 428:59–62
Raposinho PD, Pierroz DD, Broqua P, White RB, Pedrazzini T, Aubert ML (2001) Chronic administration of neuropeptide Y into the lateral ventricle of C57BL/6J male mice produces an obesity syndrome including hyperphagia, hyperleptinemia, insulin resistance, and hypogonadism. Mol Cell Endocrinol 185:195–204
Richardson RD, Boswell T, Woods SC, Wingfield JC (2000) Intracerebroventricular corticotropin-releasing factor decreases food intake in white-crowned sparrows. Physiol Behav 71:213–216
Riediger T, Bothe C, Becskei C, Lutz TA (2004) Peptide YY directly inhibits ghrelin-activated neurons of the arcuate nucleus and reverses fasting-induced c-Fos expression. Neuroendocrinology 79:317–326
Sahu A, Kalra PS, Kalra SP (1988) Food deprivation and ingestion induce reciprocal changes in neuropeptide Y concentrations in the paraventricular nucleus. Peptides 9:83–86
Sahu A, Sninsky CA, Phelps CP, Dube MG, Kalra PS, Kalra SP (1992a) Neuropeptide Y release from the paraventricular nucleus increases in association with hyperphagia in streptozotocin-induced diabetic rats. Endocrinology 131(6):2979–2985
Sahu A, White JD, Kalra PS, Kalra SP (1992b) Hypothalamic neuropeptide Y gene expression in rats on a scheduled feeding regimen. Mol Brain Res 15:15–18
Sainsbury A, Rohner-Jeanrenaud F, Cusin I, Zakrzewska KE, Halban PA, Gaillard RC, Jeanrenaud B (1997) Chronic central neuropeptide Y infusion in normal rats: status of the hypothalamo-pituitary-adrenal axis, and vagal mediation of hyperinsulinaemia. Diabetologia 40:1269–1277
Sanacora G, Kershaw M, Finkelstein JA, White JD (1990) Increased hypothalamic content of preproneuropeptide Y mRNA in genetically obese Zucker rats and its regulation by food deprivation. Endocrinology 127:730–737
Sandoval DA, Cota D, Seeley RJ (2007) The integrative role of CNS fuel-sensing mechanisms in energy balance and glucose regulation. Ann Rev Physiol 70:513–535
Schaffhauser AO, Stricker-Krongrad A, Brunner L, Cumin F, Gerald C, Whitebread S, Criscione L, Hofbauer KG (1997) Inhibition of food intake by neuropeptide Y Y5 receptor antisense oligodeoxynucleotides. Diabetes 46:1792–1798
Schaffhauser AO, Whitebread S, Haener R, Hofbauer KG, Stricker-Krongrad A (1998) Neuropeptide Y Y1 receptor antisense oligodeoxynucleotides enhance food intake in energy-deprived rats. Regul Pept 75–76:417–423
Schwartz MW, Baskin DG, Bukowski TR, Kuijper JL, Foster D, Lasser G, Prunkard DE, Porte D, Woods SC, Seeley RJ, Weigle DS (1996a) Specificity of leptin action on elevated blood glucose levels and hypothalamic neuropeptide Y gene expression in *ob/ob* mice. Diabetes 45:531–535
Schwartz MW, Figlewicz DP, Woods SC, Porte D, Baskin DG (1993a) Insulin, neuropeptide Y, and food intake. Ann New York Acad Sci 692:60–71
Schwartz MW, Marks J, Sipols AJ, Baskin DG, Woods SC, Kahn SE, Porte D Jr (1991) Central insulin administration reduces neuropeptide Y mRNA expression in the arcuate nucleus of food-deprived lean (Fa/Fa) but not obese (fa/fa) Zucker rats. Endocrinology 128:2645–2647

Schwartz MW, Seeley RJ, Campfield LA, Burn P, Baskin DG (1996b) Identification of hypothalamic targets of leptin action. J Clin Invest 98:1101–1106
Schwartz MW, Sipols AJ, Grubin CE, Baskin DG (1993b) Differential effect of fasting on hypothalamic expression of genes encoding neuropeptide Y, galanin and glutamic acid decarboxylase. Brain Res Bull 31:361–367
Schwartz MW, Sipols AJ, Marks JL, Sanacora G, White JD, Scheurinck A, Kahn SE, Baskin DG, Woods SC, Figlewicz DP, Porte D Jr (1992) Inhibition of hypothalamic neuropeptide Y gene expression by insulin. Endocrinology 130:3608–3616
Schwartz MW, Woods SC, Porte D Jr, Seeley RJ, Baskin DG (2000) Central nervous system control of food intake. Nature 404:661–671
Seeley RJ, Benoit SC, Davidson TL (1995a) The discriminative cues produced by NPY administration do not generalize to the interoceptive cues produced by food deprivation. Physiol Behav 58:1237–1241
Seeley RJ, Burklow ML, Wilmer KA, Matthews CC, Reizes O, McOsker CC, Trokhan DP, Gross MC, Sheldon RJ (2005) The effect of the melanocortin agonist, MT-II, on the defended level of body adiposity. Endocrinology 146:3732–3738
Seeley RJ, Payne CJ, Woods SC (1995b) Neuropeptide Y fails to increase intraoral intake in rats. Am J Physiol 268:R423–R427
Shibasaki T, Oda T, Imaki T, Ling N, Demura H (1993) Injection of anti-neuropeptide Y g-globulin into the hypothalamic paraventricular nucleus decreases food intake in rats. Brain Res 601:313–316
Sindelar DK, Mystkowski P, Marsh DJ, Palmiter RD, Schwartz MW (2002) Attenuation of diabetic hyperphagia in neuropeptide Y–deficient mice. Diabetes 51:778–783
Sindelar DK, Palmiter RD, Woods SC, Schwartz MW (2005) Attenuated feeding responses to circadian and palatability cues in mice lacking neuropeptide Y. Peptides 26:2597–2602
Sindelar DK, Ste Marie L, Miura GI, Palmiter RD, McMinn JE, Morton GJ, Schwartz MW (2004) Neuropeptide Y is required for hyperphagic feeding in response to neuroglucopenia. Endocrinology 145:3363–3368
Sipols AJ (1991) Neuropeptide Y stimulates food intake and causes conditioned taste aversions in rats. In: Psychology. University of Washington, Seattle
Sipols AJ, Baskin DG, Schwartz MW (1995) Effect of intracerebroventricular insulin infusion on diabetic hyperphagia and hypothalamic neuropeptide gene expression. Diabetes 44:147–151
Sipols AJ, Brief DJ, Ginter KL, Saghafi S, Woods SC (1992) Neuropeptide Y paradoxically increases food intake yet causes conditioned flavor aversions. Physiol Behav 51:1257–1260
Stanley BG, Anderson KC, Grayson MH, Leibowitz SF (1989) Repeated hypothalamic stimulation with neuropeptide Y increases daily carbohydrate and fat intake and body weight gain in female rats. Physiol Behav 46:173–177
Stanley BG, Daniel DR, Chin AS, Leibowitz SF (1985) Paraventricular nucleus injections of peptide YY and neuropeptide Y preferentially enhance carbohydrate ingestion. Peptides 6:1205–1211
Stanley BG, Kyrkouli SE, Lampert S, Leibowitz SF (1986) Neuropeptide Y chronically injected into the hypothalamus: a powerful neurochemical inducer of hyperphagia and obesity. Peptides 7:1189–1192
Stanley BG, Leibowitz SF (1984) Neuropeptide Y injected into the paraventricular hypothalamus: a powerful stimulant of feeding behavior. Proc Natl Acad Sci USA 82:3940–3943
Steinman JL, Gunion MW, Morley JE (1994) Forebrain and hindbrain involvement of neuropeptide Y in ingestive behaviors of rats. Pharmacol Biochem Behav 47:207–214
Stricker-Krongrad A, Barbanel G, Beck B, Burlet A, Nicolas JP, Burlet C (1993) K(+)-stimulated neuropeptide Y release into the paraventricular nucleus and relation to feeding behavior in free-moving rats. Neuropeptides 24:307–312
Stricker-Krongrad A, Kozak R, Burlet C, Nicolas JP, Beck B (1997) Physiological regulation of hypothalamic neuropeptide Y release in lean and obese rats. Am J Physiol 273:R2112–2116
Stricker-Krongrad A, Max JP, Musse N, Nicolas JP, Burlet C, Beck B (1994) Increased threshold concentrations of neuropeptide Y for a stimulatory effect on food intake in obese Zucker rats–changes in the microstructure of the feeding behavior. Brain Res 660:162–166

Strubbe JH, Woods SC (2004) The timing of meals. Psychol Rev 111:128–141
Szentirmai E, Krueger JM (2006) Central administration of neuropeptide Y induces wakefulness in rats. Am J Physiol Regul Integr Comp Physiol 291:R473–480
Tang-Christensen M, Kristensen P, Stidsen CE, Brand CL, Larsen PJ (1998) Central administration of Y5 receptor antisense decreases spontaneous food intake and attenuates feeding in response to exogenous neuropeptide Y. J Endocrinol 159:307–312
Tatemoto K, Carlquist M, Mutt V (1982) Neuropeptide Y–a novel brain peptide with structural similarities to peptide YY and pancreatic polypeptide. Nature 296:659–660
Taylor K, Lester E, Hudson B, Ritter S (2007) Hypothalamic and hindbrain NPY, AGRP and NE increase consummatory feeding responses. Physiol Behav 90:744–750
Teff K (2000) Nutritional implications of the cephalic-phase reflexes: endocrine responses. Appetite 34:206–213
Thiele T, van Dijik G, Yagaloff K, Fisher S, Schwartz M, Burn P, Seeley R (1998a) Central infusion of melanocortin agonist MTII in rats: assessment of c-Fos expression and taste aversion. Am J Physiol 274:248–254
Thiele TE, Marsh DJ, Ste Marie L, Bernstein IL, Palmiter RD (1998b) Ethanol consumption and resistance are inversely related to neuropeptide Y levels. Nature 396:366–369
Thiele TE, Sparta DR, Hayes DM, Fee JR (2004) A role for neuropeptide Y in neurobiological responses to ethanol and drugs of abuse. Neuropeptides 38:235–243
Thorsell A, Caberlotto L, Rimondini R, Heilig M (2002) Leptin suppression of hypothalamic NPY expression and feeding, but not amygdala NPY expression and experimental anxiety. Pharmacol Biochem Behav 71:425–430
Traebert M, Riediger T, Whitebread S, Scharrer E, Schmid HA (2002) Ghrelin acts on leptin-responsive neurones in the rat arcuate nucleus. J Neuroendocrinol 14:580–586
Vahl TP, Drazen DL, Seeley RJ, D'Alessio DA, Woods SC (2009) Meal-anticipatory glucagon-like peptide-1 secretion in rats. Endocrinology 151:569–575
Walter MJ, Scherrer JF, Flood JF, Morley JE (1994) Effects of localized injections of neuropeptide Y antibody on motor activity and other behaviors. Peptides 15:607–613
Watanabe Y, Akira A, McEwen FS (1995) Adrenal steroid regulation of neuropeptide Y (NPY) mRNA: differences between dentate hilus and locus coeruleus and arcuate nucleus. Mol Brain Res 28:135–140
Welch CC, Grace MK, Billington CJ, Levine AS (1994) Preference and diet type affect macronutrient selection after morphine, NPY, norepinephrine, and deprivation. Am J Physiol 266: R426–433
White BD, He B, Dean RG, Martin RJ (1994a) Low protein diets increase neuropeptide Y gene expression in the basomedial hypothalamus of rats. J Nutr 124:1152–1160
White BD, Dean RG, Edwards GL, Martin RJ (1994b) Type II corticosteroid receptor stimulation increases NPY gene expression in basomedial hypothalamus of rats. Am J Physiol 266: R1523–1529
Widdowson PS (1997) Regionally-selective down-regulation of NPY receptor subtypes in the obese Zucker rat. Relationship to the Y5 'feeding' receptor. Brain Res 758:17–25
Wieland HA, Engel W, Eberlein W, Rudolf K, Doods HN (1998) Subtype selectivity of the novel nonpeptide neuropeptide Y Y1 receptor antagonist BIBO 3304 and its effect on feeding in rodents. Br J Pharmacol 125:549–555
Wilding JPH, Gilbey SG, Bailey CJ, Batt RAL, Williams G, Ghatei MA, Bloom SR (1993) Increased neuropeptide-Y messenger ribonucleic acid (mRNA) and decreased neurotensin mRNA in the hypothalamus of the obese (ob/ob) mouse. Endocrinology 132:1939–1944
Williams G, Shellard L, Lewis DE, McKibbin PE, McCarthy HD, Koeslag DG, Russell JC (1992) Hypothalamic neuropeptide Y disturbances in the obese (cp/cp) JCR:LA corpulent rat. Peptides 13:537–540
Wolak ML, DeJoseph MR, Cator AD, Mokashi AS, Brownfield MS, Urban JH (2003) Comparative distribution of neuropeptide Y Y1 and Y5 receptors in the rat brain by using immunohistochemistry. J Comp Neurol 464:285–311

Woods SC (1991) The eating paradox: how we tolerate food. Psychol Rev 98:488–505

Woods SC (2009) The control of food intake: behavioral versus molecular perspectives. Cell Metab 9:489–498

Woods SC, Figlewicz DP, Madden L, Porte D Jr, Sipols AJ, Seeley RJ (1998) NPY and food intake: discrepancies in the model. Regul Pept 75–76:403–408

Wyss P, Stricker-Krongrad A, Brunner L, Miller J, Crossthwaite A, Whitebread S, Criscione L (1998) The pharmacology of neuropeptide Y (NPY) receptor-mediated feeding in rats characterizes better Y5 than Y1, but not Y2 or Y4 subtypes. Regul Pept 75–76:363–371

Xu B, Li B-H, Rowland NE, Kalra SP (1995) Neuropeptide Y injection into the fourth cerebroventricle stimulates c-Fos expression in the paraventricular nucleus and other nuclei in the forebrain: effect of food consumption. Brain Res 698:227–231

Yang L, Scott KA, Hyun J, Tamashiro KL, Tray N, Moran TH, Bi S (2009) Role of dorsomedial hypothalamic neuropeptide Y in modulating food intake and energy balance. J Neurosci 29:179–190

Yavropoulou MP, Kotsa K, Kesisoglou I, Anastasiou O, Yovos JG (2008) Intracerebroventricular infusion of neuropeptide Y increases glucose dependent-insulinotropic peptide secretion in the fasting conscious dog. Peptides 29:2281–2285

Yokosuka M, Dube MG, Kalra PS, Kalra SP (2001) The mPVN mediates blockade of NPY-induced feeding by a Y5 receptor antagonist: a c-FOS analysis. Peptides 22:507–514

Yoshihara T, Honma S, Honma K (1996) Effects of restricted daily feeding on neuropeptide Y release in the rat paraventricular nucleus. Am J Physiol 270:E589–E595

The Neuroendocrine Circuitry Controlled by POMC, MSH, and AGRP

Heike Biebermann, Peter Kühnen, Gunnar Kleinau, and Heiko Krude

Contents

Abstract Obesity is one of the most challenging health problems worldwide. Over the past few decades, our knowledge concerning mechanisms of weight regulation has increased tremendously leading to the identification of the leptin–melanocortin pathway. The filling level of energy stores is signaled to the brain, and the information is integrated by hypothalamic nuclei, resulting in a well-orchestrated response to food intake and energy expenditure to ensure constant body weight. One of the key players in this system is proopiomelanocortin (POMC), a precursor of a variety of neuropeptides. POMC-derived alpha- and beta-MSH play an important role in energy homeostasis by activating melanocortin receptors expressed in the arcuate nucleus (MC3R) and in the nucleus paraventricularis (MC4R).

H. Biebermann (✉) • P. Kühnen • G. Kleinau • H. Krude
Institut für Experimentelle Pädiatrische Endokrinologie, Charité Universitätsmedizin Berlin, Augustenburger Platz 1, 13353 Berlin, Germany
e-mail: heike.biebermann@charite.de; peter.kuehnen@charite.de; gunnar.kleinau@charite.de; heiko.krude@charite.de

H.-G. Joost (ed.), *Appetite Control*, Handbook of Experimental Pharmacology 209,
DOI 10.1007/978-3-642-24716-3_3, © Springer-Verlag Berlin Heidelberg 2012

Activation of these two G protein-coupled receptors is antagonized by agouti-related peptide (AgRP). Naturally occurring mutations in this system were identified in patients suffering from common obesity as well as in patients demonstrating a phenotype of severe early-onset obesity, adrenal insufficiency, red hair, and pale skin. Detailed understanding of the complex system of POMC-AgRP-MC3R-MC4R and their interaction with other hypothalamic as well as peripheral signals is a prerequisite to combat the obesity epidemic.

Keywords POMC • MSH • AgRP • MC3R • MC4R • Mutations

Abbreviations

ACTH	Adrenocorticotropic hormone
ARC	Arcuate nucleus
BDNF	Brain-derived neurotrophic factor
b-LPH	Beta-lipotropin
BMI	Body mass index
CLIP	Corticotrophin-like intermediate lobe peptide
CNS	Central nervous system
GPCR	G protein-coupled receptor
GHSR	Ghrelin receptor
HFRW	His-Phe-Arg-Trp
MC3R	Melanocortin receptor 3
MC4R	Melanocortin receptor 4
MRAP	Melanocortin receptor 2 accessory protein
MSH	Melanocyte-stimulating hormone
PC1	Prohormone convertase 1
PC2	Prohormone convertase 2
POMC	Proopiomelanocortin
PVN nucleus	Paraventricularis
SIM	Single-minded gene
TrkB	Tcpomyoin receptor kinase B
TSH	Thyroid-stimulating hormone

1 The Leptin–Melanocortin Pathway in Hypothalamic Weight Regulation

Although food intake and energy expenditure vary from day-to-day, body weight remains remarkably stable, indicating that homeostatic regulators of body weight exist. Intriguingly, during the past few decades, obesity has become one of the most frequent health problems, not only in industrialized societies, indicating that dysregulation of the homeostatic system is problematic, at least for parts of the human population. Our knowledge of the mechanisms that regulate body weight has increased tremendously and includes the identification of a neuroendocrine circuit, the leptin–melanocortin pathway. This pathway transmits information on nutrient supply and energy storage to hypothalamic nuclei, allowing a suitable response to the actual situation.

Genes involved in hypothalamic weight regulation encode proteins operating the neuroendocrine network which translates the information on body fat mass – signaled by the adipose tissue hormone leptin – into hypothalamic counter-regulatory responses, e.g., changes in food intake and metabolic rate. Within the arcuate nucleus of the hypothalamus (ARC), binding of leptin ((Zhang et al. 1994), see chapter by Julian Mercer/Elizabeth Cottrell) to the leptin receptor on one set of neurons activates the formation of anorexigenic acting neurotransmitters generated from the precursor protein proopiomelanocortin (POMC). On another set of neurons, orexigenic neuropeptides (neuropeptide Y (NPY) and agouti-related peptide (AgRP)) are expressed when leptin levels decrease. Activation or inhibition of the postsynaptic melanocortin receptors 3 and 4 (MC3R, MC4R) by these neuropeptides results in efferent responses to achieve the balance of energy intake and energy expenditure (Cone 1999) (Fig. 1). Recently, mutations in the human homologues of these key-regulator genes of body weight maintenance were found in cases of extreme and early-onset obesity, suggesting a high level of functional conservation of the set-point genes during evolution (Barsh et al. 2000).

In contrast to the few monogenic causes of extreme obesity, the problem of common obesity, which arose during the past two decades in postindustrialized societies, cannot be explained by these rare mutations in set-point genes. More likely, the recent epidemics of obesity results from a decompensation of the genetic counter-regulatory network within an extremely changed environment, e.g., increased caloric energy intake and minimized physical exercise. In this respect, the observation might be of importance that particularly those individuals with a formerly high normal set point (reflected by a body mass index (BMI) percentile of 90–97) now tend to develop severe obesity (Barth et al. 1997). In these individuals, it appears that the allelic combination which has determined a high normal body weight in the changed environment predisposes to obesity.

Results of several twin studies have underlined the role of genetic factors in body-weight regulation (Maes et al. 1997). Therefore, it is necessary to understand the genetic basis of common obesity and to identify those individual allelic combinations, which are limiting for an appropriate counter-regulatory capacity of the neuroendocrine network. Several independent genome screens based on familial cases of severe obesity revealed that a high level of leptin, probably reflecting leptin resistance of the hypothalamus (Considine et al. 1996), is linked to a region of chromosome 2, which includes the *POMC* gene locus (Hager et al. 1998).

In this chapter, we focus on the physiological as well as pathophysiological role of POMC, POMC-derived peptides alpha- and beta-MSH and their receptors, MC3R, MC4R, and AgRP in the hypothalamic circuit of weight regulation.

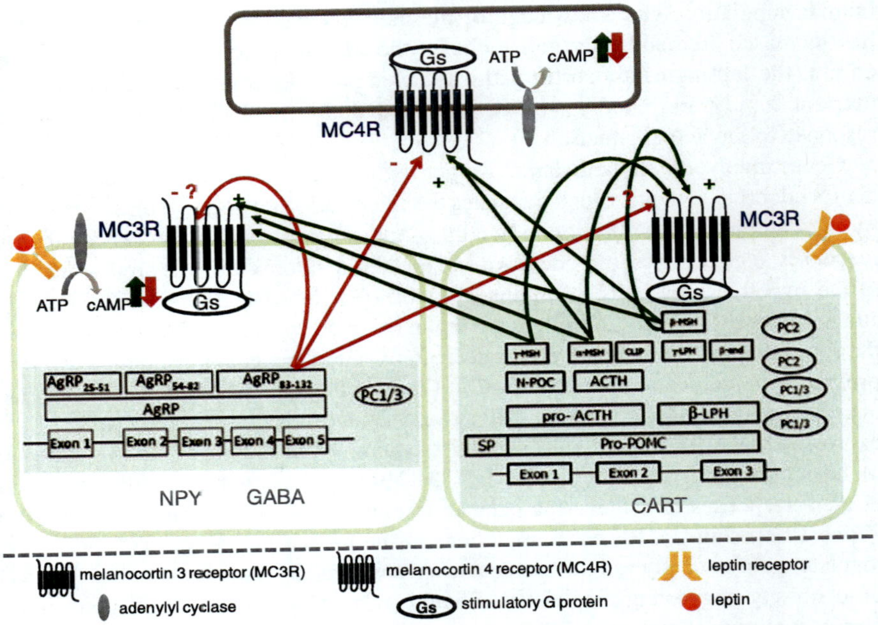

Fig. 1 Schematic illustration of POMC, MSH, and AgRP function within the leptin–melanocortin pathway. Leptin signals the filling content of adipocytes from the periphery to the brain, especially to the hypothalamic nucleus arcuatus. Arcuate neurons are indicated as light *gray squares*. Binding of leptin to the leptin receptor enhances the expression of POMC and represses the expression of AgRP. Genomic organization, transcription, and translation of POMC and AgRP are depicted (for details, see 1.1 and 1.2). Alpha- and beta-MSH are the endogenous ligands of the MC3R and MC4R. Auto- and paracrine activation (indicated as *plus*) of MC3R is indicated as *green lines*. Axonal transport of alpha- and beta-MSH to the nucleus paraventricularis (*dark gray square*) and activation (indicated as *plus*) of the MC4R are indicated by a green line. Inhibition (indicated as *minus*, if the effect is not proven, a ? is shown) of MC3R and MC4R via AgRP is indicated by *red lines*. The effect of MC3R and MC4R on Gs/adenylyl cyclase activation is depicted as *green arrow* when cAMP accumulation is stimulated and as *red arrow* when cAMP accumulation is reduced

1.1 *The Role of POMC in Weight Regulation*

POMC is expressed in the pituitary, skin, and hypothalamus and plays a key role in the regulation of energy metabolism. Several hormones, neuropeptides, and nutrients including leptin and insulin regulate the expression of POMC. Under fasting conditions, POMC expression is reduced, whereas filled energy stores enhance the expression. Leptin-induced STAT3 expression was shown to modulate expression of POMC (Munzberg et al. 2003). POMC is a 31-kDa prohormone precursor that is synthesized in the endoplasmatic reticulum. The N-terminal sequence of POMC targets the precursor to secretory granules in the regulated secretory pathway. A membrane-bound enzyme, carboxypeptidase E, acts as a sorting receptor and binds the N-terminal sequence of POMC. During trafficking of POMC, tissue-specific posttranslational processing into a variety of biologically active peptides occurs. Prohormone convertases 1/3 and 2 (PC1/3 and PC2) are responsible for endoproteolytic cleavage at dibasic amino acids. Initially, POMC is cleaved by PC1/3 into pro-ACTH and beta-lipotropin (b-LPH) (Fig. 1). Pro-ACTH is further processed into ACTH and N-terminal POMC (N-POC) by PC1/3. In rodents, but not in humans, the cleavage site for processing N-POC into gamma melanocyte-stimulating hormone (MSH) is missing.

In the hypothalamic arcuate nucleus, ACTH is processed by PC2 and carboxypeptidase E into alpha-MSH and CLIP (corticotrophin-like intermediate lobe peptide). Bioactive alpha-MSH requires N-terminal acetylation, but no substantial differences in receptor binding of the two alpha-MSH forms were found. The in vivo difference of the two forms may be related to the higher stability of acetylated alpha-MSH (Abbott et al. 2000; Kask et al. 2000). This acetylation is dynamically regulated by leptin, and it is suggested that this acetylation is an important control pathway for the melanocortic system (Mountjoy et al. 2003). Beta-LPH is cleaved by PC2 into gamma-LPH and beta-endorphin. In humans, but not in rodents, gamma-LPH is further processed to beta-MSH.

For body-weight regulation, alpha- and beta-MSH are the important neuropeptides. As early as in the 1980s, it was shown in rats that intracerebroventricular injection of alpha-MSH or of ACTH decreases food intake (Poggioli et al. 1986; Vergoni et al. 1986). For alpha- and beta-MSH as well as for ACTH, the reducing effect on food intake after intracerebroventricular or intraperitoneal injection was shown in several studies (Abbott et al. 2000; Harrold and Williams 2006; Millington et al. 2001). An additional role for gamma-MSH that specifically activates the MC3R is likely but not yet proven. The core sequence of alpha-, beta-, and gamma-MSH as well as ACTH that is needed for activation of melanocortin receptors consists of a specific motif of four amino acids: histidine, phenylalanine, arginine, and tryptophan (His-Phe-Arg-Trp motif) (Table 1) (Hruby et al. 1987).

1.2 The Role of AgRP in Weight Regulation

Agouti encodes a 131-amino acid paracrine signaling molecule, which is produced in dermal skin papilla and regulates the type of pigment in melanocytes. Agouti inhibits the MC1R by affecting cAMP response. Thereby, follicular melanocytes synthesize the yellow/red pigment pheomelanin. On the contrary, alpha-MSH stimulates the MC1R and leads to the production of the black pigment eumelanin. Pulsatile expression of agouti protein during normal hair cycles in rodents results in a yellow band in the middle of the hair (Voisey and van Daal 2002). First indications that agouti also plays a role in body-weight regulation were generated over 100 years ago when Cuenot described the lethal yellow A^y mouse (Cuenot 1909, Barsh 1999). This A^y mutation leads to ectopic activated expression of agouti leading to obesity, hyperphagia, hyperinsulinism, and yellow coat color by a

Table 1 Melanocortin receptors and their ligands

Peptide/sequence	MC1R Skin, hair follicle	MC2R Adrenals	MC3R Hypothalamus	MC4R Hypothalamus	MC5R Exocrine glands
ACTH Ser-Tyr-Ser-Met-Glu-***His-Phe-Arg-Trp***-Gly-Lys-Pro-Val-Gly-Lys-Lys-Arg-Arg-Pro-Val-Lys-Val-Tyr-Pro-Asn-Gly-Ala-Glu-Asp-Glu-Leu-Ala-Glu-Ala-Phe-Pro-Leu-Glu-Phe	Full agonist	Full agonist	Full agonist	Partial agonist	Partial agonist
Gamma-MSH Tyr-Val-Met-Gly-***His-Phe-Arg-Trp***-Asp-Arg-Phe-Gly	Partial agonist	–	Full agonist	Partial agonist	Partial agonist
Alpha-MSH Ser-Tyr-Ser-Met-Glu-***His-Phe-Arg-Trp***-Gly-Lys-Pro-Val	Full agonist	–	Full agonist	Full agonist	Full agonist
Beta-MSH Ala-Glu-Lys-Lys-Asp-Glu-Gly-Pro-Tyr-Arg-Met-Glu-***His-Phe-Arg-Trp***-Gly-Ser-Pro-Pro-Lys-Asp	Full agonist	–	Full agonist	Full agonist	Full agonist
AgRP	–	–	Antagonist	Inverse agonist	–
Agouti	Inverse agonist	Antagonist	–	–	Antagonist

For all five melanocortin receptors, the location with the highest expression is indicated. MC3R and MC4R are highlighted in gray. The sequence of all POMC-derived peptides that activate melanocortin receptors is depicted, and the amino acids of the core-binding sequence are indicated in bold italics

120–10 kb deletion that removes the entire coding region of the upstream *Raly* gene. The deletion of the *Raly* gene leads to the absence of hnRNP (heterogenous nuclear ribonucleoprotein) and to embryonic lethality in the homozygous state. Heterozygous $A^{y/+}$ animals are viable, but in the mutated allele, the *Raly* promoter drives ubiquitous expression of agouti resulting in the described phenotype (Miltenberger et al. 1999; Yen et al. 1994). Several other dominant agouti mutations have been described that cause an ectopic agouti expression like viable yellow (A^{vy}), intermediate yellow (A^{ly}), and sienna yellow (A^{sy}). A^{vy} mice result from an insertion of an intracisternal A particle (IAP) into the noncoding exon 1A (Duhl et al. 1994; Wolff et al. 1999).

Comparison of expressed sequence tag database with agouti led to the identification of AgRP (Graham et al. 1997; Ollmann et al. 1997). AgRP is exclusively expressed within the arcuate nucleus of the hypothalamus in mice and humans in which it is localized in NPY-expressing neurons.

Similar to agouti, AgRP (both containing an Arg-Phe-Phe motif) is able to antagonize MSH stimulation of MC3R and MC4R as an antagonist and inverse agonist (Fu and van den Pol 2008; Haskell-Luevano and Monck 2001; Nijenhuis et al. 2001; Oosterom et al. 2001). Thereby, the neuropeptide AgRP promotes food intake and energy expenditure. AgRP neurons, which are only expressed within the arcuate nucleus, predominantly secrete NPY and GABA and project either to neighboring POMC neurons or other hypothalamic nuclei. In this context, GABA release has been shown to inhibit POMC positive neurons in the hypothalamus (Belgardt et al. 2009).

Central AgRP administration or AgRP overexpression leads to hyperphagia and obesity (Barsh et al. 1999; Broberger and Hokfelt 2001). AgRP mRNA expression is upregulated during fasting and in *ob/ob* and *db/db* mice, respectively. Furthermore, leptin inhibits AgRP expression. These observations point to a role for leptin as a negative regulator of AgRP.

The *AgRP* knockout mouse shows a remarkably normal phenotype (Qian et al. 2002). Subsequent studies identified a modest lean phenotype with reduced body weight after 6 months, with increased metabolic rate, locomotor activity, and thyroid hormone levels (Wortley et al. 2005). To gain further insides into the role of AgRP neurons in the regulation of body weight, four different studies analyzed mouse models with ablated AgRP hypothalamic neurons. In one study, transgenic expression of the CAG expanded repeat form of ataxin-3 in *AgRP*, which induces cell death, was generated (Bewick et al. 2005). Moreover, postnatal cell death was induced by cre-mediated deletion of the mitochondrial transcription factor *Tfam* (Gropp et al. 2005). Two other groups generated AgRP neurons with expressed diphtheria toxin receptors. Thereby, ablation of AgRP neurons could be generated by administration of diphtheria toxin (Gropp et al. 2005; Luquet et al. 2005). All four studies led to ablation of AgRP neurons within the hypothalamus and to a lean and hypophagic phenotype. Thus, the phenotype differs depending on whether AgRP neurons were eliminated during the neonatal period or early postnatal. It has therefore been postulated that early loss of AgRP neurons leads to a compensatory mechanism creating a nearly unaffected phenotype (Flier 2006).

Based on previous studies, it is difficult to interpret a definite role of AgRP. Future studies will elucidate whether AgRP or other neurotransmitters like NPY or GABA are important for body-weight regulation embedded in the leptin–melanocortin signaling pathway.

1.3 Physiological Function of Melanocortin Receptors 3 and 4

Melanocortin receptors belong to the rhodopsin/beta2-adrenergic receptor-like GPCRs and consist of five members named MC1R to MC5R in order of their cloning. All five melanocortin receptors are activated by POMC-derived ligands (Table 1). MC1R (also known as MSH receptor) is mainly expressed in the skin and hair follicles and is responsible for skin, hair, or coat pigmentation. Activation of MC2R (also known as ACTH receptor), which is expressed in the adrenal glands, results in glucocorticoid production. MC3R and MC4R are widely expressed in the hypothalamus-receiving projections of POMC fibers and are responsible for the modulation of food intake and energy expenditure. MC5R is expressed in exocrine glands and promotes sebum production (for further information, see (Cone 2000)).

Interestingly, for melanocortin receptors, endogenous antagonists, agouti and AgRP, exist. Both peptides have effects on coat pigmentation and food intake. Recent studies have indicated that AgRP acts as an inverse agonist (depression of ligand-independent signaling activity) on MC4R and MC1R (Adan 2006).

The human *MC3R* is encoded by a single-exon gene located on chromosome 20q13.2–13.3 (Gantz et al. 1993b) coding for 361 amino acids. The MC3R is activated by all POMC-derived peptides with a comparable affinity (Table 1) but is the only melanocortin receptor that is fully activated by gamma-MSH. The MC3R is predominantly expressed in the arcuate nucleus of the hypothalamus (Fig. 1) (Gantz et al. 1993b) and also in the periphery in the pancreas, stomach, and placenta. The MC3R plays an important role in energy metabolism; however, its physiological role is different compared to MC4R (see below). Target deletion of the *MC3R* results in adiposity in mice, although food intake in these mice is reduced (Chen et al. 2000). It is therefore suggested that in these mice, nutrients are preferentially stored as fat. Obesity in these mice is indicated by an increased fat mass of 50–60% and a reduction in lean mass. On a high fat diet, these mice demonstrated an unusually high respiratory quotient pointing to preferential carbohydrate oxidation of carbohydrates. Moreover, *MC3R* knockout mice have reduced energy expenditure (Butler et al. 2000). Interestingly, *MC3R* null mice are protected from the development of fatty liver disease and insulin resistance despite elevated levels of body fat (Sutton et al. 2006).

The 322 amino acids of the MC4R are encoded by a single-exon gene located on chromosome 18q21.3 (Gantz et al. 1993a). The MC4R is primarily expressed in the hypothalamic nucleus paraventricularis (Fig. 1), spinal cord, sympathetic preganglionic neurons, brainstem, and penis (Van der Ploeg et al. 2002). The MC4R is fully activated by POMC-derived alpha- and beta-MSH (Table 1) and plays a profound role in weight regulation. Target disruption of the *MC4R* in mice results in maturity onset of severe obesity, hyperphagia, hyperinsulinemia, and hyperglycemia (Huszar et al. 1997). As a result of lower energy expenditure, these mice gain weight faster than their wild-type littermates when consuming the same amount of food (Ste Marie et al. 2000). Strikingly, heterozygous female *MC4R* knockout mice display higher body weight compared to wild-type littermates, indicating a gene dosage effect for MC4R function. Remarkably, it was shown that there is a divergence in MC4R function to control food intake and energy expenditure. By rescue of MC4R expression in the nucleus paraventricularis (PVN) of *MC4R* knockout mice, it was demonstrated that MC4R function in the PVN is responsible for food intake, whereas MC4R expression in other brain areas are involved in energy expenditure (Balthasar et al. 2005). The important role of MC4R in PVN is further supported by adenoviral knockdown of *MC4R* in the PVN which causes hyperphagia and obesity in rats (Garza et al. 2008). Voluntary exercise by offering running wheels to *MC4R* knockout mice impedes the course of obesity (Irani et al. 2005) and implies that even a severe monogenetic defect can be influenced by changing lifestyle habits, at least in rodents.

The phenotype of *MC3R* knockouts is different from that of the *MC4R* knockout mice indicating different and nonredundant mechanisms underlying obesity in both mouse lines. The obese phenotype of *MC3R* null mice is most likely due to aberrant feed efficiency and reduced locomotor activity, whereas *MC4R* null mice are hyperphagic and have reduced energy expenditure (Chen et al. 2000). Double knockouts of both receptors displays a more obese phenotype compared to knockout of either receptor alone (Chen et al. 2000).

MC4R downstream signaling components are inadequately understood. So far, it was recognized that injection of corticotropin-releasing hormone reduces food intake, and this is also true for *MC4R* knockout mice (Lu et al. 2003; Marsh et al. 1999). There is evidence that Sim1 (single-minded gene 1) and BDNF (brain-derived neutrophic factor) act downstream of MC4R signaling. Sim1 is expressed in the PVN and amygdala, and haploinsufficiency of Sim1 is associated with hyperphagia, obesity, and increased body length. Application of highly potent melanocortins is not sufficient to suppress food intake (Kublaoui et al. 2006). BDNF expression is mediated by nutritional state and by MC4R signaling (Xu et al. 2003). Injection of BDNF in the PVN reduces food intake and stimulates energy expenditure (Wang et al. 2010). Furthermore, peripheral injection of neutrophin-4, a ligand for the BDNF receptor, TrkB (tropomyosin receptor kinase B), suppresses appetite and weight gain in a dose-dependent manner (Tsao et al. 2008).

1.4 *Effects of Artificial MSH-Derived Ligands on MC4R Activation In Vitro and In Vivo*

Artificial alpha-MSH analogs were developed to enhance affinity and efficacy at MC4R such as NDP-alpha-MSH (Ac-Ser-Tyr-Ser-Nle-Glu-His-D-Phe-Arg-Trp-Gly-Lys-Pro-Val-NH_2) or the cyclic melanotan II (Ac-Nle-*c*[Asp-His-D-Phe-Arg-Trp-Lys]-NH_2). These substances were effective in vitro as well as in rodents; however, they are not selective for MC4R and activate also MC3R. Interestingly, substitution of the histidine residue of the core-binding motif His-Phe-Arg-Trp to proline (Ac-Nle-c[Asp-Pro-D-Phe-Arg-Trp-Lys]-Pro-Val-NH2 (PG-931) (Grieco et al. 2003) also exerts full agonism at the MC4R. In contrast, cyclic lactam analogs with a bulkier amino acid as phenylalanine at position 7, D-p-iodophenylalanine 7, exert antagonistic activity at the MC4R (Hruby et al. 1995).

Several alpha-MSH analogs were developed with the purpose to selectively activate MC4R and to pass the blood–brain barrier. On basis of the core His-Phe-Arg-Trp binding motif substance THIQ (*N*-[(3R)-1,2,3,4-tetrahydroisoquinolinium-3-ylcarbonyl]-(1R)-1-(4-chlorobenzyl)-2-[4-cyclohexyl-4-(1H-1,2,4-triazol-1ylmethyl) piperidin-1-yl]-2-oxoethylamine) and other small molecule MC4R agonists were developed (Sebhat et al. 2002; Van der Ploeg et al. 2002). Additional approaches were performed and tested in vitro and in vivo. It was demonstrated that analogs based on spiroindane amide are suitable for oral application (Guo et al. 2010; He et al. 2010a, b), and recently, it was shown that two MC4R agonists, IRC-022493 and IRC-022511, are able to efficiently activate loss-of-function MC4R mutations (Roubert et al. 2010). However, effectiveness of these substances as treatment option for human obese MC4R mutation carriers needs to be shown.

The first study to reduce body weight in humans uses ACTH4-10 which was found to be effective in healthy adults (Fehm et al. 2001). However, in patients with POMC mutations, this substance did not have any influence on body weight (Krude et al. 2003). A new substance that was highly MC4R selective and effective in mice to reduce energy intake and to induce weight loss, MK-0493 (*N*-(1S)-1-[2-(1-{[3S,4R]-tert-butyl-4-(2,4-difluorophenyl)pyrrolidin-3-yl)-5-chlorophenyl]ethyl}acetamide(hydrochloride) (Fong and Strack 2008), was used to reduce body weight in normal weight to obese healthy men. Sibutramine, a serotonin–norepinephrine reuptake inhibitor, was used as positive control. Unfortunately in humans, MK-0493 was not effective to reduce energy intake and to induce weight loss (Krishna et al. 2009). So far, these studies point to the problem of transferability of rodent model systems to the human situation.

1.5 Structural Characteristics and Signaling Properties of MC3R and MC4R

Melanocortin receptors are characterized by common structural features of the rhodopsin/beta2-adrenergic receptor family A such as highly conserved amino acid motifs at the transmembrane helices (TMH), e.g., the so-called Asn-Arg-Tyr motif in TMH3, or the Asn-Pro-x-x-Tyr motif in TMH7. Interestingly, they also lack typical family A GPCR features like a conserved cysteine residue in extracellular loop (ECL) 2 and prolines in TMH5. Additionally, they have specific features such as disulphide bridged cysteine residues in ECL3 and an extremely short ECL2 comprising four amino acids.

Melanocortin receptors are activated by POMC-derived peptides (Table 1). Thus, the binding site for the orthosteric (endogenous) ligand should be similar in all melanocortin receptors. The N-terminal receptor region as well as the extracellular loops except ECL3 (Tarnow et al. 2003) are of less importance for ligand binding (Schioth et al. 1997; Yang et al. 2000). It is known that specific amino acids in the TMH region are sensitive to interaction with the His-Phe-Arg-Trp motif (Table 1) of all POMC-derived ligands. Within the TMHs, ionic and aromatic residues are mainly responsible for high affinity ligand binding (reviewed in (Tao 2010a; Yang 2011)). In detail, residues important for ligand binding by MC3R are localized in TMH2 (Glu131), TMH3 (Asp154 and Asp158), and TMH6 (Trp292, Phe295, His298) (Chen et al. 2006). Residues important for binding by MC4R are also found in TMH2 (Glu100), TMH3 (Asp122 and Asp126), and TMH6 (Trp258, Phe261, and His264) (Chen et al. 2007; Yang et al. 2000). In the MC4R, some motifs in ECLs 2 and 3 were found to be important for AgRP 87–132 binding affinity but not for AgRP 110–117; moreover, amino acid side chains in TMHs 3 and 4 are important for binding of both AgRPs (Yang et al. 2003).

Ligand interaction with melanocortin receptors induces changes mainly in the amplitude of the Gs/adenylyl cyclase signaling pathway (Gantz et al. 1993a), whereby receptor activation by agonists leads to enhancement of intracellular cyclic AMP (cAMP), which consequently results in protein kinase A (PKA) activation. In addition to activation of the Gs-signaling pathway, it was shown in vivo that the suppression of food intake requires the induction of extracellular signal-regulated kinase (ERK) 1/2 (Sutton et al. 2005). Central application of a superpotent MC4R agonist led to phosphorylation of ERK 1/2 and cAMP-responsive element-binding protein (CREB). It is speculated that activation of the cAMP-ERK1/2-CREB cascade is a molecular integrator for converging satiety signals from the gut to the brain (Sutton et al. 2005).

For the MC4R and MC3R, further (secondary) signaling pathways like mobilization of intracellular calcium (Mountjoy et al. 2001), which is cholera toxin sensitive and pertussis toxin insensitive, were reported. Activation of the Gq/11-mediated pathway is controversially discussed, but so far, no physiological relevance of this pathway is known (Mountjoy et al. 2001; Newman et al. 2006). Additionally, pertussis toxin-sensitive MC4R signaling in hypothalamic mouse GT1-7 implicates activation of Gi/o-mediated signaling (Buch et al. 2009). Interestingly, AgRP is able to activate Gi simultaneously to antagonism of the Gs-signaling pathway, which was referred to as a biased agonist of MC4R signaling (Breit et al. 2011; Buch et al. 2009).

Di- or oligomerization of GPCRs is accepted nowadays as a relevant feature of GPCR physiology. For melanocortin receptors, it was shown that MC3R and MC4R are able to form homodimers (Biebermann et al. 2003; Elsner et al. 2006; Mandrika et al. 2005; Nickolls and Maki 2006). Additionally, heterodimerization of MC4R and MC3R with other GPCRs that are involved in weight regulation was demonstrated (Rediger et al. 2009). It has been shown that the MC4R is able to interact with GPR7, a receptor that is also assumed to be expressed in the PVN. For the MC3R, intermolecular interaction with the GHSR (known as ghrelin receptor) was determined.

For several GPCRs, accessory proteins are reported to be essential for their function. This is also true for the MC2R which depends on the interaction with melanocortin receptor 2 accessory protein (MRAP) for proper cell surface expression (Metherell et al. 2005). MRAP is able to interact with all five melanocortin receptors subtypes, but (excluding MC2R) MRAP is not mandatory for their cell surface expression. In contrast, MRAP reduces MC4R and MC5R cell surface expression (Chan et al. 2009). Moreover, a significant inhibitory role of MRAP for signaling properties for MC1R, MC3R, MC4R, and MC5R was demonstrated (Chan et al. 2009). Another protein, the attractrin-like protein (Haqq et al. 2003), interacts with the MC4R and was identified by a yeast two-hybrid approach. It is suggested that the attractrin-like protein acts as a coreceptor needed for binding of AgRP; however, so far no functional evidence for this interaction exists.

2 Naturally Occurring Mutations in POMC, MC3R, and MC4R

Shortly after the detection of the genetic defect *ob/ob*, a mutation in the leptin gene (Zhang et al. 1994), human mutations in key components of the leptin–melanocortin pathway, was identified in early-onset obese patients.

2.1 *Complete and Partial Loss of POMC Function (OMIM #609734)*

The *POMC* gene is located on chromosome 2p22 and consists of three exons with the start ATG located in exon 2. Screening for mutations in large cohorts of patients revealed few polymorphisms including a 9- and an 18-base pair insertion between amino acids 73 and 74 within the gamma-MSH gene, which occurs with a frequency of 3–5% (Hinney et al. 1998).

2.1.1 Complete Loss-of-Function Mutations

The phenotype of patients carrying mutations that lead to complete loss of POMC function is due to the lack of peptides activating their particular receptors (Krude and Grüters 2000). The pituitary-derived ACTH regulates cortisol secretion via binding to the adrenal MC2R. Hair and skin pigmentation is regulated by activation of the MC1R by alpha- and beta-MSH. Stimulation of the MC4R by both peptides and MC3R by alpha-, beta-, and gamma-MSH in the CNS is responsible for energy metabolism. The complete lack of these peptides therefore results in syndromic obesity characterized by adrenal insufficiency, pale skin, and red hair as well as early-onset obesity. In 1998, the first patients with POMC deficiency were described (Krude et al. 1998), prior to the *POMC* knockout mouse which presents a corresponding phenotype (Yaswen et al. 1999). The frequency of complete loss-of-function mutations is extremely rare. So far, only a few patients were reported (Table 2; (Clement et al. 2008; Farooqi et al. 2006; Krude et al. 1998, 2003)) as suffering from adrenal insufficiency and early-onset obesity. Deficiency in skin pigmentation has only been recognized in Caucasian patients so far. The skin and hair phenotype of patients with African or Arabian origin (Clement et al. 2008; Farooqi et al. 2006) does not differ significantly from nonmutation carriers. The reason for this might be explained by studies performed in mice that demonstrated that the genetic background has a strong influence on coat color (Slominski et al. 2005). It was speculated that the high ligand-independent activity of the melanocortin 1 receptor could ensure conversion of pheomelanin to eumelanin. An observation in one patient supported this suggestion as it was recognized that a

Table 2 Complete and partial loss of POMC function mutations

Complete loss-of-function mutations				Partial loss-of-function mutations			
Position in protein	Amino acid change	Nucleotide change	Reference	Position in protein	Amino acid change	Nucleotide change	Reference
Before start codon		g.3804 C > A	Krude et al. (1998)				
				NPOC	p.Cys28Phe	g.3897 G > T	Creemers et al. (2008)
				NPOC	p.Leu37Phe	g.3923 C > T	Creemers et al. (2008)
NPOC	p.Lys51Term	g.A6851 > T	Krude et al. (1998)				
NPOC	p.Pro69LeufsX2	g.6906delC	Farooqi et al. (2006)				
NPOC	p.Pro74ProfsX44	g.6922insC	Clement et al. (2008)				
NPOC	p.Gly99AlafsX58	g.6996delG	Krude et al. (2003)				
NPOC	p.Glu134fsX158	g.71001insGG	Krude et al. (2003)				
NPOC	p.Arg145ProfsX12	g.7134delG	Krude et al. (1998)				
				α-MSH	p.Phe144Leu	g.7132 T > C	Dubern et al. (2008)
				Before β-MSH	p.Glu180Term	g.7316 G > T	Hinney et al. (1998)
				β-MSH	p.Tyr221Cys	g.7363A > G	Lee et al. (2006), Biebermann et al. (2006)
				β-MSH/β-end cleavage	p.Arg236Gly	g.7408 C > G	Challis et al. (2002)

All so far, published mutations in order of their genomic location and their functional effects are shown
For position in the POMC gene, see Fig. 1

hair color change from red to brown in the second and third year of life occurs (Krude et al. 2003).

2.1.2 Phenotype and Treatment of Patients with POMC Mutations

After birth, patients are often identified by the diagnosis of hypoglycemia and subsequently central hypocortisolism. Patients were treated with low doses of hydrocortisone (8–10 mg/kg/m^2 body surface per day) (Krude et al. 2003). Treatment of these patients with intranasal ACTH 4–10 did not result in a reduction of food intake and weight loss. Investigation of thyroid function revealed slightly elevated TSH levels and low, sometimes reduced, total T4 but normal T3 values. Treatment of T4 in both patients did not change the course of obesity (Krude et al. 2003). The birth weight of all so far reported patients was normal. However, *POMC* mutations lead to the development of postnatal, early-onset, and severe obesity. This is combined with ACTH deficiency, pale skin, and red hair in the Caucasian population (Farooqi 2007). The dramatic increase of body weight points to the severe consequence of a dysregulated body weight set point. However, intensive lifestyle intervention, e.g., during treatment in a hospital with a sport program and assistance by a nutritionist can lead to a stabilization of body weight during this period.

2.1.3 Partial Loss of POMC Function as an Example for Common Obesity

Susceptibility to obesity might result from alleles of genes which determined a high set point of body weight in former times of restricted food supply but may now predispose for obesity within the changed environment. To understand the genetic basis of so-called common obesity, it will therefore be of interest to identify those alleles which cause a limitation for appropriate regulation of body weight homeostasis when energy intake is increased and exercise is reduced.

Patients suffering from common obesity with no other signs than increased fat mass were screened for mutations. Those variants in the *POMC* gene that predispose to common obesity were successfully identified (Table 2, (Biebermann et al. 2006; Challis et al. 2002; Dubern et al. 2008; Hinney et al. 1998; Lee et al. 2006)). However, in the beginning, no attention was given to those variants because they were located downstream of the region coding for alpha-MSH. These mutations were located in the prohormone convertase recognition site between beta-MSH and beta-endorphin (Challis et al. 2002; Hinney et al. 1998) and result in a beta-MSH/beta-endorphin fusion protein lowering the amount of functional beta-MSH. Additionally, a stop mutation before the coding region of beta-MSH was identified (Hinney et al. 1998). A missense mutation in beta-MSH gene that leads to the exchange of a highly conserved tyrosine residue for a cysteine residue causes a reduced capacity to bind to the MC4R (Biebermann et al. 2006; Lee et al. 2006). Intracerebroventricular injection of the mutated beta-MSH in contrast to wild-type beta-MSH or alpha-MSH was not sufficient to inhibit food intake (Biebermann et al. 2006). These studies point, for the first time, to the important role of beta-MSH in hypothalamic weight regulation. It was further demonstrated that beta-MSH is the predominant neuropeptide in the hypothalamic nucleus arcuatus (Biebermann et al. 2006). For a long time, it was assumed that the phenotype of POMC deficiency is exclusively caused by mutations in the region of the *POMC* gene coding for alpha-MSH. However, so far, only one missense mutation in the region coding for alpha-MSH has been reported (Dubern et al. 2008).

Additional mutations that influence POMC trafficking were detected in patients suffering from common obesity. These mutations are located in the N-terminal domain of POMC (Table 2) (Creemers et al. 2008), which is suggested to be involved in sorting POMC to the secretory pathway (Creemers et al. 2008). Therefore, both mutations lead to impaired processing of POMC to bioactive peptides representing a new molecular mechanism of POMC deficiency.

A gene dosage effect was reported due to partial loss of POMC function in heterozygous parents of complete loss-of-function mutation carriers. For these parents, a body weight in the upper normal range with a tendency to become overweight was recognized indicating that the control of energy balance is sensitive to POMC function (Farooqi et al. 2006; Krude et al. 2003).

2.2 *Naturally Occurring MC3R Mutations (OMIM # 15540)*

The physiological effect of mutations in the human MC3R contributing to obesity is still under strong debate (Calton et al. 2009; Mencarelli et al. 2011). So far, only few *MC3R* mutations with functional relevance were identified (Fig. 2a) (Lee et al. 2002; Mencarelli et al. 2011, 2008; Rached et al. 2004; Tao 2007; Zegers et al. 2011). However, few variants of the MC3R seem to have a stronger impact on the individual susceptibility of weight gain.

The first identified mutation of the MC3R Ile183Asn is located in transmembrane helix 3 and was detected in an obese child and father (Lee et al. 2002). Functional characterization confirmed a complete loss of function despite nearly normal cell surface expression and ligand binding indicating that the mutation constrained the receptor in the inactive state (Rached et al. 2004). Further studies identified additional mutations in obese subjects (Fig. 2a). These mutations display either wild-type-like receptor signaling functions (Ala293Thr, Val177Ile, X361Ser, Ala70Thr, Met134Ile) (Lee et al. 2007; Mencarelli et al. 2008; Rached et al. 2004) or reduced signaling properties (Ser17Thr, Thr280Ser, and Ile335Ser, c.397_726delins228) (Calton et al. 2009; Mencarelli et al. 2008; Rached et al. 2004). Interestingly, *MC3R* mutations were also found in normal weight individuals (Ile87Thr, Leu285Val, Arg257Ser). These mutations induce enhanced signaling capacities compared to the wild-type MC3R (Rached et al. 2004). Some mutations were reported in lean controls but have not yet been functionally characterized, e.g., Ser69Cys and Phe82Ser identified in (Calton et al. 2009) and Ile87Thr, Ala260Val, Met275Thr, and Leu297Val identified in obese patients (Calton et al. 2009).

In the past decade, numerous studies investigated association of two frequent MC3R variants with obesity or diabetes in various collectives (Feng et al. 2005; Hani et al. 2001; Li et al. 2000; Schalin-Jantti et al. 2003; Wong et al. 2002). These variants are located in the coding region and lead to amino acid exchanges Thr6Lys

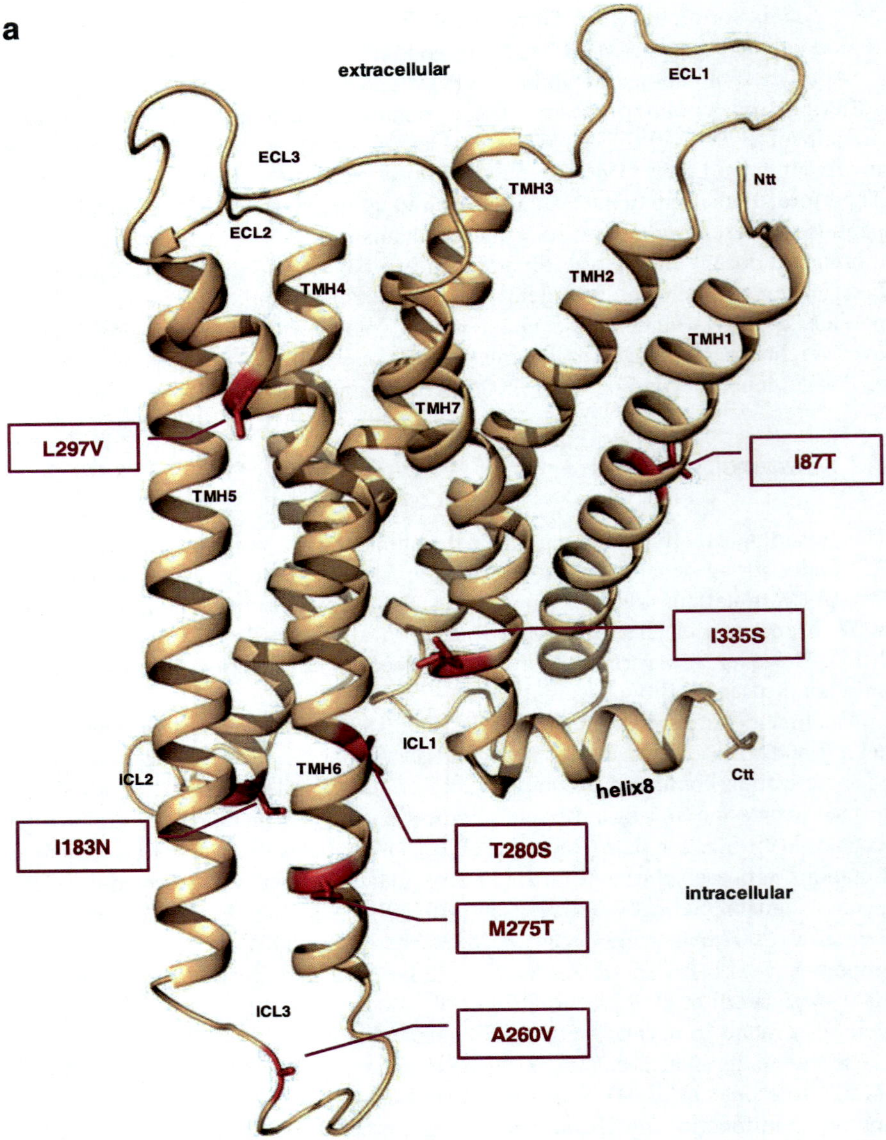

Fig. 2 (continued)

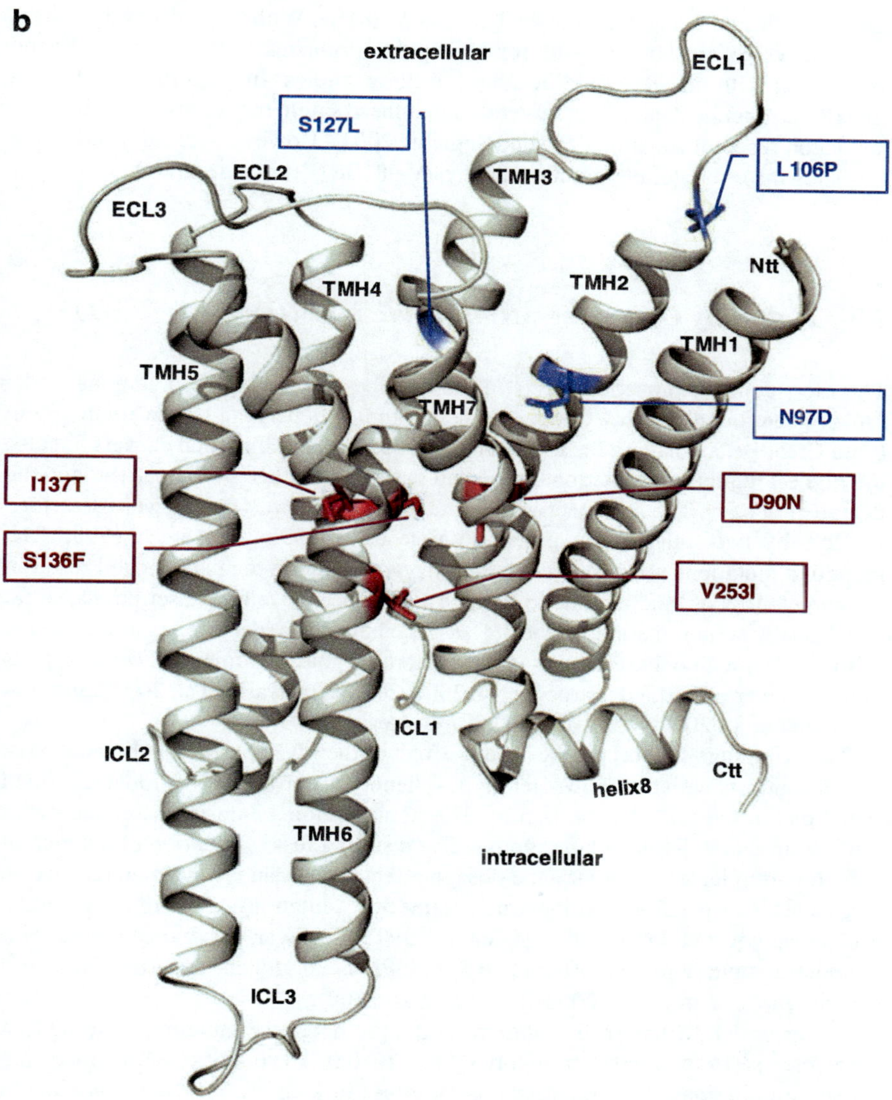

Fig. 2 Computational modeling of MC3R and MC4R. (**a**) In the MC3R model, few exemplary positions of pathogenic mutations causing reduced signal transduction capabilities identified in obese patients are highlighted (*magenta sticks*). They are located in known signaling relevant receptor components like transmembrane helices 3, 6, and 7. (**b**) In the MC4R, particular positions of pathogenic mutations known to reduce specific binding (*blue sticks*) and signaling (*magenta sticks*) are displayed. In accordance with MC3R, binding is impeded by mutations at the extracellular ends of transmembrane helices 2 and 3. Signal transduction is disrupted by modification of the interplay between helices 2, 3, and 6

and Val81Ile and the double mutant Thr6Lys/Val81Ile. With the exception of obese children which are homozygous for both polymorphisms (Feng et al. 2005), no association with obesity could be found in these studies. In vitro investigation of signal transduction capabilities suggested that these single mutations do not have an impact on receptor function (Tao and Segaloff 2004). However, partial inactivation was shown for the double mutant Thr6Lys/Val81Ile (Feng et al. 2005).

2.3 *Naturally Occurring MC4R Mutations (OMIM # 15541)*

The most common genetic defects in human obesity to date are mutations in the *MC4R* gene found in 2–6% of obese patients studied dependent on the ethnic origin (with Great Britain having the highest prevalence of obese mutation carriers but also the highest number of mutational screened patients), the age of the patient, and the degree of obesity. The first inactivating mutations in the *MC4R* gene were described in 1998 by two independent groups (Vaisse et al. 1998; Yeo et al. 1998). The identified mutations were found in a heterozygous state in the patients and result in a complete loss of function due to a frameshift. Since these first descriptions, so far over 150 different mutations in the *MC4R* gene have been identified (summarized in (Tao 2010a)) with only rare cases in which homozygous or compound heterozygous mutations were identified (Farooqi et al. 2000, 2003; Kobayashi et al. 2002; Lubrano-Berthelier et al. 2004; Ma et al. 2004; Tarnow et al. 2003).

These findings support a gene dosage effect of the MC4R in weight homeostasis, which contrasts with the known genotype–phenotype correlations of most identified mutations in the GPCR gene family. The human findings are in accordance with findings in the *MC4R* knockout mouse. The resulting loss of function of the mutant receptors implicates a similar gene dosage effect on human weight maintenance as in rodents. For the first two identified frameshift mutations, a dominant-negative effect was not found (Ho and MacKenzie 1999). However, for two complete loss-of-function mutations (Asp90Asn and Ser136Phe), a dominant-negative effect was shown (Biebermann et al. 2003; Tarnow et al. 2008).

So far, several frameshift mutations and eight nonsense mutations leading to a complete loss of function were reported (Tao 2010a). The majority of mutations are missense mutations. The mutations result either in a complete loss of function, a partial loss of function, or a function comparable to the wild-type receptor. Due to their functional consequences, the mutations were grouped into five classes (Tao 2005):

Class I: null mutations. Mutations that result in defective protein synthesis or were degraded like all nonsense mutations (Trp16X, Tyr35X, Gln43X, Glu61X, Tyr80X, Cys277X).

Class II: mutants defective in cell surface expression. These mutations are frameshift mutations or missense mutations like Asn62Ser, Gly98Arg, and Tyr302Phe, to name only a few.

Class III: mutants with binding defect. These are for example Asn97Asp, Leu106Pro, and Ser127Leu (Fig. 2b). These mutants are characterized by normal expression on the cell surface and abolished binding. Binding experiments were either performed with alpha-MSH or the highly potent analog NDP-alpha-MSH. For some mutants like Ser127Leu, discrepancies between both ligands exist, as NDP-alpha-MSH challenges exhibit properties comparable to wild type, but stimulation with alpha-MSH results in a partial loss of function.

Class IV: signaling defective mutants. These mutants are expressed on the cell surface and bind the ligand properly, but signal transduction is altered like in Asp90Asn, Ser136Phe, Ile137Thr, and Val253Ile mutants (Fig. 2b). It has been discussed whether mutations with defects that reduce the ligand-independent activity also belong to this class (Tao 2010a).

Class V: variants with unknown defects. For these mutations, no differences in comparison with the wild-type receptor were found so far using the current state-of-the-art experimental procedure. However, since all experiments are performed in heterologous expression systems, no appropriate human hypothalamic cell line is available, and one can only speculate that these mutations have no functional effect, e.g., Thr11Ile, Ser36Tyr, Ile69Thr, Asp126Tyr, and many more. An additional group of mutations that is not appropriately included in this categorization are those that lead to a higher constitutive activity. So far, only few constitutively active mutations are reported (His76Arg, Asp146Asn, and His158Arg) (Fan et al. 2008).

There is no final consensus about this nomenclature (classes I to V), but in general, the categorization is helpful to get a mechanistic idea of the receptors malfunction. However, one have to take into account that functional characterization of the mutants is performed either after stable or transient transfection. Since the MC4R has a certain degree of constitutive activity, functional characteristics depend on the degree of cell surface expression. The higher expression levels of transient overexpression systems could potentially mask a weak expression in vivo. Furthermore, different systems to measure functional properties and different ligands – endogenous and highly potent ligands – are used to study these receptors, rendering data comparison making mutant classification difficult.

So far, only one variation in the 5′ untranslated region of *MC4R* was reported (Lubrano-Berthelier et al. 2003; Valli-Jaakola et al. 2006). A deletion of two base pairs at position −439 was suggested to destroy a binding site for nescient helix loop helix 2 (Nhlh2). Nhlh2 is known to regulate the transcription of PC1 and PC2 (Fox et al. 2007), but regulation of *MC4R* transcription was not proven yet. The investigation of the MC4R from different species as well as from different ethnic groups revealed that the MC4R has been subject to high levels of continuous purification selection with codon usage bias resulting in an intriguingly low level of silent polymorphisms in humans (Hughes et al. 2009).

The mutational screening in large cohorts of obese patients revealed that approximately 2% have a mutation with a functional defect (Stutzmann et al. 2008). Screening in a normal weight cohort revealed that 0.15% have a mutation of functional relevance (Hinney et al. 2006).

Taking these findings together, it is suggested that inactivating mutations in the MC4R do not comprise a form of monogenic but of oligogenic obesity (Dempfle et al. 2004; Hinney et al. 2003), indicating that inactivating MC4R mutations can but do not always predispose for obesity.

The first patients with a mutation in the MC4R were discovered in 1998 (Vaisse et al. 1998; Yeo et al. 1998). MC4R mutations lead to early-onset obesity. However, the phenotype, with respect to their weight gain, seems to be milder compared to patients suffering from a POMC mutation. The patients develop fasting hyperinsulinemia and accelerated linear growth (Martinelli et al. 2011). This increased growth is accompanied by impaired GH secretion with an increased pulsatility and increased total GH secretion (Martinelli et al. 2011). Moreover, although obesity is one of the major causes of hypertension (Garrison et al. 1987), studies reveal a decreased prevalence of hypertension in MC4R patients. This observation seems to be due to reduced activity of the sympathetic nervous system (Greenfield et al. 2009).

2.4 Variants That Are Protective Against Obesity

Besides frequent variants that predispose to obesity, two variants were indentified that protect against obesity, Val103Ile (rs2229616) and Ile251Leu (rs52820871) (Gotoda et al. 1997; Hinney et al. 1999). Large meta-analysis has revealed that Val103Ile reduces the obesity risk by approximately 21% and the Ile251Leu allele by approximately 50%. The frequency of Ile103 in the Caucasian population is rather low, between 2 and 4% (Geller et al. 2004); therefore, large sample sizes are needed to obtain sufficient power for the identification of small effects (Heid et al. 2005). In a population-based study, a 0.8 kg/m^2 lower body mass index was reported for Ile103 (Stutzmann et al. 2007). Functional characterization of Val103Ile revealed no difference with MC4R wild-type function despite the fact that a twofold reduction in AgRP potency was observed (Xiang et al. 2006). The frequency of the Leu251 allele is lower compared to Ile103 with approximately 1–2% in the normal population and is responsible for a 0.7 kg/m^2 lower body mass index (Stutzmann et al. 2007). Interestingly, an increased constitutive activity was reported for in vitro studies of this mutation (Xiang et al. 2006). Remarkably, an increased constitutive activity was also reported for other MC4R mutations (Tao 2010a). Carriers of these mutations are obese, indicating that more studies are necessary to unravel the molecular mechanism of these protective MC4R mutations.

Large genome-wide association studies were performed for the identification of variants that are associated with obesity. These studies identified two single nucleotide polymorphisms near the MC4R locus that are highly associated with obesity, rs17782313 188 kb downstream of the MC4R gene (Loos et al. 2008) and rs129070134 150 kb downstream of the MC4R gene (Chambers et al. 2008).

Both variants are not in linkage disequilibrium with the two protective variants. So far, it is unknown how these two variants influence MC4R function. Therefore, it could only be speculated that they modulate *MC4R* gene expression.

References

Abbott CR, Rossi M, Kim M, AlAhmed SH, Taylor GM, Ghatei MA, Smith DM, Bloom SR (2000) Investigation of the melanocyte stimulating hormones on food intake. Lack Of evidence to support a role for the melanocortin-3-receptor. Brain Res 869:203–210

Adan RA (2006) Constitutive receptor activity series: endogenous inverse agonists and constitutive receptor activity in the melanocortin system. Trends Pharmacol Sci 27:183–186

Balthasar N, Dalgaard LT, Lee CE, Yu J, Funahashi H, Williams T, Ferreira M, Tang V, McGovern RA, Kenny CD, Christiansen LM, Edelstein E, Choi B, Boss O, Aschkenasi C, Zhang CY, Mountjoy K, Kishi T, Elmquist JK, Lowell BB (2005) Divergence of melanocortin pathways in the control of food intake and energy expenditure. Cell 123:493–505

Barsh G (1999) From Agouti to Pomc-100 years of fat blonde mice. Nat Med 5:984–985

Barsh GS, Farooqi IS, O'Rahilly S (2000) Genetics of body-weight regulation. Nature 404:644–651

Barsh GS, Ollmann MM, Wilson BD, Miller KA, Gunn TM (1999) Molecular pharmacology of Agouti protein in vitro and in vivo. Ann N Y Acad Sci 885:143–152

Barth N, Ziegler A, Himmelmann GW, Coners H, Wabitsch M, Hennighausen K, Mayer H, Remschmidt H, Schafer H, Hebebrand J (1997) Significant weight gains in a clinical sample of obese children and adolescents between 1985 and 1995. Int J Obes Relat Metab Disord 21:122–126

Belgardt BF, Okamura T, Bruning JC (2009) Hormone and glucose signalling in POMC and AgRP neurons. J Physiol 587:5305–5314

Bewick GA, Gardiner JV, Dhillo WS, Kent AS, White NE, Webster Z, Ghatei MA, Bloom SR (2005) Post-embryonic ablation of AgRP neurons in mice leads to a lean, hypophagic phenotype. FASEB J 19:1680–1682

Biebermann H, Castaneda TR, van Landeghem F, von Deimling A, Escher F, Brabant G, Hebebrand J, Hinney A, Tschop MH, Gruters A, Krude H (2006) A role for beta-melanocyte-stimulating hormone in human body-weight regulation. Cell Metab 3:141–146

Biebermann H, Krude H, Elsner A, Chubanov V, Gudermann T, Gruters A (2003) Autosomal-dominant mode of inheritance of a melanocortin-4 receptor mutation in a patient with severe early-onset obesity is due to a dominant-negative effect caused by receptor dimerization. Diabetes 52:2984–2988

Breit A, Buch TR, Boekhoff I, Solinski HJ, Damm E, Gudermann T (2011) Alternative G protein coupling and biased agonism: new insights into melanocortin-4 receptor signalling. Mol Cell Endocrinol 331:232–240

Broberger C, Hokfelt T (2001) Hypothalamic and vagal neuropeptide circuitries regulating food intake. Physiol Behav 74:669–682

Buch TR, Heling D, Damm E, Gudermann T, Breit A (2009) Pertussis toxin-sensitive signaling of melanocortin-4 receptors in hypothalamic GT1-7 cells defines agouti-related protein as a biased agonist. J Biol Chem 284:26411–26420

Butler AA, Kesterson RA, Khong K, Cullen MJ, Pelleymounter MA, Dekoning J, Baetscher M, Cone RD (2000) A unique metabolic syndrome causes obesity in the melanocortin-3 receptor-deficient mouse. Endocrinology 141:3518–3521

Calton MA, Ersoy BA, Zhang S, Kane JP, Malloy MJ, Pullinger CR, Bromberg Y, Pennacchio LA, Dent R, McPherson R, Ahituv N, Vaisse C (2009) Association of functionally significant

Melanocortin-4 but not Melanocortin-3 receptor mutations with severe adult obesity in a large North American case-control study. Hum Mol Genet 18:1140–1147

Challis BG, Pritchard LE, Creemers JW, Delplanque J, Keogh JM, Luan J, Wareham NJ, Yeo GS, Bhattacharyya S, Froguel P, White A, Farooqi IS, O'Rahilly S (2002) A missense mutation disrupting a dibasic prohormone processing site in pro-opiomelanocortin (POMC) increases susceptibility to early-onset obesity through a novel molecular mechanism. Hum Mol Genet 11:1997–2004

Chambers JC, Elliott P, Zabaneh D, Zhang W, Li Y, Froguel P, Balding D, Scott J, Kooner JS (2008) Common genetic variation near MC4R is associated with waist circumference and insulin resistance. Nat Genet 40:716–718

Chan LF, Webb TR, Chung TT, Meimaridou E, Cooray SN, Guasti L, Chapple JP, Egertova M, Elphick MR, Cheetham ME, Metherell LA, Clark AJ (2009) MRAP and MRAP2 are bidirectional regulators of the melanocortin receptor family. Proc Natl Acad Sci USA 106:6146–6151

Chen AS, Marsh DJ, Trumbauer ME, Frazier EG, Guan XM, Yu H, Rosenblum CI, Vongs A, Feng Y, Cao L, Metzger JM, Strack AM, Camacho RE, Mellin TN, Nunes CN, Min W, Fisher J, Gopal-Truter S, MacIntyre DE, Chen HY, Van der Ploeg LH (2000) Inactivation of the mouse melanocortin-3 receptor results in increased fat mass and reduced lean body mass. Nat Genet 26:97–102

Chen M, Aprahamian CJ, Celik A, Georgeson KE, Garvey WT, Harmon CM, Yang Y (2006) Molecular characterization of human melanocortin-3 receptor ligand-receptor interaction. Biochemistry 45:1128–1137

Chen M, Cai M, Aprahamian CJ, Georgeson KE, Hruby V, Harmon CM, Yang Y (2007) Contribution of the conserved amino acids of the melanocortin-4 receptor in [corrected] [Nle4, D-Phe7]-alpha-melanocyte-stimulating [corrected] hormone binding and signaling. J Biol Chem 282:21712–21719

Clement K, Dubern B, Mencarelli M, Czernichow P, Ito S, Wakamatsu K, Barsh GS, Vaisse C, Leger J (2008) Unexpected endocrine features and normal pigmentation in a young adult patient carrying a novel homozygous mutation in the POMC gene. J Clin Endocrinol Metab 93:4955–4962

Cone RD (1999) The ups and downs of leptin action. Endocrinology 140:4921–4922

Cone RD (2000) The melanocortin receptors. Humana, NJ

Considine RV, Sinha MK, Heiman ML, Kriauciunas A, Stephens TW, Nyce MR, Ohannesian JP, Marco CC, McKee LJ, Bauer TL et al (1996) Serum immunoreactive-leptin concentrations in normal-weight and obese humans. N Engl J Med 334:292–295

Creemers JW, Lee YS, Oliver RL, Bahceci M, Tuzcu A, Gokalp D, Keogh J, Herber S, White A, O'Rahilly S, Farooqi IS (2008) Mutations in the amino-terminal region of proopiomelanocortin (POMC) in patients with early-onset obesity impair POMC sorting to the regulated secretory pathway. J Clin Endocrinol Metab 93:4494–4499

Cuenot L (1909) Recent views of L. Cuenot on the origin of species by mutation. Science 30:768–769

Dempfle A, Hinney A, Heinzel-Gutenbrunner M, Raab M, Geller F, Gudermann T, Schafer H, Hebebrand J (2004) Large quantitative effect of melanocortin-4 receptor gene mutations on body mass index. J Med Genet 41:795–800

Dubern B, Lubrano-Berthelier C, Mencarelli M, Ersoy B, Frelut ML, Bougle D, Costes B, Simon C, Tounian P, Vaisse C, Clement K (2008) Mutational analysis of the pro-opiomelanocortin gene in French obese children led to the identification of a novel deleterious heterozygous mutation located in the alpha-melanocyte stimulating hormone domain. Pediatr Res 63:211–216

Duhl DM, Vrieling H, Miller KA, Wolff GL, Barsh GS (1994) Neomorphic agouti mutations in obese yellow mice. Nat Genet 8:59–65

Elsner A, Tarnow P, Schaefer M, Ambrugger P, Krude H, Gruters A, Biebermann H (2006) MC4R oligomerizes independently of extracellular cysteine residues. Peptides 27:372–379

Fan ZC, Sartin JL, Tao YX (2008) Molecular cloning and pharmacological characterization of porcine melanocortin-3 receptor. J Endocrinol 196:139–148

Farooqi IS, Drop S, Clements A, Keogh JM, Biernacka J, Lowenbein S, Challis BG, O'Rahilly S (2006) Heterozygosity for a POMC-null mutation and increased obesity risk in humans. Diabetes 55:2549–2553

Farooqi IS, Yeo GS, Keogh JM, Aminian S, Jebb SA, Butler G, Cheetham T, O'Rahilly S (2000) Dominant and recessive inheritance of morbid obesity associated with melanocortin 4 receptor deficiency. J Clin Invest 106:271–279

Farooqi IS, Yeo GS, O'Rahilly S (2003) Binge eating as a phenotype of melanocortin 4 receptor gene mutations. N Engl J Med 349:606–609, author reply 606–9

Farooqi S (2007) Insights from the genetics of severe childhood obesity. Horm Res 68(Suppl 5):5–7

Fehm HL, Smolnik R, Kern W, McGregor GP, Bickel U, Born J (2001) The melanocortin melanocyte-stimulating hormone/adrenocorticotropin(4–10) decreases body fat in humans. J Clin Endocrinol Metab 86:1144–1148

Feng N, Young SF, Aguilera G, Puricelli E, Adler-Wailes DC, Sebring NG, Yanovski JA (2005) Co-occurrence of two partially inactivating polymorphisms of MC3R is associated with pediatric-onset obesity. Diabetes 54:2663–2667

Flier JS (2006) AgRP in energy balance: will the real AgRP please stand up? Cell Metab 3:83–85

Fong TM, Strack AM (2008) Melanocortin-4 receptor as a new target for drug development. Handbook of obesity 3 rd edn: 389–394

Fox DL, Vella KR, Good DJ (2007) Energy balance pathways converging on the Nhlh2 transcription factor. Front Biosci 12:3983–3993

Fu LY, van den Pol AN (2008) Agouti-related peptide and MC3/4 receptor agonists both inhibit excitatory hypothalamic ventromedial nucleus neurons. J Neurosci 28:5433–5449

Gantz I, Miwa H, Konda Y, Shimoto Y, Tashiro T, Watson SJ, DelValle J, Yamada T (1993a) Molecular cloning, expression, and gene localization of a fourth melanocortin receptor. J Biol Chem 268:15174–15179

Gantz I, Tashiro T, Barcroft C, Konda Y, Shimoto Y, Miwa H, Glover T, Munzert G, Yamada T (1993b) Localization of the genes encoding the melanocortin-2 (adrenocorticotropic hormone) and melanocortin-3 receptors to chromosomes 18p11.2 and 20q13.2-q13.3 by fluorescence in situ hybridization. Genomics 18:166–167

Garrison RJ, Kannel WB, Stokes J 3rd, Castelli WP (1987) Incidence and precursors of hypertension in young adults: the Framingham Offspring Study. Prev Med 16:235–251

Garza JC, Kim CS, Liu J, Zhang W, Lu XY (2008) Adeno-associated virus-mediated knockdown of melanocortin-4 receptor in the paraventricular nucleus of the hypothalamus promotes high-fat diet-induced hyperphagia and obesity. J Endocrinol 197:471–482

Geller F, Reichwald K, Dempfle A, Illig T, Vollmert C, Herpertz S, Siffert W, Platzer M, Hess C, Gudermann T, Biebermann H, Wichmann HE, Schafer H, Hinney A, Hebebrand J (2004) Melanocortin-4 receptor gene variant I103 is negatively associated with obesity. Am J Hum Genet 74:572–581

Gotoda T, Scott J, Aitman TJ (1997) Molecular screening of the human melanocortin-4 receptor gene: identification of a missense variant showing no association with obesity, plasma glucose, or insulin. Diabetologia 40:976–979

Graham M, Shutter JR, Sarmiento U, Sarosi I, Stark KL (1997) Overexpression of Agrt leads to obesity in transgenic mice. Nat Genet 17:273–274

Greenfield JR, Miller JW, Keogh JM, Henning E, Satterwhite JH, Cameron GS, Astruc B, Mayer JP, Brage S, See TC, Lomas DJ, O'Rahilly S, Farooqi IS (2009) Modulation of blood pressure by central melanocortinergic pathways. N Engl J Med 360:44–52

Grieco P, Balse-Srinivasan P, Han G, Weinberg D, MacNeil T, Van der Ploeg LH, Hruby VJ (2003) Extensive structure-activity studies of lactam derivatives of MT-II and SHU-9119: their activity and selectivity at human melanocortin receptors 3, 4, and 5. J Pept Res 62:199–206

Gropp E, Shanabrough M, Borok E, Xu AW, Janoschek R, Buch T, Plum L, Balthasar N, Hampel B, Waisman A, Barsh GS, Horvath TL, Bruning JC (2005) Agouti-related peptide-expressing neurons are mandatory for feeding. Nat Neurosci 8:1289–1291

Guo L, Ye Z, Liu J, He S, Bakshi RK, Sebhat IK, Dobbelaar PH, Hong Q, Jian T, Dellureficio JP, Tsou NN, Ball RG, Weinberg DH, MacNeil T, Tang R, Tamvakopoulos C, Peng Q, Chen HY, Chen AS, Martin WJ, MacIntyre DE, Strack AM, Fong TM, Wyvratt MJ, Nargund RP (2010) Discovery of potent, selective, and orally bioavailable 3 H-spiro[isobenzofuran-1,4′-piperidine] based melanocortin subtype-4 receptor agonists. Bioorg Med Chem Lett 20:4895–4900

Hager J, Dina C, Francke S, Dubois S, Houari M, Vatin V, Vaillant E, Lorentz N, Basdevant A, Clement K, Guy-Grand B, Froguel P (1998) A genome-wide scan for human obesity genes reveals a major susceptibility locus on chromosome 10. Nat Genet 20:304–308

Hani EH, Dupont S, Durand E, Dina C, Gallina S, Gantz I, Froguel P (2001) Naturally occurring mutations in the melanocortin receptor 3 gene are not associated with type 2 diabetes mellitus in French Caucasians. J Clin Endocrinol Metab 86:2895–2898

Haqq AM, Rene P, Kishi T, Khong K, Lee CE, Liu H, Friedman JM, Elmquist JK, Cone RD (2003) Characterization of a novel binding partner of the melanocortin-4 receptor: attractin-like protein. Biochem J 376:595–605

Harrold JA, Williams G (2006) Melanocortin-4 receptors, beta-MSH and leptin: key elements in the satiety pathway. Peptides 27:365–371

Haskell-Luevano C, Monck EK (2001) Agouti-related protein functions as an inverse agonist at a constitutively active brain melanocortin-4 receptor. Regul Pept 99:1–7

He S, Ye Z, Dobbelaar PH, Sebhat IK, Guo L, Liu J, Jian T, Lai Y, Franklin CL, Bakshi RK, Dellureficio JP, Hong Q, Tsou NN, Ball RG, Cashen DE, Martin WJ, Weinberg DH, Macneil T, Tang R, Tamvakopoulos C, Peng Q, Miller RR, Stearns RA, Chen HY, Chen AS, Strack AM, Fong TM, Macintyre DE, Wyvratt MJ Jr, Nargund RP (2010a) Discovery of a spiroindane based compound as a potent, selective, orally bioavailable melanocortin subtype-4 receptor agonist. Bioorg Med Chem Lett 20:2106–2110

He S, Ye Z, Dobbelaar PH, Sebhat IK, Guo L, Liu J, Jian T, Lai Y, Franklin CL, Bakshi RK, Dellureficio JP, Hong Q, Weinberg DH, Macneil T, Tang R, Strack AM, Tamvakopoulos C, Peng Q, Miller RR, Stearns RA, Chen HY, Chen AS, Fong TM, Wyvratt MJ Jr, Nargund RP (2010b) Spiroindane based amides as potent and selective MC4R agonists for the treatment of obesity. Bioorg Med Chem Lett 20:4399–4405

Heid IM, Vollmert C, Hinney A, Doring A, Geller F, Lowel H, Wichmann HE, Illig T, Hebebrand J, Kronenberg F (2005) Association of the 103I MC4R allele with decreased body mass in 7937 participants of two population based surveys. J Med Genet 42:e21

Hinney A, Becker I, Heibult O, Nottebom K, Schmidt A, Ziegler A, Mayer H, Siegfried W, Blum WF, Remschmidt H, Hebebrand J (1998) Systematic mutation screening of the pro-opiomelanocortin gene: identification of several genetic variants including three different insertions, one nonsense and two missense point mutations in probands of different weight extremes. J Clin Endocrinol Metab 83:3737–3741

Hinney A, Bettecken T, Tarnow P, Brumm H, Reichwald K, Lichtner P, Scherag A, Nguyen TT, Schlumberger P, Rief W, Vollmert C, Illig T, Wichmann HE, Schafer H, Platzer M, Biebermann H, Meitinger T, Hebebrand J (2006) Prevalence, spectrum, and functional characterization of melanocortin-4 receptor gene mutations in a representative population-based sample and obese adults from Germany. J Clin Endocrinol Metab 91:1761–1769

Hinney A, Hohmann S, Geller F, Vogel C, Hess C, Wermter AK, Brokamp B, Goldschmidt H, Siegfried W, Remschmidt H, Schafer H, Gudermann T, Hebebrand J (2003) Melanocortin-4 receptor gene: case-control study and transmission disequilibrium test confirm that functionally relevant mutations are compatible with a major gene effect for extreme obesity. J Clin Endocrinol Metab 88:4258–4267

Hinney A, Schmidt A, Nottebom K, Heibult O, Becker I, Ziegler A, Gerber G, Sina M, Gorg T, Mayer H, Siegfried W, Fichter M, Remschmidt H, Hebebrand J (1999) Several mutations in the melanocortin-4 receptor gene including a nonsense and a frameshift mutation associated with dominantly inherited obesity in humans. J Clin Endocrinol Metab 84:1483–1486

Ho G, MacKenzie RG (1999) Functional characterization of mutations in melanocortin-4 receptor associated with human obesity. J Biol Chem 274:35816–35822

Hruby VJ, Lu D, Sharma SD, Castrucci AL, Kesterson RA, Al-Obeidi FA, Hadley ME, Cone RD (1995) Cyclic lactam alpha-melanotropin analogues of Ac-Nle4-cyclo[Asp5, D-Phe7, Lys10] alpha-melanocyte-stimulating hormone-(4–10)-NH2 with bulky aromatic amino acids at position 7 show high antagonist potency and selectivity at specific melanocortin receptors. J Med Chem 38:3454–3461

Hruby VJ, Wilkes BC, Hadley ME, Al-Obeidi F, Sawyer TK, Staples DJ, de Vaux AE, Dym O, Castrucci AM, Hintz MF et al (1987) alpha-Melanotropin: the minimal active sequence in the frog skin bioassay. J Med Chem 30:2126–2130

Hughes DA, Hinney A, Brumm H, Wermter AK, Biebermann H, Hebebrand J, Stoneking M (2009) Increased constraints on MC4R during primate and human evolution. Hum Genet 124:633–647

Huszar D, Lynch CA, Fairchild-Huntress V, Dunmore JH, Fang Q, Berkemeier LR, Gu W, Kesterson RA, Boston BA, Cone RD, Smith FJ, Campfield LA, Burn P, Lee F (1997) Targeted disruption of the melanocortin-4 receptor results in obesity in mice. Cell 88:131–141

Irani BG, Xiang Z, Moore MC, Mandel RJ, Haskell-Luevano C (2005) Voluntary exercise delays monogenetic obesity and overcomes reproductive dysfunction of the melanocortin-4 receptor knockout mouse. Biochem Biophys Res Commun 326:638–644

Kask A, Rago L, Wikberg JE, Schioth HB (2000) Differential effects of melanocortin peptides on ingestive behaviour in rats: evidence against the involvement of MC(3) receptor in the regulation of food intake. Neurosci Lett 283:1–4

Kobayashi H, Ogawa Y, Shintani M, Ebihara K, Shimodahira M, Iwakura T, Hino M, Ishihara T, Ikekubo K, Kurahachi H, Nakao K (2002) A Novel homozygous missense mutation of melanocortin-4 receptor (MC4R) in a Japanese woman with severe obesity. Diabetes 51:243–246

Krishna R, Gumbiner B, Stevens C, Musser B, Mallick M, Suryawanshi S, Maganti L, Zhu H, Han TH, Scherer L, Simpson B, Cosgrove D, Gottesdiener K, Amatruda J, Rolls BJ, Blundell J, Bray GA, Fujioka K, Heymsfield SB, Wagner JA, Herman GA (2009) Potent and selective agonism of the melanocortin receptor 4 with MK-0493 does not induce weight loss in obese human subjects: energy intake predicts lack of weight loss efficacy. Clin Pharmacol Ther 86:659–666

Krude H, Biebermann H, Luck W, Horn R, Brabant G, Gruters A (1998) Severe early-onset obesity, adrenal insufficiency and red hair pigmentation caused by POMC mutations in humans. Nat Genet 19:155–157

Krude H, Biebermann H, Schnabel D, Tansek MZ, Theunissen P, Mullis PE, Gruters A (2003) Obesity due to proopiomelanocortin deficiency: three new cases and treatment trials with thyroid hormone and ACTH4-10. J Clin Endocrinol Metab 88:4633–4640

Krude H, Gruters A (2000) Implications of proopiomelanocortin (POMC) mutations in humans: the POMC deficiency syndrome. Trends Endocrinol Metab 11:15–22

Kublaoui BM, Holder JL Jr, Gemelli T, Zinn AR (2006) Sim1 haploinsufficiency impairs melanocortin-mediated anorexia and activation of paraventricular nucleus neurons. Mol Endocrinol 20:2483–2492

Lee YS, Challis BG, Thompson DA, Yeo GS, Keogh JM, Madonna ME, Wraight V, Sims M, Vatin V, Meyre D, Shield J, Burren C, Ibrahim Z, Cheetham T, Swift P, Blackwood A, Hung CC, Wareham NJ, Froguel P, Millhauser GL, O'Rahilly S, Farooqi IS (2006) A POMC variant implicates beta-melanocyte-stimulating hormone in the control of human energy balance. Cell Metab 3:135–140

Lee YS, Poh LK, Kek BL, Loke KY (2007) The role of melanocortin 3 receptor gene in childhood obesity. Diabetes 56:2622–2630

Lee YS, Poh LK, Loke KY (2002) A novel melanocortin 3 receptor gene (MC3R) mutation associated with severe obesity. J Clin Endocrinol Metab 87:1423–1426

Li WD, Joo EJ, Furlong EB, Galvin M, Abel K, Bell CJ, Price RA (2000) Melanocortin 3 receptor (MC3R) gene variants in extremely obese women. Int J Obes Relat Metab Disord 24:206–210

Loos RJ, Lindgren CM, Li S, Wheeler E, Zhao JH, Prokopenko I, Inouye M, Freathy RM, Attwood AP, Beckmann JS, Berndt SI, Jacobs KB, Chanock SJ, Hayes RB, Bergmann S, Bennett AJ, Bingham SA, Bochud M, Brown M, Cauchi S, Connell JM, Cooper C, Smith GD, Day I, Dina

C, De S, Dermitzakis ET, Doney AS, Elliott KS, Elliott P, Evans DM, Sadaf Farooqi I, Froguel P, Ghori J, Groves CJ, Gwilliam R, Hadley D, Hall AS, Hattersley AT, Hebebrand J, Heid IM, Lamina C, Gieger C, Illig T, Meitinger T, Wichmann HE, Herrera B, Hinney A, Hunt SE, Jarvelin MR, Johnson T, Jolley JD, Karpe F, Keniry A, Khaw KT, Luben RN, Mangino M, Marchini J, McArdle WL, McGinnis R, Meyre D, Munroe PB, Morris AD, Ness AR, Neville MJ, Nica AC, Ong KK, O'Rahilly S, Owen KR, Palmer CN, Papadakis K, Potter S, Pouta A, Qi L, Randall JC, Rayner NW, Ring SM, Sandhu MS, Scherag A, Sims MA, Song K, Soranzo N, Speliotes EK, Syddall HE, Teichmann SA, Timpson NJ, Tobias JH, Uda M, Vogel CI, Wallace C, Waterworth DM, Weedon MN, Willer CJ, Wraight YX, Zeggini E, Hirschhorn JN, Strachan DP, Ouwehand WH, Caulfield MJ et al (2008) Common variants near MC4R are associated with fat mass, weight and risk of obesity. Nat Genet 40:768–775

Lu XY, Barsh GS, Akil H, Watson SJ (2003) Interaction between alpha-melanocyte-stimulating hormone and corticotropin-releasing hormone in the regulation of feeding and hypothalamo-pituitary-adrenal responses. J Neurosci 23:7863–7872

Lubrano-Berthelier C, Cavazos M, Le Stunff C, Haas K, Shapiro A, Zhang S, Bougneres P, Vaisse C (2003) The human MC4R promoter: characterization and role in obesity. Diabetes 52:2996–3000

Lubrano-Berthelier C, Le Stunff C, Bougneres P, Vaisse C (2004) A homozygous null mutation delineates the role of the melanocortin-4 receptor in humans. J Clin Endocrinol Metab 89:2028–2032

Luquet S, Perez FA, Hnasko TS, Palmiter RD (2005) NPY/AgRP neurons are essential for feeding in adult mice but can be ablated in neonates. Science 310:683–685

Ma L, Tataranni PA, Bogardus C, Baier LJ (2004) Melanocortin 4 receptor gene variation is associated with severe obesity in Pima Indians. Diabetes 53:2696–2699

Maes HH, Neale MC, Eaves LJ (1997) Genetic and environmental factors in relative body weight and human adiposity. Behav Genet 27:325–351

Mandrika I, Petrovska R, Wikberg J (2005) Melanocortin receptors form constitutive homo- and heterodimers. Biochem Biophys Res Commun 326:349–354

Marsh DJ, Hollopeter G, Huszar D, Laufer R, Yagaloff KA, Fisher SL, Burn P, Palmiter RD (1999) Response of melanocortin-4 receptor-deficient mice to anorectic and orexigenic peptides. Nat Genet 21:119–122

Martinelli CE, Keogh JM, Greenfield JR, Henning E, van der Klaauw AA, Blackwood A, O'Rahilly S, Roelfsema F, Camacho-Hubner C, Pijl H, Farooqi IS (2011) Obesity due to melanocortin 4 receptor (MC4R) deficiency is associated with increased linear growth and final height, fasting hyperinsulinemia, and incompletely suppressed growth hormone secretion. J Clin Endocrinol Metab 96:E181–E188

Mencarelli M, Dubern B, Alili R, Maestrini S, Benajiba L, Tagliaferri M, Galan P, Rinaldi M, Simon C, Tounian P, Hercberg S, Liuzzi A, Di Blasio AM, Clement K (2011) Rare melanocortin-3 receptor mutations with in vitro functional consequences are associated with human obesity. Hum Mol Genet 20:392–399

Mencarelli M, Walker GE, Maestrini S, Alberti L, Verti B, Brunani A, Petroni ML, Tagliaferri M, Liuzzi A, Di Blasio AM (2008) Sporadic mutations in melanocortin receptor 3 in morbid obese individuals. Eur J Hum Genet 16:581–586

Metherell LA, Chapple JP, Cooray S, David A, Becker C, Ruschendorf F, Naville D, Begeot M, Khoo B, Nurnberg P, Huebner A, Cheetham ME, Clark AJ (2005) Mutations in MRAP, encoding a new interacting partner of the ACTH receptor, cause familial glucocorticoid deficiency type 2. Nat Genet 37:166–170

Millington GW, Tung YC, Hewson AK, O'Rahilly S, Dickson SL (2001) Differential effects of alpha-, beta- and gamma(2)-melanocyte-stimulating hormones on hypothalamic neuronal activation and feeding in the fasted rat. Neuroscience 108:437–445

Miltenberger RJ, Mynatt RL, Bruce BD, Wilkison WO, Woychik RP, Michaud EJ (1999) An agouti mutation lacking the basic domain induces yellow pigmentation but not obesity in transgenic mice. Proc Natl Acad Sci USA 96:8579–8584

Mountjoy KG, Kong PL, Taylor JA, Willard DH, Wilkison WO (2001) Melanocortin receptor-mediated mobilization of intracellular free calcium in HEK293 cells. Physiol Genomics 5:11–19

Mountjoy KG, Wu CS, Cornish J, Callon KE (2003) alpha-MSH and desacetyl-alpha-MSH signaling through melanocortin receptors. Ann N Y Acad Sci 994:58–65

Munzberg H, Huo L, Nillni EA, Hollenberg AN, Bjorbaek C (2003) Role of signal transducer and activator of transcription 3 in regulation of hypothalamic proopiomelanocortin gene expression by leptin. Endocrinology 144:2121–2131

Newman EA, Chai BX, Zhang W, Li JY, Ammori JB, Mulholland MW (2006) Activation of the melanocortin-4 receptor mobilizes intracellular free calcium in immortalized hypothalamic neurons. J Surg Res 132:201–207

Nickolls SA, Maki RA (2006) Dimerization of the melanocortin 4 receptor: a study using bioluminescence resonance energy transfer. Peptides 27:380–387

Nijenhuis WA, Oosterom J, Adan RA (2001) AgRP(83–132) acts as an inverse agonist on the human-melanocortin-4 receptor. Mol Endocrinol 15:164–171

Ollmann MM, Wilson BD, Yang YK, Kerns JA, Chen Y, Gantz I, Barsh GS (1997) Antagonism of central melanocortin receptors in vitro and in vivo by agouti-related protein. Science 278:135–138

Oosterom J, Garner KM, den Dekker WK, Nijenhuis WA, Gispen WH, Burbach JP, Barsh GS, Adan RA (2001) Common requirements for melanocortin-4 receptor selectivity of structurally unrelated melanocortin agonist and endogenous antagonist, Agouti protein. J Biol Chem 276:931–936

Poggioli R, Vergoni AV, Bertolini A (1986) ACTH-(1–24) and alpha-MSH antagonize feeding behavior stimulated by kappa opiate agonists. Peptides 7:843–848

Qian S, Chen H, Weingarth D, Trumbauer ME, Novi DE, Guan X, Yu H, Shen Z, Feng Y, Frazier E, Chen A, Camacho RE, Shearman LP, Gopal-Truter S, MacNeil DJ, Van der Ploeg LH, Marsh DJ (2002) Neither agouti-related protein nor neuropeptide Y is critically required for the regulation of energy homeostasis in mice. Mol Cell Biol 22:5027–5035

Rached M, Buronfosse A, Begeot M, Penhoat A (2004) Inactivation and intracellular retention of the human I183N mutated melanocortin 3 receptor associated with obesity. Biochim Biophys Acta 1689:229–234

Rediger A, Tarnow P, Bickenbach A, Schaefer M, Krude H, Gruters A, Biebermann H (2009) Heterodimerization of hypothalamic G-protein-coupled receptors involved in weight regulation. Obes Facts 2:80–86

Roubert P, Dubern B, Plas P, Lubrano-Berthelier C, Alihi R, Auger F, Deoliveira DB, Dong JZ, Basdevant A, Thurieau C, Clement K (2010) Novel pharmacological MC4R agonists can efficiently activate mutated MC4R from obese patient with impaired endogenous agonist response. J Endocrinol 207:177–183

Schalin-Jantti C, Valli-Jaakola K, Oksanen L, Martelin E, Laitinen K, Krusius T, Mustajoki P, Heikinheimo M, Kontula K (2003) Melanocortin-3-receptor gene variants in morbid obesity. Int J Obes Relat Metab Disord 27:70–74

Schioth HB, Petersson S, Muceniece R, Szardenings M, Wikberg JE (1997) Deletions of the N-terminal regions of the human melanocortin receptors. FEBS Lett 410:223–228

Sebhat IK, Martin WJ, Ye Z, Barakat K, Mosley RT, Johnston DB, Bakshi R, Palucki B, Weinberg DH, MacNeil T, Kalyani RN, Tang R, Stearns RA, Miller RR, Tamvakopoulos C, Strack AM, McGowan E, Cashen DE, Drisko JE, Hom GJ, Howard AD, MacIntyre DE, van der Ploeg LH, Patchett AA, Nargund RP (2002) Design and pharmacology of N-[(3R)-1,2,3,4-tetrahydroisoquinolinium- 3-ylcarbonyl]-(1R)-1-(4-chlorobenzyl)- 2-[4-cyclohexyl-4-(1 H-1,2,4-triazol- 1-ylmethyl)piperidin-1-yl]-2-oxoethylamine (1), a potent, selective, melanocortin subtype-4 receptor agonist. J Med Chem 45:4589–4593

Slominski A, Plonka PM, Pisarchik A, Smart JL, Tolle V, Wortsman J, Low MJ (2005) Preservation of eumelanin hair pigmentation in proopiomelanocortin-deficient mice on a nonagouti (a/a) genetic background. Endocrinology 146:1245–1253

Ste Marie L, Miura GI, Marsh DJ, Yagaloff K, Palmiter RD (2000) A metabolic defect promotes obesity in mice lacking melanocortin-4 receptors. Proc Natl Acad Sci USA 97:12339–12344

Stutzmann F, Tan K, Vatin V, Dina C, Jouret B, Tichet J, Balkau B, Potoczna N, Horber F, O'Rahilly S, Farooqi IS, Froguel P, Meyre D (2008) Prevalence of melanocortin-4 receptor deficiency in Europeans and their age-dependent penetrance in multigenerational pedigrees. Diabetes 57:2511–2518

Stutzmann F, Vatin V, Cauchi S, Morandi A, Jouret B, Landt O, Tounian P, Levy-Marchal C, Buzzetti R, Pinelli L, Balkau B, Horber F, Bougneres P, Froguel P, Meyre D (2007) Non-synonymous polymorphisms in melanocortin-4 receptor protect against obesity: the two facets of a Janus obesity gene. Hum Mol Genet 16:1837–1844

Sutton GM, Duos B, Patterson LM, Berthoud HR (2005) Melanocortinergic modulation of cholecystokinin-induced suppression of feeding through extracellular signal-regulated kinase signaling in rat solitary nucleus. Endocrinology 146:3739–3747

Sutton GM, Trevaskis JL, Hulver MW, McMillan RP, Markward NJ, Babin MJ, Meyer EA, Butler AA (2006) Diet-genotype interactions in the development of the obese, insulin-resistant phenotype of C57BL/6 J mice lacking melanocortin-3 or −4 receptors. Endocrinology 147:2183–2196

Tao YX (2005) Molecular mechanisms of the neural melanocortin receptor dysfunction in severe early onset obesity. Mol Cell Endocrinol 239:1–14

Tao YX (2007) Functional characterization of novel melanocortin-3 receptor mutations identified from obese subjects. Biochim Biophys Acta 1772:1167–1174

Tao YX (2010a) The melanocortin-4 receptor: physiology, pharmacology, and pathophysiology. Endocr Rev 31:506–543

Tao YX (2010b) Mutations in the melanocortin-3 receptor (MC3R) gene: Impact on human obesity or adiposity. Curr Opin Investig Drugs 11:1092–1096

Tao YX, Segaloff DL (2004) Functional characterization of melanocortin-3 receptor variants identify a loss-of-function mutation involving an amino acid critical for G protein-coupled receptor activation. J Clin Endocrinol Metab 89:3936–3942

Tarnow P, Rediger A, Brumm H, Ambrugger P, Rettenbacher E, Widhalm K, Hinney A, Kleinau G, Schaefer M, Hebebrand J, Krause G, Gruters A, Biebermann H (2008) A heterozygous mutation in the third transmembrane domain causes a dominant-negative effect on signalling capability of the MC4R. Obes Facts 1:155–162

Tarnow P, Schoneberg T, Krude H, Gruters A, Biebermann H (2003) Mutationally induced disulfide bond formation within the third extracellular loop causes melanocortin 4 receptor inactivation in patients with obesity. J Biol Chem 278:48666–48673

Tsao D, Thomsen HK, Chou J, Stratton J, Hagen M, Loo C, Garcia C, Sloane DL, Rosenthal A, Lin JC (2008) TrkB agonists ameliorate obesity and associated metabolic conditions in mice. Endocrinology 149:1038–1048

Vaisse C, Clement K, Guy-Grand B, Froguel P (1998) A frameshift mutation in human MC4R is associated with a dominant form of obesity. Nat Genet 20:113–114

Valli-Jaakola K, Palvimo JJ, Lipsanen-Nyman M, Salomaa V, Peltonen L, Kontula K, Schalin-Jantti C (2006) A two-base deletion -439delGC in the melanocortin-4 receptor promoter associated with early-onset obesity. Horm Res 66:61–69

Van der Ploeg LH, Martin WJ, Howard AD, Nargund RP, Austin CP, Guan X, Drisko J, Cashen D, Sebhat I, Patchett AA, Figueroa DJ, DiLella AG, Connolly BM, Weinberg DH, Tan CP, Palyha OC, Pong SS, MacNeil T, Rosenblum C, Vongs A, Tang R, Yu H, Sailer AW, Fong TM, Huang C, Tota MR, Chang RS, Stearns R, Tamvakopoulos C, Christ G, Drazen DL, Spar BD, Nelson RJ, MacIntyre DE (2002) A role for the melanocortin 4 receptor in sexual function. Proc Natl Acad Sci USA 99:11381–11386

Vergoni AV, Poggioli R, Bertolini A (1986) Corticotropin inhibits food intake in rats. Neuropeptides 7:153–158

Voisey J, van Daal A (2002) Agouti: from mouse to man, from skin to fat. Pigment Cell Res 15:10–18

Wang C, Godar RJ, Billington CJ, Kotz CM (2010) Chronic administration of brain-derived neurotrophic factor in the hypothalamic paraventricular nucleus reverses obesity induced by high-fat diet. Am J Physiol Regul Integr Comp Physiol 298:R1320–R1332

Wolff GL, Roberts DW, Mountjoy KG (1999) Physiological consequences of ectopic agouti gene expression: the yellow obese mouse syndrome. Physiol Genomics 1:151–163

Wong J, Love DR, Kyle C, Daniels A, White M, Stewart AW, Schnell AH, Elston RC, Holdaway IM, Mountjoy KG (2002) Melanocortin-3 receptor gene variants in a Maori kindred with obesity and early onset type 2 diabetes. Diabetes Res Clin Pract 58:61–71

Wortley KE, Anderson KD, Yasenchak J, Murphy A, Valenzuela D, Diano S, Yancopoulos GD, Wiegand SJ, Sleeman MW (2005) Agouti-related protein-deficient mice display an age-related lean phenotype. Cell Metab 2:421–427

Xiang Z, Litherland SA, Sorensen NB, Proneth B, Wood MS, Shaw AM, Millard WJ, Haskell-Luevano C (2006) Pharmacological characterization of 40 human melanocortin-4 receptor polymorphisms with the endogenous proopiomelanocortin-derived agonists and the agouti-related protein (AGRP) antagonist. Biochemistry 45:7277–7288

Xu B, Goulding EH, Zang K, Cepoi D, Cone RD, Jones KR, Tecott LH, Reichardt LF (2003) Brain-derived neurotrophic factor regulates energy balance downstream of melanocortin-4 receptor. Nat Neurosci 6:736–742

Yang Y (2011) Structure, function and regulation of the melanocortin receptors. Eur J Pharmacol 660:125–130

Yang Y, Chen M, Lai Y, Gantz I, Yagmurlu A, Georgeson KE, Harmon CM (2003) Molecular determination of agouti-related protein binding to human melanocortin-4 receptor. Mol Pharmacol 64:94–103

Yang YK, Fong TM, Dickinson CJ, Mao C, Li JY, Tota MR, Mosley R, Van Der Ploeg LH, Gantz I (2000) Molecular determinants of ligand binding to the human melanocortin-4 receptor. Biochemistry 39:14900–14911

Yaswen L, Diehl N, Brennan MB, Hochgeschwender U (1999) Obesity in the mouse model of pro-opiomelanocortin deficiency responds to peripheral melanocortin. Nat Med 5:1066–1070

Yen TT, Gill AM, Frigeri LG, Barsh GS, Wolff GL (1994) Obesity, diabetes, and neoplasia in yellow A(vy)/- mice: ectopic expression of the agouti gene. FASEB J 8:479–488

Yeo GS, Farooqi IS, Aminian S, Halsall DJ, Stanhope RG, O'Rahilly S (1998) A frameshift mutation in MC4R associated with dominantly inherited human obesity. Nat Genet 20:111–112

Zegers D, Beckers S, de Freitas F, Peeters AV, Mertens IL, Verhulst SL, Rooman RP, Timmermans JP, Desager KN, Massa G, Van Gaal LF, Van Hul W (2011) Identification of three novel genetic variants in the melanocortin-3 receptor of obese children. Obesity (Silver Spring) 19:152–159

Zhang Y, Proenca R, Maffei M, Barone M, Leopold L, Friedman JM (1994) Positional cloning of the mouse obese gene and its human homologue. Nature 372:425–432

Neuropeptides Controlling Energy Balance: Orexins and Neuromedins

Joshua P. Nixon, Catherine M. Kotz, Colleen M. Novak, Charles J. Billington, and Jennifer A. Teske

Contents

J.P. Nixon • J.A. Teske
Veterans Affairs Medical Center, Research Service (151), Minneapolis, MN, USA

Department of Food Science and Nutrition, University of Minnesota, 1334 Eckles Avenue, St. Paul, MN 55108, USA

Minnesota Obesity Center, University of Minnesota, 1334 Eckles Avenue, St. Paul, MN 55108, USA

C.M. Kotz (✉)
Veterans Affairs Medical Center, GRECC (11 G), Minneapolis, MN, USA

Veterans Affairs Medical Center, Research Service (151), Minneapolis, MN, USA

Department of Food Science and Nutrition, University of Minnesota, 1334 Eckles Avenue, St. Paul, MN 55108, USA

Minnesota Obesity Center, University of Minnesota, 1334 Eckles Avenue, St. Paul, MN 55108, USA
e-mail: kotzx004@umn.edu

C.M. Novak
Department of Biological Sciences, Kent State University, Kent, OH, USA

C.J. Billington
Veterans Affairs Medical Center, Research Service (151), Minneapolis, MN, USA

Veterans Affairs Medical Center, Endocrine Unit (111 G), Minneapolis, MN, USA

Department of Food Science and Nutrition, University of Minnesota, 1334 Eckles Avenue, St. Paul, MN 55108, USA

Minnesota Obesity Center, University of Minnesota, 1334 Eckles Avenue, St. Paul, MN 55108, USA

H.-G. Joost (ed.), *Appetite Control*, Handbook of Experimental Pharmacology 209,
DOI 10.1007/978-3-642-24716-3_4,

Abstract In this chapter, we review the feeding and energy expenditure effects of orexin (also known as hypocretin) and neuromedin. Orexins are multifunctional neuropeptides that affect energy balance by participating in regulation of appetite, arousal, and spontaneous physical activity. Central orexin signaling for all functions originates in the lateral hypothalamus–perifornical area and is likely functionally differentiated based on site of action and on interacting neural influences. The effect of orexin on feeding is likely related to arousal in some ways but is nonetheless a separate neural process that depends on interactions with other feeding-related neuropeptides. In a pattern distinct from other neuropeptides, orexin stimulates both feeding and energy expenditure. Orexin increases in energy expenditure are mainly by increasing spontaneous physical activity, and this energy expenditure effect is more potent than the effect on feeding. Global orexin manipulations, such as in transgenic models, produce energy balance changes consistent with a dominant energy expenditure effect of orexin. Neuromedins are gut–brain peptides that reduce appetite. There are gut sources of neuromedin, but likely the key appetite-related neuromedin-producing neurons are in the hypothalamus and parallel other key anorectic neuropeptide expression in the arcuate to paraventricular hypothalamic projection. As with other hypothalamic feeding-related peptides, hindbrain sites are likely also important sources and targets of neuromedin anorectic action. Neuromedin increases physical activity in addition to reducing appetite, thus producing a consistent negative energy balance effect. Together with the other various neuropeptides, neurotransmitters, neuromodulators, and neurohormones, neuromedin and orexin act in the appetite network to produce changes in food intake and energy expenditure, which ultimately influences the regulation of body weight.

Keywords Brain • Feeding • Obesity • Physical activity

1 Brain Orexins and Energy Balance

1.1 Orexin

When the discovery of a novel peptide apparently limited to cell bodies in the hypothalamus was announced in 1998 (de Lecea et al. 1998; Sakurai et al. 1998), interest was high due to the possibility of its involvement with feeding. The peptide, dubbed orexin by Sakurai et al. and hypocretin by de Lecea et al., was independently discovered in two laboratories using very different methods (de Lecea et al. 1998; Sakurai et al. 1998). One group isolated the long form of orexin, orexin A (OXA), by searching for ligands for "orphaned" G protein–coupled receptors (Sakurai et al. 1998). The second group first isolated the precursor protein, prepro-orexin, in 1996 using a subtractive PCR technique to recover hypothalamus-specific proteins (Gautvik et al. 1996) but did not publish a detailed investigation of the precursor or its derivatives until early in 1998 (de Lecea et al. 1998).

The initial reports of these discoveries showed that the orexins are a family containing two peptides, the 33-amino-acid OXA (hypocretin-1) and the shorter 28-amino-acid orexin B (OXB, hypocretin-2), both derived from the precursor protein, prepro-orexin (PPO), through proteolytic processing (de Lecea et al. 1998; Sakurai et al. 1998). The PPO gene, which is highly conserved across species, has some similarities with the secretin/incretin family of peptides (de Lecea et al. 1998) and appears to have arisen early during chordate evolution through a circular mutation of an incretin gene (Alvarez and Sutcliffe 2002). Orexin has been identified in all major vertebrate taxa, including fish (Huesa et al. 2005; Kaslin et al. 2004), amphibians (Shibahara et al. 1999; Yamamoto et al. 2004; Singletary et al. 2005), reptiles (Farrell et al. 2003), birds (Ohkubo et al. 2002), and mammals (Sakurai et al. 1999). Within the central nervous system, prepro-orexin mRNA was initially reported to be limited to cell bodies in the lateral hypothalamus (LH) (Gautvik et al. 1996). While there is some evidence for orexin neurons in other brain regions, including the paraventricular hypothalamic and supraoptic nuclei, amygdala, median eminence, and ependyma (Chen et al. 1999; Ciriello et al. 2003a; Kummer et al. 2001; Nixon and Smale 2007), to date, there is no conclusive evidence of orexin mRNA in any brain region except the lateral hypothalamus.

The orexins bind to two G protein-coupled receptors: OXA binds equally to either orexin receptor 1 (OX_1R) or orexin receptor 2 (OX_2R); OXB binds to both receptors but displays moderate selectivity for OX_2R (Sakurai et al. 1998; Smart et al. 1999). Orexin A and B have been shown to increase the postsynaptic activity of GABAergic and glutamatergic cells (van den Pol et al. 1998). The orexins may also affect the presynaptic effect of Ca^{2+}-dependent transmitters by increasing calcium levels, both through mobilization of internal Ca^{2+} stores and through secondary influx of external calcium (Smart et al. 1999).

Although the total number of orexin neurons is fairly small, axonal projections from these cells extend from the LH to many regions of the rat brain and spinal cord (Chen et al. 1999; Nixon and Smale 2007; Cutler et al. 1999; Date et al. 1999;

Peyron et al. 1998; Nambu et al. 1999), and the distribution of these neurons and axonal projections is very similar across rodent strains and species (Nixon and Smale 2007). The overall distribution of orexin fibers in the brain and spinal cord allows this small population of neurons to play roles in integrating multiple autonomic and behavioral functions, primarily feeding, sleep/wake behavior, and arousal (Niimi et al. 2001a; Kotz et al. 2002; Rodgers et al. 2000; Kunii et al. 1999; Haynes et al. 2000; Mondal et al. 1999; Yamanaka et al. 2000; Tsujino and Sakurai 2009; Nunez et al. 2009; Siegel 1999; Lin et al. 1999; Piper et al. 2000; Hungs and Mignot 2001), as well as nociception, respiratory, motor, neuroendocrine, and cardiovascular systems (Nixon and Smale 2007; Cutler et al. 1999; Date et al. 1999; Peyron et al. 1998; Nambu et al. 1999; Volgin et al. 2002; Zhang and Luo 2002; Samson et al. 1999; Shirasaka et al. 2002; Zhang et al. 2005a; Berthoud et al. 2005). Disruptions or deficiencies in orexin signaling have been linked to a number of sleep/wake and endocrine disorders in humans and in animal models (Lin et al. 1999; Petersén et al. 2005; Nevsimalova et al. 2005; Thannickal et al. 2000; Nishino et al. 2000).

There is also strong evidence for an important role for orexin outside of the central nervous system. Both orexin and orexin receptors are present in peripheral tissues. Both PPO and orexin receptor mRNA are present in the gut in several species, including rats, guinea pigs, dogs, horse, deer, mice, sheep, and humans (Kirchgessner and Liu 1999; Ehrstrom et al. 2005; Dall'aglio et al. 2009, 2008, 2011; de Miguel and Burrell 2002); however, at least one report questions these findings (Baumann et al. 2008). Additionally, PPO mRNA has been identified in the heart and testicular tissue of rats (Johren et al. 2001), and orexin receptors have been found in rat lung, in the adrenal glands and gonads of both rats and sheep, and in the enteric nervous system of several species (Dall'aglio et al. 2009, 2008; Johren et al. 2001; Zhang et al. 2005b).

2 Orexin and Feeding

The hypothalamic distribution of cell bodies containing the precursor protein suggested that orexins are involved in feeding behavior. Prior to the discovery of orexin, the only other peptide known to be found in cell bodies limited to the LH was melanin-concentrating hormone (MCH), a peptide known to be involved in the regulation of feeding (Qu et al. 1996). Evidence that 48 h of fasting elicited a 2.4-fold increase in rat prepro-orexin mRNA (Sakurai et al. 1998) quickly prompted more extensive investigation into the relationship between the orexins and feeding. Early experiments showed that injections of OXA and OXB elicit ingestion in rats, although the effects of OXA appeared to be stronger than those of OXB, perhaps due to its more stable structure (Sakurai et al. 1998). This difference in the orexigenic effects of OXB in comparison to OXA has been replicated several times, with most studies suggesting that OXB is less effective than OXA in eliciting feeding or drinking behavior (Sakurai et al. 1998; Kunii et al. 1999; Edwards et al. 1999).

In some cases, OXB has been ineffective in eliciting any ingestive behavior whatsoever (Sweet et al. 1999; Lubkin and Stricker-Krongrad 1998).

Orexin effects on ingestive behavior appear to depend upon interactions with other food-related signaling systems, such as neuropeptide Y (NPY), leptin, MCH, ghrelin, galanin, and agouti-related protein (Ehrstrom et al. 2005; Broberger et al. 1998; Horvath et al. 1999a; Schwartz et al. 2000; Takenoya et al. 2005; Sakurai 2003; Rauch et al. 2000). For example, the ingestive behavior stimulated by orexin is attenuated or blocked by leptin, a potent inhibitor of food intake (reviewed in Schwartz et al. 2000). In one study, pretreatment with leptin blocked orexin-induced activity in nearly half of the orexin-responsive neurons identified in the arcuate nucleus (Arc) (Rauch et al. 2000). In addition, leptin injections in the rat are capable of blocking both OXA-induced feeding behavior as well as NPY-induced Fos immunoreactivity in OXA cells (Niimi et al. 2001a). Leptin may block the effects of orexins directly or indirectly, as some OXB cells have been shown to express leptin receptors and NPY cells in the rat and monkey Arc receiving orexin fiber contact also express leptin receptors (Horvath et al. 1999a). Orexin appears to have a reciprocal blocking effect on leptin, as OXA administered intravenously reduces plasma leptin concentrations in humans (Ehrstrom et al. 2005).

Previous studies have shown that orexin and arcuate nucleus NPY neurons have reciprocal functional connections important in feeding (Yamanaka et al. 2000; Horvath et al. 1999a; Elias et al. 1998). Orexin neurons in the LH send projections to the Arc, and these fibers form synaptic contacts with NPY-containing neurons in this nucleus (Horvath et al. 1999a). Administration of orexin increases expression of the early-active gene cFos in Arc NPY neurons (Yamanaka et al. 2000). In turn, Arc NPY neurons project to the LH where they make synaptic contact with orexin neurons (Broberger et al. 1998; Horvath et al. 1999a; Elias et al. 1998). While a large number of these NPY-orexin contacts in the LH appear to originate in the Arc, NPY neurons in other regions are also known to project to the LH and may contribute to this NPYergic input to orexin neurons (Elias et al. 1998). Injection of NPY into the perifornical LH (PeF), in which orexin neurons are found, robustly stimulates food intake (Stanley et al. 1993), and this induction of intake shows a circadian pattern of effectiveness, eliciting the greatest response during the active period (Stanley and Thomas 1993), matching the endogenous circadian patterns of cFos expression in orexin neurons (Martinez et al. 2002). Orexin neurons in the PeF are known to express NPY Y4 receptors, and cFos expression is increased in orexin neurons following application of NPY or a Y4-specific agonist (Niimi et al. 2001a; Campbell et al. 2003a). While central orexin injection increases food intake, the effect is at least partly dependent on activation of NPY neurons, as orexin-induced intake is attenuated (but not blocked) by administration of an NPY Y1 receptor antagonist (Yamanaka et al. 2000); this effect is complicated by the finding that NPY tonically pre- and postsynaptically inhibits orexin neurons via a Y1-specific pathway (Fu et al. 2004). Interestingly, NPY-induced food intake appears to be partly dependent on orexin, as treatment with an orexin antibody reduces (but does not eliminate) NPY-induced food intake (Niimi et al. 2001a), and anatomical evidence suggests orexin neurons may be a downstream target of NPY action in feeding (Broberger et al. 1998). Both orexin and NPY neurons express receptors for the hunger-signaling hormone leptin (Horvath et al. 1999a), suggesting that while NPY neurons might represent the main target of this hormone, orexin neurons are also responsive to peripheral signals of energy balance (Meister 2000).

There are several lines of evidence that suggest interactions between orexin and blood glucose levels. Insulin-induced hypoglycemia results in a rapid rise in nuclear cFos expression in OXA cells of the rat (Moriguchi et al. 1999). Orexin-containing pancreatic islet cells also contain insulin in humans, and some of these cells express orexin receptors (Ehrstrom et al. 2005). Intravenous orexin administration raises insulin levels in the blood, presumably by stimulating pancreatic cells expressing such receptors (Ehrstrom et al. 2005). In addition, there are some indications that defects in the orexin system may affect the regulation of glucose in humans. For example, in humans with narcolepsy, a condition associated with low or nonexistent levels of orexins (Nishino et al. 2000, 2001), there appears to be a higher risk of non-insulin-dependent diabetes (Honda et al. 1986).

Despite the documented relationship between the orexins and feeding and satiety systems, there is some controversy over the actual effect orexins have on feeding behavior. The administration of orexins into the central nervous system has not always reliably increased feeding behavior (reviewed in (Sutcliffe and de Lecea 2000)). Some have argued that increased ingestion following orexin administration is due solely to the increased locomotor activity caused by orexin; however, at least one study suggests that locomotor and feeding effects of orexin are independent rather than coupled (Kotz et al. 2002). While it is generally agreed that the orexins are not as potent as NPY in being a stimulator of feeding, for example, the relative strength or weakness of the orexins as compared to other peptides such as MCH has not been clearly established (Edwards et al. 1999; Lubkin and Stricker-Krongrad 1998). Indeed, in studies performed in various laboratories, the orexins elicited an ingestive response ranging from very robust (Sakurai et al. 1998; Yamanaka et al. 2000), moderate (Edwards et al. 1999; Sweet et al. 1999), to weak (Lubkin and Stricker-Krongrad 1998).

The differences in feeding behavior elicited in individual studies may be explained by several factors. First, orexins have been shown to increase both GABA and glutamate release in the rat in vitro (van den Pol et al. 1998). These peptides thus appear to have the ability to affect the fast synaptic excitatory or inhibitory activity of many parts of the hypothalamus. Therefore, the reported effects of centrally administered orexins may not be physiologically relevant, as spillover into other brain regions could activate or inhibit systems not normally involved in the feeding effects of orexin. The actual discrete local effects of naturally released peptides are presumably much more finely controlled by the brain than even the most carefully placed injection. Indirect actions or spillover effects of injected orexins have been proposed as explanations for differences seen in several studies (Edwards et al. 1999; Lubkin and Stricker-Krongrad 1998). Second, the relative degree of feeding behavior observed after introduction of orexin may be related to stress, as at least some orexin-induced ingestive responses rely upon interactions between orexin, NPY, and corticosterone levels

(Yamanaka et al. 2000; Horvath et al. 1999a; Ida et al. 2000a; Jászberényi et al. 2000). Finally, orexin-induced feeding might be time dependent. Circadian responsiveness of the feeding effect of OXA in the rat has been observed in at least one study, with an increase in food intake following OXA injection only occurring during the light phase of the cycle (Kotz et al. 2002).

Although the exact role of the orexins in feeding has yet to be established, it is possible that the orexins are involved in the coordination of locomotor activity and arousal in response to stress and variation in food availability. During short-term food deprivation, orexin receptor mRNA is upregulated (Lu et al. 2000). Orexins promote wakefulness (Piper et al. 2000; Methippara et al. 2000; Hagan et al. 1999), reliably increase locomotor activity in rats (Kotz et al. 2002, 2006; Kiwaki et al. 2004), and occasionally lead to increased searching and exploratory behavior (Ida et al. 1999; Jones et al. 2001). A decrease in food availability may thus increase arousal at times that the animal is normally quiescent, leading to increased locomotion and searching behaviors. By modifying the timing of arousal, the orexin system might increase the chance of the animal encountering a food source that is not available at other times of the day. The orexin system is clearly uniquely situated for involvement in the coordination of an interrelated suite of behaviors related to food intake and arousal.

3 Orexin and Arousal

The overall distribution of orexin fibers in the brain has suggested that the orexins play a role in a number of systems, including the maintenance of arousal (Hagan et al. 1999; Horvath et al. 1999b). Orexin fibers have been shown to project to various brain nuclei implicated in the control of sleep state (Date et al. 1999; Peyron et al. 1998; Nambu et al. 1999; Moore et al. 2001; Mintz et al. 2001). Application of OXA in the locus coeruleus (Hagan et al. 1999; Bourgin et al. 2000) and lateral preoptic area (Methippara et al. 2000) of the rat has been shown to increase wakefulness, primarily through a decrease in rapid eye movement (REM) sleep (Bourgin et al. 2000). Activity in locus coeruleus neurons increases following application of OXA (Hagan et al. 1999; Horvath et al. 1999b; Bourgin et al. 2000). In contrast to OXA, OXB does not seem to affect wakefulness (Bourgin et al. 2000).

The orexin cells also receive input from brain systems involved in regulation of sleep/wakefulness. In mammals, circadian organization of activity including sleep/wake behavior is regulated by the endogenous clock located in the suprachiasmatic nucleus (SCN) (reviewed in Weaver 1998). Orexin cell bodies receive both limited direct contact from the SCN (Abrahamson et al. 2001) as well as substantial indirect contact from the SCN via the medial preoptic area (Deurveilher and Semba 2005). Introduction of chemicals known to increase arousal in rats, such as methamphetamines or the antinarcoleptic drug modafinil, increases nuclear Fos expression in orexin cell bodies (Weaver 1998; Chemelli et al. 1999; Estabrooke et al. 2001). Furthermore, increasing the behavioral arousal of rats by sleep deprivation induced due to handling also increases the expression of nuclear Fos in OXA cells (Estabrooke et al. 2001). The orexins thus appear to be capable of both receiving information relating to the arousal state of the animal and relaying arousal information to other nuclei known to promote wakefulness. The finding that a defect in the orexin system is associated with the sleep disorder narcolepsy (Lin et al. 1999; Chemelli et al. 1999; Hungs et al. 2001; Hara et al. 2001) has strengthened the association between the orexins and arousal.

4 Orexin Actions on Endocrine and Autonomic Systems

Orexin may also be involved in the regulation of autonomic functions. There are extensive projections from orexin neurons to hindbrain nuclei that regulate cardiovascular and sympathetic processes (Date et al. 1999; Zheng et al. 2005). Several studies have shown that application of OXA increases heart rate, blood pressure, and respiration rate in rats and mice (Shirasaka et al. 2002; Zhang et al. 2005a, 2005; Ciriello et al. 2003b; de Oliveira and Ciriello 2003). Body temperature, which generally rises during active periods and decreases when animals are quiescent, increases following injection of OXA (Yoshimichi et al. 2001), but not after injection of OXB (Jones et al. 2001). The increase in body temperature following application of OXA does not appear to be a result of increased locomotor activity (Yoshimichi et al. 2001).

Finally, orexins have been implicated in modulation of the hypothalamic—pituitary–gonadal (HPG) axis at several levels. First, within the hypothalamus, orexin has been shown to stimulate the release of gonadotropin-releasing hormone (GnRH) (Russell et al. 2001). Cells containing GnRH receive direct contact from orexin fibers in rats and sheep (Campbell et al. 2003b; Iqbal et al. 2001), and in rats, GnRH neurons have also been shown to express orexin receptors (Campbell et al. 2003b). In addition, orexin projections to the hypothalamic magnocellular nuclei that project to the pituitary also appear to be important in HPG regulation. Magnocellular neurons in the Pa express orexin receptors, and these receptors are selectively upregulated during the estrous cycle and early lactation in rats (Wang et al. 2003a). At the level of the pituitary, much evidence for orexin involvement in HPG regulation has been found. Specifically, both rat and human pituitaries express orexin receptors (Johren et al. 2001; Blanco et al. 2001), and OXA acting on these receptors appears to directly block GnRH-mediated release of luteinizing hormone in proestrous female rats (Russell et al. 2001). With respect to the gonads, testicular tissue in rats expresses orexin, and both rat and sheep testicular tissues express orexin receptor mRNA (Johren et al. 2001; Zhang et al. 2005b). Although orexin receptors have been found in rat ovary, unlike in male gonads, orexin mRNA appears to be absent (Johren et al. 2001). Although the specific actions of orexin on gonadal tissue are currently unknown, the presence of orexin and orexin receptors in the gonads suggests the possibility that orexins may affect the HPG axis at all three levels.

5 Orexin, Physical Activity, and Energy Expenditure

Orexin augmentation of energy expenditure was reported shortly after initial reports describing the orexins in the literature (de Lecea et al. 1998; Sakurai et al. 1998; Lubkin and Stricker-Krongrad 1998). Orexin A infusion into the third ventricle increased metabolic rate, and the increase was more robust in the dark cycle (active phase) relative to the light cycle (resting phase) in mice (Lubkin and Stricker-Krongrad 1998). In contrast, equimolar doses of OXB were ineffective. The circadian variation in OXA-induced metabolic rate (Lubkin and Stricker-Krongrad 1998) parallels nuclear cFos immunoreactivity (an indicator of cellular activity) in orexin neurons across the light/dark cycle (Estabrooke et al. 2001), which highlights the contribution of orexins to basal metabolism. The stimulatory effect of ventricular OXA in mice (Asakawa et al. 2002) was confirmed and was later extended to rats as OXA-stimulated oxygen consumption normalized to body weight (Wang et al. 2001; Semjonous et al. 2009). Orexin augmentation of whole-body energy expenditure can be attributed to specific brain sites of action. Orexin A infusion into the arcuate nucleus increases oxygen consumption in anesthetized rats (Wang et al. 2003b) and increases thermogenesis after infusion into the hypothalamic paraventricular nucleus (PVH) and rostral lateral hypothalamus (rLH) in conscious rats (Kiwaki et al. 2004; Novak et al. 2006a, 2010; Teske et al. 2010). In contrast, OXA has no effect on oxygen consumption in anesthetized rats after infusion into the locus coeruleus (LC), paraventricular thalamic nucleus (PVT), caudal lateral hypothalamus (cLH), PVH, medial preoptic area (MPO), and the dorsomedial and ventromedial hypothalamic nuclei (Wang et al. 2003b).

Inconsistent effects of OXA infusion in the PVH and LH are likely due to differences in the anesthesia state, dose range of OXA tested, location of the injectate (rLH vs. cLH), and the endpoint reported between studies. That OXA reduces respiratory quotient (Lubkin and Stricker-Krongrad 1998) underscores the importance of measuring the change in both oxygen and carbon dioxide during indirect calorimetry experiments and reporting energy expenditure as heat production. Finally, the opposing effect of OXA in the rLH and cLH on energy expenditure demonstrates that behavioral effects of neuropeptides can be regionally specific similar to the feeding effects of OXA in the lateral hypothalamus (Thorpe et al. 2003), urocortin in the lateral septum (Wang and Kotz 2002), or inhibition of NPY-induced feeding by naltrexone in the nucleus of the solitary tract (NTS) (Kotz et al. 2000).

6 Physical Activity

Physical exertion requires ATP utilization and substrate oxidation, and as such, physical activity ranging from muscle movements during volitional motion to actions as small as postural maintenance incurs an energetic cost (Webb et al. 1980). The effects of orexin on physical activity, including increases in locomotion, rearing, grooming, and burrowing activities, require muscular contraction and thus expend energy. Consistent with this idea, acute intra-PVH OXA dose-dependently increases physical activity and energy expenditure in rodents (Kotz et al. 2006; Kiwaki et al. 2004), and chronic OXA increases physical activity and reduces body weight (Novak and Levine 2009).

Injections of OXA and, to a lesser extent OXB, have been shown to increase locomotor activity (Kotz et al. 2002, 2006; Kiwaki et al. 2004; Jones et al. 2001; Yoshimichi et al. 2001; Ida et al. 2000b; Nakamura et al. 2000) and burrowing behavior (Ida et al. 1999). Activation of orexin neurons appears to be positively correlated to the level of locomotor activity in several rodent species (Estabrooke et al. 2001; España et al. 2003; Nixon and Smale 2004). Orexin B does not appear to have as strong effect on general locomotor activity as does OXA but has been shown to be more effective than OXA in eliciting searching behavior (Ida et al. 1999; Jones et al. 2001) or exploration of novel environments (Jones et al. 2001). Importantly, the increase in locomotor activity observed after application of OXA does not appear to be related to the concurrent increase in feeding often observed following orexin injections (Kotz et al. 2002), suggesting that feeding and activity effects of orexin may be influenced by different neural mechanisms. Face washing and grooming behavior also increase in frequency following injections of OXA but not OXB (Ida et al. 1999, 2000b; Jones et al. 2001; Nakamura et al. 2000; Duxon et al. 2001). The increase in grooming following injection of OXA in the rat is blocked by prior application of a CRF antagonist, suggesting that the behavior may be linked to stress (Ida et al. 2000b). Both the grooming and locomotor effects of orexins may also involve interactions with dopaminergic (Nakamura et al. 2000) and serotonergic (Nakamura et al. 2000; Duxon et al. 2001) systems.

Orexin A stimulates several types of physical activity following ventricular (Hagan et al. 1999; Ida et al. 1999, 2000b; Zheng et al. 2005; Yoshimichi et al. 2001; Nakamura et al. 2000; Matsuzaki et al. 2002; Sunter et al. 2001; España et al. 2001; Volkoff and Peter 2000; Samson et al. 2010) or peripheral infusion (John et al. 2000). Multiple sites of action appear to underlie this stimulatory effect. Indeed, OXA infusion into the LH (Kotz et al. 2002, 2006; Teske et al. 2006), PVH (Kotz et al. 2006; Kiwaki et al. 2004, 2006a; Novak et al. 2010), substantia nigra (Kotz et al. 2006), tuberomammillary nucleus (TMN), LC, and dorsal raphe (Kotz et al. 2008) stimulates physical activity. To date, orexin A has had a stimulatory effect on locomotion after site-specific infusion with one exception. Relatively high doses of OXA (3 μg) in the PVT inhibited locomotion (Li et al. 2009). Unlike locomotion, effects of OXA on other types of physical activity appear to be site dependent. España et al. (2001) compared the effect of OXA in the lateral ventricle and into forebrain nuclei on grooming, rearing, and quadrant entries. While there was no effect of OXA in the substantia innominata on the behaviors tested. OXA in the lateral ventricle and intra-MPO increased all behaviors, while OXA in the medial septum stimulated grooming only. Finally, it is important to note that measurement duration should be considered when comparing efficacy of OXA within the same site. For example, in one study, OXA in the nucleus accumbens shell (AcbS) increased locomotor activity during the 30- to 120-min postinjection interval (Thorpe and Kotz 2005), but there was no effect of OXA 10–30 min postinjection, consistent with others who reported no effect of OXA during the 0- to 30-min postinjection time interval (Baldo and Kelley 2001). This lack of effect in the immediate postinjection period may be due to heightened physical activity due to handling during the injection process, as increased physical activity for up to 20 min postinjection in all treatment groups has been shown (Kotz et al. 2002).

Orexin stimulation of locomotor activity may rely in part on projections to the thalamic intergeniculate leaflet (IGL). The IGL is a thin structure located in the lateral geniculate complex of the thalamus, between the ventral and dorsolateral geniculate nuclei (VG and DLG, respectively) (Moore and Card 1994; Harrington 1997). NPY neurons in the IGL project directly to the SCN (Harrington et al. 1987). Patterns of cellular activation of IGL NPY neurons are correlated with activity patterns in rodents (Janik et al. 1995; Smale et al. 2001; Webb et al. 2008). Manipulations which mimic release of NPY into the SCN result in changes in patterns of physical activity in rodents, and perturbations of the IGL or of NPY cells therein block these changes in activity (Johnson et al. 1989; Rusak et al. 1989; Biello et al. 1994; Huhman and Albers 1994; Wickland and Turek 1994). Orexin neurons send moderately dense axonal projections to the IGL but little or no fibers to the VG or DLG; the presence of these fibers is highly consistent between species (Nixon and Smale 2007; Cutler et al. 1999; Date et al. 1999; Peyron et al. 1998; Mintz et al. 2001; McGranaghan and Piggins 2001). Data suggest that at least some of the mechanisms through which orexin affects physical activity might rely on these projections to the IGL. Orexin fibers form close appositions with NPY neurons in the IGL and appear to form functional contacts with these cells (Nixon and Smale 2004, 2005). Patterns of cFos activation in orexin neurons are correlated with cFos expression patterns in NPY neurons of the IGL (Webb et al. 2008; Nixon and Smale 2005), suggesting that orexin neurons influence the role of IGL NPY neurons in control of behavioral activity. Furthermore, in one study, orexin fibers in the IGL preferentially approached NPY cells, which expressed cFos in patterns correlating with physical activity (Nixon and Smale 2005), suggesting that cellular activation of these neurons is influenced by orexigenic input.

Consistent with effect of orexins on physical activity, orexin augments muscle tone, which would be expected to influence energy expenditure. Orexin A infusion into the LC (Kiyashchenko et al. 2001) or the alpha gigantocellular reticular nucleus in the medioventral medullary region (Mileykovskiy et al. 2002) increases hindlimb muscle tone. Likewise, OXA infusion into the trigeminal motor nucleus and hypoglossal motor nucleus increases EMG activity, an indicator of muscle activity, in the masseter and geniosglossus muscles (Peever et al. 2003). In contrast,

OXA microinjection into the pontine inhibitory area (Kiyashchenko et al. 2001) or the ventral gigantocellular reticular nucleus (Mileykovskiy et al. 2002) inhibits hindlimb muscle tone.

7 Orexin Effects on Sympathetic Outflow: Cardiovascular and Thermoregulatory Systems

Orexin effects on cardiovascular and thermoregulatory systems have been extensively reviewed (Shirasaka et al. 2002; Szekely et al. 2002, 2010; Ferguson and Samson 2003; Samson et al. 2005; Szekely 2006). Brain areas classically involved in thermoregulation and cardiovascular function receive orexin projections and express orexin receptors (Date et al. 1999, 2000; van den Pol 1999; van den Top et al. 2003), providing a neuroanatomical basis for orexin involvement in cardiovascular function and thermoregulation. Behavioral studies further suggest that sympathetic outflow is orexin-mediated. Orexin A infusion into the lateral ventricle increases renal sympathetic nerve activity (Shirasaka et al. 1999; Matsumura et al. 2001), plasma epinephrine (Shirasaka et al. 1999), noradrenaline release (Hirota et al. 2001), and firing rate of sympathetic nerves innervating the interscapular brown adipose tissue (iBAT) (Monda et al. 2004a, 2003, 2001), which would be expected to increase thermogenesis. Likewise, intrathecal OXA infusion (Antunes et al. 2001) and in vitro application (van den Top et al. 2003) stimulate sympathetic outflow. Tachycardia is observed following OXA in the lateral (Wang et al. 2001; Shirasaka et al. 1999; Monda et al. 2004a, 2003, 2001) and fourth ventricles (Zheng et al. 2005) or following infusion intracisternally (Chen et al. 2000) and intrathecally (Antunes et al. 2001), but not intravenously (Chen et al. 2000). Moreover, the pressor response to orexin is similar to the effect of orexin on heart rate (Samson et al. 1999; Shirasaka et al. 1999; Matsumura et al. 2001; Antunes et al. 2001; Chen et al. 2000). These sympathetic and cardiovascular effects are clearly mediated by multiple sites of action. Orexin A infusion into the rostral lateral ventral medulla (RVLM) (Chen et al. 2000), NTS (de Oliveira et al. 2003; Smith et al. 2002), Arc (Wang et al. 2003b), PVH (Monda et al. 2004a; Sato-Suzuki et al. 2002), and the diagonal band of Broca induces tachycardia (Monda et al. 2004a), and OXA in the RVLM (Chen et al. 2000) and NTS (de Oliveira et al. 2003; Smith et al. 2002) stimulate mean arterial pressure. In contrast, infusion into the nucleus ambiguous (de Oliveira and Ciriello 2003) or the subfornical organ induces bradycardia (Smith et al. 2007). While OXA in the nucleus ambiguous (de Oliveira and Ciriello 2003) has no effect on blood pressure, intrasubfornical organ OXA (Smith et al. 2007) reduces blood pressure, thereby indicating site-specific actions of orexin on cardiovascular responses. In contrast to the cardiovascular response to orexin, orexin action on temperature appears to be consistently observed independent of route of administration. Orexin A increases colonic (Monda et al. 2001), iBAT (Monda et al. 2003, 2001), cutaneous (Monda et al. 2003), and core body temperature

(Zheng et al. 2005) following infusion into the lateral and fourth ventricles (Zheng et al. 2005), Arc (Wang et al. 2003b), and the diagonal band of Broca (Monda et al. 2004b). A recent report showed that chronic infusion of the OX1R antagonist (SB-334867-A) into the lateral ventricle increased iBAT temperature during the dark cycle and UCP1 protein expression in the iBAT (Verty et al. 2010). Furthermore, there is no effect of acute OXA on colonic temperature (Hagan et al. 1999) or of chronic OXA infusion on iBAT temperature (Haynes et al. 1999). It is plausible that these discrepancies may be due to differences in dose or duration of injectate, site of administration, location of thermistor, or duration of measurement. Together, these data indicate that orexin action is sympathoexcitatory, which would supplement energy expenditure due to orexin-mediated metabolic rate and physical activity.

8 Orexin Integration of Feeding and Physical Activity

A greater understanding of how orexin may integrate information important to both energy balance and physical activity may be derived from studying the effects of gains in orexin action and loss of orexin function studies. Loss of orexin neurons, either by lesion, genetics, or postnatal ablation, affects feeding behavior and physical activity. Likewise, orexin neurons respond to changes in energy balance brought about by nutritional status and exercise, suggesting that orexins receive information relevant to both behaviors. While there is not enough existing information to understand how orexin may integrate feeding and activity, orexin neuron circuitry would lend itself well to such a role: Orexin neurons receive input from several important energy sensing areas and project mono- and multisynaptically throughout the brain to multiple brain areas with diverse functionality. A physiological state of energy deficit would confer interoceptive cues signaling appetitive drive, whereas states of energy excess could signal for energy loss, perhaps via nonexercise energy expenditure. Orexin neurons project to and excite arcuate neuropeptide Y neurons (Horvath 2005). During food restriction, this signal is robustly enhanced, demonstrating that negative energy states are sensed by orexin neurons, which respond by enhancing orexigenic tone to restore energy balance. Whether this relationship exists for physical activity is unknown. Exercise may induce an interoceptive state of temporary satiety, which is not perceived as a situation of negative energy balance by the brain. Yet existing data suggest that exercise may also stimulate orexin neurons (Nixon and Smale 2004; Wu et al. 2002). While orexin neurons are glucose sensing, and thus could respond to glucose alterations associated with physical activity, significance at the synapse level is unclear. Exercise-associated motor activity or food anticipatory-associated activity could also be two mechanisms responsible for this induction of orexin activity. The metabolic and sensory milieu following exercise vs. spontaneous physical activity is likely very different, but to date, there are no studies differentiating between these activity states and the corresponding effects on orexin signaling. As mentioned above, the induction of spontaneous physical activity by orexin A is inconsistent with an endocrine feedback loop, in which one might expect a reduction of orexin activity after motor activity. While currently there is no explanation for this relationship between orexin A and physical activity, it is likely that identification of subpopulations of orexin neurons and clarification of their functional roles will shed light on orexin and physical activity interactions.

9 Orexin Effects: Energetic Balance

As indicated above, orexin elevates both eating behavior and energy expenditure. The increase in both of these outputs does not fit the typical profile for neuromodulators of energy balance (Bray 2000), which is that of an inverse relationship between the two outputs of feeding and energy expenditure; for example, if a compound increases food intake, it concurrently reduces energy expenditure, whereas if a compound reduces food intake, it increases energy expenditure. These opposing outputs have been noted for most described neuromodulators of food intake and energy expenditure (Bray 2000). Why this is the case is unclear, but this model is attractive as it fits a classic homeostatic model of regulation and allows for the categorization of compounds (neuropeptides, neuromodulators, neurotransmitters, and neurohormones) into either the "satiety" or "obesigenic" category. In what category lies a compound that elevates both food intake and energy expenditure? This lack of ability to classify orexin as either obesity producing (via enhanced food intake) or obesity preventive (via enhanced energy expenditure) has created a confused discussion of this peptide's function. The purpose of this section is to integrate the knowledge of orexin effects on food intake and energy expenditure and clarify the role that orexin has in body weight regulation.

As suggested by multiple studies, the orexin signal in different brain areas with different functionality translates to different outcomes; orexin in one area may have a feeding effect, whereas in another area, an effect on physical activity and energy expenditure. The sum total of all these actions will influence body weight. We clearly do not have the type of comprehensive evidence that is needed to understand precisely what effect endogenously produced orexin, acting in all projection sites, has on eating behavior and energy expenditure in different physiological states, but we can start to make some assumptions about this based on receptor profiles, site functionality, and pharmacological studies and by studying obesity-prone and obesity-resistant models.

Narcolepsy in humans is accompanied by significantly reduced or absent orexin and significantly increased body mass (Thannickal et al. 2000; Nishino et al. 2001; Overeem et al. 2001), suggesting that the overall effect of orexin is obesity resistance. Animal models of orexin loss support this observation, as mice with transgenic ablation of orexin neurons become obese (Hara et al. 2001). Orexin global overexpression has mixed results on body weight, but this can be expected in studies in which the orexin signal is placed in areas in which it is not normally expressed. While there are few studies of orexin overexpression exclusively in the LH area, at least one study suggests that increasing orexin expression protects against weight gain in animals fed on a high-fat diet (Funato et al. 2009).

Clearly, orexin action is determined not just by orexin output but also by receptor expression. The consequences of increased orexin receptor expression have only just begun to be studied. Work from our laboratory shows that obesity-resistant rats have increased physical activity and that resistance to obesity is associated with increased orexin receptor expression (Kotz et al. 2002, 2006; Teske et al. 2006). Obesity-resistant rat activity is also more sensitive to orexin (Kotz et al. 2002, 2006; Teske et al. 2006; Kiwaki et al. 2004; Novak et al. 2006a, 2010), suggesting either elevated capacity for orexin action via increased receptor. Additionally, early levels of physical activity are associated with lifelong reduced adiposity in obesity-resistant rats (Teske, in press). These findings suggest that the lean phenotype of OR rats may be explained by their high level of physical activity, which appears to be mediated by orexin receptors. This finding is supported by work showing that enhanced orexin receptor expression in rats mitigates the propensity for obesity on a high-fat diet (Funato et al. 2009).

Thus, the energetic consequence of these two behavioral outputs, when added up on a caloric basis, results in negative energy balance and reduces body weight. As orexin enhances feeding behavior and physical activity in a site-specific manner, it is unclear at this point where the dominant effects on each output are taking place. Nonetheless, the calories taken in by the effects of the orexin signal are outweighed by those expended via physical activity. Based on this, orexins may be considered as potential targets for obesity therapy rather than obesigenic.

10 Neuromedins

Neuromedins are one group of gut–brain peptides that illustrate how the gut and brain communicate and act in parallel to modulate energy balance. Though less is known about neuromedins compared to other gut–brain peptides such as CCK or ghrelin, interest in the role of neuromedins in energy balance has increased markedly over the past few years. Of the neuromedins, neuromedin U (NMU) is perhaps the subject of most investigations (Brighton et al. 2004). Related peptides, including neuromedins B, C, K, L, N, and S, likely serve similar or related functions, and some exert their actions through the same receptors. Investigations into these peptides have revealed that, like other gut–brain hormones, a variety of physiological functions and behaviors are influenced by neuromedins. The peripheral actions of neuromedins have been reviewed (Brighton et al. 2004; Mori et al. 2008; Miyazato et al. 2008; Mitchell et al. 2009); here, we will focus primarily on central actions of the neuromedins, particularly of neuromedin-containing neurons found in the brain.

Neuromedin U was first identified as a spinal cord peptide that induced smooth muscle contractions in the uterus, the tissue for which it is named

(Minamino et al. 1985a, b). NMU is found at high levels within the intestine, specifically in neurons of the enteric nervous system (Brighton et al. 2004; Honzawa et al. 1990; Ballesta et al. 1988). Whether or not circulating NMU from peripheral sources acts on the brain to exert its actions is not known. Like other gut–brain peptides, there are sources of neuromedins intrinsic to the brain (Honzawa et al. 1987), possibly reflecting parallel gut–brain systems seen in other gut–brain peptides. Within the brain, most attention has been given to NMU-containing cells in hypothalamic regions important in energy balance, particularly its actions in the PVH and Arc, specifically pro-opiomelanocortin (POMC)-containing arcuate neurons (Graham et al. 2003), and also dorsomedial hypothalamus (Ballesta et al. 1988; Graham et al. 2003; Nogueiras et al. 2006). Several nonhypothalamic regions also contain NMU-immunoreactive neurons and fibers, however, including hindbrain regions important in arousal and energy balance that are also responsive to CCK (Honzawa et al. 1987; Ivanov et al. 2004). Though reports vary somewhat in the pattern of NMU cell or projection distribution, this could be secondary to cross-reactivity with other neuromedins, such as neuromedins S (NMS), which share the same receptors (Rucinski et al. 2007; Peier et al. 2009).

As would be expected from the cell and fiber distribution, NMU receptors or their mRNAs have been identified in hypothalamic brain regions associated with energy balance. Two receptor subtypes have been identified in the brain, NMR1 and NMR2, with the reports citing NMR2 as the most prevalent receptor in the central nervous system and NMR1 found primarily in peripheral tissues (Raddatz et al. 2000; Shan et al. 2000). Within the hypothalamus, NMR2 have been identified in the PVH, dorsomedial nucleus, the dorsal periventricular nuclei, surrounding the ventromedial nucleus, and in the ependymal layer of the ventricle; NMR2 are also found in the brain stem (Graham et al. 2003; Ivanov et al. 2004; Howard et al. 2000; Guan et al. 2001). Other brain systems also contain neuromedin binding or NMR2, including the hippocampus (Ivanov et al. 2004; Guan et al. 2001; Mangold et al. 2008; Zhang et al. 2010). NMR1 has also been identified in brain regions, including the amygdala (Gartlon et al. 2004). Functional studies have used Fos to identify brain regions and systems that are activated by central (i.c.v.) treatment with NMU; these include the most common hypothalamic nuclei associated with NMU – PVH, Arc, dorsomedial nucleus, and lateral hypothalamic area – but also forebrain regions associated with motivation and reward (amygdala, nucleus accumbens, frontal cortex), the supraoptic nucleus, and the hindbrain parabrachial nucleus and nucleus of the solitary tract (Gartlon et al. 2004; Ivanov et al. 2002; Niimi et al. 2001b). Though the hypothalamic regions and, to a lesser extent, the hindbrain regions, have received attention in functional behavioral or physiological studies (Ivanov et al. 2004; Novak et al. 2007, 2006b; Yokota et al. 2004; Wren et al. 2002; Thompson et al. 2004; Qiu et al. 2003, 2005), relatively little attention is paid to the potential functions of neuromedins in the forebrain.

Neuromedins are one set of neuropeptides that act on the "anorexigenic" arm regulating appetite (Semjonous et al. 2009; Howard et al. 2000; Ivanov et al. 2002; Wren et al. 2002). Activation of neuromedin receptors decreases food intake and increases energy expenditure and physical activity (Semjonous et al. 2009; Peier et al. 2009; Novak et al. 2007, 2006b; Wren et al. 2002; Nakazato et al. 2000). Moreover, commonalities in the behavioral and energetic actions of neuromedins and homologous peptides can be seen in nonmammals, even invertebrates (Maruyama et al. 2011, 2008; Bader et al. 2007; Kamisoyama et al. 2007; Tachibana et al. 2010a, b; Yayou et al. 2009). It has been hypothesized that one action of leptin is to stimulate the release of NMU, through which it exerts its actions on metabolism (Graham et al. 2003; Nogueiras et al. 2006; Wren et al. 2002; Jethwa et al. 2005). Transgenic overexpression of NMU in mice leads to hypophagia and leanness (Kowalski et al. 2005), and deletion of the gene for NMU results in obesity, hyperphagia, and decreased physical activity (Hanada et al. 2004). Some have found that mice lacking the NMUR2 gene do not show this phenotype (Zeng et al. 2006), but others have found that NMUR2-deficient mice are lean, hypophagic, and somewhat resistant to weight and fat gain on a high-fat diet (Peier et al. 2009). In humans, variants in the gene encoding pro-NMU are associated with obesity (Hainerova et al. 2006). This led to interest in neuromedins as a potential target for weight-loss therapy. Though acute NMU leads to decreased food intake and increased energy expenditure and physical activity, twice-daily intra-PVH treatment with NMU failed to induce significant weight loss (Thompson et al. 2004). In contrast, chronic central (i.c.v.) infusions of NMU using osmotic minipumps significantly suppressed energy intake, body weight, and adiposity, (Peier et al. 2009) though this effect may depend on the diet (Egecioglu et al. 2009).

The central actions of neuromedins are similar to several other neuropeptides commonly termed "anorexigenic," such as corticotrophin-releasing hormone (CRH): decreased appetite, increased energy expenditure, and increased physical activity (Novak et al. 2006b; Novak and Levine 2007; Sutton et al. 1982). In fact, neuromedins are likely to be one important component that activates brain CRH to affect behavior (Yokota et al. 2004; Wren et al. 2002; Hanada et al. 2001, 2003); CRH also appears to be necessary for some of the behavioral effects of NMU (Hanada et al. 2003). This may be one reason why neuromedins affect behaviors traditionally associated with brain CRH and, more generally, with the stress response (such as locomotion and increased grooming) (Sutton et al. 1982; Hanada et al. 2001; Jaszberenyi et al. 2007). In fact, brain neuromedins affect several other functions and behaviors besides food intake, including reproduction, the sleep/wake cycle (Ahnaou and Drinkenburg 2011), hippocampal and memory function (Zhang et al. 2010; Iwai et al. 2008), sympathetic outflow (Tanida et al. 2009), reproductive and stress axis function (Ivanov et al. 2002; Thompson et al. 2004; Hanada et al. 2004; Jaszberenyi et al. 2007; Jethwa et al. 2006; Fukue et al. 2006; Yang et al. 2010), prolactin secretion (Gartlon et al. 2004), pain sensitivity (Zeng et al. 2006), brain oxytocin and vasopressin systems (Niimi et al. 2001b; Qiu et al. 2005; Sakamoto et al. 2008, 2007), and bone remodeling (Sato et al. 2007).

11 Orexins and Neuromedins in the Appetite Network

Orexin and the neuromedins are multifunctional neuropeptides that participate in a wide variety of neuroregulatory processes. Each of these processes, including appetite, arousal, and spontaneous activity, is the result of the combined output of many neuropeptides acting in a number of brain sites, all organized into a network. Thus, there is a network of brain sites and activities for appetite, a related but distinct network for arousal, and another for physical activity. The actions of orexins throughout the brain are a particularly good illustration of this concept because orexins clearly perform different functions at different brain sites, even though the origin of orexins neurons is in a highly focused place in the lateral hypothalamus and perifornical areas. Work by Thorpe and Kotz (Kotz 2006) has shown that while there are certain brain sites, like rostral LH and paraventricular hypothalamus, where orexin increases feeding and also increases spontaneous physical activity, there are other brain sites like the locus coeruleus where orexin only affects activity.

Brain sites where orexin affects appetite are an important subset of the known sites involved in the appetite regulatory network. Orexin's role at these brain sites ultimately results in increases in feeding, but the exact mode of producing this behavioral phenotype with respect to neuronal function, and particularly the context of that neuronal function involving basal state and other neuromodulators, is incompletely defined. Considering only the effect of orexin stimulation on the behavioral phenotype of appetite, it is important to note that orexin effects do not fit into a clearly established unidirectional action pathway. The example that establishes this concept is the interaction between orexin action in the lateral hypothalamus and neuropeptide Y action in the arcuate and paraventricular nuclei. As reviewed above, orexin in the LH can activate neuropeptide Y-producing neurons in the arcuate nucleus. Neuropeptide Y action in the paraventricular nucleus can also activate orexin neurons in the lateral hypothalamus. Thus, a bidirectional stimulatory pathway involving these two orexigenic stimuli can be identified. Further, both lateral hypothalamus and paraventricular nucleus are connected through other pathways involving other neurotransmitters with other components of the appetite regulatory network. In the case of orexins itself, feeding stimulatory signals from the LH project to rostral LH, paraventricular hypothalamus, and nucleus accumbens.

A linear action pathway cannot account for the known database of brain sites and signals that participate in appetite/body weight regulation, whose action and function rely upon inputs from each other and from peripheral signals. Bidirectional information transfer, as with the example of orexin and neuropeptide Y, is a common theme in this distributed network. There are many such examples of neural interactions, and these interactions indicate that no one "regulator" is operating alone or within one brain area to determine the food intake behavioral response, but rather, a dynamic neural network of neurotransmitters at several brain sites are communicating with each other to determine this output. Therefore, the orexin effect on appetite is not a linear model with one initiation point, but it is a multipoint model. Further, there is cross talk between the networks for appetite, arousal, and activity (among others), such that each of the networks' responsiveness to orexin can be seen both as a function within that domain and as a contributor to the phenotypic output of other domains as well.

12 Orexin and Neuromedin Receptors as Therapeutic Targets

The pharmacologic efficacy of selective and dual orexin receptor antagonists has been tested. Scammell and Winrow recently reviewed the preclinical and clinical pharmacology of multiple orexin receptor antagonists and described their favorable therapeutic efficacy for insomnia (Scammell and Winrow 2011). Despite this, the efficacy for other pathologies such as obesity remains to be determined. The lack of a commercially available orexin 2 receptor antagonist and comprehensive testing of antagonists on non-sleep-related parameters that influence energy balance has hampered progress. The relevance of such testing is imperative as orexin stimulates energy intake and energy expenditure and promotes stabilization of the sleep/wake cycle, which together influence body weight regulation. Therefore, comprehensive testing in addition to distinguishing the functional specificity of the orexin receptors is necessary as it is unclear whether stimulation of one or both receptors is necessary and/or sufficient to elicit a behavioral response and thus a given pharmacologic effect. Based on the behavioral effects of orexin receptor antagonists (reviewed below), it appears that an orexin-based obesity treatment must promote satiety, stimulate energy expenditure, and stabilize sleep/wake parameters.

12.1 Orexin 1 Receptor Antagonists

The first commercially available selective orexin 1 receptor antagonist, SB-334867, has been shown to reverse orexin A-induced feeding (Haynes et al. 2000), physical activity-induced thermogenesis (Kiwaki et al. 2004), grooming (Duxon et al. 2001), sympathetic activity (Hirota et al. 2003), and arousal (Smith et al. 2003), as well as the delay in the normal transition from eating to resting (behavioral satiety sequence) induced by orexin A (Rodgers et al. 2001). Selective blockade of orexin 1 receptors by SB-334867 also attenuated orexin B-stimulated physical activity (Jones et al. 2001). In contrast, SB-408124 had no effect on sleep, physical activity, or body temperature after peripheral administration (Dugovic et al. 2009). Other selective orexin 1 receptor antagonists including SB-410220 (Langmead et al. 2004) and diaryl urea analogues of SB-334867 (Perrey et al. 2011) have been described pharmacologically; however, behavioral effects have yet to determined.

12.2 Orexin 2 Receptor Antagonists

Two proprietary selective orexin 2 receptor antagonists have been reported. N-ethyl-2-[(6-methoxy-pyridin-3-yl)-(toluene-2-sulphonyl)-amino]-*N*-pyridin-3-ylmethyl-acetamide (EMPA) reduced dark cycle basal physical activity, and [Ala^{11}, D-Leu^{15}], orexin B-stimulated physical activity (Malherbe et al. 2009). Peripheral infusion of JNJ-10397049 promoted sleep and reduced basal physical activity and body temperature in the light/dark cycle (Dugovic et al. 2009).

12.3 Dual Receptor Antagonists

Actelion described the first dual orexin receptor antagonist as being most effective during the active phase of the light/dark cycle in rats, dogs, and humans (Brisbare-Roch et al. 2007). Oral almorexant (ACT-078573) was shown to promote sleep and reduce home cage activity despite no effect on body temperature during the active dark period in rats. Parallel effects were observed in dogs with efficacious sleep promotion and physical activity reduction effects observed during the day but absent when almorexant was administered prior to sleep. In humans, oral administration in the morning reduced clinical and subjective alertness and promoted sleep demonstrated by reduced latency with no adverse effects. Recently, Li et al. reported that almorexant reduced oxygen consumption and promoted sleep in rats after oral gavage in the dark cycle only and had no effect on body temperature (Li and Nattie 2010). Additional biocomparison, tolerability, pharmacokinetic, and pharmacodynamic tests in humans suggest that almorexant may be promising for treatment of insomnia (Hoch et al. 2011; Hoever et al. 2010).

Several dual receptor antagonists developed by Winrow and colleagues at Merck have been described. Suvorexant (MK-4305) has been shown to reduce physical activity and promote sleep in rats, dogs, and rhesus monkeys (Winrow et al. 2011). Like its predecessor (Bergman et al. 2008), DORA-1 promoted sleep, reduced basal physical activity, and reduced physical activity stimulated by [Ala^{11}, D-Leu^{15}] orexin B and amphetamine (Winrow et al. 2010). In a similar manner, another dual orexin receptor antagonist based on a 1,4-diazepane central scaffold reduced basal dark cycle physical activity (Whitman et al. 2009). From these studies, DORA-5 was developed and was shown to reduce home cage physical activity and increase sleep after oral administration (Whitman et al. 2009). The clinical efficacy of another dual receptor antagonist, SB-674042, has also been demonstrated for insomnia (Reviewed in Scammell and Winrow 2011).

Some progress has also been made to exploit the brain neuromedin system for potential pharmacological treatment for obesity. An antagonist, R-PSOP, has been described (Liu et al. 2009). This antagonist binds competitively to the NMUR2 with high affinity and significantly attenuates the nociceptive response induced by NMU-23 treatment. However, there have yet to be reports on the development of agents that might act to stem obesity by targeting central NM receptors to alter behavior or energy expenditure.

13 Summary

Orexin affects energy balance in several ways, notably by increasing feeding in some behavioral contexts and more potently by stimulating energy expenditure mainly through increasing the level of spontaneous activity. The brain site in which orexin engages its receptor principally determines the regulatory functions of orexin, but taken together, orexin mainly exerts negative energy balance influence. The two forms of orexin and the two receptor subtypes likely also play a role in differentiating function, as do the contexts of the receiving neurons and the network in which the signaling is taking place.

Neuromedin is an anorexic peptide with expression for the peptide and receptor in gut, hypothalamus, and brain stem. There is evidence that neuromedin may play a major role in signaling certain kinds of satiety signals, such as leptin-based signals. Neuromedin also contributes to negative energy balance influences by increasing physical activity and thereby energy expenditure.

References

Abrahamson EE, Leak RK, Moore RY (2001) The suprachiasmatic nucleus projects to posterior hypothalamic arousal systems. Neuroreport 12(2):435–440

Ahnaou A, Drinkenburg WH (2011) Neuromedin U(2) receptor signaling mediates alteration of sleep-wake architecture in rats. Neuropeptides 45(2):165–174

Alvarez CE, Sutcliffe JG (2002) Hypocretin is an early member of the incretin gene family. Neurosci Lett 324(3):169–172

Antunes VR, Brailoiu GC, Kwok EH, Scruggs P, Dun NJ (2001) Orexins/hypocretins excite rat sympathetic preganglionic neurons in vivo and in vitro. Am J Physiol Regul Integr Comp Physiol 281(6):R1801–R1807

Asakawa A, Inui A, Goto K, Yuzuriha H, Takimoto Y, Inui T et al (2002) Effects of agouti-related protein, orexin and melanin-concentrating hormone on oxygen consumption in mice. Int J Mol Med 10(4):523–525

Bader R, Colomb J, Pankratz B, Schrock A, Stocker RF, Pankratz MJ (2007) Genetic dissection of neural circuit anatomy underlying feeding behavior in Drosophila: distinct classes of hugin-expressing neurons. J Comp Neurol 502(5):848–856

Baldo BA, Kelley AE (2001) Amylin infusion into rat nucleus accumbens potently depresses motor activity and ingestive behavior. Am J Physiol Regul Integr Comp Physiol 281(4): R1232–R1242

Ballesta J, Carlei F, Bishop AE, Steel JH, Gibson SJ, Fahey M et al (1988) Occurrence and developmental pattern of neuromedin U-immunoreactive nerves in the gastrointestinal tract and brain of the rat. Neuroscience 25(3):797–816

Baumann CR, Clark EL, Pedersen NP, Hecht JL, Scammell TE (2008) Do enteric neurons make hypocretin? Regul Pept 147(1–3):1–3

Bergman JM, Roecker AJ, Mercer SP, Bednar RA, Reiss DR, Ransom RW et al (2008) Proline bis-amides as potent dual orexin receptor antagonists. Bioorg Med Chem Lett 18(4):1425–1430

Berthoud HR, Patterson LM, Sutton GM, Morrison C, Zheng H (2005) Orexin inputs to caudal raphe neurons involved in thermal, cardiovascular, and gastrointestinal regulation. Histochem Cell Biol 123(2):147–156

Biello SM, Janik D, Mrosovsky N (1994) Neuropeptide Y and behaviorally induced phase shifts. Neuroscience 62(1):273–279

Blanco M, Lopez M, Garcia-Caballero T, Gallego R, Vazquez-Boquete A, Morel G et al (2001) Cellular localization of orexin receptors in human pituitary. J Clin Endocrinol Metab 86 (4):1616–1619

Bourgin P, Huitron-Resendiz S, Spier AD, Fabre V, Morte B, Criado JR et al (2000) Hypocretin-1 modulates rapid eye movement sleep through activation of locus coeruleus neurons. J Neurosci 20(20):7760–7765

Bray GA (2000) Reciprocal relation of food intake and sympathetic activity: experimental observations and clinical implications. Int J Obes Relat Metab Disord 24(Suppl 2):S8–S17

Brighton PJ, Szekeres PG, Willars GB (2004) Neuromedin U and its receptors: structure, function, and physiological roles. Pharmacol Rev 56(2):231–248

Brisbare-Roch C, Dingemanse J, Koberstein R, Hoever P, Aissaoui H, Flores S et al (2007) Promotion of sleep by targeting the orexin system in rats, dogs and humans. Nat Med 13 (2):150–155

Broberger C, De Lecea L, Sutcliffe JG, Hokfelt T (1998) Hypocretin/orexin- and melanin-concentrating hormone-expressing cells form distinct populations in the rodent lateral hypothalamus: relationship to the neuropeptide Y and agouti gene-related protein systems. J Comp Neurol 402(4):460–474

Campbell RE, Smith MS, Allen SE, Grayson BE, Ffrench-Mullen JM, Grove KL (2003a) Orexin neurons express a functional pancreatic polypeptide Y4 receptor. J Neurosci 23(4):1487–1497

Campbell RE, Grove KL, Smith MS (2003b) Gonadotropin-releasing hormone neurons coexpress orexin 1 receptor immunoreactivity and receive direct contacts by orexin fibers. Endocrinology 144(4):1542–1548

Chemelli RM, Willie JT, Sinton CM, Elmquist JK, Scammell T, Lee C et al (1999) Narcolepsy in orexin knockout mice: molecular genetics of sleep regulation. Cell 98(4):437–451

Chen CT, Dun SL, Kwok EH, Dun NJ, Chang JK (1999) Orexin A-like immunoreactivity in the rat brain. Neurosci Lett 260(3):161–164

Chen CT, Hwang LL, Chang JK, Dun NJ (2000) Pressor effects of orexins injected intracisternally and to rostral ventrolateral medulla of anesthetized rats. Am J Physiol Regul Integr Comp Physiol 278(3):R692–R697

Ciriello J, Rosas-Arellano MP, Solano-Flores LP, de Oliveira CV (2003a) Identification of neurons containing orexin-B (hypocretin-2) immunoreactivity in limbic structures. Brain Res 967 (1–2):123–131

Ciriello J, Li Z, de Oliveira CV (2003b) Cardioacceleratory responses to hypocretin-1 injections into rostral ventromedial medulla. Brain Res 991(1–2):84–95

Cutler DJ, Morris R, Sheridhar V, Wattam TA, Holmes S, Patel S et al (1999) Differential distribution of orexin-A and orexin-B immunoreactivity in the rat brain and spinal cord. Peptides 20(12):1455–1470

Dall'Aglio C, Pascucci L, Mercati F, Giontella A, Pedini V, Scocco P et al (2008) Identification of orexin A- and orexin type 2 receptor-positive cells in the gastrointestinal tract of neonatal dogs. Eur J Histochem 52(4):229–235

Dall'aglio C, Pascucci L, Mercati F, Giontella A, Pedini V, Ceccarelli P (2009) Immunohistochemical identification and localization of orexin A and orexin type 2 receptor in the horse gastrointestinal tract. Res Vet Sci 86(2):189–193

Dall'aglio C, Pascucci L, Mercati F, Boiti C, Ceccarelli P (2011) Localization of the orexin system in the gastrointestinal tract of fallow deer. Acta Histochem. 2011 March 11 [Epub ahead of print]

Date Y, Ueta Y, Yamashita H, Yamaguchi H, Matsukura S, Kangawa K et al (1999) Orexins, orexigenic hypothalamic peptides, interact with autonomic, neuroendocrine and neuroregulatory systems. Proc Natl Acad Sci USA 96(2):748–753

Date Y, Mondal MS, Matsukura S, Nakazato M (2000) Distribution of orexin-A and orexin-B (hypocretins) in the rat spinal cord. Neurosci Lett 288(2):87–90

de Lecea L, Kilduff TS, Peyron C, Gao X, Foye PE, Danielson PE et al (1998) The hypocretins: hypothalamus-specific peptides with neuroexcitatory activity. Proc Natl Acad Sci USA 95 (1):322–327

de Miguel MJ, Burrell MA (2002) Immunocytochemical detection of orexin A in endocrine cells of the developing mouse gut. J Histochem Cytochem 50(1):63–69

de Oliveira CV, Ciriello J (2003) Cardiovascular responses to hypocretin-1 in nucleus ambiguus of the ovariectomized female rat. Brain Res 986(1–2):148–156

de Oliveira CV, Rosas-Arellano MP, Solano-Flores LP, Ciriello J (2003) Cardiovascular effects of hypocretin-1 in nucleus of the solitary tract. Am J Physiol Heart Circ Physiol 284(4): H1369–H1377

Deurveilher S, Semba K (2005) Indirect projections from the suprachiasmatic nucleus to major arousal-promoting cell groups in rat: implications for the circadian control of behavioural state. Neuroscience 130(1):165–183

Dugovic C, Shelton JE, Aluisio LE, Fraser IC, Jiang X, Sutton SW et al (2009) Blockade of orexin-1 receptors attenuates orexin-2 receptor antagonism-induced sleep promotion in the rat. J Pharmacol Exp Ther 330(1):142–151

Duxon MS, Stretton J, Starr K, Jones DN, Holland V, Riley G et al (2001) Evidence that orexin-A-evoked grooming in the rat is mediated by orexin- 1 (OX1) receptors, with downstream 5-HT2C receptor involvement. Psychopharmacology (Berl) 153(2):203–209

Edwards CM, Abusnana S, Sunter D, Murphy KG, Ghatei MA, Bloom SR (1999) The effect of the orexins on food intake: comparison with neuropeptide Y, melanin-concentrating hormone and galanin. J Endocrinol 160(3):R7–R12

Egecioglu E, Ploj K, Xu X, Bjursell M, Salome N, Andersson N et al (2009) Central NMU signaling in body weight and energy balance regulation: evidence from NMUR2 deletion and chronic central NMU treatment in mice. Am J Physiol Endocrinol Metab 297(3):E708–E716

Ehrstrom M, Gustafsson T, Finn A, Kirchgessner A, Gryback P, Jacobsson H et al (2005) Inhibitory effect of exogenous orexin A on gastric emptying, plasma leptin and the distribution of orexin and orexin receptors in the gut and pancreas in man. J Clin Endocrinol Metab 90 (4):2370–2377

Elias CF, Saper CB, Maratos-Flier E, Tritos NA, Lee C, Kelly J et al (1998) Chemically defined projections linking the mediobasal hypothalamus and the lateral hypothalamic area. J Comp Neurol 402(4):442–459

España RA, Baldo BA, Kelley AE, Berridge CW (2001) Wake-promoting and sleep-suppressing actions of hypocretin (orexin): basal forebrain sites of action. Neuroscience 106(4):699–715

España RA, Valentino RJ, Berridge CW (2003) Fos immunoreactivity in hypocretin-synthesizing and hypocretin-1 receptor-expressing neurons: effects of diurnal and nocturnal spontaneous waking, stress and hypocretin-1 administration. Neuroscience 121(1):201–217

Estabrooke IV, McCarthy MT, Ko E, Chou TC, Chemelli RM, Yanagisawa M et al (2001) Fos expression in orexin neurons varies with behavioral state. J Neurosci 21(5):1656–1662

Farrell WJ, Delville Y, Wilczynski W (2003) Immunocytochemical localization of orexin in the brain of the green ancle lizard (*Anolis carolinensis*). Soc Neurosci Abs 33:828

Ferguson AV, Samson WK (2003) The orexin/hypocretin system: a critical regulator of neuroendocrine and autonomic function. Front Neuroendocrinol 24(3):141–150

Fu LY, Acuna-Goycolea C, van den Pol AN (2004) Neuropeptide Y inhibits hypocretin/orexin neurons by multiple presynaptic and postsynaptic mechanisms: tonic depression of the hypothalamic arousal system. J Neurosci 24(40):8741–8751

Fukue Y, Sato T, Teranishi H, Hanada R, Takahashi T, Nakashima Y et al (2006) Regulation of gonadotropin secretion and puberty onset by neuromedin U. FEBS Lett 580(14):3485–3488

Funato H, Tsai AL, Willie JT, Kisanuki Y, Williams SC, Sakurai T et al (2009) Enhanced orexin receptor-2 signaling prevents diet-induced obesity and improves leptin sensitivity. Cell Metab 9(1):64–76, PMCID: 2630400

Gartlon J, Szekeres P, Pullen M, Sarau HM, Aiyar N, Shabon U et al (2004) Localisation of NMU1R and NMU2R in human and rat central nervous system and effects of neuromedin-U following central administration in rats. Psychopharmacology (Berl) 177(1–2):1–14

Gautvik KM, de Lecea L, Gautvik VT, Danielson PE, Tranque P, Dopazo A et al (1996) Overview of the most prevalent hypothalamus-specific mRNAs, as identified by directional tag PCR subtraction. Proc Natl Acad Sci USA 93(16):8733–8738

Graham ES, Turnbull Y, Fotheringham P, Nilaweera K, Mercer JG, Morgan PJ et al (2003) Neuromedin U and Neuromedin U receptor-2 expression in the mouse and rat hypothalamus: effects of nutritional status. J Neurochem 87(5):1165–1173

Guan XM, Yu H, Jiang Q, Van Der Ploeg LH, Liu Q (2001) Distribution of neuromedin U receptor subtype 2 mRNA in the rat brain. Brain Res Gene Expr Patterns 1(1):1–4

Hagan JJ, Leslie RA, Patel S, Evans ML, Wattam TA, Holmes S et al (1999) Orexin A activates locus coeruleus cell firing and increases arousal in the rat. Proc Natl Acad Sci USA 96 (19):10911–10916

Hainerova I, Torekov SS, Ek J, Finkova M, Borch-Johnsen K, Jorgensen T et al (2006) Association between neuromedin U gene variants and overweight and obesity. J Clin Endocrinol Metab 91 (12):5057–5063

Hanada R, Nakazato M, Murakami N, Sakihara S, Yoshimatsu H, Toshinai K et al (2001) A role for neuromedin U in stress response. Biochem Biophys Res Commun 289(1):225–228

Hanada T, Date Y, Shimbara T, Sakihara S, Murakami N, Hayashi Y et al (2003) Central actions of neuromedin U via corticotropin-releasing hormone. Biochem Biophys Res Commun 311 (4):954–958

Hanada R, Teranishi H, Pearson JT, Kurokawa M, Hosoda H, Fukushima N et al (2004) Neuromedin U has a novel anorexigenic effect independent of the leptin signaling pathway. Nat Med 10(10):1067–1073

Hara J, Beuckmann CT, Nambu T, Willie JT, Chemelli RM, Sinton CM et al (2001) Genetic ablation of orexin neurons in mice results in narcolepsy, hypophagia, and obesity. Neuron 30 (2):345–354

Harrington ME (1997) The ventral lateral geniculate nucleus and the intergeniculate leaflet: interrelated structures in the visual and circadian systems. Neurosci Biobehav Rev 21 (5):705–727

Harrington ME, Nance DM, Rusak B (1987) Double-labeling of neuropeptide Y-immunoreactive neurons which project from the geniculate to the suprachiasmatic nuclei. Brain Res 410 (2):275–282

Haynes AC, Jackson B, Overend P, Buckingham RE, Wilson S, Tadayyon M et al (1999) Effects of single and chronic intracerebroventricular administration of the orexins on feeding in the rat. Peptides 20(9):1099–1105

Haynes AC, Jackson B, Chapman H, Tadayyon M, Johns A, Porter RA et al (2000) A selective orexin-1 receptor antagonist reduces food consumption in male and female rats. Regul Pept 96(1–2):45–51

Hirota K, Kushikata T, Kudo M, Kudo T, Lambert DG, Matsuki A (2001) Orexin A and B evoke noradrenaline release from rat cerebrocortical slices. Br J Pharmacol 134(7):1461–1466

Hirota K, Kushikata T, Kudo M, Kudo T, Smart D, Matsuki A (2003) Effects of central hypocretin-1 administration on hemodynamic responses in young-adult and middle-aged rats. Brain Res 981(1–2):143–150

Hoch M, Hoever P, Haschke M, Krahenbuhl S, Dingemanse J (2011) Food effect and biocomparison of two formulations of the dual orexin receptor antagonist almorexant in healthy male subjects. J Clin Pharmacol 51(7):1116–1121

Hoever P, de Haas S, Winkler J, Schoemaker RC, Chiossi E, van Gerven J et al (2010) Orexin receptor antagonism, a new sleep-promoting paradigm: an ascending single-dose study with almorexant. Clin Pharmacol Ther 87(5):593–600

Honda Y, Doi Y, Ninomiya R, Ninomiya C (1986) Increased frequency of non-insulin-dependent diabetes mellitus among narcoleptic patients. Sleep 9(1):254–259
Honzawa M, Sudoh T, Minamino N, Tohyama M, Matsuo H (1987) Topographic localization of neuromedin U-like structures in the rat brain: an immunohistochemical study. Neuroscience 23 (3):1103–1122
Honzawa M, Sudoh T, Minamino N, Kangawa K, Matsuo H (1990) Neuromedin U-like immunoreactivity in rat intestine: regional distribution and immunohistochemical study. Neuropeptides 15(1):1–9
Horvath TL (2005) The hardship of obesity: a soft-wired hypothalamus. Nat Neurosci 8 (5):561–565
Horvath TL, Diano S, van den Pol AN (1999a) Synaptic interaction between hypocretin (orexin) and neuropeptide Y cells in the rodent and primate hypothalamus: a novel circuit implicated in metabolic and endocrine regulations. J Neurosci 19(3):1072–1087
Horvath TL, Peyron C, Diano S, Ivanov A, Aston-Jones G, Kilduff TS et al (1999b) Hypocretin (orexin) activation and synaptic innervation of the locus coeruleus noradrenergic system. J Comp Neurol 415(2):145–159
Howard AD, Wang R, Pong SS, Mellin TN, Strack A, Guan XM et al (2000) Identification of receptors for neuromedin U and its role in feeding. Nature 406(6791):70–74
Huesa G, van den Pol AN, Finger TE (2005) Differential distribution of hypocretin (orexin) and melanin-concentrating hormone in the goldfish brain. J Comp Neurol 488(4):476–491
Huhman KL, Albers HE (1994) Neuropeptide Y microinjected into the suprachiasmatic region phase shifts circadian rhythms in constant darkness. Peptides 15(8):1475–1478
Hungs M, Mignot E (2001) Hypocretin/orexin, sleep and narcolepsy. Bioessays 23(5):397–408
Hungs M, Fan J, Lin L, Lin X, Maki RA, Mignot E (2001) Identification and functional analysis of mutations in the hypocretin (orexin) genes of narcoleptic canines. Genome Res 11(4):531–539
Ida T, Nakahara K, Katayama T, Murakami N, Nakazato M (1999) Effect of lateral cerebroventricular injection of the appetite- stimulating neuropeptide, orexin and neuropeptide Y, on the various behavioral activities of rats. Brain Res 821(2):526–529
Ida T, Nakahara K, Kuroiwa T, Fukui K, Nakazato M, Murakami T et al (2000a) Both corticotropin releasing factor and neuropeptide Y are involved in the effect of orexin (hypocretin) on the food intake in rats. Neurosci Lett 293(2):119–122
Ida T, Nakahara K, Murakami T, Hanada R, Nakazato M, Murakami N (2000b) Possible involvement of orexin in the stress reaction in rats. Biochem Biophys Res Commun 270(1):318–323
Iqbal J, Pompolo S, Sakurai T, Clarke IJ (2001) Evidence that orexin-containing neurones provide direct input to gonadotropin-releasing hormone neurones in the ovine hypothalamus. J Neuroendocrinol 13(12):1033–1041
Ivanov TR, Lawrence CB, Stanley PJ, Luckman SM (2002) Evaluation of neuromedin U actions in energy homeostasis and pituitary function. Endocrinology 143(10):3813–3821
Ivanov TR, Le Rouzic P, Stanley PJ, Ling WY, Parello R, Luckman SM (2004) Neuromedin U neurones in the rat nucleus of the tractus solitarius are catecholaminergic and respond to peripheral cholecystokinin. J Neuroendocrinol 16(7):612–619
Iwai T, Iinuma Y, Kodani R, Oka J (2008) Neuromedin U inhibits inflammation-mediated memory impairment and neuronal cell-death in rodents. Neurosci Res 61(1):113–119
Janik D, Mikkelsen JD, Mrosovsky N (1995) Cellular colocalization of Fos and neuropeptide Y in the intergeniculate leaflet after nonphotic phase-shifting events. Brain Res 698(1–2):137–145
Jaszberenyi M, Bagosi Z, Thurzo B, Foldesi I, Telegdy G (2007) Endocrine and behavioral effects of neuromedin S. Horm Behav 52(5):631–639
Jászberényi M, Bujdosó E, Pataki I, Telegdy G (2000) Effects of orexins on the hypothalamic-pituitary-adrenal system. J Neuroendocrinol 12(12):1174–1178
Jethwa PH, Small CJ, Smith KL, Seth A, Darch SJ, Abbott CR et al (2005) Neuromedin U has a physiological role in the regulation of food intake and partially mediates the effects of leptin. Am J Physiol Endocrinol Metab 289(2):E301–E305

Jethwa PH, Smith KL, Small CJ, Abbott CR, Darch SJ, Murphy KG et al (2006) Neuromedin U partially mediates leptin-induced hypothalamo-pituitary adrenal (HPA) stimulation and has a physiological role in the regulation of the HPA axis in the rat. Endocrinology 147 (6):2886–2892

John J, Wu MF, Siegel JM (2000) Systemic administration of hypocretin-1 reduces cataplexy and normalizes sleep and waking durations in narcoleptic dogs. Sleep Res Online 3(1):23–28

Johnson RF, Moore RY, Morin LP (1989) Lateral geniculate lesions alter circadian activity rhythms in the hamster. Brain Res Bull 22(2):411–422

Johren O, Neidert SJ, Kummer M, Dendorfer A, Dominiak P (2001) Prepro-orexin and orexin receptor mRNAs are differentially expressed in peripheral tissues of male and female rats. Endocrinology 142(8):3324–3331

Jones DN, Gartlon J, Parker F, Taylor SG, Routledge C, Hemmati P et al (2001) Effects of centrally administered orexin-B and orexin-A: a role for orexin-1 receptors in orexin-B-induced hyperactivity. Psychopharmacology (Berl) 153(2):210–218

Kamisoyama H, Honda K, Saneyasu T, Sugahara K, Hasegawa S (2007) Central administration of neuromedin U suppresses food intake in chicks. Neurosci Lett 420(1):1–5

Kaslin J, Nystedt JM, Ostergard M, Peitsaro N, Panula P (2004) The orexin/hypocretin system in zebrafish is connected to the aminergic and cholinergic systems. J Neurosci 24(11):2678–2689

Kirchgessner AL, Liu M (1999) Orexin synthesis and response in the gut. Neuron 24(4):941–951

Kiwaki K, Kotz CM, Wang C, Lanningham-Foster L, Levine JA (2004) Orexin A (hypocretin 1) injected into hypothalamic paraventricular nucleus and spontaneous physical activity in rats. Am J Physiol Endocrinol Metab 286(4):E551–E559

Kiyashchenko LI, Mileykovskiy BY, Lai YY, Siegel JM (2001) Increased and decreased muscle tone with orexin (hypocretin) microinjections in the locus coeruleus and pontine inhibitory area. J Neurophysiol 85(5):2008–2016

Kotz CM (2006) Integration of feeding and spontaneous physical activity: role for orexin. Physiol Behav 88(3):294–301

Kotz CM, Glass MJ, Levine AS, Billington CJ (2000) Regional effect of naltrexone in the nucleus of the solitary tract in blockade of NPY-induced feeding. Am J Physiol Regul Integr Comp Physiol 278(2):R499–R503

Kotz CM, Teske JA, Levine JA, Wang C (2002) Feeding and activity induced by orexin A in the lateral hypothalamus in rats. Regul Pept 104(1–3):27–32

Kotz CM, Wang C, Teske JA, Thorpe AJ, Novak CM, Kiwaki K et al (2006) Orexin A mediation of time spent moving in rats: neural mechanisms. Neuroscience 142(1):29–36

Kotz CM, Teske JA, Billington CJ (2008) Neuroregulation of nonexercise activity thermogenesis and obesity resistance. Am J Physiol Regul Integr Comp Physiol 294(3):R699–R710

Kowalski TJ, Spar BD, Markowitz L, Maguire M, Golovko A, Yang S et al (2005) Transgenic overexpression of neuromedin U promotes leanness and hypophagia in mice. J Endocrinol 185 (1):151–164

Kummer M, Neidert SJ, Johren O, Dominiak P (2001) Orexin (hypocretin) gene expression in rat ependymal cells. Neuroreport 12(10):2117–2120

Kunii K, Yamanaka A, Nambu T, Matsuzaki I, Goto K, Sakurai T (1999) Orexins/hypocretins regulate drinking behaviour. Brain Res 842(1):256–261

Langmead CJ, Jerman JC, Brough SJ, Scott C, Porter RA, Herdon HJ (2004) Characterisation of the binding of [3 H]-SB-674042, a novel nonpeptide antagonist, to the human orexin-1 receptor. Br J Pharmacol 141(2):340–346, PMCID: 1574197

Li A, Nattie E (2010) Antagonism of rat orexin receptors by almorexant attenuates central chemoreception in wakefulness in the active period of the diurnal cycle. J Physiol 588(Pt 15):2935–2944, PMCID: 2956908

Li Y, Li S, Sui N, Kirouac GJ (2009) Orexin-A acts on the paraventricular nucleus of the midline thalamus to inhibit locomotor activity in rats. Pharmacol Biochemand Behav 93(4):506–514

Lin L, Faraco J, Li R, Kadotani H, Rogers W, Lin X et al (1999) The sleep disorder canine narcolepsy is caused by a mutation in the hypocretin (orexin) receptor 2 gene. Cell 98 (3):365–376

Liu JJ, Payza K, Huang J, Liu R, Chen T, Coupal M et al (2009) Discovery and pharmacological characterization of a small-molecule antagonist at neuromedin U receptor NMUR2. J Pharmacol Exp Ther 330(1):268–275

Lu XY, Bagnol D, Burke S, Akil H, Watson SJ (2000) Differential distribution and regulation of OX1 and OX2 orexin/hypocretin receptor messenger RNA in the brain upon fasting. Horm Behav 37(4):335–344

Lubkin M, Stricker-Krongrad A (1998) Independent feeding and metabolic actions of orexins in mice. Biochem Biophys Res Commun 253(2):241–245

Malherbe P, Borroni E, Gobbi L, Knust H, Nettekoven M, Pinard E et al (2009) Biochemical and behavioural characterization of EMPA, a novel high-affinity, selective antagonist for the OX (2) receptor. Br J Pharmacol 156(8):1326–1341

Mangold C, Ksiazek I, Yun SW, Berger E, Binkert C (2008) Distribution of neuromedin U binding sites in the rat CNS revealed by in vitro receptor autoradiography. Neuropeptides 42 (4):377–386

Martinez GS, Smale L, Nuñez AA (2002) Diurnal and nocturnal rodents show rhythms in orexinergic neurons. Brain Res 955(1–2):1–7

Maruyama K, Konno N, Ishiguro K, Wakasugi T, Uchiyama M, Shioda S et al (2008) Isolation and characterisation of four cDNAs encoding neuromedin U (NMU) from the brain and gut of goldfish, and the inhibitory effect of a deduced NMU on food intake and locomotor activity. J Neuroendocrinol 20(1):71–78

Maruyama K, Kaiya H, Miyazato M, Konno N, Wakasugi T, Uchiyama M et al (2011) Isolation and characterisation of two cDNAs encoding the neuromedin U receptor from goldfish brain. J Neuroendocrinol 23(3):282–291

Matsumura K, Tsuchihashi T, Abe I (2001) Central orexin-A augments sympathoadrenal outflow in conscious rabbits. Hypertension 37(6):1382–1387

Matsuzaki I, Sakurai T, Kunii K, Nakamura T, Yanagisawa M, Goto K (2002) Involvement of the serotonergic system in orexin-induced behavioral alterations in rats. Regul Pept 104 (1–3):119–123

McGranaghan PA, Piggins HD (2001) Orexin A-like immunoreactivity in the hypothalamus and thalamus of the Syrian hamster (*Mesocricetus auratus*) and Siberian hamster (*Phodopus sungorus*), with special reference to circadian structures. Brain Res 904(2):234–244

Meister B (2000) Control of food intake via leptin receptors in the hypothalamus. Vitam Horm 59:265–304

Methippara MM, Alam MN, Szymusiak R, McGinty D (2000) Effects of lateral preoptic area application of orexin-A on sleep- wakefulness. Neuroreport 11(16):3423–3426

Mileykovskiy BY, Kiyashchenko LI, Siegel JM (2002) Muscle tone facilitation and inhibition after orexin-a (hypocretin-1) microinjections into the medial medulla. J Neurophysiol 87(5):2480–2489

Minamino N, Kangawa K, Matsuo H (1985a) Neuromedin U-8 and U-25: novel uterus stimulating and hypertensive peptides identified in porcine spinal cord. Biochem Biophys Res Commun 130(3):1078–1085

Minamino N, Sudoh T, Kangawa K, Matsuo H (1985b) Neuromedins: novel smooth-muscle stimulating peptides identified in porcine spinal cord. Peptides 6(Suppl 3):245–248

Mintz EM, van den Pol AN, Casano AA, Albers HE (2001) Distribution of hypocretin-(orexin) immunoreactivity in the central nervous system of Syrian hamsters (Mesocricetus auratus). J Chem Neuroanat 21(3):225–238

Mitchell JD, Maguire JJ, Davenport AP (2009) Emerging pharmacology and physiology of neuromedin U and the structurally related peptide neuromedin S. Br J Pharmacol 158 (1):87–103, PMCID: 2795236

Miyazato M, Mori K, Ida T, Kojima M, Murakami N, Kangawa K (2008) Identification and functional analysis of a novel ligand for G protein-coupled receptor, Neuromedin S. Regul Pept 145(1–3):37–41

Monda M, Viggiano A, Mondola P, De Luca V (2001) Inhibition of prostaglandin synthesis reduces hyperthermic reactions induced by hypocretin-1/orexin A. Brain Res 909(1–2):68–74

Monda M, Viggiano A, De Luca V (2003) Paradoxical [correction of parodoxical] effect of orexin A: hypophagia induced by hyperthermia. Brain Res 961(2):220–228

Monda M, Viggiano AN, Viggiano AL, Fuccio F, De Luca V (2004a) Cortical spreading depression blocks the hyperthermic reaction induced by orexin A. Neuroscience 123 (2):567–574

Monda M, Viggiano A, Viggiano A, Fuccio F, De Luca V (2004b) Injection of orexin A into the diagonal band of Broca induces sympathetic and hyperthermic reactions. Brain Res 1018 (2):265–271

Mondal MS, Nakazato M, Date Y, Murakami N, Yanagisawa M, Matsukura S (1999) Widespread distribution of orexin in rat brain and its regulation upon fasting. Biochem Biophys Res Commun 256(3):495–499

Moore RY, Card JP (1994) Intergeniculate leaflet: an anatomically and functionally distinct subdivision of the lateral geniculate complex. J Comp Neurol 344(3):403–430

Moore RY, Abrahamson EA, Van Den Pol A (2001) The hypocretin neuron system: an arousal system in the human brain. Arch Ital Biol 139(3):195–205

Mori K, Miyazato M, Kangawa K (2008) Neuromedin S: discovery and functions. Results Probl Cell Differ 46:201–212

Moriguchi T, Sakurai T, Nambu T, Yanagisawa M, Goto K (1999) Neurons containing orexin in the lateral hypothalamic area of the adult rat brain are activated by insulin-induced acute hypoglycemia. Neurosci Lett 264(1–3):101–104

Nakamura T, Uramura K, Nambu T, Yada T, Goto K, Yanagisawa M et al (2000) Orexin-induced hyperlocomotion and stereotypy are mediated by the dopaminergic system. Brain Res 873 (1):181–187

Nakazato M, Hanada R, Murakami N, Date Y, Mondal MS, Kojima M et al (2000) Central effects of neuromedin U in the regulation of energy homeostasis. Biochem Biophys Res Commun 277 (1):191–194

Nambu T, Sakurai T, Mizukami K, Hosoya Y, Yanagisawa M, Goto K (1999) Distribution of orexin neurons in the adult rat brain. Brain Res 827(1–2):243–260

Nevsimalova S, Vankova J, Stepanova I, Seemanova E, Mignot E, Nishino S (2005) Hypocretin deficiency in Prader-Willi syndrome. Eur J Neurol 12(1):70–72

Niimi M, Sato M, Taminato T (2001a) Neuropeptide Y in central control of feeding and interactions with orexin and leptin. Endocrine 14(2):269–273

Niimi M, Murao K, Taminato T (2001b) Central administration of neuromedin U activates neurons in ventrobasal hypothalamus and brainstem. Endocrine 16(3):201–206

Nishino S, Ripley B, Overeem S, Lammers GJ, Mignot E (2000) Hypocretin (orexin) deficiency in human narcolepsy. Lancet 355(9197):39–40

Nishino S, Ripley B, Overeem S, Nevsimalova S, Lammers GJ, Vankova J et al (2001) Low cerebrospinal fluid hypocretin (Orexin) and altered energy homeostasis in human narcolepsy. Ann Neurol 50(3):381–388

Nixon JP, Smale L (2004) Individual differences in wheel-running rhythms are related to temporal and spatial patterns of activation of orexin A and B cells in a diurnal rodent (*Arvicanthis niloticus*). Neuroscience 127(1):25–34

Nixon JP, Smale L (2005) Orexin fibers form appositions with Fos expressing neuropeptide-Y cells in the grass rat intergeniculate leaflet. Brain Res 1053(1–2):33–37

Nixon JP, Smale L (2007) A comparative analysis of the distribution of immunoreactive orexin A and B in the brains of nocturnal and diurnal rodents. Behav Brain Funct 3:28, PMCID: 1913054

Nogueiras R, Tovar S, Mitchell SE, Barrett P, Rayner DV, Dieguez C et al (2006) Negative energy balance and leptin regulate neuromedin-U expression in the rat pars tuberalis. J Endocrinol 190 (2):545–553

Novak CM, Levine JA (2007) Central neural and endocrine mechanisms of non-exercise activity thermogenesis and their potential impact on obesity. J Neuroendocrinol 19(12):923–940

Novak CM, Levine JA (2009) Daily intraparaventricular orexin-A treatment induces weight loss in rats. Obesity (Silver Spring) 7(8):1493–1498

Novak CM, Kotz CM, Levine JA (2006a) Central orexin sensitivity, physical activity, and obesity in diet-induced obese and diet-resistant rats. Am J Physiol Endocrinol Metab 290(2): E396–E403

Novak CM, Zhang M, Levine JA (2006b) Neuromedin U in the paraventricular and arcuate hypothalamic nuclei increases non-exercise activity thermogenesis. J Neuroendocrinol 18 (8):594–601

Novak CM, Zhang M, Levine JA (2007) Sensitivity of the hypothalamic paraventricular nucleus to the locomotor-activating effects of neuromedin U in obesity. Brain Res 1169:57–68

Novak CM, Escande C, Burghardt PR, Zhang M, Barbosa MT, Chini EN et al (2010) Spontaneous activity, economy of activity, and resistance to diet-induced obesity in rats bred for high intrinsic aerobic capacity. Horm Behav 58(3):355–367, PMCID: 2923555

Nunez A, Rodrigo-Angulo ML, Andres ID, Garzon M (2009) Hypocretin/Orexin neuropeptides: participation in the control of sleep-wakefulness cycle and energy homeostasis. Curr Neuropharmacol 7(1):50–59, PMCID: 2724663

Ohkubo T, Boswell T, Lumineau S (2002) Molecular cloning of chicken prepro-orexin cDNA and preferential expression in the chicken hypothalamus. Biochim Biophys Acta 1577(3):476–480

Overeem S, Mignot E, Gert van Dijk J, Lammers GJ (2001) Narcolepsy: clinical features, new pathophysiologic insights, and future perspectives. J Clin Neurophysiol 18(2):78–105

Peever JH, Lai YY, Siegel JM (2003) Excitatory effects of hypocretin-1 (orexin-A) in the trigeminal motor nucleus are reversed by NMDA antagonism. J Neurophysiol 89(5):2591–2600

Peier A, Kosinski J, Cox-York K, Qian Y, Desai K, Feng Y et al (2009) The antiobesity effects of centrally administered neuromedin U and neuromedin S are mediated predominantly by the neuromedin U receptor 2 (NMUR2). Endocrinology 150(7):3101–3109, PMCID: 2703546

Perrey DA, Gilmour BP, Runyon SP, Thomas BF, Zhang Y (2011) Diaryl urea analogues of SB-334867 as orexin-1 receptor antagonists. Bioorg Med Chem Lett 21(10):2980–2985, PMCID: 3085582

Petersén Å, Gil J, Maat-Schieman ML, Björkqvist M, Tanila H, Araújo IM et al (2005) Orexin loss in Huntington's disease. Hum Mol Genet 14(1):39–47

Peyron C, Tighe DK, van den Pol AN, de Lecea L, Heller HC, Sutcliffe JG et al (1998) Neurons containing hypocretin (orexin) project to multiple neuronal systems. J Neurosci 18 (23):9996–10015

Piper DC, Upton N, Smith MI, Hunter AJ (2000) The novel brain neuropeptide, orexin-A, modulates the sleep-wake cycle of rats. Eur J Neurosci 12(2):726–730

Qiu DL, Chu CP, Shirasaka T, Nabekura T, Kunitake T, Kato K et al (2003) Neuromedin U depolarizes rat hypothalamic paraventricular nucleus neurons in vitro by enhancing IH channel activity. J Neurophysiol 90(2):843–850

Qiu DL, Chu CP, Tsukino H, Shirasaka T, Nakao H, Kato K et al (2005) Neuromedin U receptor-2 mRNA and HCN channels mRNA expression in NMU-sensitive neurons in rat hypothalamic paraventricular nucleus. Neurosci Lett 374(1):69–72

Qu D, Ludwig DS, Gammeltoft S, Piper M, Pelleymounter MA, Cullen MJ et al (1996) A role for melanin-concentrating hormone in the central regulation of feeding behaviour. Nature 380 (6571):243–247

Raddatz R, Wilson AE, Artymyshyn R, Bonini JA, Borowsky B, Boteju LW et al (2000) Identification and characterization of two neuromedin U receptors differentially expressed in peripheral tissues and the central nervous system. J Biol Chem 275(42):32452–32459

Rauch M, Riediger T, Schmid HA, Simon E (2000) Orexin A activates leptin-responsive neurons in the arcuate nucleus. Pflugers Arch 440(5):699–703

Rodgers RJ, Halford JC, Nunes de Souza RL, Canto de Souza AL, Piper DC, Arch JR et al (2000) Dose-response effects of orexin-A on food intake and the behavioural satiety sequence in rats. Regul Pept 96(1–2):71–84

Rodgers RJ, Halford JC, Nunes de Souza RL, Canto de Souza AL, Piper DC, Arch JR et al (2001) SB-334867, a selective orexin-1 receptor antagonist, enhances behavioural satiety and blocks the hyperphagic effect of orexin-A in rats. Eur J Neurosci 13(7):1444–1452

Rucinski M, Ziolkowska A, Neri G, Trejter M, Zemleduch T, Tyczewska M et al (2007) Expression of neuromedins S and U and their receptors in the hypothalamus and endocrine glands of the rat. Int J Mol Med 20(2):255–259

Rusak B, Meijer JH, Harrington ME (1989) Hamster circadian rhythms are phase-shifted by electrical stimulation of the geniculo-hypothalamic tract. Brain Res 493(2):283–291

Russell SH, Small CJ, Kennedy AR, Stanley SA, Seth A, Murphy KG et al (2001) Orexin A interactions in the hypothalamo-pituitary gonadal axis. Endocrinology 142(12):5294–5302

Sakamoto T, Mori K, Nakahara K, Miyazato M, Kangawa K, Sameshima H et al (2007) Neuromedin S exerts an antidiuretic action in rats. Biochem Biophys Res Commun 361(2):457–461

Sakamoto T, Mori K, Miyazato M, Kangawa K, Sameshima H, Nakahara K et al (2008) Involvement of neuromedin S in the oxytocin release response to suckling stimulus. Biochem Biophys Res Commun 375(1):49–53

Sakurai T (2003) Orexin: a link between energy homeostasis and adaptive behaviour. Curr Opin Clin Nutr Metab Care 6(4):353–360

Sakurai T, Amemiya A, Ishii M, Matsuzaki I, Chemelli RM, Tanaka H et al (1998) Orexins and orexin receptors: a family of hypothalamic neuropeptides and G protein-coupled receptors that regulate feeding behavior. Cell 92(4):573–585

Sakurai T, Moriguchi T, Furuya K, Kajiwara N, Nakamura T, Yanagisawa M et al (1999) Structure and function of human prepro-orexin gene. J Biol Rhythms 274(25):17771–17776

Samson WK, Gosnell B, Chang JK, Resch ZT, Murphy TC (1999) Cardiovascular regulatory actions of the hypocretins in brain. Brain Res 831(1–2):248–253

Samson WK, Taylor MM, Ferguson AV (2005) Non-sleep effects of hypocretin/orexin. Sleep Med Rev 9(4):243–252

Samson WK, Bagley SL, Ferguson AV, White MM (2010) Orexin receptor subtype activation and locomotor behaviour in the rat. Acta Physiol (Oxf) 198(3):313–324

Sato S, Hanada R, Kimura A, Abe T, Matsumoto T, Iwasaki M et al (2007) Central control of bone remodeling by neuromedin U. Nat Med 13(10):1234–1240

Sato-Suzuki I, Kita I, Seki Y, Oguri M, Arita H (2002) Cortical arousal induced by microinjection of orexins into the paraventricular nucleus of the rat. Behav Brain Res 128(2):169–177

Scammell TE, Winrow CJ (2011) Orexin receptors: pharmacology and therapeutic opportunities. Annu Rev Pharmacol Toxicol 51:243–266

Schwartz MW, Woods SC, Porte D Jr, Seeley RJ, Baskin DG (2000) Central nervous system control of food intake. Nature 404(6778):661–671

Semjonous NM, Smith KL, Parkinson JR, Gunner DJ, Liu YL, Murphy KG et al (2009) Coordinated changes in energy intake and expenditure following hypothalamic administration of neuropeptides involved in energy balance. Int J Obes (Lond) 33(7):775–785, PMCID: 2711051

Shan L, Qiao X, Crona JH, Behan J, Wang S, Laz T et al (2000) Identification of a novel neuromedin U receptor subtype expressed in the central nervous system. J Biol Chem 275 (50):39482–39486

Shibahara M, Sakurai T, Nambu T, Takenouchi T, Iwaasa H, Egashira SI et al (1999) Structure, tissue distribution, and pharmacological characterization of Xenopus orexins. Peptides 20 (10):1169–1176

Shirasaka T, Nakazato M, Matsukura S, Takasaki M, Kannan H (1999) Sympathetic and cardiovascular actions of orexins in conscious rats. Am J Physiol 277(6 Pt 2):R1780–R1785

Shirasaka T, Kunitake T, Takasaki M, Kannan H (2002) Neuronal effects of orexins: relevant to sympathetic and cardiovascular functions. Regul Pept 104(1–3):91–95

Siegel JM (1999) Narcolepsy: a key role for hypocretins (orexins). Cell 98(4):409–412

Singletary KG, Delville Y, Farrell WJ, Wilczynski W (2005) Distribution of orexin/hypocretin immunoreactivity in the nervous system of the green Treefrog, Hyla cinerea. Brain Res 1041 (2):231–236

Smale L, McElhinny T, Nixon J, Gubik B, Rose S (2001) Patterns of wheel running are related to Fos expression in neuropeptide- Y-containing neurons in the intergeniculate leaflet of Arvicanthis niloticus. J Biol Rhythms 16(2):163–172

Smart D, Jerman JC, Brough SJ, Rushton SL, Murdock PR, Jewitt F et al (1999) Characterization of recombinant human orexin receptor pharmacology in a Chinese hamster ovary cell-line using FLIPR. Br J Pharmacol 128(1):1–3
Smith PM, Connolly BC, Ferguson AV (2002) Microinjection of orexin into the rat nucleus tractus solitarius causes increases in blood pressure. Brain Res 950(1–2):261–267
Smith MI, Piper DC, Duxon MS, Upton N (2003) Evidence implicating a role for orexin-1 receptor modulation of paradoxical sleep in the rat. Neurosci Lett 341(3):256–258
Smith PM, Samson WK, Ferguson AV (2007) Cardiovascular actions of orexin-A in the rat subfornical organ. J Neuroendocrinol 19(1):7–13
Stanley BG, Thomas WJ (1993) Feeding responses to perifornical hypothalamic injection of neuropeptide Y in relation to circadian rhythms of eating behavior. Peptides 14(3):475–481
Stanley BG, Magdalin W, Seirafi A, Thomas WJ, Leibowitz SF (1993) The perifornical area: the major focus of (a) patchily distributed hypothalamic neuropeptide Y-sensitive feeding system (s). Brain Res 604(1–2):304–317
Sunter D, Morgan I, Edwards CM, Dakin CL, Murphy KG, Gardiner J et al (2001) Orexins: effects on behavior and localisation of orexin receptor 2 messenger ribonucleic acid in the rat brainstem. Brain Res 907(1–2):27–34
Sutcliffe JG, de Lecea L (2000) The hypocretins: excitatory neuromodulatory peptides for multiple homeostatic systems, including sleep and feeding. J Neurosci Res 62(2):161–168
Sutton RE, Koob GF, Le Moal M, Rivier J, Vale W (1982) Corticotropin releasing factor produces behavioural activation in rats. Nature 297(5864):331–333
Sweet DC, Levine AS, Billington CJ, Kotz CM (1999) Feeding response to central orexins. Brain Res 821(2):535–538
Szekely M (2006) Orexins, energy balance, temperature, sleep-wake cycle. Am J Physiol Regul Integr Comp Physiol 291(3):R530–R532
Szekely M, Petervari E, Balasko M, Hernadi I, Uzsoki B (2002) Effects of orexins on energy balance and thermoregulation. Regul Pept 104(1–3):47–53
Szekely M, Petervari E, Balasko M (2010) Thermoregulation, energy balance, regulatory peptides: recent developments. Front Biosci (Schol Ed) 2:1009–1046
Tachibana T, Matsuda K, Khan MS, Ueda H, Cline MA (2010a) Feeding and drinking response following central administration of neuromedin S in chicks. Comp Biochem Physiol A Mol Integr Physiol 157(1):63–67
Tachibana T, Matsuda K, Khan SI, Ueda H, Cline MA (2010b) Feeding and drinking response following central administrations of bombesin-like peptides in chicks. Comp Biochem Physiol A Mol Integr Physiol 156(4):394–399
Takenoya F, Hirayama M, Kageyama H, Funahashi H, Kita T, Matsumoto H et al (2005) Neuronal interactions between galanin-like-peptide- and orexin- or melanin-concentrating hormone-containing neurons. Regul Pept 126(1–2):79–83
Tanida M, Satomi J, Shen J, Nagai K (2009) Autonomic and cardiovascular effects of central neuromedin U in rats. Physiol Behav 96(2):282–288
Teske JA, Levine AS, Kuskowski M, Levine JA, Kotz CM (2006) Elevated hypothalamic orexin signaling, sensitivity to orexin A, and spontaneous physical activity in obesity-resistant rats. Am J Physiol Regul Integr Comp Physiol 291(4):R889–R899
Teske JA, Billington CJ, Kotz CM (2010) Hypocretin/Orexin and Energy Expenditure. Acta Physiol (Oxf) 198:303–312
Thannickal TC, Moore RY, Nienhuis R, Ramanathan L, Gulyani S, Aldrich M et al (2000) Reduced number of hypocretin neurons in human narcolepsy. Neuron 27(3):469–474
Thompson EL, Murphy KG, Todd JF, Martin NM, Small CJ, Ghatei MA et al (2004) Chronic administration of NMU into the paraventricular nucleus stimulates the HPA axis but does not influence food intake or body weight. Biochem Biophys Res Commun 323(1):65–71
Thorpe AJ, Kotz CM (2005) Orexin A in the nucleus accumbens stimulates feeding and locomotor activity. Brain Res 1050(1–2):156–162

Thorpe AJ, Mullett MA, Wang C, Kotz CM (2003) Peptides that regulate food intake: regional, metabolic, and circadian specificity of lateral hypothalamic orexin A feeding stimulation. Am J Physiol Regul Integr Comp Physiol 284(6):R1409–R1417

Tsujino N, Sakurai T (2009) Orexin/Hypocretin: a neuropeptide at the interface of sleep, energy homeostasis, and reward system. Pharmacol Rev 61(2):162–176

van den Pol AN (1999) Hypothalamic hypocretin (orexin): robust innervation of the spinal cord. J Neurosci 19(8):3171–3182

van den Pol AN, Gao XB, Obrietan K, Kilduff TS, Belousov AB (1998) Presynaptic and postsynaptic actions and modulation of neuroendocrine neurons by a new hypothalamic peptide, hypocretin/orexin. J Neurosci 18(19):7962–7971

van den Top M, Nolan MF, Lee K, Richardson PJ, Buijs RM, Davies CH et al (2003) Orexins induce increased excitability and synchronisation of rat sympathetic preganglionic neurones. J Physiol 549(Pt 3):809–821

Verty AN, Allen AM, Oldfield BJ (2010) The endogenous actions of hypothalamic peptides on brown adipose tissue thermogenesis in the rat. Endocrinology 151(9):4236–4246

Volgin DV, Saghir M, Kubin L (2002) Developmental changes in the orexin 2 receptor mRNA in hypoglossal motoneurons. Neuroreport 13(4):433–436

Volkoff H, Peter RE (2000) Effects of CART peptides on food consumption, feeding and associated behaviors in the goldfish, *Carassius auratus*: actions on neuropeptide Y- and orexin A-induced feeding. Brain Res 887(1):125–133

Wang C, Kotz C (2002) Urocortin in the lateral septal area modulates feeding induced by orexin A in the lateral hypothalamus. Am J Physiol Regul Integr Comp Physiol 283(2):R358–R367

Wang J, Osaka T, Inoue S (2001) Energy expenditure by intracerebroventricular administration of orexin to anesthetized rats. Neurosci Lett 315(1–2):49–52

Wang JB, Murata T, Narita K, Honda K, Higuchi T (2003a) Variation in the expression of orexin and orexin receptors in the rat hypothalamus during the estrous cycle, pregnancy, parturition, and lactation. Endocrine 22(2):127–134

Wang J, Osaka T, Inoue S (2003b) Orexin-A-sensitive site for energy expenditure localized in the arcuate nucleus of the hypothalamus. Brain Res 971(1):128–134

Weaver DR (1998) The suprachiasmatic nucleus: a 25-year retrospective. J Biol Rhythms 13 (2):100–112

Webb P, Annis JF, Troutman SJ Jr (1980) Energy balance in man measured by direct and indirect calorimetry. Am J Clin Nutr 33(6):1287–1298

Webb IC, Patton DF, Hamson DK, Mistlberger RE (2008) Neural correlates of arousal-induced circadian clock resetting: hypocretin/orexin and the intergeniculate leaflet. Eur J Neurosci 27 (4):828–835

Whitman DB, Cox CD, Breslin MJ, Brashear KM, Schreier JD, Bogusky MJ et al (2009) Discovery of a potent, CNS-penetrant orexin receptor antagonist based on an n, n-disubstituted-1,4-diazepane scaffold that promotes sleep in rats. ChemMedChem 4(7):1069–1074

Wickland C, Turek FW (1994) Lesions of the thalamic intergeniculate leaflet block activity-induced phase shifts in the circadian activity rhythm of the golden hamster. Brain Res 660 (2):293–300

Winrow CJ, Tanis KQ, Reiss DR, Rigby AM, Uslaner JM, Uebele VN et al (2010) Orexin receptor antagonism prevents transcriptional and behavioral plasticity resulting from stimulant exposure. Neuropharmacology 58(1):185–194

Winrow CJ, Gotter AL, Cox CD, Doran SM, Tannenbaum PL, Breslin MJ et al (2011) Promotion of sleep by suvorexant-a novel dual orexin receptor antagonist. J Neurogenet 25(1–2):52–61

Wren AM, Small CJ, Abbott CR, Jethwa PH, Kennedy AR, Murphy KG et al (2002) Hypothalamic actions of neuromedin U. Endocrinology 143(11):4227–4234

Wu MF, John J, Maidment N, Lam HA, Siegel JM (2002) Hypocretin release in normal and narcoleptic dogs after food and sleep deprivation, eating, and movement. Am J Physiol Regul Integr Comp Physiol 283(5):R1079–R1086

Yamamoto T, Suzuki H, Uemura H, Yamamoto K, Kikuyama S (2004) Localization of orexin-A-like immunoreactivity in prolactin cells in the bullfrog (Rana catesbeiana) pituitary. Gen Comp Endocrinol 135(2):186–192

Yamanaka A, Kunii K, Nambu T, Tsujino N, Sakai A, Matsuzaki I et al (2000) Orexin-induced food intake involves neuropeptide Y pathway. Brain Res 859(2):404–409

Yang G, Su J, Yao Y, Lei Z, Zhang G, Li X (2010) The regulatory mechanism of neuromedin S on luteinizing hormone in pigs. Anim Reprod Sci 122(3–4):367–374

Yayou K, Kitagawa S, Ito S, Kasuya E, Sutoh M (2009) Effects of intracerebroventricular administration of neuromedin U or neuromedin S in steers. Gen Comp Endocrinol 163 (3):324–328

Yokota M, Ozaki Y, Sakamoto F, Yamada S, Saito J, Fujihara H et al (2004) Fos expression in CRF-containing neurons in the rat paraventricular nucleus after central administration of neuromedin U. Stress 7(2):109–112

Yoshimichi G, Yoshimatsu H, Masaki T, Sakata T (2001) Orexin-A regulates body temperature in coordination with arousal status. Exp Biol Med (Maywood) 226(5):468–476

Zeng H, Gragerov A, Hohmann JG, Pavlova MN, Schimpf BA, Xu H et al (2006) Neuromedin U receptor 2-deficient mice display differential responses in sensory perception, stress, and feeding. Mol Cell Biol 26(24):9352–9363

Zhang J, Luo P (2002) Orexin B immunoreactive fibers and terminals innervate the sensory and motor neurons of jaw-elevator muscles in the rat. Synapse 44(2):106–110

Zhang W, Fukuda Y, Kuwaki T (2005a) Respiratory and cardiovascular actions of orexin-A in mice. Neurosci Lett 385(2):131–136

Zhang S, Blache D, Vercoe PE, Adam CL, Blackberry MA, Findlay PA et al (2005b) Expression of orexin receptors in the brain and peripheral tissues of the male sheep. Regul Pept 124 (1–3):81–87

Zhang Y, Jiang D, Zhang J, Wang F, Jiang X, Tao J (2010) Activation of neuromedin U type 1 receptor inhibits L-type Ca2+ channel currents via phosphatidylinositol 3-kinase-dependent protein kinase C epsilon pathway in mouse hippocampal neurons. Cell Signal 22 (11):1660–1668

Zheng H, Patterson LM, Berthoud HR (2005) Orexin-A projections to the caudal medulla and orexin-induced c-Fos expression, food intake, and autonomic function. J Comp Neurol 485 (2):127–142

The Central Insulin System and Energy Balance

Denovan P. Begg and Stephen C. Woods

Contents

Abstract Insulin acts throughout the body to reduce circulating energy and to increase energy storage. Within the brain, insulin produces a net catabolic effect by reducing food intake and increasing energy expenditure; this is evidenced by the hypophagia and increased brown adipose tissue sympathetic nerve activity induced by central insulin infusion. Reducing the activity of the brain insulin system via administration of insulin antibodies, receptor antisense treatment, or receptor knockdown results in hyperphagia and increased adiposity. However, despite decades of research into the role of central insulin in food intake, many questions remain to be answered, including the underlying mechanism of action.

Keywords Energy balance • Food intake • Glucose homeostasis • Insulin • Leptin • NPY/AgRP • POMC

Supported by NIH DK17844; DPB supported by NHMRC Early Career Fellowship GNT1013264.

D.P. Begg
Department of Psychiatry and Behavioral Neuroscience, University of Cincinnati, 2170 East Galbraith Road, Cincinnati, OH 45237, USA

Department of Medicine, University of Melbourne, Melbourne, VIC 3010, Australia

S.C. Woods (✉)
Department of Psychiatry and Behavioral Neuroscience, University of Cincinnati, 2170 East Galbraith Road, Cincinnati, OH 45237, USA
e-mail: steve.woods@uc.edu

H.-G. Joost (ed.), *Appetite Control*, Handbook of Experimental Pharmacology 209, DOI 10.1007/978-3-642-24716-3_5, © Springer-Verlag Berlin Heidelberg 2012

Although the pancreatic islets were first described by Langerhans (1869), it was decades before their function was identified when Minkowski and von Mering reported that removal of the pancreas caused a diabetic phenotype (von Mering and Minkowski 1889). Numerous researchers subsequently found that crude pancreatic extracts produced moderate reductions in glycosuria, and in 1921, Banting and Best isolated insulin from pancreatic islets and administered it in the first successful clinical treatment of type 1 diabetes (Banting et al. 1922). Shortly thereafter, insulin purified from animal sources became commercially available for the treatment of diabetes (Swann 1986).

Insulin is a 51-amino acid peptide hormone processed by proteases from proinsulin in the secretory vesicles of pancreatic beta cells (Sanger and Tuppy 1951a, b). The first synthetic insulin was produced in the 1960s (Volvin et al. 1964; Katsoyannis et al. 1966), and in 1978, biosynthetic human insulin was produced by Genentech and soon became commercially available for treatment of diabetes as Humulin (Banga and Chien 1988). Numerous forms of insulin and insulin analogs have been engineered for the treatment of diabetes, including rapid acting analogs (e.g., insulin aspart, insulin glulisine) as well as extended release and long-acting insulin (e.g., insulin glargine, insulin detemir) (Zib and Raskin 2006).

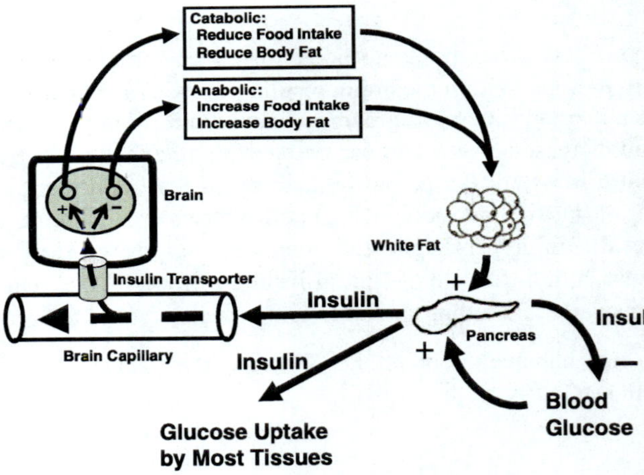

Fig. 1 The hormone insulin is secreted into the blood from the pancreas in response to circulating glucose and in direct proportion to the level of fat stored in white adipose tissue. One of its primary functions is to reduce blood glucose by increasing glucose uptake into most tissues (e.g., liver, skeletal muscle). As it circulates through brain capillaries, a small amount of insulin is transported into the brain, where it acts on insulin receptors on neurons with either net anabolic (e.g., NPY, AgRP) or net catabolic (e.g., POMC) activity, for example, in the hypothalamic arcuate nucleus. These neurons in turn influence energy homeostasis (food intake, energy expenditure) and ultimately the amount of fat stored in the body, by exerting a net catabolic or anabolic action. Increased insulin entering the brain activates catabolic circuits and inhibits anabolic circuits, leading to hypophagia and weight loss

The insulin receptor (IR) is a transmembrane tyrosine kinase receptor and is the only target receptor for insulin (Joost 1995). The receptor contains two alpha and two beta subunits that are encoded by the INSR gene (De Meyts et al. 1990). Despite decades of research, no antagonist to the IR has been identified, making it impossible to determine the effects of directly blocking insulin without the confounding factors introduced by insulin antibody administration (Strubbe and Mein 1977), insulin receptor antisense treatment (Obici et al. 2002), or knocking out the receptor (Brüning et al. 2000). There is considerable cross-reactivity between the insulin system and insulin-like growth factors (IGFs) and their receptors (Kitamura et al. 2003). IGF1 has a strong affinity for the IGF1 receptor (IGF1R) (Morgan et al. 1986; Ullrich et al. 1986) and also binds with the IR, albeit with far less affinity than insulin (Lawrence et al. 2007). IGF2 is the only known endogenous ligand that activates the IGF2 receptor (IGF2R) (Morgan et al. 1987) and has strong affinities for IGF1R and IR (Morrione et al. 1997); for example, IGF2, rather than insulin, acts via the IR to promote embryonic development (Louvi et al. 1997). Another receptor, the insulin receptor-related receptor (IRR), has recently been identified in this family (Shier and Watt 1989). Neither insulin nor IGF binds with the IRR, and no endogenous ligand has been confirmed to act at the receptor, although it has been suggested that IRR is an alkali sensor in the renal system (Deyev et al. 2011).

While insulin has many biological actions, it is best known for its peripheral role of reducing circulating glucose. This is a result of several functions including acting on insulin-sensitive tissues to increase their rate of uptake of glucose and to use it as fuel or else store it as glycogen (Cooney et al. 1985). In the liver, besides increasing glycogen synthesis (Rossetti and Giaccari 1990), insulin inhibits gluconeogenesis (Yki-Jarvinen et al. 1989) and glycogenolysis (Petersen et al. 1998). Therefore, peripheral administration of insulin results in hypoglycemia, a decrease of blood glucose.

Neurons require a continuous source of energy from the blood to remain functional, and most of this energy is derived from blood glucose (Hoyer 1990). If hypoglycemia occurs, for example, after insulin is administered, receptors in the brain detect it and trigger reflexes to increase glucose secretion from the liver while simultaneously stimulating food intake (Ritter et al. 1981). For this reason, the hypoglycemia that occurs with insulin treatment has been proposed as the basis of the increased food intake and weight gain associated with insulin treatment in diabetics (Rosenbloom and Giordano 1977; McFarlane 2009; MacKay et al. 1940; Lotter and Woods 1977). The effects of insulin on circulating energy are not limited to increased glucose uptake. Insulin additionally stimulates the uptake of lipids into adipocytes and inhibits free fatty acid release from adipose tissue (Prior and Smith 1982; Large et al. 2004; Wang et al. 2008). In the absence of insulin, adipocytes cannot store lipid (Benoit et al. 2004). In skeletal muscle and other tissues, insulin increases the cellular uptake of amino acids (Pozefsky et al. 1969; Biolo et al. 1995).

The primary stimulant of insulin secretion is an increase of blood glucose, and this typically occurs mainly when food is consumed (Woods 1990). In the absence of an increase of glucose, only a small, stable secretion of insulin occurs, called basal insulin. In a normal, nondiabetic human or animal, blood glucose is low and relatively constant throughout much of the day, rising when meals containing carbohydrates are eaten (Woods 1991). Insulin follows a similar pattern, being low and steady except during meals when it is recruited to return prandial glucose to basal levels (called glucose tolerance) (Woods 1990). Another important influence is body fat. Both basal and stimulated insulin secretion occur in direct proportion to adiposity; i.e., leaner individuals have lower basal and stimulated insulin secretion, and fatter individuals have higher basal and stimulated insulin secretion (Bagdade and Bagdade 1968; Bagdade et al. 1967; Stephan et al. 1972). Because of this, insulin is considered to be an adiposity signal to the brain (Woods and Seeley 2001; Moran and Ladenheim 2011). A hormone secreted by white adipocytes, leptin, is also an adiposity signal since its secretion is also directly proportional to body fat and since, like insulin, it is transported into the brain where it acts on leptin receptors on neurons influencing energy homeostasis (Woods et al. 2000; Figlewicz 2003; Belgardt and Brüning 2010).

1 The Central Insulin System

The central insulin system is a concept that derives from the multiple biological effects caused by centrally administered exogenous insulin, including reductions in food intake (Woods et al. 1979) and increases in energy expenditure (Muller et al. 1997) (see Fig. 1). Most insulin is made in the pancreas, and there is no convincing evidence that insulin can be synthesized in the brain (Schwartz et al. 1992). Because insulin is a large peptide hormone, for many years, it was considered unable to easily penetrate the blood brain barrier (Havrankova et al. 1978a). However, around 35 years ago, it was first proposed that insulin passes from the blood to the cerebrospinal fluid (CSF) and then to the brain, providing a hormonal indicator of peripheral adiposity (Woods and Porte Jr 1976). The original suggestion was that insulin enters the CSF from plasma via the choroid plexus, and then passes through the ependymal lining to act at insulin receptors on adjacent neurons (Woods and Porte Jr 1976). This hypothesis was compelling given that there are dense insulin binding sites in the choroid plexus and that several nuclei in the ventral hypothalamus, located close to the wall of the third ventricle, contain high numbers of insulin receptors (Havrankova et al. 1978b, 1981; van Houten and Posner 1979; van Houten et al. 1980; Young et al. 1980; Baskin et al. 1983a). Further, the hypothesis was consistent with the observation that when exogenous insulin is administered directly into the third ventricle, it has a net catabolic effect (Woods et al. 1979; Benoit et al. 2002; Air et al. 2002a, b, c). However, subsequent studies assessing the dynamics of insulin uptake into the brain and CSF determined that rather than entering the CNS via the choroid plexus and CSF, insulin is transported into the brain via an insulin receptor-mediated, saturable pathway in brain capillary endothelial cells (Baura et al. 1993). Thus, the normal movement of insulin vis-à-vis the brain best fits a three-compartment model (plasma–brain interstitial fluid–CSF)

(Schwartz et al. 1992). The passage of insulin through brain capillary endothelial cells has since been confirmed by various state-of-the-art methods (Banks 2004; Banks et al. 1997). The insulin that can be measured in the CSF has likely passed through the brain and will be removed by the IR on the choroid plexus. In order to engage neuronal insulin receptors, insulin that is experimentally or therapeutically administered into the CSF must therefore penetrate into the brain against the normal flow of fluid.

Insulin receptors are abundant and widely distributed throughout both the developing and adult CNS (Hill et al. 1986). The distribution of insulin and its receptor has been well characterized by immunohistochemistry (Baskin et al. 1983b) and autoradiography (van Houten and Posner 1979; van Houten et al. 1979) as well as by in situ hybridization for insulin receptor mRNA (Marks et al. 1990). Relatively high levels of IR are found in the olfactory bulbs and the arcuate (ARC) nucleus of the hypothalamus. They are also abundant in several other regions including the cerebral cortex, cerebellum, hippocampus, and choroid plexus, as well as other hypothalamic areas and regions of the lower brainstem (van Houten and Posner 1979; Baskin et al. 1983a, b; van Houten et al. 1979; Marks et al. 1990; Marks and Eastman 1990). The widespread distribution of insulin and its receptors within the CNS (Havrankova et al. 1981) provides an indication of the diverse actions of insulin in the brain, including influencing energy homeostasis (food intake and energy expenditure) (Brüning et al. 2000; Woods et al. 1979; Benoit et al. 2002), systemic glucose responses (Fisher et al. 2005), reproductive processes (Brüning et al. 2000), cognitive function (Stockhorst et al. 2004, 2011), and many other processes (van der Heide et al. 2006).

The ARC, which comprises the floor of the third ventricle in the region of the ventral hypothalamus, is the region most commonly associated with the effects of insulin on food intake. As reviewed previously (Belgardt and Brüning 2010; Elmquist et al. 2005; Schwartz et al. 2000; Seeley and Woods 2003), the ARC contains two major types of neurons that project to other hypothalamic as well as other brain areas. The first of these neurons are found mainly in the ventromedial ARC and synthesize and secrete neuropeptide Y (NPY) and agouti-related protein (AgRP). Primary targets for their axonal projections are the hypothalamic paraventricular nuclei (PVN) and the lateral hypothalamic area (LH). Stimulation of these NPY/AgRP neurons elicits an anabolic response including a rapid and prolonged increase of food intake, and one endogenous stimulant of ARC NPY/AgRP neurons is the gastric hormone, ghrelin (Horvath et al. 2001; Tschöp et al. 2000). Administration of exogenous ghrelin, or of NPY or AgRP, elicits hyperphagia. The adiposity signals insulin, and leptin inhibits these neurons. The second group of ARC neurons, which are found more laterally, synthesize the peptides proopiomelanocorticotropin (POMC) and cocaine- and amphetamine-regulated transcript (CART). POMC is a large prohormone that can be cleaved into any of several possible active products depending upon the cell(s) in which it is synthesized. Some ARC neurons cleave POMC into the active neuropeptide called α-melanocyte-stimulating hormone (αMSH). POMC/CART neurons in the ARC project to many of the same areas as NPY/AgRP neurons, including the PVN and LH, where αMSH acts on melanocortin receptors (MC3 and MC4) to reduce food intake. When POMC/CART neurons are activated, they elicit a net catabolic response including inhibiting food intake and increasing energy expenditure (Belgardt and Brüning 2010; Morton et al. 2006). Insulin and leptin act on POMC/CART neurons to release αMSH, and the ability of both insulin and leptin to reduce food intake via the ARC requires signaling through the melanocortin receptor system (Niswender and Schwartz 2003; Seeley et al. 1997). The balance of activity of NPY/AgRP and POMC/CART neurons in the ARC determines whether there is a net anabolic or catabolic tone in the ventral hypothalamus.

In sum, when any of the neuropeptides found endogenously in the ARC are administered into the third ventricle [or directly into the PVN, for example, αMSH (Kim et al. 2002)], they produce robust effects on energy homeostasis, relating to both food intake and energy expenditure (Benoit et al. 2000; Morton and Schwartz 2001). Importantly, both types of ARC neurons express receptors for peripheral hormones which enter the brain and signal peripheral energy stores (Ellacott and Cone 2004), including insulin and leptin which cause a net catabolic response (Benoit et al. 2002; Halaas et al. 1995) and ghrelin which elicits a net anabolic response (Tschöp et al. 2000; Wren et al. 2000).

Insulin and leptin stimulate overlapping signal transduction pathways in ARC neurons. When insulin binds to its receptor, tyrosine kinase is activated, causing phosphorylation of intracellular tyrosine residues on the receptor (Rees-Jones et al. 1984; Plum et al. 2005). Insulin-receptor substrate (IRS) proteins then bind to the phosphorylated residues and are activated; the IRS proteins then activate phosphatidylinositol 3-kinase (PI3K), consisting of a regulatory subunit (p85) and a catalytic subunit (p110) (Esposito et al. 2001). PI3K can then catalyze phosphorylation of phosphatidylinositol (4,5)-bisphosphate (PIP2) to phosphatidylinositol (3,4,5)-trisphosphate (PIP3) (Rodriguez-Escudero et al. 2005). PIP3 then acts on downstream targets including 3-phosphoinositide-dependent kinase-1 (PDK1) (Standaert et al. 2001), glycogen synthase kinase-3 (GSK3) (Brady et al. 1998), and protein kinase B (PKB/Akt) (Su et al. 2006). The interaction between insulin and leptin may occur, at least in part, through leptin-mediated PI3K activation.

2 Central Insulin Administration and Food Intake

Central infusion of insulin was first demonstrated to reduce food intake in baboons in the late 1970s (Woods et al. 1979). In that report, chronic treatment with insulin administered into the lateral cerebral ventricle caused a dose-dependent decrease in food intake over a period of 2–3 weeks. The decrease in food intake occurred within days and was accompanied by a reduction in body weight; the reduction in food intake was reversed following cessation of the infusions. Numerous studies since have examined the effects of central insulin infusion on food intake and body weight, primarily in rats (Air et al. 2002a, b; Ikeda et al. 1986; McGowan et al. 1992); the data have supported the initial finding, and higher doses of insulin cause greater decreases in both food intake and body weight (Brief and Davis 1984). It has been confirmed that the anorexia and weight loss that occur are not secondary to a conditioned aversion formed from an illness following infusion (Chavez et al. 1995a), and the food intake reduction has been demonstrated to not be the result of reduced mobility of animals following insulin infusion (Chavez et al. 1995b). Further evidence of insulin's role in food intake comes from data demonstrating that when antibodies to insulin are injected into the ventral hypothalamus, animals have greater food intake compared with control days (Strubbe and Mein 1977; McGowan et al. 1992).

Hypophagia is observed following either acute (bolus) administration of insulin into the third ventricle (mainly of rats) or with chronic central insulin infusion, and both are dose-dependent (Woods et al. 1979; Brief and Davis 1984; Riedy et al. 1995). Injection of insulin directly into the ARC produces hypophagia at doses that are subthreshold when injected into the ventricle (McGowan et al. 1992); these data implicate the ARC as a key region involved in the reduced food intake following insulin infusion.

Central insulin-induced hypophagia has been observed in baboons (Woods et al. 1979), rats (Benoit et al. 2002; Air et al. 2002a, b; Ikeda et al. 1986; Brief and Davis 1984; Clegg et al. 2003; Plata-Salaman and Oomura 1986), mice (Brown et al. 2006; Jaillard et al. 2009), chicks (Shiraishi et al. 2011), and sheep (Foster et al. 1991). Interestingly, recent data indicate that humans also have reduced food intake following an increase in central insulin. Using the novel approach of intranasal insulin administration to increase central insulin levels, it was found that there is a dose-dependent reduction of food intake during an ad libitum breakfast buffet (Hallschmid et al. 2004), and follow-up experiments found that men were more sensitive to insulin's anorexigenic action than women (Benedict et al. 2008). These data are consistent with findings in the rat, with males more sensitive to the anorexigenic effects of insulin than females and females being more sensitive to the anorexigenic effect of leptin (Clegg et al. 2003, 2006). In the blood, females have higher leptin and lower insulin than comparably obese males (Clegg et al. 2006). These observations are consistent with the sex difference in fat distribution. Females tend to have more subcutaneous fat than males, and males tend to have more visceral fat than females (Bjorntorp 1996; Nedungadi and Clegg 2009). Leptin is secreted disproportionately more from subcutaneous than from visceral fat (Montague et al. 1997; Dusserre et al. 2000), and insulin secretion is proportional to visceral as opposed to subcutaneous fat (Pouliot et al. 1992). Thus, the fat-to-adiposity signal-to-hypothalamic sensitivity system is tuned to leptin in females and to insulin in males. This has important clinical implications since high visceral fat and high plasma insulin predispose to the metabolic syndrome and its sequelae of obesity, cardiovascular problems, hyperlipidemia, and several cancers (Nedungadi and Clegg 2009; Matsuzawa et al. 1994, 1995). The sex differences observed in hypophagia following central administration of adiposity signals are due, at least in part, to the actions of estrogen in the hypothalamus (Musatov et al. 2007).

In some instances, insulin and leptin act at their individual receptors on the same neurons, and some of their intracellular signaling pathways are the same, at least in the ARC. For example, the hypophagic effects of both i3vt insulin and leptin can be blocked by administration of PI3 kinase inhibitors (Niswender and Schwartz 2003) as well as by melanocortin 3/4 receptor antagonists (Benoit et al. 2002; Seeley et al. 1997).

There is evidence that the reduced food intake and weight loss caused by i3vt insulin are dependent on central leptin signaling since insulin is ineffective at reducing food intake in obese Zucker rats lacking functional leptin receptors (Ikeda et al. 1986). The response to central insulin infusion is also reduced when animals are maintained on a high-fat diet (Arase et al. 1988; Chavez et al. 1996; Clegg et al. 2005), implying that these animals have insulin resistance both peripherally and in the central nervous system. Within the brain, it could also be secondary to the central leptin resistance that has been observed in obese animals (Koch et al. 2010). When insulin and leptin are both administered i3vt, there are dose-dependent interactions (Air et al. 2002b). At most doses, the two are additive in reducing food intake, whereas at others, the combined reduction of food intake is actually significantly less than the sum of their individual actions, implying that there are likely differential effects of the two hormones on different neuronal populations.

To summarize, insulin receptors are expressed on some hypothalamic ARC neurons. When exogenous insulin is administered into the cerebroventricular system, it inhibits NPY/AgRP neurons while stimulating POMC neurons (Benoit et al. 2002; Sato et al. 2005) and results in increased release and activity of αMSH onto neurons in the PVN, LH, and elsewhere, resulting in a reduction of food intake (Benoit et al. 2000) and, if given chronically, body weight as well (Woods et al. 1979, 1998; Schwartz et al. 2000). Leptin has similar actions in the ARC (Heisler et al. 2003), and leptin activity appears to be necessary for insulin to exert its anorectic effect (Ikeda et al. 1986). A key question is whether endogenous insulin has the same action as exogenous insulin. What is known is that mice lacking insulin receptors on all neurons (NIRKO mice) weigh more than wild-type controls (Brüning et al. 2000).

In contrast to the hypophagic action of centrally administered insulin (Woods et al. 1979; Clegg et al. 2003), peripherally administered insulin generally leads to overeating and significant weight gain (Ryan et al. 2008), and this is a detrimental clinical problem for diabetics being treated with insulin as it reduces compliance (Bryden et al. 1999). However, the long-acting human insulin analog, insulin detemir, which has an attached myristyric acid molecule, has the same ability to improve glycemia but increases body weight less and often induces weight loss in diabetic patients (Hermansen et al. 2007; Hermansen and Mortensen 2007). It has been suggested that the beneficial effects of insulin detemir on weight may be the result of increased penetration of insulin into the brain (Hermansen and Davies 2007).

3 Interaction with Satiation Peptides

Few, if any, meals are initiated in response to low glucose or other indicators of low or dwindling energy. Rather, meals occur because of habit, time of day, opportunity, the social situation, and other psychological factors (Woods 1991, 2009).

Physiologic controls that help determine body weight are therefore exerted on how much food is eaten once a meal begins. There is compelling evidence that, in addition to psychological factors, the feeling of fullness (satiation) is determined by the gradual buildup, during the meal, of hormones and other signals secreted by the gastrointestinal tract (Woods 2004; Woods and D'Alessio 2008). The best known of these signals is the duodenal hormone, cholecystokinin (CCK) (Moran and Kinzig 2004). CCK is secreted in response to consumed nutrients entering the intestine (especially lipids), and it stimulates CCK receptors on adjacent sensory nerves that relay the signal to the brain. CCK is one of several gastrointestinal peptides implicated in satiation, others including glucagon-like peptide-1 (GLP-1) (Turton et al. 1996; Tang-Christensen et al. 1996; Barrera et al. 2011) and peptide YY (PYY) (Batterham et al. 2002). If people or animals are administered CCK or other satiation peptides as a meal is beginning, they eat fewer calories than in a control condition, and when antagonists to these compounds are given, individuals eat larger meals. The cumulated effect of CCK and the other satiation signals during a meal ultimately increases to the point that eating stops. Insulin and leptin act in the brain to increase its sensitivity to CCK and other satiation signals (Riedy et al. 1995; Matson et al. 1997, 2000; Emond et al. 1999, 2001). Thus, when an individual has lost some weight, insulin and leptin levels in the blood (and consequently in the brain as well) are reduced. This makes the brain less sensitive to signals from CCK, and the consequence is that larger meals will be consumed until weight returns to normal. Likewise, an increase of weight is associated with an increase of circulating insulin and leptin and heightened sensitivity to CCK such that smaller meals are eaten. This negative feedback mechanism helps to maintain body weight relatively constant.

4 Genetic Manipulation of the Central Insulin System and Food Intake

In recent years, it has become possible to apply molecular genetic techniques to manipulate selected populations of insulin receptors in the brain. Total body knockouts of either insulin or the insulin receptor are lethal soon after birth, the mice dying of ketoacidosis (Accili et al. 1996). Tissue-specific knockout of the insulin-receptor gene has thus provided a more useful tool to examine the roles of insulin signaling in various tissues. Knocking out (KO) the insulin receptor in skeletal muscle results in dyslipidemia (Brüning et al. 1998); liver insulin-receptor KO leads to moderate insulin resistance (Michael et al. 2000); selective KO of the insulin receptor in adipose tissue leads to protection from obesity (Bluher et al. 2002); and selective KO in pancreatic beta cells causes impaired glucose tolerance (Kulkarni et al. 1999), as does combined muscle and adipose tissue knockout (Lauro et al. 1998). Of particular interest with consideration of the role of insulin in the control of food intake, knocking out the insulin receptors selectively in neurons (NIRKO mice) leads to numerous metabolic abnormalities (Brüning et al. 2000; Fisher et al. 2005; Schubert et al. 2004; Diggs-Andrews et al. 2010).

Female NIRKO mice have increased food intake compared to wild-type controls (Brüning et al. 2000), and both sexes are obesity prone, having increased weight gain compared with controls when placed on an energy-dense diet. However, given that there are concurrent effects related to reduced fertility in these animals (Brüning et al. 2000), it has been suggested that the obesity could be secondary to hypogonadism, especially in the females (Woods and Seeley 2001). NIRKO mice also have a reduced capacity to mount a sympathetic adrenal response, achieving plasma epinephrine and norepinephrine that are only 50% and 90% of control responses, respectively, following insulin-induced hypoglycemia (Fisher et al. 2005). The lack of central insulin signaling also leads to reduced hypothalamic GLUT4 expression and a reduced hypothalamic neuronal sensing of glucose in response to hypoglycemia (Diggs-Andrews et al. 2010).

Because insulin receptors are expressed widely throughout the brain (van Houten et al. 1979; Corp et al. 1986), including areas involved with caloric homeostasis (Abizaid and Horvath 2008), with hedonics and rewards (Lattemann 2008), and with memory and cognitive behavior (McNay 2007), more recent reports have used sophisticated molecular genetic techniques to knock the insulin receptor out of specific populations of neurons in the brain. For example, selective knockout in brainstem catecholaminergic neurons results in hyperphagia and increased body weight and body fat compared with controls (Konner et al. 2011). Conversely, knockout of the insulin receptor uniquely in steroidogenic factor-1 expressing neurons that are localized to the ventromedial hypothalamus protects against diet-induced obesity, leptin resistance, and glucose intolerance (Klockener et al. 2011). Knockout of both insulin and leptin receptors specifically in POMC neurons results in insulin resistance (Hill et al. 2010), and this did not occur with knockout of either of these receptor types alone, demonstrating the important interactions between these two adiposity signals. Interestingly, deletion of insulin receptors specifically in the POMC neurons did not produce hyperphagia or induce weight gain (Hill et al. 2010). This may indicate that the actions of central insulin on food intake are dependent on reducing the anabolic NPY/AgRP signaling, rather than increasing catabolic POMC activity. This series of experiments from the Brüning lab also nicely demonstrates the complex calculus of insulin's actions across a large number of brain areas. The observation that administering insulin into the ventricular system of normal animals reduces food intake and body weight therefore implies that this catabolic action represents the integrated output of many neuronal circuits.

Further evidence for the involvement of insulin in obesity comes from protein tyrosine phosphatase-1B (PTP-1B) knockout mice. PTP-1B is implicated in insulin signaling and is thought to impart insulin resistance (Kennedy 1999; Byon et al. 1998). PTP-1B−/− mice are obesity resistant and remain insulin sensitive, even when maintained on a high-fat diet (Elchebly et al. 1999). They also have higher energy expenditure than wild-type mice (Klaman et al. 2000). Interestingly, the catabolic actions of removing PTP-1B are consistent with observations following central insulin infusion (Rahmouni et al. 2004). Another factor that reduces insulin receptor activity is plasma cell membrane glycoprotein 1 (PC-1). PC-1 acts via a direct interaction with the alpha subunit of the insulin receptor to reduce tyrosine kinase activity and therefore to compromise subsequent insulin-initiated cellular signaling (Maddux and Goldfine 2000). Polymorphism of the PC-1 gene is associated with obesity in a number of human populations (Hamaguchi et al. 2004; Grarup et al. 2006; Matsuoka et al. 2006; Wang et al. 2011), and it has been suggested that PC-1 may be involved in development of insulin resistance in obesity. While these data are of interest, it is important to note that unlike the NIRKO studies, these data relate to alteration of both central and peripheral insulin signaling.

References

Abizaid A, Horvath TL (2008) Brain circuits regulating energy homeostasis. Regul Pept 149:3–10

Accili D, Drago J, Lee EJ, Johnson MD, Cool MH, Salvatore P, Asico LD, Jose PA, Taylor SI, Westphal H (1996) Early neonatal death in mice homozygous for a null allele of the insulin receptor gene. Nat Genet 12:106–109

Air EL, Benoit SC, Blake Smith KA, Clegg DJ, Woods SC (2002a) Acute third ventricular administration of insulin decreases food intake in two paradigms. Pharmacol Biochem Behav 72:423–429

Air EL, Benoit SC, Clegg DJ, Seeley RJ, Woods SC (2002b) Insulin and leptin combine additively to reduce food intake and body weight in rats. Endocrinology 143:2449–2452

Air EL, Strowski MZ, Benoit SC, Conarello SL, Salituro GM, Guan XM, Liu K, Woods SC, Zhang BB (2002c) Small molecule insulin mimetics reduce food intake and body weight and prevent development of obesity. Nat Med 8:179–183

Arase K, Fisler JS, Shargill NS, York DA, Bray GA (1988) Intracerebroventricular infusions of 3-OHB and insulin in a rat model of dietary obesity. Am J Physiol 255:R974–R981

Bagdade JD, Bagdade JD (1968) Basal insulin and obesity. Lancet 2:630–631

Bagdade JD, Bierman EL, Porte D Jr (1967) The significance of basal insulin levels in the evaluation of the insulin response to glucose in diabetic and nondiabetic subjects. J Clin Invest 46:1549–1557

Banga AK, Chien YW (1988) Systemic delivery of therapeutic peptides and proteins. Int J Pharm 48:15–50

Banks WA (2004) The source of cerebral insulin. Eur J Pharmacol 490:5–12

Banks WA, Jaspan JB, Huang W, Kastin AJ (1997) Transport of insulin across the blood-brain barrier: saturability at euglycemic doses of insulin. Peptides 18:1423–1429

Banting FG, Best CH, Collip JB, Campbell WR, Fletcher AA (1922) Pancreatic extracts in the treatment of Diabetes Mellitus. Can Med Assoc J 12:141–146

Barrera JG, Jones KR, Herman JP, D'Alessio DA, Woods SC, Seeley RJ (2011) Hyperphagia and increased fat accumulation in two models of chronic CNS glucagon-like peptide-1 loss of function. J Neurosci 31:3904–3913

Baskin DG, Porte D Jr, Guest K, Dorsa DM (1983a) Regional concentrations of insulin in the rat brain. Endocrinology 112:898–903

Baskin DG, Woods SC, West DB, van HM, Posner BI, Dorsa DM, Porte D Jr (1983b) Immunocytochemical detection of insulin in rat hypothalamus and its possible uptake from cerebrospinal fluid. Endocrinology 113:1818–1825

Batterham RL, Cowley MA, Small CJ, Herzog H, Cohen MA, Dakin CL, Wren AM, Brynes AE, Low MJ, Ghatei MA, Cone RD, Bloom SR (2002) Gut hormone PYY(3–36) physiologically inhibits food intake. Nature 418:650–654

Baura G, Foster D, Porte D Jr, Kahn SE, Bergman RN, Cobelli C, Schwartz MW (1993) Saturable transport of insulin from plasma into the central nervous system of dogs in vivo: a mechanism for regulated insulin delivery to the brain. J Clin Invest 92:1824–1830

Belgardt BF, Brüning JC (2010) CNS leptin and insulin action in the control of energy homeostasis. Ann NY Acad Sci 1212:97–113

Benedict C, Kern W, Schultes B, Born J, Hallschmid M (2008) Differential sensitivity of men and women to anorexigenic and memory-improving effects of intranasal insulin. J Clin Endocrinol Metab 93:1339–1344

Benoit SC, Schwartz MW, Baskin DG, Woods SC, Seeley RJ (2000) CNS melanocortin system involvement in the regulation of food intake and body weight. Horm Behav 37:299–308

Benoit SC, Air EL, Coolen LM, Strauss R, Jackman A, Clegg DJ, Seeley RJ, Woods SC (2002) The catabolic action of insulin in the brain is mediated by melanocortins. J Neurosci 22:9048–9052

Benoit SC, Clegg DJ, Seeley RJ, Woods SC (2004) Insulin and leptin as adiposity signals. Recent Prog Horm Res 59:267–285

Biolo G, Declan Fleming RY, Wolfe RR (1995) Physiologic hyperinsulinemia stimulates protein synthesis and enhances transport of selected amino acids in human skeletal muscle. J Clin Invest 95:811–819

Bjorntorp P (1996) The regulation of adipose tissue distribution in humans. Int J Obes Relat Metab Disord 20:291–302

Bluher M, Michael MD, Peroni OD, Ueki K, Carter N, Kahn BB, Kahn CR (2002) Adipose tissue selective insulin receptor knockout protects against obesity and obesity-related glucose intolerance. Dev Cell 3:25–38

Brady MJ, Bourbonais FJ, Saltiel AR (1998) The activation of glycogen synthase by insulin switches from kinase inhibition to phosphatase activation during adipogenesis in 3T3-L1 cells. J Biol Chem 273:14063–14066

Brief DJ, Davis JD (1984) Reduction of food intake and body weight by chronic intraventricular insulin infusion. Brain Res Bull 12:571–575

Brown LM, Clegg DJ, Benoit SC, Woods SC (2006) Intraventricular insulin and leptin reduce food intake and body weight in C57BL/6 J mice. Physiol Behav 30:687–691

Brüning JC, Michael MD, Winnay JN, Hayashi T, Horsch D, Accili D, Goodyear LJ, Kahn CR (1998) A muscle-specific insulin receptor knockout exhibits features of the metabolic syndrome of NIDDM without altering glucose tolerance. Mol Cell 2:559–569

Brüning JC, Gautam D, Burks DJ, Gillette J, Schubert M, Orban PC, Klein R, Krone W, Muller-Wieland D, Kahn CR (2000) Role of brain insulin receptor in control of body weight and reproduction. Science 289:2122–2125

Bryden KS, Neil A, Mayou RA, Peveler RC, Fairburn CG, Dunger DB (1999) Eating habits, body weight, and insulin misuse. A longitudinal study of teenagers and young adults with type 1 diabetes. Diabetes Care 22:1956–1960

Byon JC, Kusari AB, Kusari J (1998) Protein-tyrosine phosphatase-1B acts as a negative regulator of insulin signal transduction. Mol Cell Biochem 182:101–108

Chavez M, Seeley RJ, Woods SC (1995a) A comparison between effects of intraventricular insulin and intraperitoneal lithium chloride on three measures sensitive to emetic agents. Behav Neurosci 109:547–550

Chavez M, Kaiyala K, Madden LJ, Schwartz MW, Woods SC (1995b) Intraventricular insulin and the level of maintained body weight in rats. Behav Neurosci 109:528–531

Chavez M, Riedy CA, van Dijk G, Woods SC (1996) Central insulin and macronutrient intake in the rat. Am J Physiol 271:R727–R731

Clegg DJ, Riedy CA, Smith KA, Benoit SC, Woods SC (2003) Differential sensitivity to central leptin and insulin in male and female rats. Diabetes 52:682–687

Clegg DJ, Benoit SC, Reed JA, Woods SC, Dunn-Meynell A, Levin BE (2005) Reduced anorexic effects of insulin in obesity-prone rats fed a moderate-fat diet. Am J Physiol Regul Integr Comp Physiol 288:R981–R986

Clegg DJ, Brown LM, Woods SC, Benoit SC (2006) Gonadal hormones determine sensitivity to central leptin and insulin. Diabetes 55:978–987

Cooney GJ, Caterson ID, Newsholme EA (1985) The effect of insulin and noradrenaline on the uptake of 2-[1-14C]deoxyglucose in vivo by brown adipose tissue and other glucose-utilising tissues of the mouse. FEBS Lett 188:257–261

Corp ES, Woods SC, Porte D Jr, Dorsa DM, Figlewicz DP, Baskin DG (1986) Localization of ^{125}I-insulin binding sites in the rat hypothalamus by quantitative autoradiography. Neurosci Lett 70:17–22

De Meyts P, Gu JL, Shymko RM, Kaplan BE, Bell GI, Whittaker J (1990) Identification of a ligand-binding region of the human insulin receptor encoded by the second exon of the gene. Mol Endocrinol 4:409–416

Deyev IE, Sohet F, Vassilenko KP, Serova OV, Popova NV, Zozulya SA, Burova EB, Houillier P, Rzhevsky DI, Berchatova AA, Murashev AN, Chugunov AO, Efremov RG, Nikol'sky NN, Bertelli E, Eladari D, Petrenko AG (2011) Insulin receptor-related receptor as an extracellular alkali sensor. Cell Metab 13:679–689

Diggs-Andrews KA, Zhang X, Song Z, Daphna-Iken D, Routh VH, Fisher SJ (2010) Brain insulin action regulates hypothalamic glucose sensing and the counterregulatory response to hypoglycemia. Diabetes 59:2271–2280

Dusserre E, Moulin P, Vidal H (2000) Differences in mRNA expression of the proteins secreted by the adipocytes in human subcutaneous and visceral adipose tissues. Biochim Biophys Acta 1500:88–96

Elchebly M, Payette P, Michaliszyn E, Cromlish W, Collins S, Loy AL, Normandin D, Cheng A, Himms-Hagen J, Chan CC, Ramachandran C, Gresser MJ, Tremblay ML, Kennedy BP (1999) Increased insulin sensitivity and obesity resistance in mice lacking the protein tyrosine phosphatase-1B gene. Science 283:1544–1548

Ellacott KL, Cone RD (2004) The central melanocortin system and the integration of short- and long-term regulators of energy homeostasis. Recent Prog Horm Res 59:395–408

Elmquist JK, Coppari R, Balthasar N, Ichinose M, Lowell BB (2005) Identifying hypothalamic pathways controlling food intake, body weight, and glucose homeostasis. J Comp Neurol 493:63–71

Emond M, Schwartz GJ, Ladenheim EE, Moran TH (1999) Central leptin modulates behavioral and neural responsivity to CCK. Am J Physiol 276:R1545–R1549

Emond M, Ladenheim EE, Schwartz GJ, Moran TH (2001) Leptin amplifies the feeding inhibition and neural activation arising from a gastric nutrient preload. Physiol Behav 72:123–128

Esposito DL, Li Y, Cama A, Quon MJ (2001) Tyr(612) and Tyr(632) in human insulin receptor substrate-1 are important for full activation of insulin-stimulated phosphatidylinositol 3-kinase activity and translocation of GLUT4 in adipose cells. Endocrinology 142:2833–2840

Figlewicz DP (2003) Adiposity signals and food reward: expanding the CNS roles of insulin and leptin. Am J Physiol Regul Integr Comp Physiol 284:R882–R892

Fisher SJ, Brüning JC, Lannon S, Kahn CR (2005) Insulin signaling in the central nervous system is critical for the normal sympathoadrenal response to hypoglycemia. Diabetes 54:1447–1451

Foster LA, Ames NK, Emery RS (1991) Food intake and serum insulin responses to intraventricular infusions of insulin and IGF-I. Physiol Behav 50(4):745–749

Grarup N, Urhammer SA, Ek J, Albrechtsen A, Glumer C, Borch-Johnsen K, Jorgensen T, Hansen T, Pedersen O (2006) Studies of the relationship between the ENPP1 K121Q polymorphism and type 2 diabetes, insulin resistance and obesity in 7,333 Danish white subjects. Diabetologia 49:2097–2104

Halaas JL, Gajiwala KS, Maffei M, Cohen SL, Chait BT, Rabinowitz D, Lallone RL, Burley SK, Friedman JM (1995) Weight-reducing effects of the plasma protein encoded by the obese gene. Science 269:543–546

Hallschmid M, Benedict C, Schultes B, Fehm HL, Born J, Kern W (2004) Intranasal insulin reduces body fat in men but not in women. Diabetes 53:3024–3029
Hamaguchi K, Terao H, Kusuda Y, Yamashita T, Hazoury Bahles JA, Cruz LM, Brugal VL, Jongchong WB, Yoshimatsu H, Sakata T (2004) The PC-1 Q121 allele is exceptionally prevalent in the Dominican Republic and is associated with type 2 diabetes. J Clin Endocrinol Metab 89:1359–1364
Havrankova J, Schmechel D, Roth J, Brownstein MJ (1978a) Identification of insulin in the rat brain. Proc Natl Acad Sci USA 75:5737–5741
Havrankova J, Roth J, Browstein M (1978b) Insulin receptors are widely distributed in the central nervous system of the rat. Nature 272:827–829
Havrankova J, Brownstein M, Roth J (1981) Insulin and insulin receptors in the rodent brain. Diabetologia 20:268–273
Heisler LK, Cowley MA, Kishi T, Tecott LH, Fan W, Low MJ, Smart JL, Rubinstein M, Tatro JB, Zigman JM, Cone RD, Elmquist JK (2003) Central serotonin and melanocortin pathways regulating energy homeostasis. Ann NY Acad Sci 994:169–174
Hermansen K, Davies M (2007) Does insulin detemir have a role in reducing risk of insulin-associated weight gain? Diabetes Obes Metab 9:209–217
Hermansen K, Mortensen LS (2007) Bodyweight changes associated with antihyperglycaemic agents in type 2 diabetes mellitus. Drug Saf 30:1127–1142
Hermansen K, Lund P, Clemmensen K, Breum L, Kleis Moller M, Mette Rosenfalck A, Christiansen E (2007) 3-Month results from Denmark within the globally prospective and observational study to evaluate insulin Detemir treatment in Type 1 and Type 2 Diabetes: the PREDICTIVE study. Rev Diabet Stud 4:89–97
Hill JM, Lesniak MA, Pert CB, Roth J (1986) Autoradiographic localization of insulin receptors in rat brain: prominence in olfactory and limbic areas. Neuroscience 17:1127–1138
Hill JW, Elias CF, Fukuda M, Williams KW, Berglund ED, Holland WL, Cho YR, Chuang JC, Xu Y, Choi M, Lauzon D, Lee CE, Coppari R, Richardson JA, Zigman JM, Chua S, Scherer PE, Lowell BB, Brüning JC, Elmquist JK (2010) Direct insulin and leptin action on pro-opiomelanocortin neurons is required for normal glucose homeostasis and fertility. Cell Metab 11:286–297
Horvath TL, Diano S, Sotonyi P, Heiman M, Tschop M (2001) Minireview: ghrelin and the regulation of energy balance – a hypothalamic perspective. Endocrinology 142:4163–4169
Hoyer S (1990) Brain glucose and energy metabolism during normal aging. Aging (Milano) 2:245–258
Ikeda H, West DB, Pustek JJ, Figlewicz DP, Greenwood MR, Porte D Jr, Woods SC (1986) Intraventricular insulin reduces food intake and body weight of lean but not obese Zucker rats. Appetite 7:381–386
Jaillard T, Roger M, Galinier A, Guillou P, Benani A, Leloup C, Casteilla L, Penicaud L, Lorsignol A (2009) Hypothalamic reactive oxygen species are required for insulin-induced food intake inhibition: an NADPH oxidase-dependent mechanism. Diabetes 58:1544–1549
Joost HG (1995) Structural and functional heterogeneity of insulin receptors. Cell Signal 7:85–91
Katsoyannis PG, Tometsko A, Zalut C (1966) Insulin peptides. XII. Human insulin generation by combination of synthetic A and B chains. J Am Chem Soc 88:166–167
Kennedy BP (1999) Role of protein tyrosine phosphatase-1B in diabetes and obesity. Biomed Pharmacother 53:466–470
Kim EM, Grace MK, O'Hare E, Billington CJ, Levine AS (2002) Injection of alpha-MSH, but not beta-endorphin, into the PVN decreases POMC gene expression in the ARC. Neuroreport 13:497–500
Kitamura T, Kahn CR, Accili D (2003) Insulin receptor knockout mice. Annu Rev Physiol 65:313–332
Klaman LD, Boss O, Peroni OD, Kim JK, Martino JL, Zabolotny JM, Moghal N, Lubkin M, Kim YB, Sharpe AH, Stricker-Krongrad A, Shulman GI, Neel BG, Kahn BB (2000) Increased energy expenditure, decreased adiposity, and tissue-specific insulin sensitivity in protein-tyrosine phosphatase 1B-deficient mice. Mol Cell Biol 20:5479–5489

Klockener T, Hess S, Belgardt BF, Paeger L, Verhagen LA, Husch A, Sohn JW, Hampel B, Dhillon H, Zigman JM, Lowell BB, Williams KW, Elmquist JK, Horvath TL, Kloppenburg P, Brüning JC (2011) High-fat feeding promotes obesity via insulin receptor/PI3K-dependent inhibition of SF-1 VMH neurons. Nat Neurosci 14:911–918

Koch C, Augustine RA, Steger J, Ganjam GK, Benzler J, Pracht C, Lowe C, Schwartz MW, Shepherd PR, Anderson GM, Grattan DR, Tups A (2010) Leptin rapidly improves glucose homeostasis in obese mice by increasing hypothalamic insulin sensitivity. J Neurosci 30:16180–16187

Konner AC, Hess S, Tovar S, Mesaros A, Sanchez-Lasheras C, Evers N, Verhagen LA, Bronneke HS, Kleinridders A, Hampel B, Kloppenburg P, Brüning JC (2011) Role for insulin signaling in catecholaminergic neurons in control of energy homeostasis. Cell Metab 13:720–728

Kulkarni RN, Brüning JC, Winnay JN, Postic C, Magnuson MA, Kahn CR (1999) Tissue-specific knockout of the insulin receptor in pancreatic beta cells creates an insulin secretory defect similar to that in type 2 diabetes. Cell 96:329–339

Langerhans P (1869) Beiträge zur mikroskopischen Anatomie der Bauchspeicheldrüse. Gustav Lange, Berlin

Large V, Peroni O, Letexier D, Ray H, Beylot M (2004) Metabolism of lipids in human white adipocyte. Diabetes Metab 30:294–309

Lattemann DF (2008) Endocrine links between food reward and caloric homeostasis. Appetite 51:452–455

Lauro D, Kido Y, Castle AL, Zarnowski MJ, Hayashi H, Ebina Y, Accili D (1998) Impaired glucose tolerance in mice with a targeted impairment of insulin action in muscle and adipose tissue. Nat Genet 20:294–298

Lawrence MC, McKern NM, Ward CW (2007) Insulin receptor structure and its implications for the IGF-1 receptor. Curr Opin Struct Biol 17:699–705

Lotter EC, Woods SC (1977) Injections of insulin and changes of body weight. Physiol Behav 18:293–297

Louvi A, Accili D, Efstratiadis A (1997) Growth-promoting interaction of IGF-II with the insulin receptor during mouse embryonic development. Dev Biol 189:33–48

MacKay EM, Calloway JW, Barnes RH (1940) Hyperalimentation in normal animals produced by protamine insulin. J Nutr 20:59–66

Maddux BA, Goldfine ID (2000) Membrane glycoprotein PC-1 inhibition of insulin receptor function occurs via direct interaction with the receptor alpha-subunit. Diabetes 49:13–19

Marks JL, Eastman CJ (1990) Ontogeny of insulin binding in different regions of the rat brain. Dev Neurosci 12:349–358

Marks JL, Porte D Jr, Stahl WL, Baskin DG (1990) Localization of insulin receptor mRNA in rat brain by in situ hybridization. Endocrinology 127:3236

Matson CA, Wiater MF, Kuijper JL, Weigle DS (1997) Synergy between leptin and cholecystokinin (CCK) to control daily caloric intake. Peptides 18:1275–1278

Matson CA, Reid DF, Cannon TA, Ritter RC (2000) Cholecystokinin and leptin act synergistically to reduce body weight. Am J Physiol 278:R882–R890

Matsuoka N, Patki A, Tiwari HK, Allison DB, Johnson SB, Gregersen PK, Leibel RL, Chung WK (2006) Association of K121Q polymorphism in ENPP1 (PC-1) with BMI in Caucasian and African-American adults. Int J Obes (Lond) 30:233–237

Matsuzawa Y, Shimomura I, Nakamura T, Keno Y, Tokunaga K (1994) Pathophysiology and pathogenesis of visceral fat obesity. Diabetes Res Clin Pract 24(Suppl):S111–S116

Matsuzawa Y, Shimomura I, Nakamura T, Keno Y, Kotani K, Tokunaga K (1995) Pathophysiology and pathogenesis of visceral fat obesity. Obes Res 3(Suppl 2):187S–194S

McFarlane SI (2009) Insulin therapy and type 2 diabetes: management of weight gain. J Clin Hypertens (Greenwich) 11:601–607

McGowan MK, Andrews KM, Grossman SP (1992) Chronic intrahypothalamic infusions of insulin or insulin antibodies alter body weight and food intake in the rat. Physiol Behav 51:753–766

McNay EC (2007) Insulin and ghrelin: peripheral hormones modulating memory and hippocampal function. Curr Opin Pharmacol 7:628–632

Michael MD, Kulkarni RN, Postic C, Previs SF, Shulman GI, Magnuson MA, Kahn CR (2000) Loss of insulin signaling in hepatocytes leads to severe insulin resistance and progressive hepatic dysfunction. Mol Cell 6:87–97

Montague CT, Prins JB, Sanders L, Digby JE, O'Rahilly S (1997) Depot- and sex-specific differences in human leptin mRNA expression: implications for the control of regional fat distribution. Diabetes 46:342–347

Moran TH, Kinzig KP (2004) Gastrointestinal satiety signals. II. Cholecystokinin. Am J Physiol 286:G183–G188

Moran TH, Ladenheim EE (2011) Adiposity signaling and meal size control. Physiol Behav 103:21–24

Morgan DO, Jarnagin K, Roth RA (1986) Purification and characterization of the receptor for insulin-like growth factor I. Biochemistry 25:5560–5564

Morgan DO, Edman JC, Standring DN, Fried VA, Smith MC, Roth RA, Rutter WJ (1987) Insulin-like growth factor II receptor as a multifunctional binding protein. Nature 329:301–307

Morrione A, Valentinis B, Xu SQ, Yumet G, Louvi A, Efstratiadis A, Baserga R (1997) Insulin-like growth factor II stimulates cell proliferation through the insulin receptor. Proc Natl Acad Sci USA 94:3777–3782

Morton GJ, Schwartz MW (2001) The NPY/AgRP neuron and energy homeostasis. Int J Obes Relat Metab Disord 25:S56–S62

Morton GJ, Cummings DE, Baskin DG, Barsh GS, Schwartz MW (2006) Central nervous system control of food intake and body weight. Nature 443:289–295

Muller C, Voirol MJ, Stefanoni N, Surmely JF, Jequier E, Gaillard RC, Tappy L (1997) Effect of chronic intracerebroventricular infusion of insulin on brown adipose tissue activity in fed and fasted rats. Int J Obes Relat Metab Disord 21:562–566

Musatov S, Chen W, Pfaff DW, Mobbs CV, Yang XJ, Clegg DJ, Kaplitt MG, Ogawa S (2007) Silencing of estrogen receptor alpha in the ventromedial nucleus of hypothalamus leads to metabolic syndrome. Proc Natl Acad Sci USA 104:2501–2506

Nedungadi TP, Clegg DJ (2009) Sexual dimorphism in body fat distribution and risk for cardiovascular diseases. J Cardiovasc Transl Res 2:321–327

Niswender KD, Schwartz MW (2003) Insulin and leptin revisited: adiposity signals with overlapping physiological and intracellular signaling capabilities. Front Neuroendocrinol 24:1–10

Obici S, Feng Z, Karkanias G, Baskin DG, Rossetti L (2002) Decreasing hypothalamic insulin receptors causes hyperphagia and insulin resistance in rats. Nat Neurosci 5:566–572

Petersen KF, Laurent D, Rothman DL, Cline GW, Shulman GI (1998) Mechanism by which glucose and insulin inhibit net hepatic glycogenolysis in humans. J Clin Invest 101:1203–1209

Plata-Salaman CR, Oomura Y (1986) Effect of intra-third ventricular administration of insulin on food intake after food deprivation. Physiol Behav 37:735–739

Plum L, Schubert M, Brüning JC (2005) The role of insulin receptor signaling in the brain. Trends Endocrinol Metab 16:59–65

Pouliot MC, Despres JP, Nadeau A, Moorjani S, Prud'Homme D, Lupien PJ, Tremblay A, Bouchard C (1992) Visceral obesity in men. Associations with glucose tolerance, plasma insulin, and lipoprotein levels. Diabetes 41:826–834

Pozefsky T, Felig P, Tobin JD, Soeldner JS, Cahill GF Jr (1969) Amino acid balance across tissues of the forearm in postabsorptive man. Effects of insulin at two dose levels. J Clin Invest 48:2273–2282

Prior RL, Smith SB (1982) Hormonal effects on partitioning of nutrients for tissue growth: role of insulin. Fed Proc 41:2545–2549

Rahmouni K, Morgan DA, Morgan GM, Liu X, Sigmund CD, Mark AL, Haynes WG (2004) Hypothalamic PI3K and MAPK differentially mediate regional sympathetic activation to insulin. J Clin Invest 114:652–658

Rees-Jones RW, Hendricks SA, Quarum M, Roth J (1984) The insulin receptor of rat brain is coupled to tyrosine kinase activity. J Biol Chem 249:3470–3474
Riedy CA, Chavez M, Figlewicz DP, Woods SC (1995) Central insulin enhances sensitivity to cholecystokinin. Physiol Behav 58:755–760
Ritter RC, Slusser PG, Stone S (1981) Glucoreceptors controlling feeding and blood glucose: location in the hindbrain. Science 213:451–452
Rodriguez-Escudero I, Roelants FM, Thorner J, Nombela C, Molina M, Cid VJ (2005) Reconstitution of the mammalian PI3K/PTEN/Akt pathway in yeast. Biochem J 390:613–623
Rosenbloom AL, Giordano BP (1977) Chronic overtreatment with insulin in children and adolescents. Am J Dis Child 131:881–885
Rossetti L, Giaccari A (1990) Relative contribution of glycogen synthesis and glycolysis to insulin-mediated glucose uptake. A dose-response euglycemic clamp study in normal and diabetic rats. J Clin Invest 85:1785–1792
Ryan M, Livingstone MB, Ducluzeau PH, Salle A, Genaitay M, Ritz P (2008) Is a failure to recognize an increase in food intake a key to understanding insulin-induced weight gain? Diabetes Care 31:448–450
Sanger F, Tuppy H (1951a) The amino-acid sequence in the phenylalanyl chain of insulin. I. The identification of lower peptides from partial hydrolysates. Biochem J 49:463–481
Sanger F, Tuppy H (1951b) The amino-acid sequence in the phenylalanyl chain of insulin. 2. The investigation of peptides from enzymic hydrolysates. Biochem J 49:481–490
Sato I, Arima H, Ozaki N, Watanabe M, Goto M, Hayashi M, Banno R, Nagasaki H, Oiso Y (2005) Insulin inhibits neuropeptide Y gene expression in the arcuate nucleus through GABAergic systems. J Neurosci 25:8657–8664
Schubert M, Gautam D, Surjo D, Ueki K, Baudler S, Schubert D, Kondo T, Alber J, Galldiks N, Kustermann E, Arndt S, Jacobs AH, Krone W, Kahn CR, Brüning JC (2004) Role for neuronal insulin resistance in neurodegenerative diseases. Proc Natl Acad Sci USA 101:3100–3105
Schwartz MW, Figlewicz DP, Baskin DG, Woods SC, Porte D Jr (1992) Insulin in the brain: a hormonal regulator of energy balance. Endocr Rev 13:387–414
Schwartz MW, Woods SC, Porte D Jr, Seeley RJ, Baskin DG (2000) Central nervous system control of food intake. Nature 404:661–671
Seeley RJ, Woods SC (2003) Monitoring of stored and available fuel by the CNS: implications for obesity. Nat Rev Neurosci 4:901–909
Seeley RJ, Yagaloff KA, Fisher SL, Burn P, Thiele TE, van Dijk G, Baskin DG, Schwartz MW (1997) Melanocortin receptors in leptin effects. Nature 390:349
Shier P, Watt VM (1989) Primary structure of a putative receptor for a ligand of the insulin family. J Biol Chem 264:14605–14608
Shiraishi J, Yanagita K, Fukumori R, Sugino T, Fujita M, Kawakami S, McMurtry JP, Bungo T (2011) Comparisons of insulin related parameters in commercial-type chicks: evidence for insulin resistance in broiler chicks. Physiol Behav 103:233–239
Standaert ML, Bandyopadhyay G, Kanoh Y, Sajan MP, Farese RV (2001) Insulin and PIP3 activate PKC-zeta by mechanisms that are both dependent and independent of phosphorylation of activation loop (T410) and autophosphorylation (T560) sites. Biochemistry 40:249–255
Stephan F, Reville P, Thierry R, Schlienger JL (1972) Correlations between plasma insulin and body weight in obesity, anorexia nervosa and diabetes mellitus. Diabetologia 8:196–201
Stockhorst U, de Fries D, Steingrueber HJ, Scherbaum WA (2004) Insulin and the CNS: effects on food intake, memory, and endocrine parameters and the role of intranasal insulin administration in humans. Physiol Behav 83:47–54
Stockhorst U, Huenig A, Ziegler D, Scherbaum W (2011) Unconditioned and conditioned effects of intravenous insulin and glucose on heart rate variability in healthy men. Physiol Behav 103:31–38
Strubbe JH, Mein CG (1977) Increased feeding in response to bilateral injection of insulin antibodies in the VMH. Physiol Behav 19:309–313

Su X, Lodhi IJ, Saltiel AR, Stahl PD (2006) Insulin-stimulated Interaction between insulin receptor substrate 1 and p85alpha and activation of protein kinase B/Akt require Rab5. J Biol Chem 281:27982–27990

Swann JP (1986) Insulin: a case study in the emergence of collaborative pharmacomedical research. Pharm Hist 28:65–74

Tang-Christensen M, Larsen PJ, Goke R, Fink-Jensen A, Jessop DS, Moller M, Sheikh SP (1996) Central administration of GLP-1-(7-36) amide inhibits food and water intake in rats. Am J Physiol 271:R848–R856

Tschöp M, Smiley DL, Heiman ML (2000) Ghrelin induces adiposity in rodents. Nature 407:908–913

Turton MD, O'Shea D, Gunn I, Beak SA, Edwards CM, Meeran K, Choi SJ, Taylor GM, Heath MM, Lambert PD, Wilding JP, Smith DM, Ghatei MA, Herbert J, Bloom SR (1996) A role for glucagon-like peptide-1 in the central regulation of feeding. Nature 379:69–72

Ullrich A, Gray A, Tam AW, Yang-Feng T, Tsubokawa M, Collins C, Henzel W, Le Bon T, Kathuria S, Chen E et al (1986) Insulin-like growth factor I receptor primary structure: comparison with insulin receptor suggests structural determinants that define functional specificity. EMBO J 5:2503–2512

van der Heide LP, Ramakers GM, Smidt MP (2006) Insulin signaling in the central nervous system: learning to survive. Prog Neurobiol 79:205–221

van Houten M, Posner BI (1979) Insulin binds to brain blood vessels in vivo. Nature 282:623–625

van Houten M, Posner BI, Kopriwa BM, Brawer JR (1979) Insulin-binding sites in the rat brain: in vivo localization to the circumventricular organs by quantitative radioautography. Endocrinology 105:666–673

van Houten M, Posner BI, Kopriwa BM, Brawer JR (1980) Insulin binding sites localized to nerve terminals in rat median eminence and arcuate nucleus. Science 207:1081–1083

Volvin P, Chambaut AM, Eboue Bonis D, Clauser H, Brinkhoff O, Bremer H, Meienhofer J, Zahn H (1964) Biological activity of natural and synthetic Insulin A-chain preparations on the isolated rat diaphragm. Nature 203:408–409

von Mering JV, Minkowski O (1889) Diabetes Mellitus nach Pankreasextirpation. Zentralblatt Klin Med 10:393–394

Wang P, Mariman E, Renes J, Keijer J (2008) The secretory function of adipocytes in the physiology of white adipose tissue. J Cell Physiol 216:3–13

Wang R, Zhou D, Xi B, Ge X, Zhu P, Wang B, Zhou M, Huang Y, Liu J, Yu Y, Wang C (2011) ENPP1/PC-1 gene K121Q polymorphism is associated with obesity in European adult populations: evidence from a meta-analysis involving 24,324 subjects. Biomed Environ Sci 24:200–206

Woods SC (1990) On blood glucose and eating. Intl J Obesity 14:33–34

Woods SC (1991) The eating paradox: how we tolerate food. Psychol Rev 98:488–505

Woods SC (2004) Gastrointestinal satiety signals I. An overview of gastrointestinal signals that influence food intake. Am J Physiol 286:G7–G13

Woods SC (2009) The control of food intake: behavioral versus molecular perspectives. Cell Metab 9:489–498

Woods SC, D'Alessio DA (2008) Central control of body weight and appetite. J Clin Endocrinol Metab 93:S37–S50

Woods SC, Porte D Jr (1976) Insulin and the set-point regulation of body weight. In: Hunger: basic mechanisms and clinical implications. Raven, New York, pp 273–280

Woods SC, Seeley RJ (2001) Insulin as an adiposity signal. Int J Obes Relat Metab Disord 25: S35–S38

Woods SC, Lotter EC, McKay LD, Porte D Jr (1979) Chronic intracerebroventricular infusion of insulin reduces food intake and body weight of baboons. Nature 282:503–505

Woods SC, Seeley RJ, Porte D Jr, Schwartz MW (1998) Signals that regulate food intake and energy homeostasis. Science 280:1378–1383

Woods SC, Schwartz MW, Baskin DG, Seeley RJ (2000) Food intake and the regulation of body weight. Annu Rev Psychol 51:255–277

Wren AM, Small CJ, Ward HL, Murphy KG, Dakin CL, Taheri S, Kennedy AR, Roberts GH, Morgan DG, Ghatei MA, Bloom SR (2000) The novel hypothalamic peptide ghrelin stimulates food intake and growth hormone secretion. Endocrinology 141:4325–4328

Yki-Jarvinen H, Helve E, Sane T, Nurjhan N, Taskinen MR (1989) Insulin inhibition of overnight glucose production and gluconeogenesis from lactate in NIDDM. Am J Physiol 256: E732–E739

Young WSI, Kuhar MJ, Roth J, Brownstein MJ (1980) Radiohistochemical localization of insulin receptors in the adult and developing rat brain. Neuropeptides 1:15–22

Zib I, Raskin P (2006) Novel insulin analogues and its mitogenic potential. Diabetes Obes Metab 8:611–620

Peripheral Signals Modifying Food Reward

John R.W. Menzies, Karolina P. Skibicka, Emil Egecioglu, Gareth Leng, and Suzanne L. Dickson

Contents

Abstract The pleasure derived from eating may feel like a simple emotion, but the decision to eat, and perhaps more importantly what to eat, involves central pathways linking energy homeostasis and reward and their regulation by metabolic

J.R.W. Menzies (✉) • G. Leng
Centre for Integrative Physiology, School of Biomedical Sciences, University of Edinburgh, Scotland, UK
e-mail: john.menzies@ed.ac.uk

K.P. Skibicka • E. Egecioglu • S.L. Dickson
Department of Physiology/Endocrinology, Institute of Neuroscience and Physiology
The Sahlgrenska Academy at the University of Gothenburg, Gothenburg, Sweden
e-mail: suzanne.dickson@neuro.gu.se

H.-G. Joost (ed.), *Appetite Control*, Handbook of Experimental Pharmacology 209,
DOI 10.1007/978-3-642-24716-3_6, © Springer-Verlag Berlin Heidelberg 2012

and endocrine factors. Evidence is mounting that modulation of the hedonic aspects of energy balance is under the control of peripheral neuropeptides conventionally associated with homeostatic appetite control. Here, we describe the significance of reward in feeding, the neural substrates underlying the reward pathway and their modification by peptides released into the circulation from peripheral tissues.

Keywords Dopamine • Food • Ghrelin • Insulin • Leptin • Oxytocin • Peptide YY (3-36) • Reward

Abbreviations

α-MSH	α-Melanocyte-stimulating hormone
AgRP	Agouti-related peptide
Amy	Amygdala
ARC	Arcuate nucleus
CCK	Cholecystokinin
DR	Dorsal raphe
GHS-R1A	Growth hormone secretagogue receptor 1A
GIP	Gastric inhibitory polypeptide
GLP-1	Glucagon-like peptide 1
JAK-STAT	Janus kinase-signal transducer and activator of transcription
LDTg	Laterodorsal tegmental area
LH	Lateral hypothalamus
NAcc	Nucleus accumbens
NPY	Neuropeptide Y
NTS	Nucleus tractus solitarius
ObRb	Leptin (ob) receptor b
OLETF	Otsuka Long-Evans Tokushima fatty
OXT	Oxytocin
PBn	Parabrachial nucleus
PFC	Prefrontal cortex
PVN	Paraventricular nucleus
PYY	Peptide YY
SON	Supraoptic nucleus
V1aR	Vasopressin receptor 1a
VPall	Ventral pallidum
VTA	Ventral tegmental area

1 Introduction

We eat not only to obtain energy and nutrients but also for pleasure. Over evolutionary time, hedonic neural systems have developed to promote the intake of energy-dense food. The evolutionary advantage of feeding as a rewarding behaviour is self-evident; it promotes the accumulation of energy stores during unpredictable periods of increased food availability. But such is the influence of the reward pathway, homeostatic systems that regulate food intake and energy expenditure can be inundated by environmental cues related to the rewarding properties of food. With few internal forces opposing the desire to eat, obesity may be an inevitable consequence of reward-driven feeding, at least for those of us in the developed world where food is abundant and inexpensive. The incidence of obesity is high in Europe and in the USA, especially in children (Kipping et al. 2008), and for many individuals, this must reflect a strong drive to eat even when homeostatic mechanisms indicate a positive energy balance.

2 Food Reward

2.1 Feeding as an Addictive Behaviour

Food intake is rewarding and reinforcing. The pleasure associated with feeding influences future behaviour, making an organism more likely to engage in the consumption of pleasurable food again. Repeated consumption can lead to a learned association between stimuli associated with food, such as its taste and smell, and a predicted reward (Berridge 2009). Repeated palatable food intake results in an increase in self-administration, in other words, hyperphagia and consequent obesity. Obese individuals show a stronger preference for palatable food than lean individuals (Blundell et al. 2005), presumably due either to a strengthened experience of reward or an effort to achieve a similar level of reward after a blunting of the normal reward response. In support of the latter, in obese humans, the activation of brain regions associated with reward is attenuated during feeding (Stice et al. 2009).

Behaviours relating to overeating are reminiscent of behavioural responses to commonly abused psychoactive substances. Alcohol and centrally acting drugs like opiates can rapidly and powerfully influence behaviour. The reward associated with their use often leads to compulsive and repetitive reuse, even after detrimental social or health outcomes. It is becoming clear that the reward associated with "natural" behaviours like eating involves the same brain regions that mediate reward from drugs of abuse (Larsson and Engel 2004). Perhaps, just as tolerance and dependence in drug use occur, a form of tolerance to and dependence upon consumption of palatable foods may be commonplace in humans exposed to a high-energy diet. A clue to this comes from experiments using rats fed an unrestricted palatable diet (Johnson and Kenny 2010). They quickly developed a reduced sensitivity to electrical self-stimulation of the reward pathway (potentially reflecting tolerance) and continued to consume large quantities of palatable food even when conditioned to expect a painful foot shock during the feeding session (potentially reflecting an increase in motivation and dependence).

2.2 *Reward Is Required for Normal Feeding*

Activation of the reward pathway increases food intake and promotes a preference for food high in calories, sugar, salt and fat. In excess, these preferences are rightly regarded as harmful, so reward-driven behaviours could be fairly considered a major culprit behind the increase in obesity and its accompanying cardiovascular and metabolic disorders. However, reward is necessary for normal eating. Gustatory pathways allow the selection of desired foods by detecting specific components and attributes such as sugars and fats. But there is nothing intrinsically pleasurable about the sweetness of sugar or the mouthfeel of fat in butter or meat. Instead, our brains have evolved to derive pleasure from these food components solely as a consequence of their nutritional content. Mice deficient in the trpm5 gene cannot taste sweetness (de Araujo et al. 2008), yet they preferentially consume sucrose over sucralose (a less calorific sweetener). Sucrose consumption activates the reward pathway in these animals, even in the absence of what one might think of as a pleasurable taste. Thus, there may be functional links not only between the sensory systems that detect the palatability of food and the reward pathway but also between gastrointestinal systems that sense the energy content of foods and the reward pathway.

The major neurotransmitter in the reward pathway is dopamine (see Sect. 3.1). Chemical lesion of dopamine neurones in the reward pathway leads to decreases in reward-related food intake (Papp and Bal 1987), while electrical stimulation of the same neurones can lead to an increase in food consumption (Berridge and Valenstein 1991). Rats fed a high-fat diet show decreased dopamine activity in the reward pathway, even before they become obese (Davis et al. 2008). Transgenic dopamine-deficient mice suckle normally in the neonatal period, but around the time one would expect them to wean, they become aphagic—they stop eating and subsequently die of starvation within a few days (Zhou and Palmiter 1995). These mice can be rescued from starvation by L-dopa treatment or by viral injection of tyrosine hydroxylase into specific brain nuclei involved in the reward pathway (Szczypka et al. 1999). Deletion of D1 and D2 dopamine receptors results in a similar phenotype—death by starvation after 2-3 weeks even in animals nursed by hand (Kobayashi et al. 2004). These data strongly suggest that improper functioning of the reward pathway via a lack of dopamine signalling abolishes the motivation to find and consume food. Observations such as these may also have ramifications for disorders featuring underlying *hypophagia* such as anorexia nervosa (Kaye et al. 2009, but see Scheurink et al. 2010 for an alternative view based on the rewarding properties of starvation-induced physical activity).

Thus, it is clear that the hedonic aspects of food intake are important in obesity but also essential for the minimal food intake required for survival. But given that the experience of reward seems able to overwhelm systems involved in the homeostatic control of appetite to promote the intake of high-energy foods in overweight and obese individuals, the reward pathway should be considered a potential target in the therapeutic modification of food intake.

3 The Reward Pathway

The first evidence for a reward centre in the brain was presented decades ago (Olds and Milner 1954). In rats, electrical self-stimulation of specific brain regions, particularly the septum, was observed to produce behaviours comparable to or exceeding those seen in response to natural rewards such as food. These animals would self-stimulate repeatedly and for long periods with some animals spending over 90% of the experimental period activating the stimulus. Human subjects behave similarly (Heath 1963), though it is not clear whether they derived pleasure from self-stimulation or simply felt an intrinsic compulsion to self-stimulate (Green et al. 2010). Nonetheless, direct electrical stimulation of these regions in rats is a more powerful reinforcer than a "real" reward. Given the choice, rats will select self-stimulation over food, eventually resulting in self-starvation (Routtenberg and Lindy 1965).

Subsequent research on the mechanisms underlying reward has focused on a dopaminergic pathway in the midbrain as the neural substrate of reward (Palmiter 2007). A low level of dopamine in this pathway predicts hyperphagia, and an increase in dopamine causes hypophagia (Goldfield et al. 2007). The mesolimbic pathway may be dysregulated in obesity where the existence of a state of reward deficiency has been proposed.

3.1 *The Neuroanatomy of Reward*

Dopamine neurones arising in part of the substantia nigra termed the ventral tegmental area (VTA) project to the nucleus accumbens (NAcc), ventral pallidum, the prefrontal cortex and amygdala. The NAcc projects to the amygdala and several regions of the prefrontal cortex and also to the brainstem, specifically to the parabrachial nucleus (Berridge and Kringelbach 2008). The prefrontal cortex provides an important inhibitory input to the NAcc (Volkow and Wise, 2005). The VTA-NAcc-prefrontal cortex pathway is generally accepted to be the midbrain pathway underlying the motivational behaviours associated with reward (Fig. 1).

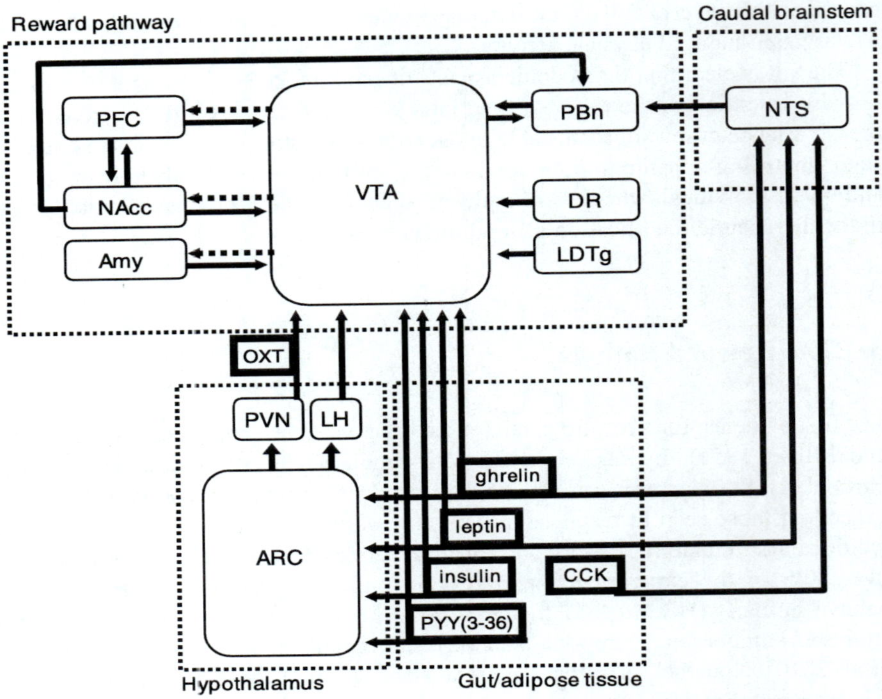

Fig. 1 Schematic representation of the mesolimbic reward pathway and its sensitivity to peripheral peptides. Dopamine neurones originate in the ventral tegmental area (VTA) and project (*dashed line*) to the nucleus accumbens (NAcc), prefrontal cortex (PFC) and amygdala (Amy). The VTA receives input from these areas and also from the dorsal raphe (DR) and laterodorsal tegmental area (LDTg). The NAcc projects to the prefrontal cortex (PFC) and to the parabrachial nucleus (PBn), a brainstem nucleus that receives cholecystokinin (CCK)- and leptin-sensitive input from the nucleus tractus solitarius (NTS). The arcuate nucleus of the hypothalamus (ARC) projects to the paraventricular nucleus (PVN) and lateral hypothalamus (LH). The ARC and VTA express receptors for several circulating metabolic signals including ghrelin, leptin, insulin and peptide YY (3-36) (PYY (3-36)). Only the primary targets of ghrelin (ARC, VTA and NTS) are illustrated here. Fig. 2 shows central ghrelin-sensitive reward-related pathways in greater detail

The VTA receives opioid-sensitive inputs from numerous regions including reciprocal connections from areas to which it projects. Electrophysiological studies indicate that dopaminergic VTA neurones respond to rewarding food intake and also to cues predictive of reward (Schultz 1998). There are at least five distinct output pathways from the VTA, but the most important in reward is the mesolimbic pathway projecting to the NAcc, the amygdala and the limbic cortex (Spanagel and Weiss 1999). The NAcc (subdivided into the core and shell) is a striatal structure containing a predominantly GABAergic neuronal population. The major input to the NAcc is dopaminergic VTA neurones, but the amygdala and hippocampus also provide input. The NAcc expresses D1 and D2 dopamine receptors, and their activation is important in reward (Bernal et al. 2008). Most rewarding behaviours are characterised by an increase in dopamine in the NAcc whether stimulated by food, drugs of abuse or electrical self-stimulation (Hernandez and Hoebel 1988), and dopamine receptor antagonism can prevent the food-seeking behaviour seen as a consequence of reward (Wise et al. 1978). The major output of the NAcc is to the ventral pallidum and thence to the frontal cortices via the thalamus. The shell of the NAcc also sends inhibitory inputs to the lateral hypothalamus that are important in the control of feeding (Maldonado-Irizarry et al. 1995). A further target of the NAcc is the aforementioned parabrachial nucleus, an important centre in taste processing. Gustatory information is relayed to the parabrachial nucleus from the nucleus of the solitary tract en route to the thalamus, lateral hypothalamus and amygdala, indicating a relationship between the taste of some foods and their rewarding properties.

3.2 *The Psychology of Reward*

The neuroanatomy of the reward pathway is fairly well understood, but how does activity in this pathway relate to behaviour? Even when the neural correlates of a certain behaviour are known with some confidence, the processes spanning the gulf between neuronal activation and behaviour are unknown in most cases. It is not at all clear how behaviour emerges from the activation or inhibition of a certain neuronal pathway. However, in the field of motivation and reward, testable ideas are now beginning to be developed on how neuronal activity can be modelled as an information processing system whose output is the implementation of behaviour (Anselme 2010; Zhang et al. 2009). What can be stated more certainly, from physiological, pharmacological and behavioural studies in rodents, is that the mesolimbic pathway is involved in the *motivational* aspects of behaviours propelled by the prospect of reward (described as "incentive salience"), whereas the *pleasurable* aspects of reward are mediated via opioid- and cannabinoid-sensitive brainstem projections to the NAcc and limbic forebrain. The motivational and pleasurable aspects are typically termed "wanting" and "liking", respectively. These terms may seem to overlap (surely, one wants what one likes and likes what one wants?). However, in models of rewarding behaviour, these two phenomena can be measured separately (but see Havermans 2011), are likely to be mediated by different parts of the brain and reflect different components of reward (Berridge 2009).

3.2.1 "Wanting"

It is important to contrast reward-related "wanting" with cognitive wanting. The former ascribes to certain stimuli, a conditioned association with reward (incentive salience). Taking food as an example, cooking smells (or even food packaging) can prompt a desire to consume the food. When a reward stimulus is associated with incentive salience, the stimulus holds more appeal and becomes a stronger target for motivation. The organism will give greater attention to and actively pursue that goal. Cognitive wanting is more intuitively understandable and may be best described as a conscious imagining of and desire to achieve a specific goal or aspiration. Incentive salience "wanting" does not require any cognitive processing such as conscious decision-making. In contrast, it is mediated by structures outwith the cortex, namely the midbrain mesolimbic pathways described in Sect. 3.1. Nor is reward-related "wanting" directly related to pleasure: to want a reward is not simply to like it. "Wanting" can occur without "liking", anyone who has acutely overeaten has probably experienced this. Indeed, it may be true that "wanting" can occur with *disliking*, for example, where an individual's use of a rewarding substance has immediate unpleasant effects.

"Wanting" is relatively easy to study in animal models. Some behaviours produce an agreeable effect, and such behaviours are likely to be repeated even when obstructions are present. Animals are willing to do work to overcome these obstructions and achieve their goal. Quantifying this willingness is a means of measuring the rewarding properties associated with a behaviour. Willingness is usually measured by operant procedures that require the animal to work harder and harder to achieve the same reward (until they give up). These approaches have been used successfully to examine the rewarding properties of food and their modulation by endogenous signals (e.g. Mello and Negus 1998).

Thus, "wanting" reflects awareness of the attractive qualities of a rewarding stimulus, qualities that demand to be attended to and sought out. "Wanting" is not decision-making or enjoyment. In bypassing both cognitive processing and "liking", it is possible that the "wanting" of food may underlie a detrimental drive to eat that is almost entirely independent not only of homeostatic needs but also the associated sense of enjoyment. Selective modification of "wanting" has the potential to reduce the impulse to eat without affecting homeostatic drives or the hedonic satisfaction of eating.

3.2.2 "Liking"

Rewards are desired because they produce pleasurable sensations. "Liking" is the pleasurable reaction to a reward: its hedonic impact. "Liking" can be thought of as a mainly psychological phenomenon (albeit with a neural basis) and as such is more difficult to quantify. The measurement of the hedonic value of food has been validated in adult humans using computer-based rating and forced-choice protocols (Finlayson et al. 2007), but other approaches need to be used in animal models. Human infants execute much of their communication using facial expressions that can be understood by other infants and adult humans. When giving a new food to taste, it is easy for a parent to determine whether the infant gets pleasure from the food by his or her facial expression. These responses are not unique to humans; primates and some rodents can demonstrate changes in facial expressions that correlate with our understanding of how "enjoyable" we would expect the food to be. In rats, for example, sweet tastes evoke positive responses (such as licking), while bitter tastes evoke negative responses (such as gaping and head shaking and rubbing). Careful analysis of these facial expressions provides a means of quantifying "liking" in commonly used animal models (Berridge 2000).

Can "liking" occur without "wanting", and under what circumstances would this be sought after? Dopamine receptor antagonism reduces "wanting". In such experiments, rat will not work for food or other rewards, but hedonic "liking" is unaffected (Berridge 2009). Evidently, this is undesirable, but a balance of hedonic happiness (tempering "wanting" without affecting "liking"?) has been suggested to be a desirable aim for humans (Kringelbach and Berridge 2009).

4 Peripheral Signals Modulating Reward in Feeding

The reward pathway is not only involved in feeding. It is also strongly implicated in sexual behaviour (Balfour et al. 2004), parental care (Champagne et al. 2004) and disorders of mood (Nestler and Carlezon 2006). Since divergent behaviours are under the influence of the mesolimbic reward pathway, a fuller understanding of the effects (if any) specific to food intake is necessary. Thus, attention has turned to neuropeptide signals originating in the gut (and other tissues involved in energy metabolism) as important modulators of rewarding behaviour.

4.1 Peripheral Peptides and Appetite

The most familiar context for a discussion of the effects of peripheral peptides involved in the control of appetite is the homeostatic control system of the hypothalamus rather than the motivational/hedonic system of the mesolimbic pathway. Since the discovery of circulating peptide hormones capable of crossing the blood-brain barrier and acting at specific receptors in the brain, much has been accomplished in the understanding of the circuitry of hypothalamic homeostatic control systems and their modification by peripheral peptide signalling. Work has concentrated on the arcuate nucleus of the hypothalamus (ARC). The ARC is extremely complex, containing neuroendocrine neurones involved in the control of feeding, reproduction, growth and lactation (Smith and Grove 2002). Within this nucleus, there are at least two distinct populations of reciprocally linked neurones, one orexigenic (mediated by neuropeptide Y (NPY)-, Agouti-related protein (AgRP)- and GABA-containing neurones) and one anorexigenic (mediated by products of the POMC gene, predominantly α-melanocyte-stimulating hormone (α-MSH)).

In terms of therapeutic interventions to treat obesity, the homeostatic control systems of the ARC and adjacent hypothalamic nuclei were initially considered to be pertinent targets. We now have a reasonably good understanding of these homeostatic systems in health, obesity and other physiological conditions including pregnancy (Augustine et al. 2008), but translational research has harvested less than originally hoped. It quickly became clear that only very few obese individuals had tractable conditions, typically genetic deficiencies in certain elements such as the melanocortin pathway (Farooqi and O'Rahilly 2008). Furthermore, genetic deletion experiments in rodents showed that homeostatic control systems contain ample redundancy. For example, transgenic mice lacking both NPY and AgRP (the major peptide neurotransmitters released by orexigenic ARC neurones) still express a normal phenotype with respect to food intake (Qian et al. 2002). Another problem is that pathways regulating appetite also influence other physiological systems. For example, the anorexigenic POMC neurones are the major source of α-MSH. As well as being a very potent inhibitor of feeding, α-MSH is a potent stimulator of sexual behaviour (Caquineau et al. 2006). It is possible that eating and sexual behaviour, being mutually exclusive, are regulated reciprocally (Caquineau et al. 2012). Thus, an attempt to intervene pharmacologically at the level of homeostatic control of appetite may have unintended and unwanted consequences for other behaviours. It seems clear that we should broaden our attention from the homeostatic systems that monitor and control energy intake and expenditure to include other systems involved in appetite.

In humans, one of the most successful treatments for morbid obesity is bariatric surgery (Sjöström et al. 2004). Not only do patients consume less energy and lose fat mass, but they become normotensive, their glucose tolerance improves, their cholesterolaemia can be reduced, and they tend not to regain weight in the longer term. Initially, it was believed that weight loss observed following surgery was a simple result of a reduction in the amount of food the patient could eat and an accompanying malabsorptive syndrome. However, it has become clear that surgery results in striking changes in the plasma profile of many gut-brain axis peptides, and the basis for these favourable changes is neural. For example, alterations in the circulating levels of the peripheral orexigenic peptide ghrelin and the anorexigenic peptides peptide YY (PYY), glucagon-like peptide-1 (GLP-1) and gastric inhibitory polypeptide (GIP) have been observed after surgery (Diniz et al. 2010).

Many peripherally released peptides involved in central homeostatic control of appetite also target the reward system. Two key peptides involved in regulating food intake are ghrelin, an orexigenic hormone secreted from the stomach in response to fasting, and leptin, an anorexigenic hormone secreted from adipocytes to serve as a signal of total body fat mass.

4.2 Ghrelin and Reward

The gut-derived peptide ghrelin (Kojima et al. 1999) was first identified as an endogenous agonist of the growth hormone secretagogue receptor (GHS-R1A). After this initial characterisation, ghrelin quickly proved to be important in the control of food intake and is currently the only circulating hormone attributed with a central orexigenic effect. Ghrelin levels increase preprandially in association with feelings of hunger and meal initiation (Cummings et al. 2001, 2004), and studies in rodents have shown orexigenic effects after acute central or peripheral administration (Wren et al. 2000, 2001). Ghrelin increases body fat by a central mechanism that includes decreased energy expenditure and, potentially, also increased food intake, although this is not reproducible in all studies (Tschöp et al. 2000; Dornonville de la Cour et al. 2005; Theander-Carrillo et al. 2006; Salomé et al. 2009). A great deal is known about the target circuits for these orexigenic and adipogenic effects. Ghrelin and synthetic ghrelin mimetics target cells in the ARC (Dickson et al. 1993; Hewson and Dickson 2000), notably the orexigenic NPY/AgRP/GABA neurones in this region (Dickson and Luckman 1997; Kamegai et al. 2001). The ghrelin receptor, GHS-R1A, is expressed in hypothalamic areas involved in appetite regulation, most prominently in the ARC and ventromedial nucleus, but also in the paraventricular nucleus (PVN) and lateral hypothalamus.

Ghrelin not only stimulates homeostatic food intake but may also be involved in the hedonic enjoyment of palatable foods (Fig. 2). In support of involvement in the reward pathway, GHS-R1A is expressed in tegmental and mesolimbic areas involved in reward, such as the VTA and laterodorsal tegmental areas (LDTg; Guan et al. 1997; Zigman et al. 2006). It now seems clear that ghrelin activates a key reward circuit, the "cholinergic-dopaminergic reward link". This circuit has been suggested to mediate the reinforcement of natural rewards and addictive drugs (Yeomans et al. 1993;

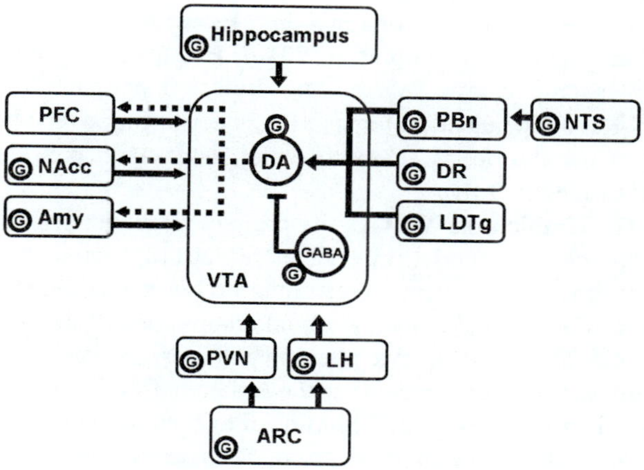

Fig. 2 Schematic representation of the ghrelin receptor-expressing brain regions comprising and converging on the reward pathway. Several brain nuclei involved in the control of homeostatic and hedonic food intake express the ghrelin receptor (GHS-R1A). Mesolimbic dopamine pathways (*dashed line*) are regulated by input from hindbrain structures (including the parabrachial nucleus (PBn) and nucleus tractus solitarius (NTS), the dorsal raphe (DR) and laterodorsal tegmental area (LDTg)), areas involved in learning and memory such as the hippocampus and regions of the hypothalamus involved in homeostatic energy balance and arousal, including the arcuate nucleus (ARC), the paraventricular nucleus (PVN) and the lateral hypothalamus (LH). At the cellular level, GHS-R1A is expressed postsynaptically on dopamine neurones in the VTA as well as on GABAergic interneurones. The central pathways targeted by ghrelin have been reviewed recently (Skibicka and Dickson, 2011)

Larsson and Engel 2004). This link includes a cholinergic afferent projection from the LDTg onto the VTA dopamine cells. Intra-VTA or intra-LDTg administration of ghrelin increases accumbal dopamine release and locomotor activity, effects abolished by nicotinic cholinergic receptor blockade (Jerlhag et al. 2006, 2007). GHS-R1A is co-localised both in dopamine (tyrosine hydroxylase)-containing cells in the VTA (Abizaid et al. 2006) and with cholinergic (choline acetyl transferase)-containing cells in the LDTg (Dickson et al. 2010). These findings, that ghrelin activates the midbrain dopamine system, including circuits implicitly involved in reward, have emerged as having direct relevance for reward from addictive drugs that include alcohol, cocaine and amphetamine (Jerlhag et al. 2009, 2010), as well as from palatable food (Egecioglu et al. 2010; Perello et al. 2010). In other words, the actions of ghrelin may influence responses not only to "natural" rewards like food but also to "unnatural" rewards like drugs of abuse.

Ghrelin signalling at the VTA appears to be important for these feeding effects as intra-VTA injection of ghrelin increases food intake (Naleid et al. 2005) with marked effects on the intake of palatable food (Egecioglu et al. 2010). The cholinergic-dopaminergic reward link is implicated in ghrelin-induced feeding as nicotinic blockade suppresses ghrelin-induced and fasting-induced feeding and also suppresses the ability of food to condition a place preference (Dickson et al. 2010). Consistent with this, blockade of the ghrelin receptor by peripheral treatment with a GHS-R1A antagonist decreases preference for palatable food, suppresses the ability of sweet food to condition a place preference (Egecioglu et al. 2010; Fig. 3) and suppresses motivated behaviour for rewarding foods, both high sugar (Skibicka et al. 2011b) and high fat (Perello et al. 2010). However, it is not the case that ghrelin promotes only calorie intake. Mice injected peripherally with ghrelin exhibit a GHS-R1A-dependent preference for sweet-tasting over bland food even when artificial (low calorie) sweeteners, rather than sugar, provide the hedonic "sweetness" (Disse et al. 2010).

Recently, we identified the VTA as the primary target for ghrelin's effects on motivated behaviour for food (Skibicka et al. 2011a). Interestingly, although ghrelin induces food intake when injected into the NAcc (Naleid et al. 2005), this does not appear to be coupled to increased food motivation (Skibicka et al. 2011b). In humans, ghrelin has been shown to alter the brain response to visual food cues in relevant reward targets areas, including the striatum (Malik et al. 2008), where dopamine signalling has previously been linked to food motivation (Volkow et al. 2002). Elevated plasma ghrelin levels have been associated with binge eating (a behaviour linked to reward dysfunction) in patient groups including those suffering from bulimia nervosa, anorexia nervosa and also Prader-Willi syndrome (a genetic disorder affecting cognitive function that is associated with obesity; Cummings et al. 2002; Tanaka et al. 2002; Nacmias et al. 2004).

Collectively, these data support the idea that an important physiological role of ghrelin is to increase the incentive motivation for natural rewards such as food. Thus, the ghrelin receptor (or, potentially, enzymes involved in the bioactivation of ghrelin such as ghrelin O-acyltransferase (Barnett et al. 2010)) may emerge as therapeutic targets to suppress food reward and motivated behaviour for food, with

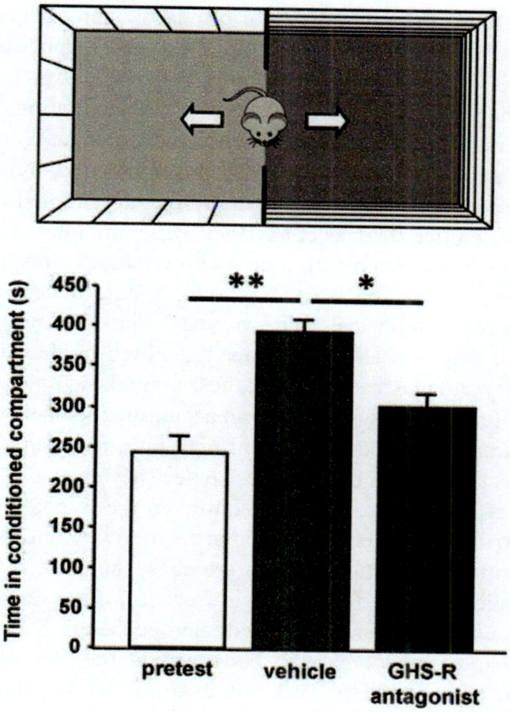

Fig. 3 The ability of rewarding food to condition a place preference is suppressed by peripheral injection of a ghrelin receptor (GHS-R1A) antagonist (JMV2959, 1 mg/kg i.p.). In this test, rats are introduced to a two-compartment chamber with distinct tactile and visual cues. In a pretest, initial preference for one chamber over the other is established. The rats are given access to palatable food in the least-preferred compartment for 20 min per day during an 8-day period of conditioning. On the test day, preference for the palatable food-paired compartment is assessed. *$p < 0.01$, **$p < 0.001$ one way ANOVA followed by Bonferroni. Adapted from Egecioglu et al. (2010), with permission from Wiley-Blackwell

possibilities for treating eating disorders, including those that lead to obesity. It is also important to explore further the extent and mechanism through which obesity may lead to ghrelin resistance (Lindqvist et al. 2005; Briggs et al. 2010; Andrews 2011) and whether the central ghrelin signalling system has a role in the pathophysiology of obesity.

4.3 Leptin and Reward

Leptin is a polypeptide hormone discovered in 1994 (Zhang et al. 1994). It is secreted from adipose tissue and released into the circulation to act via specific leptin receptors at numerous sites including the brain, bone, heart and immune cells

(Ahima and Osei 2004). Circulating leptin provides a direct communication link between adipose tissue and the brain centres involved in appetite control. Leptin inhibits the electrophysiological activity of orexigenic NPY/AgRP/GABA neurones in the ARC in vitro (Spanswick et al. 1997) and iv administration of leptin in fasted mice reduces food intake (Rentsch et al. 1995). Leptin-deficient (*ob/ob*) mice have a high body weight and large fat stores, and are hyperphagic, hyperlipidaemic and hyperglycaemic. Exogenous leptin administration reduces food intake, body weight and adiposity in these mice (Halaas et al. 1995; Pelleymounter et al. 1995). As one might predict, obese humans have high circulating leptin levels and high levels of leptin in CSF (Schwartz et al. 1996). However, a state of leptin resistance may develop in obesity (cf. insulin resistance in type 2 diabetes and ghrelin resistance reported in obesity; Briggs et al. 2010). Mice fed a high-fat diet show the expected increase in adiposity and plasma leptin, but these animals do not reduce their energy intake in response to leptin's signalling of an accumulating energy store (Frederich et al. 1995). Presumably, the coexistence of high plasma leptin but no decrease in energy intake reflects the onset of leptin resistance (Myers et al. 2010).

As well as its effects on homeostatic control in the hypothalamus, leptin also appears to act upon reward systems involved in the motivational and hedonic control of food intake. Leptin-deficient *ob/ob* mice are obese, but if *ob/ob* mice are crossed with dopamine-deficient mice (described in Sect. 2.2), the dopamine/leptin-deficient offspring are not obese—in fact, they do not feed at all (Szczypka et al. 2000). Thus, the hyperphagia and increase in body weight seen in mice lacking leptin require mesolimbic dopamine. However, administration of leptin to the VTA in vivo results in a decrease in food intake, and chronic transgenic knockdown of VTA leptin receptor (ObRb) results in an increase in food intake (Hommel et al. 2006). Furthermore, leptin suppresses dopamine signalling in the mesolimbic system via leptin receptor-expressing LH neurones (a target of NAcc output) projecting to the VTA (Krügel et al. 2003). Selective knockout of the ObRb receptor in the lateral hypothalamus of rats fed a high-fat diet causes an increase in food intake, body weight and adiposity and restores the palatable diet-induced suppression of NAcc dopamine content (Davis et al. 2011). Administration of exogenous leptin attenuates self-stimulation of NAcc dopamine release (Fulton et al. 2004).

The leptin receptor is the product of a single gene, and multiple splice variants have been described. The full-length functional isoform couples to the JAK-STAT signalling pathway (Frühbeck 2006). Dopamine neurones in the VTA express ObRb receptors (Figlewicz et al. 2003), and the JAK-STAT pathway is activated in these neurones by leptin (Fulton et al. 2006). Leptin inhibits VTA neurones in vivo and in brain slices in vitro (Hommel et al. 2006), suggesting a decreased delivery of dopamine to target areas. These findings are compatible with a leptin-driven *attenuation* of the rewarding properties of food.

In the conditioned place preference test, where rats are trained to associate a certain test area with a rewarding stimulus, starved or normally fed rats both express a preference for an area associated with a food reward. This effect is prevented by dopamine antagonism and also by leptin administration (Figlewicz et al. 2001). Similarly, in vivo leptin administration in the VTA reduces food intake and elevates the threshold for reward in a reinforcing self-stimulation paradigm (Bruijnzeel et al. 2011). However, it should be noted that the effect of leptin in the conditioned place preference test disappears in rats switched to a high-fat diet (Figlewicz et al. 2006). This suggests that, in response to a high-energy diet, a form of leptin resistance can develop in the reward pathway that echoes that proposed to occur in the hypothalamus (Matheny et al. 2011).

Exaggerated reward responses are seen in humans suffering from a rare genetic deficiency of leptin. Food-related visual stimuli evoked fMRI responses in the ventral striatum when the subjects were fasted, but, unlike controls, these responses persisted even when subjects were sated (Farooqi et al. 2007). In other words, just as in *ob*/*ob* mice, the absence of leptin led to enduring activation of the reward pathway.

In contrast to the data supporting the idea that leptin *restrains* reward, administration of dopamine agonists to *ob*/*ob* mice has been shown to reverse hyperphagia (Scislowski et al., 1999). Furthermore, despite there being no change in the density of tyrosine hydroxylase-immunopositive neurones in the VTA compared to controls, electrically stimulated dopamine release in the NAcc is markedly reduced in *ob*/*ob* mice (Fulton et al. 2006). Thus, there seems to be a decline in the activity of the reward system in leptin-deficient mice. Reduced dopamine release and reduced dopamine receptor binding in the NAcc in the absence of leptin support the notion that obesity in these animals may be partly based on an insufficiency in the dopamine-driven "wanting" of food; food intake is not found sufficiently rewarding and so hyperphagia occurs.

Nevertheless, most data suggest that leptin acts at the mesolimbic system to restrain responses to food rewards, thereby suppressing the rewarding properties of feeding and contributing in a reduction in the overall drive to eat. Taken with its strong anorexigenic effect in circuits of the hypothalamus, leptin could be considered to signal to distinct brain homeostatic and reward pathways to promote integrated anorexic or feeding behaviours in the appropriate circumstances.

4.4 Insulin and Reward

Insulin is secreted from β-cells of the pancreas in response to an acute increase in plasma glucose. Chronic insulin plasma levels mirror fat mass. Insulin promotes glucose uptake and glucagon storage in skeletal muscle, adipose tissue and the liver and is also involved in the homeostatic control of food intake. It can cross the blood-brain barrier and act at hypothalamic neurones to reduce food intake and body weight (McGowan et al. 1993). In the ARC, both NPY/AgRP/GABA and POMC neurones express insulin receptors (Marks et al. 1990) and are directly sensitive to glucose (via K_{ATP} channels) and insulin (Belgardt et al. 2009). Mice lacking central insulin receptors show increased food intake and mild diet-induced obesity (Brüning et al. 2000). The peripheral insulin resistance diagnostic of late-onset type 2 diabetes may also have a central correlate: unlike controls, obese humans given intranasal insulin do not respond with a decrease in body weight (Hallschmid et al. 2008), and a reduction of ventral striatal activity is seen in insulin-resistant patients (Anthony et al. 2006).

Insulin has a bidirectional effect on dopamine release in the rat striatum. At low concentrations, insulin increases dopamine release but inhibits release at higher concentrations (Potter et al. 1999). As observed with leptin, central administration of insulin can reduce sucrose intake in rats (Figlewicz et al. 2006). Control conditioned place preference behaviour in response to a high-fat diet disappears in fed rats given intraventricular insulin (Figlewicz et al. 2004). Insulin activates immediate-early gene expression in the substantia nigra (Warne et al. 2007) and exerts its functional effect at the VTA via the ARC and in a fashion dependent on opioid receptor activation (Figlewicz et al. 2008). However, exposure to a high-energy diet increases sucrose self-administration and prevents the ability of centrally administered insulin to reduce sucrose intake. The switch to insulin-independent reward-driven sucrose intake can be observed even before the development of overt obesity (Figlewicz et al. 2006).

Thus, insulin (and leptin) normally stifles the reward response to food. This is in line with both polypeptides' anorexigenic effect on homeostatic control of appetite. But if central resistance to insulin and leptin develops in obesity, this may eventually contribute to behavioural changes driven by homeostatic and hedonic systems.

4.5 PYY(3-36) and Reward

Peptide YY (PYY) is secreted from L-cells in the mucosa of the distal gastrointestinal tract. Plasma levels are low during fasting but increase during feeding in proportion to the calories consumed. PYY inhibits gastric, pancreatic and intestinal secretion, gastrointestinal motility and is also involved in regulating glucose homeostasis (Vincent and le Roux 2008). The central effects of PYY are mediated by the truncated form PYY(3-36). PYY(3-36) is anorexigenic and reduces food intake in rats via a mechanism dependent on the activation of inhibitory presynaptic NPY Y2 autoreceptors expressed on orexigenic NPY/AgRP/GABA neurones in the ARC (Batterham et al. 2002). PYY(3-36) reduces food intake in both lean and obese humans (Batterham et al. 2003).

The dynamics of PYY(3-36) release may depend on feeding patterns. Indeed in humans, longer duration meals incorporating breaks from eating result in a lower and slower release of PYY(3-36) compared to the same meal eaten as a "binge" (Lemmens et al. 2011). However, no difference in "liking" or "wanting" (assessed in these subjects by computer-based preference and effort tests) was observed despite different PYY(3-36) profiles.

Human subjects given a PYY(3-36) infusion to mimic satiety showed activation of the parabrachial nucleus, the VTA, limbic regions, the ventral striatum and certain frontal cortical regions as assessed by BOLD imaging (Batterham et al. 2007).

But the most marked change was seen in the orbitofrontal cortex, a region implicated in reward processing and known to be less active in people who have successfully followed a weight-reduction diet (DelParigi et al. 2007). In contrast, the greatest change in subjects given saline was in the homeostatic centre: the hypothalamus. Brain activity in both groups predicted the size of a subsequent meal. Both groups reported the meal to be equally pleasant, but those given PYY(3-36) consumed fewer calories than those given saline. Thus, PYY(3-36) may have two effects: a homeostatic brake on food intake and a simultaneous activation of the hedonic pathway increasing pleasure during the latter stages of feeding. The effect of the latter of these parallel processes may contribute to a sense of satiation.

5 Centrally Released Peptides Modulating Reward in Feeding

In addition to the growing number of peripheral peptides thought to modulate reward, peptides traditionally thought of as peripheral signals have been shown to be synthesised and released in the brain. When released centrally, many of these peptides mediate physiological processes or behaviours that complement their peripheral effects. However, some evoke unanticipated behaviours quite unrelated to their peripheral actions.

5.1 Oxytocin and Vasopressin

Oxytocin and vasopressin are two closely related peptides synthesised in the brain and released peripherally. Classically, they are involved in maternal physiology and the control of water homeostasis. Both neuropeptides can be released centrally from dendrites to act as neurohormonal messengers (Ludwig and Leng 2006). Recently, they have been shown to have an unexpectedly strong link with the control of appetite (Douglas et al. 2007; Leng et al. 2008). Hypothalamic oxytocin and vasopressin gene expression are both markedly downregulated in fasting and upregulated by leptin (Tung et al. 2008). One of the major sources of oxytocin and vasopressin is the PVN. The PVN is a source of several neuropeptides associated with the termination of feeding and is considered an important region in satiety (Shor-Posner et al. 1985). The PVN is innervated by both NPY- and α-MSH-containing fibres from the ARC, both oxytocin neurones and vasopressin neurones express MC4 receptors. Vasopressin (V1aR) receptor knockout mice have altered glucose homeostasis and suffered from diet-induced obesity (Aoyagi et al, 2007). Central administration of oxytocin inhibits feeding, even in fasted rats (Lokrantz et al. 1997), oxytocin receptor antagonists reduce the anorexigenic effect of CCK (Olson et al. 1991), and oxytocin receptor-deficient mice show a preference for sugar (Amico et al. 2005) and display an obese phenotype in adulthood (Takayanagi et al. 2008). Oxytocinergic axons arising in the PVN synapse onto mesolimbic neurones in the caudal part of the VTA (Succu et al. 2008) and exogenous oxytocin can reduce dopamine turnover in the striatum and NAcc (Qi et al. 2008). So, as well as being an important mediator of satiety, oxytocin may also have a specific role in limiting the intake of rewarding palatable food.

The consumption of sugar is universal in animals and is partly driven by its hedonic qualities. Sucrose consumption leads to increased dopamine levels in the NAcc (Hajnal et al. 2004) and an increase in oxytocin gene expression (Olszewski et al. 2009). The anorexigenic oxytocin system is preferentially activated by prolonged (but not short-term) intake of sucrose compared to a less palatable alternative (Mitra et al. 2010). Furthermore, an increase in oxytocin neurone activity is seen to coincide with the termination of sugar consumption (Olszewski et al. 2010). Therefore, oxytocin may be involved in signalling satiety after the desired level of reward during feeding has been achieved. Indeed, satiety could be thought of as the termination of reward. Satiety may involve multiple anorexigenic peptides and depend not only on the total amount of food consumed but on the intake of specific components like sugars and fats (Chaudhri et al. 2008). Regular intake of high levels of sugar reduces oxytocin activity in response to all foods (not just sugar) and may blunt the normal satiety response pushing the organism to obesity.

5.2 *Cholecystokinin*

Cholecystokinin (CCK) is a peptide secreted from I-cells in the mucosa of the duodenum in response to the presence of nutrients in the gut lumen. CCK acts at the stomach, pancreas and gallbladder as a key signal inhibiting gastric emptying and stimulating enzymatic secretion during digestion. CCK-mediated feedback in the gut involves the vagal afferent pathway. CCK also acts centrally to inhibit food intake via an effect on meal size and duration. CCK acts by stimulation of the vagus nerve and activation of neurones in the nucleus tractus solitarius (NTS), a brainstem region which receives vagal input and projects to the hypothalamic PVN and parabrachial nuclei among others (Dockray 2009).

CCK is co-expressed by some dopaminergic neurones in the VTA (Lança et al. 1998). Functional CCK-2 receptors are present in the NAcc, and their activation results in a decreased dopamine release in that region (Voigt et al. 1985). CCK-2 receptor agonists act in the NAcc to inhibit the rewarding effects of drug self-administration (Bush et al. 1999). Peripherally administered CCK suppresses operant lever pressing in a paradigm to test food reward (Hsiao and Deupree 1983; Babcock et al. 1985), and the reduced operant response is equivalent to that obtained with feeding prior to the test. Blocking effects of endogenous CCK with a CCK-2 receptor antagonist potentiates the development of conditioned behaviour related to a food reward (Josselyn and Vaccarino 1995) and increases sucrose intake when injected into the NAcc (Sills and Vaccarino 1996).

Otsuka Long-Evans Tokushima fatty (OLETF) rats have a spontaneous genetic deficiency in CCK-1 receptors. They are hyperphagic, obese and express a diabetic phenotype featuring hyperglycaemia and insulin resistance (Moran and Bi 2006). OLETF rats work harder to achieve reward-driven sucrose consumption than controls (Hajnal et al. 2007). This effect is reduced by D2 receptor antagonism but only in older, obese OLETF rats. These data reinforce the importance of CCK in curbing continued activation of the reward pathway and also point to a dysregulation of the dopamine system more readily observable in obesity.

6 Conclusion

Powerful neurobiological forces determine the frequency and composition of meals, fulfilling not only energy requirements but also hedonic desires. The midbrain VTA dopamine system has a primary role in food motivation and reinforcement, receiving integrated information from the environment through visual food cues and orosensory experiences. This food reward network is regulated by numerous endocrine and metabolic signals, and many circulating peptides have functions in reward that seem to compliment their function in homeostasis. It is important to understand the role of these signals in the aetiology of anorexia, bulimia and obesity with a view to developing novel therapeutic strategies.

Currently, the most successful and widely applied treatment for obesity is dietary restriction in combination with exercise or bariatric surgery, rather than drug therapy. Despite an improving understanding of the signalling molecules and receptors involved in homeostatic and hedonic control of energy balance (and in contrast to the successful pharmacological control of sequalae of obesity such as hypertension and type 2 diabetes), pharmacological approaches to prevent or reverse obesity have had a chequered history (Christensen et al. 2007; Scheen 2010). A case can be made for more refined and diverse strategies involving manipulation of centrally active peripheral signalling (the gut-brain axis). Approaches include tailoring macronutrient content to enhance satiety (Abete et al. 2010), understanding the role of the gut as an endocrine organ (Tolhurst et al. 2009) and the mechanisms underlying changes in gut peptide secretion (Holst et al. 2008), examining the importance of environmental cues to feeding and objective assessment of their impact on hedonic enjoyment (van der Laan et al. 2011; Delzenne et al. 2010), and the effects of the pre- and postnatal environment on nutrition during later life (Bouret and Simerly 2007; Naef et al. 2011).

There is a potential problem in targeting the reward pathway as a treatment for obesity. The successful or repeated performance of many essential behaviours depends on enjoyment of or motivation to perform them. This is pointedly illustrated by the indifference to feeding and eventual starvation seen in transgenic dopamine-deficient mice. If pharmacological blockade (or, more realistically, attenuation) of the reward pathway caused a reduction in enjoyment associated with a behaviour, the enjoyable behaviour would be performed less frequently or not at all. It seems almost tautological to state that there would be no reason to undertake an unenjoyable behaviour, but it is worth emphasising the point because the neural correlates of "wanting" and "liking" are observed in numerous behaviours. Not only in the more obvious like feeding and sexual behaviour, but also in behaviours perhaps less anticipated to activate the central reward pathway, like listening to music (Salimpoor et al. 2011) or physical exercise (Greenwood et al. 2011). Intuitively then, it would seem unprofitable to use drugs to dampen the reward response; too many behaviours would be negatively affected.

We understand relatively little about how food intake is acutely limited (even obese individuals eat less than they *could* in any given day). In other words, the involvement of reward in reaching satiation and meal termination is not at all clear. Nonetheless, perhaps it is more reasonable to aim to design anti-obesity drugs that stimulate slightly the reward response so that the thresholds for satiety and hedonic satisfaction are reached with fewer mouthfuls. One possible danger is that the exaggeration of pleasure or motivation by a reward-stimulating drug would mean that satiation never comes, resulting in enhanced, even compulsive, eating.

Acknowledgements Supported by the EU Seventh Framework programme under grant agreements FP7-KBBE-2009-245009 (NeuroFAST), FP7-HEALTH-2009–241592 (EurOCHIP) and FP7-KBBE-2010-266408 (Full4Health), The Swedish Medical Research Council (K2007-54X-20328–013), ALF Göteborg (SU7601), the Swedish Foundation for Strategic Research to Sahlgrenska Center for Cardiovascular and Metabolic Research (A305-188) and the Swedish Institute.

References

Abete I, Astrup A, Martínez JA, Thorsdottir I, Zulet MA (2010) Obesity and the metabolic syndrome: role of different dietary macronutrient distribution patterns and specific nutritional components on weight loss and maintenance. Nutr Rev 68:214–231

Abizaid A, Liu ZW, Andrews ZB, Shanabrough M, Borok E, Elsworth JD, Roth RH, Sleeman MW, Picciotto MR, Tschöp MH, Gao XB, Horvath TL (2006) Ghrelin modulates the activity and synaptic input organization of midbrain dopamine neurons while promoting appetite. J Clin Invest 116:3229–3239

Ahima RS, Osei SY (2004) Leptin signaling. Physiol Behav 81:223–241

Amico JA, Vollmer RR, Cai HM, Miedlar JA, Rinaman L (2005) Enhanced initial and sustained intake of sucrose solution in mice with an oxytocin gene deletion. Am J Physiol Regul Integr Comp Physiol 289:R1798–1806

Andrews ZB (2011) The extra-hypothalamic actions of ghrelin on neuronal function. Trends Neurosci 34:31–40

Anselme P (2010) The uncertainty processing theory of motivation. Behav Brain Res 208:291–310

Anthony K, Reed LJ, Dunn JT, Bingham E, Hopkins D, Marsden PK, Amiel SA (2006) Attenuation of insulin-evoked responses in brain networks controlling appetite and reward in insulin resistance: the cerebral basis for impaired control of food intake in metabolic syndrome? Diabetes 55:2986–2992

Aoyagi T, Birumachi J, Hiroyama M, Fujiwara Y, Sanbe A, Yamauchi J, Tanoue A (2007) Alteration of glucose homeostasis in V1a vasopressin receptor-deficient mice. Endocrinology 148:2075–2084

Augustine RA, Ladyman SR, Grattan DR (2008) From feeding one to feeding many: hormone-induced changes in bodyweight homeostasis during pregnancy. J Physiol 586:387–97

Babcock AM, Livosky M, Avery DD (1985) Cholecystokinin and bombesin suppress operant responding for food reward. Pharmacol Biochem Behav 22:893–895

Balfour ME, Yu L, Coolen LM (2004) Sexual behavior and sex-associated environmental cues activate the mesolimbic system in male rats. Neuropsychopharmacology 29:718–30

Barnett BP, Hwang Y, Taylor MS, Kirchner H, Pfluger PT, Bernard V, Lin YY, Bowers EM, Mukherjee C, Song WJ, Longo PA, Leahy DJ, Hussain MA, Tschöp MH, Boeke JD, Cole PA (2010) Glucose and weight control in mice with a designed ghrelin O-acyltransferase inhibitor. Science 330:1689–1692

Batterham RL, Cowley MA, Small CJ, Herzog H, Cohen MA, Dakin CL, Wren AM, Brynes AE, Low MJ, Ghatei MA, Cone RD, Bloom SR (2002) Gut hormone PYY(3-36) physiologically inhibits food intake. Nature 418:650–654

Batterham RL, Cohen MA, Ellis SM, Le Roux CW, Withers DJ, Frost GS, Ghatei MA, Bloom SR (2003) Inhibition of food intake in obese subjects by peptide YY3-36. N Engl J Med 349:941–948

Batterham RL, ffytche DH, Rosenthal JM, Zelaya FO, Barker GJ, Withers DJ, Williams SC (2007) PYY modulation of cortical and hypothalamic brain areas predicts feeding behaviour in humans. Nature 450:106–109

Belgardt BF, Okamura T, Brüning JC (2009) Hormone and glucose signalling in POMC and AgRP neurons. J Physiol 587:5305–5014

Bernal SY, Dostova I, Kest A, Abayev Y, Kandova E, Touzani K, Sclafani A, Bodnar RJ (2008) Role of dopamine D1 and D2 receptors in the nucleus accumbens shell on the acquisition and expression of fructose-conditioned flavor–flavor preferences in rats. Behav Brain Res 190:59–66

Berridge KC (2000) Measuring hedonic impact in animals and infants: microstructure of affective taste reactivity patterns. Neurosci Biobehav Rev 24:173–198

Berridge KC (2009) 'Liking' and 'wanting' food rewards: brain substrates and roles in eating disorders. Physiol Behav 97:537–550

Berridge KC, Kringelbach ML (2008) Affective neuroscience of pleasure: reward in humans and animals. Psychopharmacology (Berl) 199:457–480

Berridge KC, Valenstein ES (1991) What psychological process mediates feeding evoked by electrical stimulation of the lateral hypothalamus? Behav Neurosci 105:3–14

Blundell JE, Stubbs RJ, Golding C, Croden F, Alam R, Whybrow S, Le Noury J, Lawton CL (2005) Resistance and susceptibility to weight gain: individual variability in response to a high-fat diet. Physiol Behav 86:614–622

Bouret SG, Simerly RB (2007) Development of leptin-sensitive circuits. J Neuroendocrinol 19:575–582

Briggs DI, Enriori PJ, Lemus MB, Cowley MA, Andrews ZB (2010) Diet-induced obesity causes ghrelin resistance in arcuate NPY/AgRP Neurons. Endocrinology 151:4745–4755

Brüning JC, Gautam D, Burks DJ, Gillette J, Schubert M, Orban PC, Klein R, Krone W, Müller-Wieland D, Kahn CR (2000) Role of brain insulin receptor in control of body weight and reproduction. Science 289:2122–2125

Bruijnzeel AW, Corrie LW, Rogers JA, Yamada H (2011) Effects of insulin and leptin in the ventral tegmental area and arcuate hypothalamic nucleus on food intake and brain reward function in female rats. Behav Brain Res. doi:10.1016/j.bbr.2011.01.020

Bush DE, DeSousa NJ, Vaccarino FJ (1999) Self-administration of intravenous amphetamine: effect of nucleus accumbens CCKB receptor activation on fixed-ratio responding. Psychopharmacology (Berl) 147:331–334

Caquineau C, Leng G, Guan XM, Jiang M, Van der Ploeg L, Douglas AJ (2006) Effects of alpha-melanocyte-stimulating hormone on magnocellular oxytocin neurones and their activation at intromission in male rats. J Neuroendocrinol 18:685–691

Caquineau C, Leng G, Douglas AJ (2012) Sexual behaviour and neuronal activation in the vomeronasal pathway and hypothalamus of food-deprived male rats. J Neuroendocrinol, in press.

Champagne FA, Chretien P, Stevenson CW, Zhang TY, Gratton A, Meaney MJ (2004) Variations in nucleus accumbens dopamine associated with individual differences in maternal behavior in the rat. J Neurosci 24:4113–4123

Chaudhri OB, Field BC, Bloom SR (2008) Gastrointestinal satiety signals. Int J Obes (Lond) 32: S28–31

Christensen R, Kristensen PK, Bartels EM, Bliddal H, Astrup A (2007) Efficacy and safety of the weight-loss drug rimonabant: a meta-analysis of randomised trials. Lancet 370:1706–1713

Cummings DE, Clement K, Purnell JQ, Vaisse C, Foster KE, Frayo RS, Schwartz MW, Basdevant A, Weigle DS (2002) Elevated plasma ghrelin levels in Prader–Willi syndrome. Nat Med 8:643–644

Cummings DE, Frayo RS, Marmonier C, Aubert R, Chapelot D (2004) Plasma ghrelin levels and hunger scores in humans initiating meals voluntarily without time- and food-related cues. Am J Physiol Endocrinol Metab 287:E297–304

Cummings DE, Purnell JQ, Frayo RS, Schmidova K, Wisse BE, Weigle DS (2001) A preprandial rise in plasma ghrelin levels suggests a role in meal initiation in humans. Diabetes 50:1714–1719

Davis JF, Tracy AL, Schurdak JD, Tschöp MH, Lipton JW, Clegg DJ, Benoit SC (2008) Exposure to elevated levels of dietary fat attenuates psychostimulant reward and mesolimbic dopamine turnover in the rat. Behav Neurosci 122:1257–1263

Davis JF, Choi DL, Schurdak JD, Fitzgerald MF, Clegg DJ, Lipton JW, Figlewicz DP, Benoit SC (2011) Leptin regulates energy balance and motivation through action at distinct neural circuits. Biol Psychiatry 69:668–674

de Araujo IE, Oliveira-Maia AJ, Sotnikova TD, Gainetdinov RR, Caron MG, Nicolelis MA, Simon SA (2008) Food reward in the absence of taste receptor signaling. Neuron 57:930–941

DelParigi A, Chen K, Salbe AD, Hill JO, Wing RR, Reiman EM, Tataranni PA (2007) Successful dieters have increased neural activity in cortical areas involved in the control of behavior. Int J Obes (Lond) 31:440–448

Delzenne N, Blundell J, Brouns F, Cunningham K, De Graaf K, Erkner A, Lluch A, Mars M, Peters HP, Westerterp-Plantenga M (2010) Gastrointestinal targets of appetite regulation in humans. Obes Rev 11:234–250

Dickson SL, Hrabovszky E, Hansson C, Jerlhag E, Alvarez-Crespo M, Skibicka KP, Molnar CS, Liposits Z, Engel JA, Egecioglu E (2010) Blockade of central nicotine acetylcholine receptor signaling attenuate ghrelin-induced food intake in rodents. Neuroscience 171:1180–1186

Dickson SL, Leng G, Robinson ICAF (1993) Systemic administration of growth hormone-releasing peptide activates hypothalamic arcuate neurons. Neuroscience 53:303–306

Dickson SL, Luckman SM (1997) Induction of c-fos messenger ribonucleic acid in neuropeptide Y and growth hormone (GH)-releasing factor neurons in the rat arcuate nucleus following systemic injection of the GH secretagogue, GH-releasing peptide-6. Endocrinology 138:771–777

Diniz M, Azeredo Passos VM, Diniz MT (2010) Bariatric surgery and the gut-brain communication–the state of the art three years later. Nutrition 26:925–931

Disse E, Bussier AL, Veyrat-Durebex C, Deblon N, Pfluger PT, Tschöp MH, Laville M, Rohner-Jeanrenaud F (2010) Peripheral ghrelin enhances sweet taste food consumption and preference, regardless of its caloric content. Physiol Behav 101:277–281

Dockray GJ (2009) Cholecystokinin and gut-brain signalling. Regul Pept 155:6–10

Dornonville de la Cour C, Lindqvist A, Egecioglu E, Tung YC, Surve V, Ohlsson C, Jansson JO, Erlanson-Albertsson C, Dickson SL, Håkanson R (2005) Ghrelin treatment reverses the reduction in weight gain and body fat in gastrectomised mice. Gut 54:907–913

Douglas AJ, Johnstone LE, Leng G (2007) Neuroendocrine mechanisms of change in food intake during pregnancy: a potential role for brain oxytocin. Physiol Behav 91:352–365

Egecioglu E, Jerlhag E, Salomé N, Skibicka KP, Haage D, Bohlooly-Y M, Andersson D, Bjursell M, Perrissoud D, Engel JA, Dickson SL (2010) Ghrelin increases intake of rewarding food in rodents. Addict Biol 15:304–311
Farooqi IS, Bullmore E, Keogh J, Gillard J, O'Rahilly S, Fletcher PC (2007) Leptin regulates striatal regions and human eating behavior. Science 317:1355
Farooqi IS, O'Rahilly S (2008) Mutations in ligands and receptors of the leptin-melanocortin pathway that lead to obesity. Nat Clin Pract Endocrinol Metab 4:569–577
Figlewicz DP, Higgins MS, Ng-Evans SB, Havel PJ (2001) Leptin reverses sucrose-conditioned place preference in food-restricted rats. Physiol Behav 73:229–234
Figlewicz DP, Evans SB, Murphy J, Hoen M, Baskin DG (2003) Expression of receptors for insulin and leptin in the ventral tegmental area/substantia nigra (VTA/SN) of the rat. Brain Res 964:107–115
Figlewicz DP, Bennett J, Evans SB, Kaiyala K, Sipols AJ, Benoit SC (2004) Intraventricular insulin and leptin reverse place preference conditioned with high-fat diet in rats. Behav Neurosci 118:479–487
Figlewicz DP, Bennett JL, Naleid AM, Davis C, Grimm JW (2006) Intraventricular insulin and leptin decrease sucrose self-administration in rats. Physiol Behav 89:611–616
Figlewicz DP, Bennett JL, Aliakbari S, Zavosh A, Sipols AJ (2008) Insulin acts at different CNS sites to decrease acute sucrose intake and sucrose self-administration in rats. Am J Physiol Regul Integr Comp Physiol 295:R388–394
Finlayson G, King N, Blundell JE (2007) Is it possible to dissociate 'liking' and 'wanting' for foods in humans? A novel experimental procedure. Physiol Behav 90:36–42
Frederich RC, Hamann A, Anderson S, Löllmann B, Lowell BB, Flier JS (1995) Leptin levels reflect body lipid content in mice: evidence for diet-induced resistance to leptin action. Nat Med 1:1311–1314
Frühbeck G (2006) Intracellular signalling pathways activated by leptin. Biochem J 393:7–20
Fulton S, Richard D, Woodside B, Shizgal P (2004) Food restriction and leptin impact brain reward circuitry in lean and obese Zucker rats. Behav Brain Res 155:319–329
Fulton S, Pissios P, Manchon RP, Stiles L, Frank L, Pothos EN, Maratos-Flier E, Flier JS (2006) Leptin regulation of the mesoaccumbens dopamine pathway. Neuron 51:811–822
Goldfield GS, Lorello C, Doucet E (2007) Methylphenidate reduces energy intake and dietary fat intake in adults: a mechanism of reduced reinforcing value of food? Am J Clin Nutr 86:308–315
Green AL, Pereira EA, Aziz TZ (2010) Deep brain stimulation and pleasure. In: Kringelbach ML, Berridge KC (eds) Pleasures of the brain. Oxford University Press, New York
Greenwood BN, Foley TE, Le TV, Strong PV, Loughridge AB, Day HE, Fleshner M (2011) Long-term voluntary wheel running is rewarding and produces plasticity in the mesolimbic reward pathway. Behav Brain Res 217:354–362
Guan XM, Yu H, Palyha OC, McKee KK, Feighner SD, Sirinathsinghji DJ, Smith RG, Van der Ploeg LH, Howard AD (1997) Distribution of mRNA encoding the growth hormone secretagogue receptor in brain and peripheral tissues. Brain Res Mol Brain Res 48:23–29
Hajnal A, Smith GP, Norgren R (2004) Oral sucrose stimulation increases accumbens dopamine in the rat. Am J Physiol Regul Integr Comp Physiol 286:R31–7
Hajnal A, Acharya NK, Grigson PS, Covasa M, Twining RC (2007) Obese OLETF rats exhibit increased operant performance for palatable sucrose solutions and differential sensitivity to D2 receptor antagonism. Am J Physiol Regul Integr Comp Physiol 293:R1846–1854
Halaas JL, Gajiwala KS, Maffei M, Cohen SL, Chait BT, Rabinowitz D, Lallone RL, Burley SK, Friedman JM (1995) Weight-reducing effects of the plasma protein encoded by the obese gene. Science 269:543–546
Hallschmid M, Benedict C, Schultes B, Born J, Kern W (2008) Obese men respond to cognitive but not to catabolic brain insulin signaling. Int J Obes (Lond) 32:275–282
Havermans RC (2011) You say it's liking, I say it's wanting... On the difficulty of disentangling food reward in man. Appetite 57:286–294

Heath RG (1963) Electrical self-stimulation of the brain in man. Am J Psychiatry 120:571–577
Hernandez L, Hoebel BG (1988) Food reward and cocaine increase extracellular dopamine in the nucleus accumbens as measured by microdialysis. Life Sci 42:1705–1712
Hewson AK, Dickson SL (2000) Systemic administration of ghrelin induces Fos and Egr-1 proteins in the hypothalamic arcuate nucleus of fasted and fed rats. J Neuroendocrinol 12:1047–1049
Holst JJ, Deacon CF, Vilsbøll T, Krarup T, Madsbad S (2008) Glucagon-like peptide-1, glucose homeostasis and diabetes. Trends Mol Med 14:161–168
Hommel JD, Trinko R, Sears RM, Georgescu D, Liu ZW, Gao XB, Thurmon JJ, Marinelli M, DiLeone RJ (2006) Leptin receptor signaling in midbrain dopamine neurons regulates feeding. Neuron 51:801–810
Hsiao S, Deupree D (1983) Cholecystokinin and bombesin effects on rewarded and nonrewarded operants. Peptides 4:1–3
Jerlhag E, Egecioglu E, Dickson SL, Andersson M, Svensson L, Engel JA (2006) Ghrelin stimulates locomotor activity and accumbal dopamine-overflow via central cholinergic systems in mice: implications for its involvement in brain reward. Addict Biol 11:45–54
Jerlhag E, Egecioglu E, Dickson SL, Douhan A, Svensson L, Engel JA (2007) Ghrelin administration into tegmental areas stimulates locomotor activity and increases extracellular concentration of dopamine in the nucleus accumbens. Addict Biol 12:6–16
Jerlhag E, Egecioglu E, Dickson SL, Engel JA (2010) Ghrelin receptor antagonism attenuates cocaine- and amphetamine-induced locomotor stimulation, accumbal dopamine release, and conditioned place preference. Psychopharmacology (Berl) 211:415–422
Jerlhag E, Egecioglu E, Landgren S, Salomé N, Heilig M, Moechars D, Datta R, Perrissoud D, Dickson SL, Engel JA (2009) Requirement of central ghrelin signaling for alcohol reward. Proc Natl Acad Sci USA 106:11318–11323
Johnson PM, Kenny PJ (2010) Dopamine D2 receptors in addiction-like reward dysfunction and compulsive eating in obese rats. Nat Neurosci 13:635–641
Josselyn SA, Vaccarino FJ (1995) Interaction of CCKB receptors with amphetamine in responding for conditioned rewards. Peptides 16:959–964
Kamegai J, Tamura H, Shimizu T, Ishii S, Sugihara H, Wakabayashi I (2001) Chronic central infusion of ghrelin increases hypothalamic neuropeptide Y and Agouti-related protein mRNA levels and body weight in rats. Diabetes 50:2438–2443
Kaye WH, Fudge JL, Paulus M (2009) New insights into symptoms and neurocircuit function of anorexia nervosa. Nat Rev Neurosci 10:573–584
Kipping RR, Jago R, Lawlor DA (2008) Obesity in children. Part 1: epidemiology, measurement, risk factors, and screening. BMJ 337:922–927
Kobayashi M, Iaccarino C, Saiardi A, Heidt V, Bozzi Y, Picetti R, Vitale C, Westphal H, Drago J, Borrelli E (2004) Simultaneous absence of dopamine D1 and D2 receptor-mediated signaling is lethal in mice. Proc Natl Acad Sci USA 101:11465–11470
Kojima M, Hosoda H, Date Y, Nakazato M, Matsuo H, Kangawa K (1999) Ghrelin is a growth-hormone-releasing acylated peptide from stomach. Nature 402:656–660
Kringelbach ML, Berridge KC (2009) Towards a functional neuroanatomy of pleasure and happiness. Trends Cogn Sci 13:479–487
Krügel U, Schraft T, Kittner H, Kiess W, Illes P (2003) Basal and feeding-evoked dopamine release in the rat nucleus accumbens is depressed by leptin. Eur J Pharmacol 482:185–187
Lança AJ, De Cabo C, Arifuzzaman AI, Vaccarino FJ (1998) Cholecystokinergic innervation of nucleus accumbens subregions. Peptides 19:859–868
Larsson A, Engel JA (2004) Neurochemical and behavioral studies on ethanol and nicotine interactions. Neurosci Biobehav Rev 27:713–720
Lindqvist A, de la Cour CD, Stegmark A, Håkanson R, Erlanson-Albertsson C (2005) Overeating of palatable food is associated with blunted leptin and ghrelin responses. Regul Pept 130:123–132
Lokrantz CM, Uvnäs-Moberg K, Kaplan JM (1997) Effects of central oxytocin administration on intraoral intake of glucose in deprived and nondeprived rats. Physiol Behav 62:347–352

Lemmens SG, Martens EA, Born JM, Martens MJ, Westerterp-Plantenga MS (2011) Staggered meal consumption facilitates appetite control without affecting postprandial energy intake. J Nutr 141:482–488

Leng G, Onaka T, Caquineau C, Sabatier N, Tobin VA, Takayanagi Y (2008) Oxytocin and appetite. Prog Brain Res 170:137–151

Ludwig M, Leng G (2006) Dendritic peptide release and peptide-dependent behaviours. Nat Rev Neurosci 7:126–136

Maldonado-Irizarry CS, Swanson CJ, Kelley AE (1995) Glutamate receptors in the nucleus accumbens shell control feeding behavior via the lateral hypothalamus. J Neurosci 15:6779–6788

Malik S, McGlone F, Bedrossian D, Dagher A (2008) Ghrelin modulates brain activity in areas that control appetitive behavior. Cell Metab 7:400–409

Marks JL, Porte D Jr, Stahl WL, Baskin DG (1990) Localization of insulin receptor mRNA in rat brain by in situ hybridization. Endocrinology 127:3234–3236

Matheny M, Shapiro A, Tümer N, Scarpace PJ (2011) Region-specific diet-induced and leptin-induced cellular leptin resistance includes the ventral tegmental area in rats. Neuropharmacology 60:480–487

McGowan MK, Andrews KM, Fenner D, Grossman SP (1993) Chronic intrahypothalamic insulin infusion in the rat: behavioral specificity. Physiol Behav 54:1031–1034

Mello NK, Negus SS (1998) Effects of kappa opioid agonists on cocaine- and food-maintained responding by rhesus monkeys. J Pharmacol Exp Ther 286:812–824

Mitra A, Gosnell BA, Schiöth HB, Grace MK, Klockars A, Olszewski PK, Levine AS (2010) Chronic sugar intake dampens feeding-related activity of neurons synthesizing a satiety mediator, oxytocin. Peptides 31:1346–1352

Moran TH, Bi S (2006) Hyperphagia and obesity in OLETF rats lacking CCK-1 receptors. Philos Trans R Soc Lond B Biol Sci 361:1211–1218

Myers MG Jr, Leibel RL, Seeley RJ, Schwartz MW (2010) Obesity and leptin resistance: distinguishing cause from effect. Trends Endocrinol Metab 21:643–651

Nacmias B, Cellini E, Ricca V, Tedde A, Bagnoli S, Di Bernardo M, Sorbi S (2004) Ghrelin gene polymorphisms in patients with anorexia and bulimia nervosa. Eur Psychiatry 19(suppl 1):224

Naef L, Moquin L, Dal Bo G, Giros B, Gratton A, Walker CD (2011) Maternal high-fat intake alters presynaptic regulation of dopamine in the nucleus accumbens and increases motivation for fat rewards in the offspring. Neuroscience 176:225–236

Naleid AM, Grace MK, Cummings DE, Levine AS (2005) Ghrelin induces feeding in the mesolimbic reward pathway between the ventral tegmental area and the nucleus accumbens. Peptides 26:2274–2279

Nestler EJ, Carlezon WA Jr (2006) The mesolimbic dopamine reward circuit in depression. Biol Psychiatry 59:1151–1159

Olds J, Milner P (1954) Positive reinforcement produced by electrical stimulation of septal area and other regions of rat brain. J Comp Physiol Psychol 47:419–427

Olson BR, Drutarosky MD, Stricker EM, Verbalis JG (1991) Brain oxytocin receptor antagonism blunts the effects of anorexigenic treatments in rats: evidence for central oxytocin inhibition of food intake. Endocrinology 129:785–791

Olszewski PK, Shaw TJ, Grace MK, Höglund CE, Fredriksson R, Schiöth HB, Levine AS (2009) Complexity of neural mechanisms underlying overconsumption of sugar in scheduled feeding: involvement of opioids, orexin, oxytocin and NPY. Peptides 30:226–233

Olszewski PK, Klockars A, Olszewska AM, Fredriksson R, Schiöth HB, Levine AS (2010) Molecular, immunohistochemical, and pharmacological evidence of oxytocin's role as inhibitor of carbohydrate but not fat intake. Endocrinology 151:4736–4744

Palmiter RD (2007) Is dopamine a physiologically relevant mediator of feeding behavior? Trends Neurosci 30:375–381

Papp M, Bal A (1987) Separation of the motivational and motor consequences of 6-hydroxydopamine lesions of the mesolimbic or nigrostriatal system in rats. Behav Brain Res 23:221–229

Pelleymounter MA, Cullen MJ, Baker MB, Hecht R, Winters D, Boone T, Collins F (1995) Effects of the obese gene product on body weight regulation in ob/ob mice. Science 269:540–543

Perello M, Sakata I, Birnbaum S, Chuang JC, Osborne-Lawrence S, Rovinsky SA, Woloszyn J, Yanagisawa M, Lutter M, Zigman JM (2010) Ghrelin increases the rewarding value of high-fat diet in an orexin-dependent manner. Biol Psychiatry 67:880–886

Potter GM, Moshirfar A, Castonguay TW (1999) Insulin affects dopamine overflow in the nucleus accumbens and the striatum. Physiol Behav 65:811–816

Qi J, Yang JY, Song M, Li Y, Wang F, Wu CF (2008) Inhibition by oxytocin of methamphetamine-induced hyperactivity related to dopamine turnover in the mesolimbic region in mice. Naunyn Schmiedebergs Arch Pharmacol 376:441–448

Qian S, Chen H, Weingarth D, Trumbauer ME, Novi DE, Guan X, Yu H, Shen Z, Feng Y, Frazier E, Chen A, Camacho RE, Shearman LP, Gopal-Truter S, MacNeil DJ, Van der Ploeg LH, Marsh DJ (2002) Neither agouti-related protein nor neuropeptide Y is critically required for the regulation of energy homeostasis in mice. Mol Cell Biol 22:5027–5035

Rentsch J, Levens N, Chiesi M (1995) Recombinant ob-gene product reduces food intake in fasted mice. Biochem Biophys Res Commun 214:131–136

Routtenberg A, Lindy J (1965) Effects of the availability of rewarding septal and hypothalamic stimulation on bar pressing for food under conditions of deprivation. J Comp Physiol Psychol 60:158–161

Salimpoor VN, Benovoy M, Larcher K, Dagher A, Zatorre RJ (2011) Anatomically distinct dopamine release during anticipation and experience of peak emotion to music. Nat Neurosci 14:257–262

Salomé N, Hansson C, Taube M, Gustafsson-Ericson L, Egecioglu E, Karlsson-Lindahl L, Fehrentz JA, Martinez J, Perrissoud D, Dickson SL (2009) On the central mechanism underlying ghrelin's chronic pro-obesity effects in rats: new insights from studies exploiting a potent ghrelin receptor antagonist. J Neuroendocrinol 21:777–785

Scheen AJ (2010) Cardiovascular risk-benefit profile of sibutramine. Am J Cardiovasc Drugs 10:321–334

Scheurink AJ, Boersma GJ, Nergårdh R, Södersten P (2010) Neurobiology of hyperactivity and reward: agreeable restlessness in anorexia nervosa. Physiol Behav 100:490–495

Schultz W (1998) Predictive reward signal of dopamine neurons. J Neurophysiol 80:1–27

Schwartz MW, Peskind E, Raskind M, Boyko EJ, Porte D Jr (1996) Cerebrospinal fluid leptin levels: relationship to plasma levels and to adiposity in humans. Nat Med 2:589–593

Scislowski PW, Tozzo E, Zhang Y, Phaneuf S, Prevelige R, Cincotta AH (1999) Biochemical mechanisms responsible for the attenuation of diabetic and obese conditions in ob/ob mice treated with dopaminergic agonists. Int J Obes Relat Metab Disord 23:425–431

Shor-Posner G, Azar AP, Insinga S, Leibowitz SF (1985) Deficits in the control of food intake after hypothalamic paraventricular nucleus lesions. Physiol Behav 35:883–890

Sills TL, Vaccarino FJ (1996) Individual differences in the feeding response to CCKB antagonists: role of the nucleus accumbens. Peptides 17:593–599

Sjöström L, Lindroos AK, Peltonen M, Torgerson J, Bouchard C, Carlsson B, Dahlgren S, Larsson B, Narbro K, Sjöström CD, Sullivan M, Wedel H, Swedish Obese Subjects Study Scientific Group (2004) Lifestyle, diabetes, and cardiovascular risk factors 10 years after bariatric surgery. N Engl J Med 351:2683–2693

Skibicka KP, Dickson SL (2011). Ghrelin and food reward: The story of potential underlying substrates. Peptides, in press

Skibicka KP, Hansson C, Alvarez-Crespo M, Friberg A, Dickson SL (2011a) Ghrelin directly targets the mesolimbic pathway to change food motivation in rats. Neuroscience 180:129–137

Skibicka KP, Hansson C, Egecioglu E, Dickson SL (2011b) Role of ghrelin in food reward: impact of ghrelin on sucrose self-administration and mesolimbic dopamine and acetylcholine receptor gene expression. Addict Biol. doi:10.1111/j.1369-1600.2010.00294

Smith MS, Grove KL (2002) Integration of the regulation of reproductive function and energy balance: lactation as a model. Front Neuroendocrinol 23:225–256

Spanagel R, Weiss F (1999) The dopamine hypothesis of reward: past and current status. Trends Neurosci 22:521–527

Spanswick D, Smith MA, Groppi VE, Logan SD, Ashford ML (1997) Leptin inhibits hypothalamic neurons by activation of ATP-sensitive potassium channels. Nature 390:521–525

Stice E, Spoor S, Ng J, Zald DH (2009) Relation of obesity to consummatory and anticipatory food reward. Physiol Behav 97:551–560

Succu S, Sanna F, Cocco C, Melis T, Boi A, Ferri GL, Argiolas A, Melis MR (2008) Oxytocin induces penile erection when injected into the ventral tegmental area of male rats: role of nitric oxide and cyclic GMP. Eur J Neurosci 28:813–821

Szczypka MS, Rainey MA, Kim DS, Alaynick WA, Marck BT, Matsumoto AM, Palmiter RD (1999) Feeding behavior in dopamine-deficient mice. Proc Natl Acad Sci USA 96:12138–12143

Szczypka MS, Rainey MA, Palmiter RD (2000) Dopamine is required for hyperphagia in Lep(ob/ob) mice. Nat Genet 25:102–104

Takayanagi Y, Kasahara Y, Onaka T, Takahashi N, Kawada T, Nishimori K (2008) Oxytocin receptor-deficient mice developed late-onset obesity. Neuroreport 19:951–955

Tanaka M, Naruo T, Muranaga T, Yasuhara D, Shiiya T, Nakazato M, Matsukura S, Nozoe S (2002) Increased fasting plasma ghrelin levels in patients with bulimia nervosa. Eur J Endocrinol 146:R1–3

Theander-Carrillo C, Wiedmer P, Cettour-Rose P, Nogueiras R, Perez-Tilve D, Pfluger P, Castaneda TR, Muzzin P, Schürmann A, Szanto I, Tschöp MH, Rohner-Jeanrenaud F (2006) Ghrelin action in the brain controls adipocyte metabolism. J Clin Invest 116:1983–1993

Tolhurst G, Reimann F, Gribble FM (2009) Nutritional regulation of glucagon-like peptide-1 secretion. J Physiol 587:27–32

Tschöp M, Smiley DL, Heiman ML (2000) Ghrelin induces adiposity in rodents. Nature 407:908–913

Tung YC, Ma M, Piper S, Coll A, O'Rahilly S, Yeo GS (2008) Novel leptin-regulated genes revealed by transcriptional profiling of the hypothalamic paraventricular nucleus. J Neurosci 28:12419–12426

van der Laan LN, de Ridder DT, Viergever MA, Smeets PA (2011) The first taste is always with the eyes: a meta-analysis on the neural correlates of processing visual food cues. Neuroimage 55:296–303

Vincent RP, le Roux CW (2008) The satiety hormone peptide YY as a regulator of appetite. J Clin Pathol 61:548–552

Voigt MM, Wang RY, Westfall TC (1985) The effects of cholecystokinin on the in vivo release of newly synthesized [3 H]dopamine from the nucleus accumbens of the rat. J Neurosci 5:2744–2749

Volkow ND, Wang GJ, Fowler JS, Logan J, Jayne M, Franceschi D, Wong C, Gatley SJ, Gifford AN, Ding YS, Pappas N (2002) "Nonhedonic" food motivation in humans involves dopamine in the dorsal striatum and methylphenidate amplifies this effect. Synapse 44:175–180

Volkow ND, Wise RA (2005) How can drug addiction help us understand obesity? Nat Neurosci 8:555–560

Warne JP, Horneman HF, Ginsberg AB, Pecoraro NC, Foster MT, Akana SF, Dallman MF (2007) Mapping brain c-Fos immunoreactivity after insulin-induced voluntary lard intake: insulin- and lard-associated patterns. J Neuroendocrinol 19:794–808

Wise RA, Spindler J, deWit H, Gerberg GJ (1978) Neuroleptic-induced "anhedonia" in rats: pimozide blocks reward quality of food. Science 201:262–264

Wren AM, Small CJ, Abbott CR, Dhillo WS, Seal LJ, Cohen MA, Batterham RL, Taheri S, Stanley SA, Ghatei MA, Bloom SR (2001) Ghrelin causes hyperphagia and obesity in rats. Diabetes 50:2540–2547

Wren AM, Small CJ, Ward HL, Murphy KG, Dakin CL, Taheri S, Kennedy AR, Roberts GH, Morgan DG, Ghatei MA, Bloom SR (2000) The novel hypothalamic peptide ghrelin stimulates food intake and growth hormone secretion. Endocrinology 141:4325–4328

Yeomans JS, Mathur A, Tampakeras M (1993) Rewarding brain stimulation: role of tegmental cholinergic neurons that activate dopamine neurons. Behav Neurosci 107:1077–1087

Zhang Y, Proenca R, Maffei M, Barone M, Leopold L, Friedman JM (1994) Positional cloning of the mouse obese gene and its human homologue. Nature 372:425–432

Zhang J, Berridge KC, Tindell AJ, Smith KS, Aldridge JW (2009) A neural computational model of incentive salience. PLoS Comput Biol 5:e1000437

Zhou QY, Palmiter RD (1995) Dopamine-deficient mice are severely hypoactive, adipsic, and aphagic. Cell 83:1197–1209

Zigman JM, Jones JE, Lee CE, Saper CB, Elmquist JK (2006) Expression of ghrelin receptor mRNA in the rat and the mouse brain. J Comp Neurol 494:528–548

Part II
Gastrointestinal Hormones and Factors

The Role of Ghrelin in the Control of Energy Balance

Henriette Kirchner, Kristy M. Heppner, and Matthias H. Tschöp

Contents

Abstract Ghrelin is the only potent orexigenic peptide in circulation. It stimulates food intake and leads to positive energy balance, adipogenesis, and body weight gain. However, the physiological significance of ghrelin in the regulation of energy homeostasis is controversial, since loss of ghrelin function in rodents does not necessarily lead to anorexia and weight loss. In this chapter, we discuss the metabolic function of ghrelin and are highlighting recent findings including the discovery and function of ghrelin-acylating enzyme ghrelin *O*-acyltransferase (GOAT). Based on available published data, we conclude that ghrelin is a

H. Kirchner • K.M. Heppner
Department of Internal Medicine and Endocrinology, Metabolic Diseases Institute, Obesity Research Center, University of Cincinnati, 2180 E. Galbraith Rd, Cincinnati, OH 54237, USA

M.H. Tschöp (✉)
Department of Internal Medicine and Endocrinology, Metabolic Diseases Institute, Obesity Research Center, University of Cincinnati, Institute for Diabetes and Obesity, Helmholtz Center Munich, 2180 E. Galbraith Rd, Cincinnati, OH 54237, USA
e-mail: Matthias.tschoep@uc.edu

H.-G. Joost (ed.), *Appetite Control*, Handbook of Experimental Pharmacology 209,
DOI 10.1007/978-3-642-24716-3_7,

principally important endogenous regulator of energy balance, which however may affect both food intake and systemic metabolism via independent mechanisms. Importantly, ghrelin, when acylated by GOAT, might represent a key molecular link between the sensing of consumed calories and the neuroendocrine control of energy homeostasis. Thus, agents antagonizing the action of ghrelin may have therapeutic potential in the therapy of obesity.

Keywords Ghrelin • Ghrelin-*O*-acyltransferase • Ghrelin activation • Stimulation of food intake

Abbreviations

AgRP	Agouti-related protein
AMPK	5′ AMP-activated protein kinase
BW	Body weight
CNS	Central nervous system
CoA	Coenzyme A
Dap	Diaminopropionic acid
ELISA	Enzyme-linked immune sorbent assay
FE	Food efficiency
FI	Food intake
GH	Growth hormone
GHRH	Growth hormone–releasing hormone
GHSR-1a	Growth hormone secretagogue receptor 1a
GI	Gastrointestinal
GOAT	Ghrelin *O*-acyltransferase
HFD	High-fat diet
KO	Knockout
MBOAT4	Membrane-bound *O*-acyltransferase 4
MCFA	Medium-chain fatty acids
MCT	Medium-chain triglycerides
mRNA	Messenger ribonucleic acid
NPY	Neuropeptide Y
POMC	Pro-opiomelanocortin
RYGB	Roux-en-Y gastric bypass
UCP2	Uncoupling protein 2

1 Introduction

Ghrelin is a gastrointestinal peptide hormone that is mainly produced in the stomach. The peptide was discovered in 1999 by the group of Kojima et al. as the endogenous ligand of the growth hormone secretagogue receptor 1a (GHSR-1a), which was later renamed to ghrelin receptor. Ghrelin potently stimulates the release of growth hormone (GH) from primary pituitary cells (Kojima et al. 1999) and acts synergistically with growth hormone-releasing hormone (GHRH) to stimulate GH secretion in humans (Arvat et al. 2001). Kojima et al. created the name "ghrelin" in order to refer to this function. The word ghrelin originates from *ghre*, the Proto-Indo-European root of the word "grow," and *-lin* indicates ghrelin's secretagogue function. Only 1 year after its discovery, Tschöp et al. described that ghrelin induces food intake and increases adiposity in rodents and humans, a major breakthrough for the field of metabolic research (Tschop et al. 2000, 2001a). Intriguingly, ghrelin remains the only so far known circulating peptide that powerfully stimulates food intake and increases fat mass. Therefore, today, the endogenous ghrelin system is less of a target for GH-related therapies, but rather an important basis for development of potential drugs that regulate energy metabolism and body mass.

2 Chemical Structure of Ghrelin and Ghrelin Synthesis

The human ghrelin gene is located on chromosome 3 (3p25–26) and contains 5 exons (Kojima et al. 1999). Only exons 1 and 2 encode the 28 amino acids of the functional ghrelin peptide. The ghrelin gene has two transcriptional starting sites, which results in two distinct ghrelin transcripts, transcript-A and transcript-B. Transcript-A is the main form produced from ghrelin messenger ribonucleic acid (mRNA). The putative promoter region of the human ghrelin gene contains predicted binding sites for several transcription factors including AP2, NF-IL6, NF-B, and half-sites for estrogen- and glucocorticoid-binding elements (Kanamoto et al. 2004; Kishimoto et al. 2003; Tanaka et al. 2001). Translation of ghrelin mRNA leads to formation of a ghrelin precursor molecule, which is not yet the functional 28-amino-acid ghrelin peptide. This ghrelin precursor consists of 117 amino acids, and the amino acid sequence is well conserved among mammals. Rat and mouse ghrelin for example are identical and differ by only two amino acids from the human ghrelin (Arg11 and Val12 in human are replaced by Lys11 and Ala12 in rat/mouse ghrelin). The ghrelin precursor, which is called pre-proghrelin, contains a signal peptide, which is directly followed by the 28-amino-acid ghrelin sequence. Additionally to ghrelin, a 23-amino-acid long peptide called obestatin is encoded by the ghrelin gene (Zhang et al. 2005). Obestatin was believed to have ghrelin-opposing effects resulting in a decrease in food intake (Zhang et al. 2005). However, these findings could not be confirmed and currently no obestatin receptor is known. Therefore, the function and physiological role of obestatin remains mostly unknown.

In the first step of processing pre-proghrelin to ghrelin, the signal sequence is removed to produce proghrelin (Fig. 1). Next, proghrelin is cleaved between arginine and alanine of the C-terminus by the prohormone convertase PC1/3 (Zhu et al. 2006). The cleavage at this proline–arginine recognition site is rather

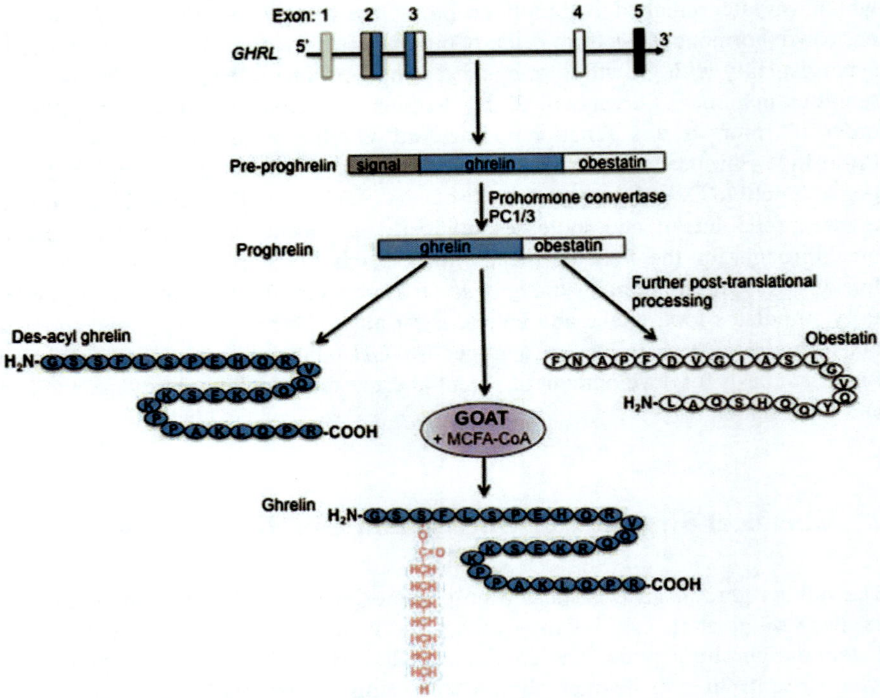

Fig. 1 From the ghrelin gene to acyl ghrelin. The ghrelin gene *GHRL* is located in humans on chromosome 3 and includes 5 exons. Transcription results in synthesis of pre-proghrelin, which is further cleaved by prohormone convertase 1/3 to proghrelin. During more posttranslational processes, proghrelin is cleaved into mature des-acyl ghrelin and obestatin. In order to form acyl ghrelin, proghrelin is acylated at its serin-3 residue with a coenzyme A–activated medium-chain fatty acid (MCFA-CoA) by the ghrelin *O*-acyltransferase (GOAT). This step is assumed to take place in the endoplasmic reticulum

uncommon for propeptide processing; however, it is the same for all mammalian ghrelins (Seidah and Chretien 1999; Steiner 1998). In another posttranslational step, the hydroxyl group of serine-3 is acylated with n-octanoic acid or another medium-chain fatty acid (MCFA), which results in the formation of octanoyl ghrelin that is generally called acyl ghrelin. This ghrelin modification is unique among proteins and necessary to activate the ghrelin receptor (Kojima et al. 1999). However, it is not entirely clear yet at what stage of its processing *n*-octanoylation of ghrelin takes place. In vitro, the presence of both ghrelin *O*-acyltransferase (GOAT) and prohormone convertase PC1/3 are necessary to produce acyl ghrelin. This finding suggests that proghrelin becomes acylated before final cleavage to ghrelin (Takahashi et al. 2009). The final acylated ghrelin peptide is predominantly produced in the stomach by X/A-like cells, which are located within the gastric oxyntic mucosa (Date et al. 2000). Further, a certain amount of ghrelin is produced along the gastrointestinal tract and the pancreas (Date et al. 2000) as well as to a smaller amount in numerous other tissues including the brain (Cowley et al. 2003;

Mondal et al. 2005), testis (Tena-Sempere 2008), pituitary (Korbonits et al. 2001), kidney (Mori et al. 2000), thyroid gland (Raghay et al. 2006), and placenta (Gualillo et al. 2001). Under physiological conditions, ghrelin secretion oscillates in a unique rhythmic pattern with the circadian light–dark cycle (LeSauter et al. 2009). Ghrelin secretion surges immediately before a meal (Liu et al. 2008) and is suppressed by incoming nutrients after food intake (Tschop et al. 2001a).

Generally speaking, serum C-esterases are the predominant enzymes that are responsible for ghrelin des-acylation, whereas peptidases complete the ghrelin-molecule breakdown. However, it should be noted that ghrelin des-acylation and peptide degradation differ highly among species. Therefore, the half-life of acyl ghrelin varies among species as well. For instance, the half-life of acyl ghrelin is about 240 min in human serum, whereas it is only 30 min in rat serum (De Vriese et al. 2004). Further, the mechanism of ghrelin cleavage and the type of enzymes responsible for ghrelin des-acylation and degradation are remarkably different between rodents and humans. In human serum, butyrylcholinesterase is one of the major enzymes responsible for ghrelin des-acylation. In contrast, des-octanoylation of ghrelin in rodents is nine times faster and predominantly catalyzed by carboxylesterases (De Vriese et al. 2004). The ghrelin backbone has five cleavage sites for molecule breakdown, which differ between species and tissue within one species. Most commonly, the cleavage sites are between the serine-2 and glutamic acid-8 residues, which leave relatively long C-terminal ghrelin fragments (De Vriese et al. 2004). These fragments are probably inactive, because the minimal structural requirement to activate the ghrelin receptor is defined by the N-terminal region and requires N-acylation of serine-3 (Matsumoto et al. 2001).

Although octanoylation at serine-3 is essential for bioactivity of ghrelin, i.e., the ghrelin receptor activation, the vast majority (80–90%) of circulating ghrelin is nonacylated. Since no specific receptor for des-acyl ghrelin is known (Kojima et al. 1999), the general consensus appears to be that without this acyl modification, ghrelin does not induce relevant effects on energy balance. However, some controversial studies exist which reported orexin-mediated effects of des-acyl ghrelin on food intake (Toshinai et al. 2006), specific binding sites in cardiomyocytes (Lear et al. 2010), and ghrelin receptor–independent effects on energy and glucose homeostasis (Thompson et al. 2004; Toshinai et al. 2006; Zhang et al. 2008).

3 Ghrelin Activation by the Ghrelin *O*-Acyltransferase

After ghrelin was discovered in 1999, an extensive search for the ghrelin-activating enzyme started. Finally, in 2008, two independent laboratories showed that the membrane-bound *O*-acyltransferase 4 (MBOAT4) is essential to produce octanoyl ghrelin in vitro. Thus, MBOAT4 was renamed to GOAT (Gutierrez et al. 2008; Yang et al. 2008a). As shown by GOAT-knockout mice, GOAT is the only enzyme that acylates ghrelin (Gutierrez et al. 2008; Kirchner et al. 2009).

GOAT is, similar to ghrelin, conserved across vertebrates, and functional GOAT activity has been shown in humans, rats, mice, and zebra fish (Gutierrez et al. 2008;

Yang et al. 2008a). Further, sequences with significant sequence similarity to GOAT exist in other vertebrates, consistent with the conservation of octanoylated ghrelin across vertebrates (Gutierrez et al. 2008). GOAT has eight predicted membrane-bound regions and is presumably located in the membrane of the endoplasmic reticulum (Yang et al. 2008a). In humans, mRNA of GOAT is most abundant, in stomach and pancreas. In mice, GOAT mRNA is mostly present in cells also expressing ghrelin (Sakata et al. 2009). To smaller amounts, GOAT is also expressed in various other tissues including heart, liver, and colon (Gutierrez et al. 2008). Biochemically, GOAT has two critical substrates, proghrelin and short- to medium-chain fatty acids that need to be conjugated with coenzyme A (CoA). In vitro data from cells that express both ghrelin and GOAT show that GOAT can acyl-modify ghrelin at its critical serine-3 residue with fatty acids that range in chain length from acetate (C2) to tetradecanoic acid (C14) (Gutierrez et al. 2008). Nevertheless, most ghrelin in circulation is octanoyl and to a lesser degree decanoyl modified, although medium-chain triglycerides (MCT) are not abundant in circulation. Importantly, octanoyl- and decanoyl-modified ghrelin forms are the optimal ligands for activation of the ghrelin receptor (Yang et al. 2008a). The biochemical mechanisms involved in the generation of these medium-chain fatty acid substrates for GOAT in the ghrelin-producing cells are still unknown.

Cell-culture studies show that GOAT acylates proghrelin before it is cleaved to the final ghrelin peptide (Takahashi et al. 2009; Yang et al. 2008b). Furthermore, des-acyl ghrelin can be acylated in vitro in the presence of microsomes that contain fatty acid CoA esters and GOAT (Ohgusu et al. 2009; Takahashi et al. 2009). These studies conclusively demonstrate that GOAT requires its substrates to be activated as CoA thioesters. Furthermore, these studies contributed to the knowledge that the short amino acid sequence GXSFX, where G, X, S, and F correspond to unblocked amino terminal glycine (G), any amino acid (X), serine (S), and phenylalanine (F), respectively, is sufficient as a recognition site for acylation of GOAT (Ohgusu et al. 2009; Yang et al. 2008b). This recognition sequence appears to be specific for ghrelin and may suggest that ghrelin is GOAT's only peptide substrate. Most recent studies that compare the in vitro selectivity of hexanoyl- and octanoyl-CoA substrates indicate that GOAT might actually prefer hexanoyl-CoA to octanoyl-CoA substrates. These studies highlight again the importance of studying the origin of these fatty acids and their metabolism in the acyl ghrelin-producing cells (Ohgusu et al. 2009).

4 Ghrelin in the Regulation of Energy Homeostasis

With the discovery of the adipokine leptin, an energy balance and body weight regulation model was introduced in which caloric intake, nutrient partitioning, fuel utilization, and energy storage are controlled by neuroendocrine circuits of the hypothalamus and the hindbrain. According to this model, energy homeostasis is maintained by a constant flow of information from the gastrointestinal system and the adipose tissue to specific agouti-related protein/neuropeptide Y (AgRP/NPY) and pro-opiomelanocortin (POMC) neurons in the arcuate nucleus of the hypothalamus, thereby providing feedback about the current energy status and requirements. In return, neuropeptides and neurotransmitters in the CNS are released and respond via neurons of the autonomic nervous system and the endocrine system to the target organs. As a result, energy homeostasis is constantly monitored and dynamically regulated by compensation mechanisms of multiple energy balance components. Ghrelin is one of these components that contribute to the regulation of energy homeostasis of the whole organism. Ghrelin has many known functions including a role in metabolic, behavioral, cardiovascular, reproductive, and immunologic processes [for review see (Kojima and Kangawa 2008; van der Lely et al. 2004)], and yet additional effects of ghrelin are still being discovered. However, ghrelin is mostly associated with its growth hormone-stimulating activity and its orexigenic effect. This chapter will focus on ghrelin's effects on energy homeostasis.

4.1 Ghrelin and Food Intake

Ghrelin plays an important role in the regulation of energy balance since it has proven orexigenic effects that lead to adipogenesis and body weight gain. A number of discoveries established ghrelin's involvement in the regulation of food intake: Central ghrelin administration in rodents induces food intake as potently as NPY (Tschop et al. 2000). Peripheral ghrelin administration reliably induces the sensation of hunger and increases food intake in lean and obese as well as in healthy and malnourished individuals (Wren et al. 2001). Ghrelin increases feeding frequency without affecting the meal size and stimulates gastric motility (Cummings et al. 2007, 2001; Tschop et al. 2000). Intravenous administration of ghrelin in healthy volunteers increases neural activity in response to food pictures in specific brain regions that are associated with reward (Malik et al. 2008). Furthermore, ghrelin secretion is highest before an anticipated meal (Cummings et al. 2001; Tschop et al. 2001a), and incoming nutrients suppress ghrelin levels after a meal (Tschop et al. 2001a). The magnitude of the ghrelin-lowering effect of ingested food is in proportion with the caloric load and macronutrient content, with ingested lipids being the least effective ghrelin suppressors (Foster-Schubert et al. 2008; Monteleone et al. 2003).

Nonetheless, the mechanism by which ghrelin contributes to the regulation of energy homeostasis might be independent of long-term food intake. Pharmacological doses of ghrelin reliably increase food intake in humans and rodents (Tschop et al. 2000; Wren et al. 2001). This ghrelin-induced hyperphagia seems to be most profound when ghrelin is delivered centrally, a very unnatural route of delivery, and all ghrelin-induced effects on food intake lasts only for few hours. There is no evidence that physiological doses of ghrelin modulate energy homeostasis by sustained increase in food intake. Concomitant, mouse models with altered ghrelin, ghrelin receptor, or GOAT function do not show significant changes in food intake despite changed energy balance in comparison to wild-type mice (Kirchner et al.

2009; Pfluger et al. 2008; Sun et al. 2003; Wortley et al. 2005; Zigman et al. 2005). Pharmacological ghrelin antagonists which were developed as antiobesity drugs failed to reduce food intake (Zorrilla et al. 2006), and long-term fasting in humans and rodents, a condition in which hunger-inducing hormones should be highest, does not increase acyl ghrelin levels (Kirchner et al. 2009; Liu et al. 2008). Therefore, it can be concluded that ghrelin only increases short-term food intake. Its significant role in regulating energy homeostasis, however, is likely achieved by mechanisms other than regulating food intake.

4.2 Ghrelin Increases Body Weight and Fat Mass

In addition to its orexigenic effect, ghrelin strongly induces adipogenesis. Chronic ghrelin administration promotes weight gain and adiposity in rodents (Tschop et al. 2000) in the absence of overfeeding (Theander-Carrillo et al. 2006). The body weight gain induced by ghrelin specifically reflects accumulation of fat mass without changes in longitudinal skeletal growth or in lean mass (Tschop et al. 2000). This effect can be explained by direct actions on energy partitioning, which leads to a switch from fat to carbohydrates as major fuel source (Longo et al. 2008; Pfluger et al. 2008). Moreover, central infusion of ghrelin in rodents induces lipogenesis in white adipose tissue through inhibition of the hypothalamic melanocortin system (Nogueiras et al. 2007; Sangiao-Alvarellos et al. 2009). Simultaneous genetic deletion of ghrelin and the ghrelin receptor leads to increased energy expenditure and locomotor activity in mice, suggesting that ghrelin has a dampening effect on energy expenditure and activity level (Pfluger et al. 2008). Thereby, ghrelin defends against energy deficiency beyond its role in mediating meal-to-meal food intake.

In humans, circulating ghrelin levels are inversely associated with obesity, insulin resistance, and weight gain (McLaughlin et al. 2004; Tschop et al. 2001b). Conversely, they are positively correlated with exercise-induced weight loss, low-calorie diet, mixed lifestyle modification, or pathologic conditions such as anorexia nervosa and cachexia (Cummings et al. 2002; Nagaya et al. 2001). The lower levels of ghrelin in obese subjects may reflect a compensatory adaptation mechanism that aims to reduce a hunger stimulus and might suggest that ghrelin itself is not the key cause of obesity. In contrast, patients with Prader–Willi syndrome and hyperphagia have very high circulating ghrelin levels (DelParigi et al. 2002). Rare gene mutations exist that lead to alterations either in ghrelin itself, its precursor pre-proghrelin, or its receptor (Hinney et al. 2002; Korbonits et al. 2002; Pantel et al. 2006; Ukkola et al. 2001, 2002). However, these alterations show only a weak correlation with human obesity and occur much less frequently than mutations in known obesity-related genes such as the melanocortin receptor subtype 4 and its pro-opiomelanocortin ligands. Thus, the significance of these rare ghrelin mutations in the pathogenesis of human obesity is questionable, although this topic needs further investigation.

4.3 Central Pathways of Ghrelin Action

The above described orexigenic and adipogenic actions of ghrelin are predominantly mediated via pathways in the central nervous system that control food intake, energy expenditure, and nutrient partitioning. Ghrelin is the only blood-derived hormone capable of activating hypothalamic NPY and AgRP neurons (Cowley et al. 2003; Dickson et al. 1993), which are currently regarded as very, if not the most, important players in regulating feeding and energy metabolism (Gropp et al. 2005; Luquet et al. 2005). The orexigenic effect of ghrelin is specifically modulated through the ghrelin receptor, which is mainly expressed in the hypothalamus. Peripheral and central ghrelin administration in rodents increases the number of stimulatory synapses on NPY/AgRP neurons (Nakazato et al. 2001), while it increases the number of inhibitory synapses on POMC neurons (Cowley et al. 2003). NPY and AgRP are necessary for ghrelin's actions on feeding behavior, since ghrelin fails to increase food intake in mice lacking both NPY and AgRP (Chen et al. 2004). A crucial regulator for the expression of NPY and AgRP is the hypothalamic homeobox domain transcription factor Bsx (Sakkou et al. 2007), which is also essential for the control of the spontaneous physical activity and the hyperphagic response of ghrelin (Sakkou et al. 2007).

The ghrelin receptor most effectively couples to the $\alpha_{q/11}$ subunit of the G protein, resulting in activation of phospholipase C and release of inositol phosphate and diacylglycerol (Holst et al. 2004). Further, signaling via Gαi and Gαs has been suggested (Carreira et al. 2004; Dezaki et al. 2007). Calcium release is mediated by the inositol phosphates or possibly through the $\beta\gamma$-subunit of the G protein (Holst et al. 2005). The "energy sensor," AMP-activated kinase (AMPK), is the best characterized kinase activated by ghrelin. Ghrelin-induced increase in NPY/AgRP expression in the hypothalamus is associated with activation of AMPK which further increases the expression of uncoupling protein 2 (UCP2) and upregulation of mitochondrial respiration (Andrews et al. 2008). Yet, the exact pathway whereby ghrelin modulates AMPK is not fully understood.

4.4 Ghrelin in Starvation

Due to ghrelin's unique secretion pattern and the fact that it strongly stimulates food intake, it has been suggested that ghrelin is a meal-initiation factor (Cummings et al. 2001) or "the hunger hormone," which is secreted during nutrient deficiency and leads to food intake. However, there are a number of findings that do not seem to fit this theory. It holds true that plasma concentrations of both acyl and des-acyl ghrelin, measured with a sophisticated in-house sandwich ELISA, increase before scheduled meals and decrease during the intermeal intervals in humans (Liu et al. 2008). Nonetheless, these characteristic ghrelin peaks might be rather a learned food-anticipatory response since they are absent in rats fed ad libitum but are inducible by scheduled meal feeding (Drazen et al. 2006). More strikingly, circulating acyl ghrelin is not increased during prolonged starvation in humans or mice (Kirchner et al. 2009; Liu et al. 2008) as it was previously believed. These findings are paralleled by gastric GOAT and ghrelin expression. After long-term fasting, GOAT mRNA levels are significantly downregulated, and ghrelin mRNA remains unchanged (Gonzalez et al. 2008; Kirchner et al. 2009). Only des-acyl ghrelin in circulation increased significantly during the long-term fast (Kirchner et al. 2009; Liu et al. 2008), which might have led to the previous assumption that ghrelin increases with fasting. Inadequate sample preparation and poor ghrelin quantification methods might have further contributed to the false assumption that ghrelin increases during starvation. Another observation that is consistent with low acyl ghrelin levels during starvation is that GOAT activity, and thus ghrelin acylation, is markedly influenced by dietary lipids (Kirchner et al. 2009; Nishi et al. 2005). Supplementation of MCT in the diet significantly increases the amount of circulating octanoyl ghrelin (Kirchner et al. 2009). Moreover, transgenic mice overexpressing both ghrelin and GOAT do not produce acyl ghrelin in abundance (Kirchner et al. 2009): When fed with standard diet, these mice have normal blood concentration of acyl ghrelin compared with wild-type mice. Dietary supplementation of MCT is required to stimulate the ghrelin acylation, which then results in the expected increase in blood acyl ghrelin, leading to increased body weight and fat mass in comparison to wild-type mice (Kirchner et al. 2009). These data clearly show that the regulation of GOAT and its product acyl ghrelin are dependent on the presence of food rather than on the absence of food. Therefore, it is unlikely that the expression of both GOAT and ghrelin are key components for the generation of a hunger signal.

4.5 Ghrelin Deficiency and Altered Ghrelin Function

Although the here described functions of ghrelin clearly show that ghrelin has an effect on food intake, body weight, and adiposity, the importance of ghrelin in regulating energy homeostasis might be controversial. The major criticism is fueled by the finding that *knockout* (KO) mice with loss of ghrelin, ghrelin receptor, or GOAT function have normal food intake, body weight, and fat mass, when fed with standard diet (Kirchner et al. 2009; Sun et al. 2003; Wortley et al. 2004); for review (Barnett et al. 2010). Further, loss of ghrelin or its receptor does not lead to growth retardation (Sun et al. 2003), and ghrelin ablation in leptin-deficient mice does not ameliorate the obesity phenotype (Sun et al. 2006). However, when ghrelin-deficient mice are chronically exposed to high-fat diet (HFD) directly after weaning, a clear metabolic benefit from ghrelin-deficiency emerges: Despite similar food intake, ghrelin-KO mice gain less body weight and body fat than wild-type mice. Moreover, respiratory quotient is decreased in ghrelin-KO mice that are fed with HFD, which is indicative for a change in metabolic fuel preference to favor fat oxidation (Wortley et al. 2005). Similarly, ghrelin receptor–KO mice are protected from diet-induced obesity, possibly due to mild hypophagia (Longo et al. 2008; Zigman et al. 2005). Expression of ghrelin receptor antisense RNA in the arcuate nucleus of rats leads to hypophagia as well as decreased body weight and body fat (Shuto et al. 2002). GOAT-KO mice that are fed with a diet that is rich in the GOAT substrate, medium-chain-triglycerides (MCT), gain less body weight and body fat and have increased energy expenditure compared with wild-type mice (Kirchner et al. 2009). In contrast, ghrelin and GOAT overproducing transgenic mice have significantly increased body weight and body fat when fed with this MCT diet (Kirchner et al. 2009). The most profound phenotype is achieved by simultaneous deletion of ghrelin and its receptor. The ghrelin–ghrelin receptor double-KO mouse shows improved metabolic benefits such as increase energy expenditure and locomotor activity together with decreased body weight, body fat, and cholesterol levels even when fed with a low-fat standard diet (Pfluger et al. 2008). Nonetheless, the observed phenotypes of mice with altered ghrelin, GOAT, and ghrelin receptor function are rather mild. It was therefore concluded that ghrelin belongs to a redundant orexigenic signaling system in which additional food intake-promoting systems exist that may compensate for the loss of GOAT, ghrelin, or its receptor. In fact, such compensation may have occurred during early development of these mice, possibly via enhanced activity of hypothalamic orexigenic neurons (Wortley et al. 2004). However, basal gene expression of orexigenic neuropeptides such as AgRP, POMC, NPY, and melanin-concentrating hormone are not changed in ghrelin-KO mice (Wortley et al. 2004). Taken together, ghrelin, GOAT, or ghrelin receptor deficiency ameliorates the development of obesity after chronic exposure to HFD or MCT diet. Furthermore, the simultaneous deletion of ghrelin and its receptor results in lower body weight and fat mass on standard diet independently of the amount of food intake. These results clearly demonstrated that ghrelin does play an important role in regulating energy balance that cannot be fully substituted by other factors.

4.6 *Ghrelin in the Treatment of Obesity*

4.6.1 Ghrelin and Ghrelin Receptor Antagonists

Due to its function to increase body weight and body fat, ghrelin is one of the most interesting targets for pharmacotherapy. One possibility to counteract ghrelin's effects is to antagonize the ghrelin receptor. Indeed, several ghrelin receptor antagonists have been developed and tested in rodents and humans as potential antiobesity drugs (Table 1). The structure of the early ghrelin receptor antagonists was mostly on a peptide basis, resulting in relatively large molecules. Rodent studies using those molecules showed that the ghrelin receptor antagonists need to cross the blood–brain barrier since for most compounds intracerebroventricular (icv) administration is required to efficiently decrease food intake and body weight (Asakawa et al. 2003; Nakazato et al. 2001; Xin et al. 2006). [D-Lys-3]-GHRP-6,

Table 1 Pharmacological modifications of the ghrelin system

Compound	Function	Effect	Test system	Selected references
BIM-28163	Selective GHSR-1a antagonist	↑ BW	Cell culture, rats	Halem et al. (2004)
Capromorelin	GHSR-1a agonist	↑ GH	Cell culture, mice	Kitazawa et al. (2005)
CP-464709-18	GHSR-1a agonist	↑ FI ↑ Memory ↑ Anxiety	Rats, mice	Atcha et al. (2009), Carpino et al. (2002)
[Dap[3]]octanoyl ghrelin	GOAT inhibitor	↓ Acyl ghrelin	Cell culture	Yang et al. (2008b)
[D-Lys-3]GHRP-6	Nonselective GHSR-1a antagonist	↓ FI ↓ BW	Rats, mice	Asakawa et al. (2003), Rudolph et al. (2007)
Ghr1, Ghr2, and Ghr3	Ghrelin immunoconjugates (vaccination)	↓ BW gain ↓ FM	Rats	Zorrilla et al. (2006)
GHR-11E11	Ghrelin antibody catalyst	↑ BMR ↓ FI after fasting	Mice	Mayorov et al. (2008)
GHRP-6	Synthetic GHS	↑ FI, improved heart function	Cell culture, mice	Kitazawa et al. (2005)
GO-CoA-Tat	GOAT inhibitor	↓ BW gain ↓ FM	Cell culture, mice	Barnett et al. (2010)
GSK894490A	GHSR-1a agonist	↑ BW ↑ Memory	Rats	Atcha et al. (2009)
Hexarelin	Synthetic GHS	↑ FI, improved heart function	Humans	Arvat et al. (2001)
L-NOX-B11	GHSR–RNA spiegelmer	↓ Acute FI ↓ GH	Cell culture, rats	Helmling et al. (2004)
MK-677	GHSR-1a agonist	↑ GH	Cell culture, rats, humans	Holst et al. (2009), Smith et al. (1997)
RC-1139	Ghrelin analogue	↑ Gastric emptying	Rats	Poitras et al. (2005)
RC-1291	Ghrelin mimetic	↑ BW	Humans	Garcia and Polvino (2007)
Triazole derivates	Competitive GHSR-1a antagonists and partial GHSR-1a agonists	↓ FI	Cell culture, rats	Demange et al. (2007), Moulin et al. (2007), Moulin et al. (2008)
TZP-101	GHSR-1a agonist	↑ Gastric emptying	Rats, humans	Ejskjaer et al. (2010), Venkova et al. (2007), Wo et al. (2011)

↑ increase, ↓ decrease, *BMR* basal metabolic rate, *BW* body weight, *FE* food efficiency, *FI* food intake, *GH* growth hormone

also a peptide-based GHSR antagonist and analogue of one of the synthetic growth hormone secretagogues, exhibits some effects on food intake and body weight when administered peripheral in mice (Asakawa et al. 2003). Nevertheless, it is not entirely clear if the decrease in food intake and body weight observed after [D-Lys-3]-GHRP-6 administration is exclusively due to ghrelin receptor blockage given that [D-Lys-3]-GHRP-6 also binds to melanocortin receptors (Schioth et al. 1997), which are known to play a crucial role in the regulation of energy balance. The selective ghrelin receptor antagonist BIM-28163, which is a full-length human ghrelin analogue, decreases the amplitude of GH secretion when administered subcutaneous or centrally but surprisingly fails to significantly decrease food intake in rats (Halem et al. 2005). Later on, some small-molecule ghrelin receptor antagonists, mostly on nonpeptidyl basis (Cheng et al. 1993; Xin et al. 2006), have been designed [for review see (Zhao and Liu 2006) and (Zizzari et al. 2011)]. Similar to the peptide-based GHSR antagonists, most of them lack significant effects on body weight and food intake in vivo, mostly due to poor bioavailability and limited brain permeability (Xin et al. 2006; Zhao and Liu 2006).

Currently, ghrelin antagonists are in preclinical development or early phases of clinical development, and it is presently unclear whether they will work as well in humans as the physiology of the ghrelin hormone suggests. In addition, ghrelin-neutralizing strategies which include vaccination against ghrelin (Zorrilla et al. 2006) and inhibition of ghrelin gene transcription (Helmling et al. 2004) have been developed. However, similar to ghrelin receptor antagonists, these approaches to antagonize ghrelin action had only limited success in the prevention or reversal of obesity (Shearman et al. 2006). It is difficult to interpret the somewhat disappointing results from pharmacological modification of ghrelin and ghrelin receptor modification, especially since many of the compounds have an effect on GH release but not on food intake and body weight. Besides the high constitutive activity of the ghrelin receptor, possible explanations could be that the ghrelin receptor is known to signal through multiple signal transduction pathways (Carreira et al. 2004; Holst et al. 2005). Further, it is postulated that other yet unknown receptor ligands and ghrelin receptors exist (Pfluger et al. 2008).

4.6.2 Ghrelin and Bariatric Surgery

Ghrelin is the only known circulating peptide that promotes adiposity, and it is the only gastrointestinal (GI) peptide that is predominantly produced in the stomach. Thus, the question arose what happens to ghrelin after bariatric surgery where the ghrelin-producing cells in the stomach are either bypassed or even removed. In theory, one would expect that total ghrelin levels drastically fall after some types of bariatric surgery, because the food does not contact the ghrelin-producing cells in the stomach and duodenum anymore. One exemption would be gastric banding where food uptake into the stomach is hindered but where the anatomy of the GI tract is unchanged. In fact, weight loss after gastric banding is not as pronounced as after the invasive surgical procedures such as the Roux-en-Y gastric bypass (RYGB), which led to the theory that changes in ghrelin might be important for the success of those surgeries. One of the first studies about effects of RYGB on ghrelin concentrations reported that ghrelin significantly decreased after the surgery compared to both obese controls and a nonoperated control group that lost weight by dieting (Cummings et al. 2002). The diurnal secretion pattern of ghrelin was completely lost after RYGB, and ghrelin concentrations remained at a minimum throughout a 24-h observation period (Cummings et al. 2002). Further, it could be shown that subjects who were still losing weight after RYGB had higher levels of ghrelin than subjects who had already achieved a steady nadir weight (Faraj et al. 2003). In contrast, a recent study reported that neither fasting plasma concentration of ghrelin nor ghrelin area under the curve during a meal challenge change after RYGB surgery (Korner et al. 2009). Subjects who underwent the less efficient gastric banding respond with a significantly increased ghrelin area under the curve after the meal challenge (Korner et al. 2009). The diversity of existing data, including the two mentioned studies, might be due to inhomogeneous study conditions that vary from fasted, nonfasted, and meal-challenged. Secondly, often only plasma concentration of total ghrelin is reported without differentiating between the active acyl and the des-acyl form. In addition, it has been shown previously that it is difficult to interpret total ghrelin plasma values because it is possible that the ratio between acyl and des-acyl ghrelin changes with no alteration of the total ghrelin concentration (Kirchner et al. 2009; Nishi et al. 2005). Nevertheless, results of studies in which acyl ghrelin is measured seem to be contradictory as well. Reports about increased, unchanged, and decreased acyl ghrelin concentrations after various types of bariatric surgery including gastric bypass, gastric band, biliopancreatic diversion, and sleeve gastrectomy exist (Holdstock et al. 2003; Kotidis et al. 2006; Langer et al. 2005). It should be mentioned at this point that most studies quantifying circulating ghrelin levels with antighrelin antibodies might be critically flawed, because appropriate blood sampling regimen preventing deacylation of ghrelin are often neglected. Moreover, commonly used immunoassays are either not sufficiently specific for acyl ghrelin or have not been thoroughly calibrated by comparison with mass spectrometry or other specific methods. In summary, it is still unclear whether ghrelin levels change after bariatric surgery and if so, in which direction. However, by resecting a large part of the stomach alone and thereby removing a significant portion of ghrelin-producing cells, gastric bypass surgery may lead to a decrease in ghrelin, which might contribute to the weight loss and reduction in fat mass seen after the surgery (Cummings et al. 2002). Only the development of new ghrelin assays and the correct re-collection of plasma samples will allow researchers to reliably answer whether or not ghrelin plays a role in the weight loss seen after bariatric procedures.

4.6.3 Pharmacological GOAT Inhibition

Theoretically, modification of GOAT activity would be an elegant way to alter ghrelin action and energy balance, especially since ghrelin seems to be the only peptide substrate of GOAT. Mouse models with altered GOAT function show that GOAT plays a significant role in the regulation of energy homeostasis (Kirchner et al. 2009). Thus, pharmacologic GOAT inhibition might have some potential to treat obesity. Indeed, GOAT-inhibiting molecules were developed recently (Barnett et al. 2010; Yang et al. 2008b). Yang et al. designed GOAT-inhibiting peptides with the use of a newly developed biochemical assay to study GOAT activity in vitro

(Yang et al. 2008b). First, they could show that GOAT activity is enhanced by the presence of long-chain fatty acid residues coupled to CoA. Secondly, the substrate recognition sequence of GOAT was identified as being glycine-1, serine-3, and phenylalanine-4. These two findings were crucial for the development of potential GOAT agonists and antagonists, since replacement of one of these amino acids leads to inhibitory effects on GOAT. Additionally, Yang and colleagues designed a GOAT inhibitory ghrelin-like pentapeptide that corresponds to the GOAT recognition sequence but is amidated at the C-terminus (Yang et al. 2008b). Interestingly, this amidated ghrelin-like pentapeptide is able to be octanoylated by GOAT, and the inhibitory effects on GOAT activity are likely due to end-product inhibition. The most efficient inhibition of GOAT activity, however, is achieved by the above described amidated ghrelin-like pentapeptide in which serine-3 is replaced by octanoylated (S)-2,3-diaminopropionic acid (Dap). Importantly, the octanoyl residue is linked to Dap through an amide bond and not the ghrelin-typical ester bond (Yang et al. 2008b).

Another GOAT inhibitor named GO-CoA-Tat shows impressive effects on energy and glucose homeostasis in vivo (Barnett et al. 2010). GO-CoA-Tat is a substrate analogue that was designed based on the fact that GOAT generates acyl ghrelin through a ternary complex mechanism by using two substrates, medium-chain fatty acids and proghrelin. The GOAT-inhibiting action of GO-CoA-Tat was achieved by linking the two GOAT substrates with a noncleavable bridge so that the binding energies of the individual ligands are combined without the entropic loss that is normally associated with forming the physiological ternary complex. The GOAT inhibitory effect of GO-CoA-Tat was studied in vitro in cell models that stably express GOAT and pre-proghrelin. GO-CoA-Tat inhibits the production of acyl ghrelin with maximal efficiency 24 h after treatment (Barnett et al. 2010). The observed delay in inhibition of acyl ghrelin production is presumably caused by preformed acyl ghrelin stores within the cells. To follow up on this idea, the effects of GO-CoA-Tat on acyl ghrelin production were tested in vitro with recombinant microsomal GOAT. In this model, nearly complete GOAT inhibition is achieved only 5 min after GO-CoA-Tat treatment (Barnett et al. 2010). Consequently, it is very likely that significant amounts of preformed acyl ghrelin exist within the ghrelin-producing cells. The effects of GO-CoA-Tat on GOAT inhibition were further tested in wild-type mice. GO-CoA-Tat treatment successfully reduces acyl ghrelin concentrations in serum with maximal inhibition after 6 h (Barnett et al. 2010). GO-CoA-Tat does not change serum concentrations of des-acyl ghrelin in mice. The inhibition of GOAT, which results in substantial decreased acyl ghrelin production, leads to a significant reduction in body weight and fat mass independently of food intake in mice that were chronically treated with GO-CoA-Tat in comparison to mice treated with control substance. Importantly, GO-CoA-Tat is very likely to be GOAT specific since treatment of ghrelin-KO mice with GO-CoA-Tat does not change body weight and fat mass in comparison to ghrelin-KO mice that are treated with vehicle (Barnett et al. 2010). Interestingly, wild-type mice that are treated with GO-CoA-Tat have decreased blood glucose and lower IGF-1 levels compared to mice treated with control substance. Consequently, GO-CoA-Tat

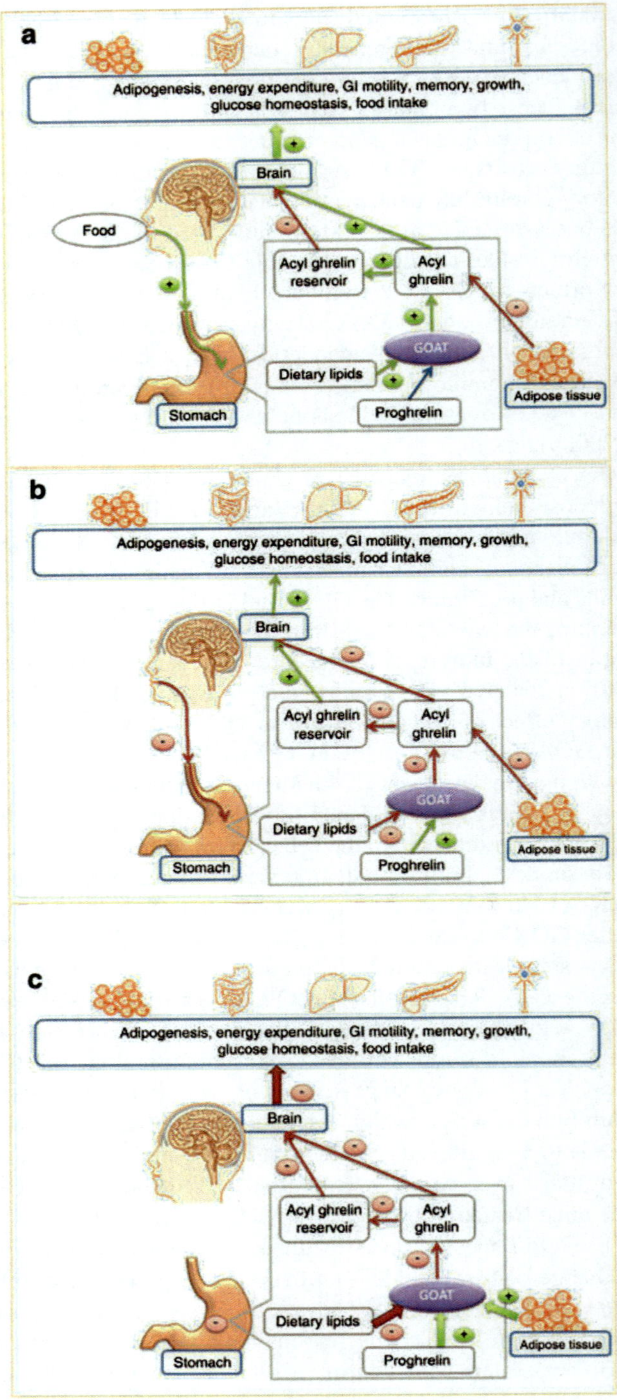

a
Adipogenesis, energy expenditure, GI motility, memory, growth, glucose homeostasis, food intake
Brain
Food
Acyl ghrelin reservoir
Acyl ghrelin
GOAT
Dietary lipids
Proghrelin
Adipose tissue
Stomach
b
Adipogenesis, energy expenditure, GI motility, memory, growth, glucose homeostasis, food intake
Brain
Acyl ghrelin reservoir
Acyl ghrelin
GOAT
Dietary lipids
Proghrelin
Adipose tissue
Stomach
c
Adipogenesis, energy expenditure, GI motility, memory, growth, glucose homeostasis, food intake
Brain
Acyl ghrelin reservoir
Acyl ghrelin
GOAT
Dietary lipids
Proghrelin
Adipose tissue
Stomach

treatment decreases glucose excursion curve by increased insulin secretion during a glucose tolerance test in mice. Again, studies in ghrelin-KO mice could prove the specificity of GO-CoA-Tat for GOAT inhibition since glucose tolerance and insulin secretion is not improved in ghrelin-KO mice treated with GO-CoA-Tat. Similar to the wild-type mouse studies, human islet cells that were pretreated with GO-CoA-Tat before a glucose challenge show a significantly increased insulin secretion compared with cells treated with the control substance prior to the glucose challenge. A possible mechanism for the glucose-lowering effect of GOAT inhibition could be that GO-CoA-Tat decreases expression of uncoupling protein 2 (UCP2), which is known to suppress insulin secretion, by 20-fold in islets of GO-CoA-Tat-treated mice (Barnett et al. 2010). Overall, the development of the GOAT-inhibiting molecules (Barnett et al. 2010; Yang et al. 2008b) contributed largely to the understanding of the mechanisms of GOAT activity. Moreover, pharmacological GOAT inhibition with GO-CoA-Tat resulted in promising metabolic benefits in vivo and will guide the development of GOAT-inhibiting molecules that could potentially be used for the treatment of obesity and type 2 diabetes mellitus in the future.

5 Summary and Conclusion

The peptide hormone ghrelin is involved in the regulation of energy balance by increasing food intake, inducing lipogenesis, reducing energy expenditure, and modulating nutrient partitioning. The promotion of weight gain and adipogenesis, however, are not driven by increased food intake alone. Since ghrelin acylation seems to be substantially modified by food intake and fatty acid composition of the ingested food, ghrelin might be a gastric lipid sensor that signals the caloric intake to the brain (Kirchner et al. 2009). In addition, ghrelin might induce an anticipatory activity to optimally prepare the organism for the incoming nutrients (Kirchner et al. 2009; LeSauter et al. 2009). Ghrelin secretion and acylation appear to represent two parallel processes, which are regulated independently as indicated by the existence of ghrelin reservoirs in cells (Barnett et al. 2010). However, if synchronized, these two processes can promote the efficiency of the ghrelin pathway by linking both entrained meal patterns via ghrelin's specific secretory profile and the abundance of dietary lipids as sensed by GOAT activity (Fig. 2). Therefore,

Fig. 2 Physiological regulation of ghrelin acylation. **(a)** During food intake, sufficient dietary lipids are available to acylate ghrelin by GOAT. Part of the newly synthesized acyl ghrelin is stored in a reservoir. Further, a significant amount is released into the circulation to signal to the brain the abundance of calories. As a result, the organism maximizes food intake and optimizes the storage of nutrients by promoting adipogenesis and shifting the metabolism toward carbohydrate oxidation. Energy-consuming processes such as growth are started. **(b)** During intermeal intervals, dietary lipids are spared. Preformed acyl ghrelin is released from its reservoir to optimally prepare the organisms for the anticipated incoming nutrients. **(c)** After long-term starvation, the acyl ghrelin reservoir is emptied. The fasting-induced lack of dietary lipids further prevents ghrelin acylation, and GOAT gene expression is downregulated. A baseline production of acyl ghrelin is maintained by using endogenous lipids as GOAT substrate. *Green arrows* indicate increase. *Red arrows* indicate decrease

when meal-entrained ghrelin secretion occurs during a phase where ghrelin is acylated by dietary lipids via GOAT, this pathway reaches its maximum efficiency: In times of abundant calorie availability, the activated ghrelin pathway serves its intended purpose of maximizing the intake and optimizing the storage of nutrients (Fig. 2a). When ghrelin is activated not only directly by diet-derived lipids but also by circulating and adipocyte-derived fatty acids, the system may still be able to ensure an optimum balance between storing versus utilizing lipid-derived calories.

References

Andrews ZB, Liu ZW, Wallingford N, Erion DM, Borok E, Friedman JM, Tschop MH, Shanabrough M, Cline G, Shulman GI, Coppola A, Gao XB, Horvath TL, Diano S (2008) UCP2 mediates ghrelin's action on NPY/AgRP neurons by lowering free radicals. Nature 454:846–51

Arvat E, Maccario M, Di Vito L, Broglio F, Benso A, Gottero C, Papotti M, Muccioli G, Dieguez C, Casanueva FF, Deghenghi R, Camanni F, Ghigo E (2001) Endocrine activities of ghrelin, a natural growth hormone secretagogue (GHS), in humans: comparison and interactions with hexarelin, a nonnatural peptidyl GHS, and GH-releasing hormone. J Clin Endocrinol Metab 86:1169–74

Asakawa A, Inui A, Kaga T, Katsuura G, Fujimiya M, Fujino MA, Kasuga M (2003) Antagonism of ghrelin receptor reduces food intake and body weight gain in mice. Gut 52:947–52

Atcha Z, Chen WS, Ong AB, Wong FK, Neo A, Browne ER, Witherington J, Pemberton DJ (2009) Cognitive enhancing effects of ghrelin receptor agonists. Psychopharmacology (Berl) 206:415–27

Barnett BP, Hwang Y, Taylor MS, Kirchner H, Pfluger PT, Bernard V, Lin YY, Bowers EM, Mukherjee C, Song WJ, Longo PA, Leahy DJ, Hussain MA, Tschop MH, Boeke JD, Cole PA (2010) Glucose and weight control in mice with a designed ghrelin O-acyltransferase inhibitor. Science 330:1689–92

Carpino PA, Lefker BA, Toler SM, Pan LC, Hadcock JR, Murray MC, Cook ER, DiBrino JN, DeNinno SL, Chidsey-Frink KL, Hada WA, Inthavongsay J, Lewis SK, Mangano FM, Mullins MA, Nickerson DF, Ng O, Pirie CM, Ragan JA, Rose CR, Tess DA, Wright AS, Yu L, Zawistoski MP, Pettersen JC, DaSilva-Jardine PA, Wilson TC, Thompson DD (2002) Discovery and biological characterization of capromorelin analogues with extended half-lives. Bioorg Med Chem Lett 12:3279–82

Carreira MC, Camina JP, Smith RG, Casanueva FF (2004) Agonist-specific coupling of growth hormone secretagogue receptor type 1a to different intracellular signaling systems. Role of adenosine. Neuroendocrinology 79:13–25

Chen HY, Trumbauer ME, Chen AS, Weingarth DT, Adams JR, Frazier EG, Shen Z, Marsh DJ, Feighner SD, Guan XM, Ye Z, Nargund RP, Smith RG, Van der Ploeg LH, Howard AD, MacNeil DJ, Qian S (2004) Orexigenic action of peripheral ghrelin is mediated by neuropeptide Y and agouti-related protein. Endocrinology 145:2607–12

Cheng K, Chan WW, Butler B, Wei L, Smith RG (1993) A novel non-peptidyl growth hormone secretagogue. Horm Res 40:109–15

Cowley MA, Smith RG, Diano S, Tschop M, Pronchuk N, Grove KL, Strasburger CJ, Bidlingmaier M, Esterman M, Heiman ML, Garcia-Segura LM, Nillni EA, Mendez P, Low MJ, Sotonyi P, Friedman JM, Liu H, Pinto S, Colmers WF, Cone RD, Horvath TL (2003) The distribution and mechanism of action of ghrelin in the CNS demonstrates a novel hypothalamic circuit regulating energy homeostasis. Neuron 37:649–61

Cummings DE, Purnell JQ, Frayo RS, Schmidova K, Wisse BE, Weigle DS (2001) A preprandial rise in plasma ghrelin levels suggests a role in meal initiation in humans. Diabetes 50:1714–9

Cummings DE, Weigle DS, Frayo RS, Breen PA, Ma MK, Dellinger EP, Purnell JQ (2002) Plasma ghrelin levels after diet-induced weight loss or gastric bypass surgery. N Engl J Med 346:1623–30

Cummings DE, Naleid AM, Figlewicz Lattemann DP (2007) Ghrelin: a link between energy homeostasis and drug abuse? Addict Biol 12:1–5

Date Y, Kojima M, Hosoda H, Sawaguchi A, Mondal MS, Suganuma T, Matsukura S, Kangawa K, Nakazato M (2000) Ghrelin, a novel growth hormone-releasing acylated peptide, is synthesized in a distinct endocrine cell type in the gastrointestinal tracts of rats and humans. Endocrinology 141:4255–61

De Vriese C, Gregoire F, Lema-Kisoka R, Waelbroeck M, Robberecht P, Delporte C (2004) Ghrelin degradation by serum and tissue homogenates: identification of the cleavage sites. Endocrinology 145:4997–5005

DelParigi A, Tschop M, Heiman ML, Salbe AD, Vozarova B, Sell SM, Bunt JC, Tataranni PA (2002) High circulating ghrelin: a potential cause for hyperphagia and obesity in Prader-Willi syndrome. J Clin Endocrinol Metab 87:5461–4

Demange L, Boeglin D, Moulin A, Mousseaux D, Ryan J, Berge G, Gagne D, Heitz A, Perrissoud D, Locatelli V, Torsello A, Galleyrand JC, Fehrentz JA, Martinez J (2007) Synthesis and pharmacological in vitro and in vivo evaluations of novel triazole derivatives as ligands of the ghrelin receptor. 1. J Med Chem 50:1939–57

Dezaki K, Kakei M, Yada T (2007) Ghrelin uses Galphai2 and activates voltage-dependent K+ channels to attenuate glucose-induced Ca2+ signaling and insulin release in islet beta-cells: novel signal transduction of ghrelin. Diabetes 56:2319–27

Dickson SL, Leng G, Robinson IC (1993) Systemic administration of growth hormone-releasing peptide activates hypothalamic arcuate neurons. Neuroscience 53:303–6

Drazen DL, Vahl TP, D'Alessio DA, Seeley RJ, Woods SC (2006) Effects of a fixed meal pattern on ghrelin secretion: evidence for a learned response independent of nutrient status. Endocrinology 147:23–30

Ejskjaer N, Dimcevski G, Wo J, Hellstrom PM, Gormsen LC, Sarosiek I, Softeland E, Nowak T, Pezzullo JC, Shaughnessy L, Kosutic G, McCallum R (2010) Safety and efficacy of ghrelin agonist TZP-101 in relieving symptoms in patients with diabetic gastroparesis: a randomized, placebo-controlled study. Neurogastroenterol Motil 22:1069–e281

Faraj M, Havel PJ, Phelis S, Blank D, Sniderman AD, Cianflone K (2003) Plasma acylation-stimulating protein, adiponectin, leptin, and ghrelin before and after weight loss induced by gastric bypass surgery in morbidly obese subjects. J Clin Endocrinol Metab 88:1594–602

Foster-Schubert KE, Overduin J, Prudom CE, Liu J, Callahan HS, Gaylinn BD, Thorner MO, Cummings DE (2008) Acyl and total ghrelin are suppressed strongly by ingested proteins, weakly by lipids, and biphasically by carbohydrates. J Clin Endocrinol Metab 93:1971–9

Garcia JM, Polvino WJ (2007) Effect on body weight and safety of RC-1291, a novel, orally available ghrelin mimetic and growth hormone secretagogue: results of a phase I, randomized, placebo-controlled, multiple-dose study in healthy volunteers. Oncologist 12:594–600

Gonzalez CR, Vazquez MJ, Lopez M, Dieguez C (2008) Influence of chronic undernutrition and leptin on GOAT mRNA levels in rat stomach mucosa. J Mol Endocrinol 41:415–21

Gropp E, Shanabrough M, Borok E, Xu AW, Janoschek R, Buch T, Plum L, Balthasar N, Hampel B, Waisman A, Barsh GS, Horvath TL, Bruning JC (2005) Agouti-related peptide-expressing neurons are mandatory for feeding. Nat Neurosci 8:1289–91

Gualillo O, Caminos J, Blanco M, Garcia-Caballero T, Kojima M, Kangawa K, Dieguez C, Casanueva F (2001) Ghrelin, a novel placental-derived hormone. Endocrinology 142:788–94

Gutierrez JA, Solenberg PJ, Perkins DR, Willency JA, Knierman MD, Jin Z, Witcher DR, Luo S, Onyia JE, Hale JE (2008) Ghrelin octanoylation mediated by an orphan lipid transferase. Proc Natl Acad Sci USA 105:6320–5

Halem HA, Taylor JE, Dong JZ, Shen Y, Datta R, Abizaid A, Diano S, Horvath T, Zizzari P, Bluet-Pajot MT, Epelbaum J, Culler MD (2004) Novel analogs of ghrelin: physiological and clinical implications. Eur J Endocrinol 151(Suppl 1):S71–5

Halem HA, Taylor JE, Dong JZ, Shen Y, Datta R, Abizaid A, Diano S, Horvath TL, Culler MD (2005) A novel growth hormone secretagogue-1a receptor antagonist that blocks ghrelin-induced growth hormone secretion but induces increased body weight gain. Neuroendocrinology 81:339–49

Helmling S, Maasch C, Eulberg D, Buchner K, Schroder W, Lange C, Vonhoff S, Wlotzka B, Tschop MH, Rosewicz S, Klussmann S (2004) Inhibition of ghrelin action in vitro and in vivo by an RNA-Spiegelmer. Proc Natl Acad Sci USA 101:13174–9

Hinney A, Hoch A, Geller F, Schafer H, Siegfried W, Goldschmidt H, Remschmidt H, Hebebrand J (2002) Ghrelin gene: identification of missense variants and a frameshift mutation in extremely obese children and adolescents and healthy normal weight students. J Clin Endocrinol Metab 87:2716

Holdstock C, Engstrom BE, Ohrvall M, Lind L, Sundbom M, Karlsson FA (2003) Ghrelin and adipose tissue regulatory peptides: effect of gastric bypass surgery in obese humans. J Clin Endocrinol Metab 88:3177–83

Holst B, Holliday ND, Bach A, Elling CE, Cox HM, Schwartz TW (2004) Common structural basis for constitutive activity of the ghrelin receptor family. J Biol Chem 279:53806–17

Holst B, Brandt E, Bach A, Heding A, Schwartz TW (2005) Nonpeptide and peptide growth hormone secretagogues act both as ghrelin receptor agonist and as positive or negative allosteric modulators of ghrelin signaling. Mol Endocrinol 19:2400–11

Holst B, Frimurer TM, Mokrosinski J, Halkjaer T, Cullberg KB, Underwood CR, Schwartz TW (2009) Overlapping binding site for the endogenous agonist, small-molecule agonists, and ago-allosteric modulators on the ghrelin receptor. Mol Pharmacol 75:44–59

Kanamoto N, Akamizu T, Tagami T, Hataya Y, Moriyama K, Takaya K, Hosoda H, Kojima M, Kangawa K, Nakao K (2004) Genomic structure and characterization of the 5'-flanking region of the human ghrelin gene. Endocrinology 145:4144–53

Kirchner H, Gutierrez JA, Solenberg PJ, Pfluger PT, Czyzyk TA, Willency JA, Schurmann A, Joost HG, Jandacek RJ, Hale JE, Heiman ML, Tschop MH (2009) GOAT links dietary lipids with the endocrine control of energy balance. Nat Med 15:741–5

Kishimoto M, Okimura Y, Nakata H, Kudo T, Iguchi G, Takahashi Y, Kaji H, Chihara K (2003) Cloning and characterization of the 5(')-flanking region of the human ghrelin gene. Biochem Biophys Res Commun 305:186–92

Kitazawa T, De Smet B, Verbeke K, Depoortere I, Peeters TL (2005) Gastric motor effects of peptide and non-peptide ghrelin agonists in mice in vivo and in vitro. Gut 54:1078–84

Kojima M, Kangawa K (2008) Structure and function of ghrelin. Results Probl Cell Differ 46:89–115

Kojima M, Hosoda H, Date Y, Nakazato M, Matsuo H, Kangawa K (1999) Ghrelin is a growth-hormone-releasing acylated peptide from stomach. Nature 402:656–60

Korbonits M, Bustin SA, Kojima M, Jordan S, Adams EF, Lowe DG, Kangawa K, Grossman AB (2001) The expression of the growth hormone secretagogue receptor ligand ghrelin in normal and abnormal human pituitary and other neuroendocrine tumors. J Clin Endocrinol Metab 86:881–7

Korbonits M, Gueorguiev M, O'Grady E, Lecoeur C, Swan DC, Mein CA, Weill J, Grossman AB, Froguel P (2002) A variation in the ghrelin gene increases weight and decreases insulin secretion in tall, obese children. J Clin Endocrinol Metab 87:4005–8

Korner J, Inabnet W, Febres G, Conwell IM, McMahon DJ, Salas R, Taveras C, Schrope B, Bessler M (2009) Prospective study of gut hormone and metabolic changes after adjustable gastric banding and Roux-en-Y gastric bypass. Int J Obes (Lond) 33:786–95

Kotidis EV, Koliakos G, Papavramidis TS, Papavramidis ST (2006) The effect of biliopancreatic diversion with pylorus-preserving sleeve gastrectomy and duodenal switch on fasting serum ghrelin, leptin and adiponectin levels: is there a hormonal contribution to the weight-reducing effect of this procedure? Obes Surg 16:554–9

Langer FB, Reza Hoda MA, Bohdjalian A, Felberbauer FX, Zacherl J, Wenzl E, Schindler K, Luger A, Ludvik B, Prager G (2005) Sleeve gastrectomy and gastric banding: effects on plasma ghrelin levels. Obes Surg 15:1024–9

Lear PV, Iglesias MJ, Feijoo-Bandin S, Rodriguez-Penas D, Mosquera-Leal A, Garcia-Rua V, Gualillo O, Ghe C, Arnoletti E, Muccioli G, Dieguez C, Gonzalez-Juanatey JR, Lago F (2010) Des-Acyl ghrelin has specific binding sites and different metabolic effects from ghrelin in cardiomyocytes. Endocrinology 151(7):3286–3298

LeSauter J, Hoque N, Weintraub M, Pfaff DW, Silver R (2009) Stomach ghrelin-secreting cells as food-entrainable circadian clocks. Proc Natl Acad Sci USA 106:13582–7

Liu J, Prudom CE, Nass R, Pezzoli SS, Oliveri MC, Johnson ML, Veldhuis P, Gordon DA, Howard AD, Witcher DR, Geysen HM, Gaylinn BD, Thorner MO (2008) Novel ghrelin assays provide evidence for independent regulation of ghrelin acylation and secretion in healthy young men. J Clin Endocrinol Metab 93:1980–7

Longo KA, Charoenthongtrakul S, Giuliana DJ, Govek EK, McDonagh T, Qi Y, DiStefano PS, Geddes BJ (2008) Improved insulin sensitivity and metabolic flexibility in ghrelin receptor knockout mice. Regul Pept 150:55–61

Luquet S, Perez FA, Hnasko TS, Palmiter RD (2005) NPY/AgRP neurons are essential for feeding in adult mice but can be ablated in neonates. Science 310:683–5

Malik S, McGlone F, Bedrossian D, Dagher A (2008) Ghrelin modulates brain activity in areas that control appetitive behavior. Cell Metab 7:400–9

Matsumoto M, Hosoda H, Kitajima Y, Morozumi N, Minamitake Y, Tanaka S, Matsuo H, Kojima M, Hayashi Y, Kangawa K (2001) Structure-activity relationship of ghrelin: pharmacological study of ghrelin peptides. Biochem Biophys Res Commun 287:142–6

Mayorov AV, Amara N, Chang JY, Moss JA, Hixon MS, Ruiz DI, Meijler MM, Zorrilla EP, Janda KD (2008) Catalytic antibody degradation of ghrelin increases whole-body metabolic rate and reduces refeeding in fasting mice. Proc Natl Acad Sci USA 105:17487–92

McLaughlin T, Abbasi F, Lamendola C, Frayo RS, Cummings DE (2004) Plasma ghrelin concentrations are decreased in insulin-resistant obese adults relative to equally obese insulin-sensitive controls. J Clin Endocrinol Metab 89:1630–5

Mondal MS, Date Y, Yamaguchi H, Toshinai K, Tsuruta T, Kangawa K, Nakazato M (2005) Identification of ghrelin and its receptor in neurons of the rat arcuate nucleus. Regul Pept 126:55–9

Monteleone P, Bencivenga R, Longobardi N, Serritella C, Maj M (2003) Differential responses of circulating ghrelin to high-fat or high-carbohydrate meal in healthy women. J Clin Endocrinol Metab 88:5510–4

Mori K, Yoshimoto A, Takaya K, Hosoda K, Ariyasu H, Yahata K, Mukoyama M, Sugawara A, Hosoda H, Kojima M, Kangawa K, Nakao K (2000) Kidney produces a novel acylated peptide, ghrelin. FEBS Lett 486:213–6

Moulin A, Demange L, Berge G, Gagne D, Ryan J, Mousseaux D, Heitz A, Perrissoud D, Locatelli V, Torsello A, Galleyrand JC, Fehrentz JA, Martinez J (2007) Toward potent ghrelin receptor ligands based on trisubstituted 1,2,4-triazole structure. 2. Synthesis and pharmacological in vitro and in vivo evaluations. J Med Chem 50:5790–806

Moulin A, Demange L, Ryan J, M'Kadmi C, Galleyrand JC, Martinez J, Fehrentz JA (2008) Trisubstituted 1,2,4-triazoles as ligands for the ghrelin receptor: on the significance of the orientation and substitution at position 3. Bioorg Med Chem Lett 18:164–8

Nagaya N, Uematsu M, Kojima M, Date Y, Nakazato M, Okumura H, Hosoda H, Shimizu W, Yamagishi M, Oya H, Koh H, Yutani C, Kangawa K (2001) Elevated circulating level of ghrelin in cachexia associated with chronic heart failure: relationships between ghrelin and anabolic/catabolic factors. Circulation 104:2034–8

Nakazato M, Murakami N, Date Y, Kojima M, Matsuo H, Kangawa K, Matsukura S (2001) A role for ghrelin in the central regulation of feeding. Nature 409:194–8

Nishi Y, Hiejima H, Hosoda H, Kaiya H, Mori K, Fukue Y, Yanase T, Nawata H, Kangawa K, Kojima M (2005) Ingested medium-chain fatty acids are directly utilized for the acyl modification of ghrelin. Endocrinology 146:2255–64

Nogueiras R, Wiedmer P, Perez-Tilve D, Veyrat-Durebex C, Keogh JM, Sutton GM, Pfluger PT, Castaneda TR, Neschen S, Hofmann SM, Howles PN, Morgan DA, Benoit SC, Szanto I, Schrott B, Schurmann A, Joost HG, Hammond C, Hui DY, Woods SC, Rahmouni K, Butler

AA, Farooqi IS, O'Rahilly S, Rohner-Jeanrenaud F, Tschop MH (2007) The central melanocortin system directly controls peripheral lipid metabolism. J Clin Invest 117:3475–88
Ohgusu H, Shirouzu K, Nakamura Y, Nakashima Y, Ida T, Sato T, Kojima M (2009) Ghrelin O-acyltransferase (GOAT) has a preference for n-hexanoyl-CoA over n-octanoyl-CoA as an acyl donor. Biochem Biophys Res Commun 386:153–8
Pantel J, Legendre M, Cabrol S, Hilal L, Hajaji Y, Morisset S, Nivot S, Vie-Luton MP, Grouselle D, de Kerdanet M, Kadiri A, Epelbaum J, Le Bouc Y, Amselem S (2006) Loss of constitutive activity of the growth hormone secretagogue receptor in familial short stature. J Clin Invest 116:760–8
Pfluger PT, Kirchner H, Gunnel S, Schrott B, Perez-Tilve D, Fu S, Benoit SC, Horvath T, Joost HG, Wortley KE, Sleeman MW, Tschop MH (2008) Simultaneous deletion of ghrelin and its receptor increases motor activity and energy expenditure. Am J Physiol Gastrointest Liver Physiol 294:G610–8
Poitras P, Polvino WJ, Rocheleau B (2005) Gastrokinetic effect of ghrelin analog RC-1139 in the rat. Effect on post-operative and on morphine induced ileus. Peptides 26:1598–601
Raghay K, Garcia-Caballero T, Nogueiras R, Morel G, Beiras A, Dieguez C, Gallego R (2006) Ghrelin localization in rat and human thyroid and parathyroid glands and tumours. Histochem Cell Biol 125:239–46
Rudolph J, Esler WP, O'Connor S, Coish PD, Wickens PL, Brands M, Bierer DE, Bloomquist BT, Bondar G, Chen L, Chuang CY, Claus TH, Fathi Z, Fu W, Khire UR, Kristie JA, Liu XG, Lowe DB, McClure AC, Michels M, Ortiz AA, Ramsden PD, Schoenleber RW, Shelekhin TE, Vakalopoulos A, Tang W, Wang L, Yi L, Gardell SJ, Livingston JN, Sweet LJ, Bullock WH (2007) Quinazolinone derivatives as orally available ghrelin receptor antagonists for the treatment of diabetes and obesity. J Med Chem 50:5202–16
Sakata I, Yang J, Lee CE, Osborne-Lawrence S, Rovinsky SA, Elmquist JK, Zigman JM (2009) Colocalization of ghrelin O-acyltransferase and ghrelin in gastric mucosal cells. Am J Physiol Endocrinol Metab 297:E134–41
Sakkou M, Wiedmer P, Anlag K, Hamm A, Seuntjens E, Ettwiller L, Tschop MH, Treier M (2007) A role for brain-specific homeobox factor Bsx in the control of hyperphagia and locomotory behavior. Cell Metab 5:450–63
Sangiao-Alvarellos S, Vazquez MJ, Varela L, Nogueiras R, Saha AK, Cordido F, Lopez M, Dieguez C (2009) Central ghrelin regulates peripheral lipid metabolism in a growth hormone-independent fashion. Endocrinology 150:4562–74
Schioth HB, Muceniece R, Wikberg JE (1997) Characterization of the binding of MSH-B, HB-228, GHRP-6 and 153N-6 to the human melanocortin receptor subtypes. Neuropeptides 31:565–71
Seidah NG, Chretien M (1999) Proprotein and prohormone convertases: a family of subtilases generating diverse bioactive polypeptides. Brain Res 848:45–62
Shearman LP, Wang SP, Helmling S, Stribling DS, Mazur P, Ge L, Wang L, Klussmann S, Macintyre DE, Howard AD, Strack AM (2006) Ghrelin neutralization by a ribonucleic acid-SPM ameliorates obesity in diet-induced obese mice. Endocrinology 147:1517–26
Shuto Y, Shibasaki T, Otagiri A, Kuriyama H, Ohata H, Tamura H, Kamegai J, Sugihara H, Oikawa S, Wakabayashi I (2002) Hypothalamic growth hormone secretagogue receptor regulates growth hormone secretion, feeding, and adiposity. J Clin Invest 109:1429–36
Smith RG, Van der Ploeg LH, Howard AD, Feighner SD, Cheng K, Hickey GJ, Wyvratt MJ Jr, Fisher MH, Nargund RP, Patchett AA (1997) Peptidomimetic regulation of growth hormone secretion. Endocr Rev 18:621–45
Steiner DF (1998) The proprotein convertases. Curr Opin Chem Biol 2:31–9
Sun Y, Ahmed S, Smith RG (2003) Deletion of ghrelin impairs neither growth nor appetite. Mol Cell Biol 23:7973–81
Sun Y, Asnicar M, Saha PK, Chan L, Smith RG (2006) Ablation of ghrelin improves the diabetic but not obese phenotype of ob/ob mice. Cell Metab 3:379–86
Takahashi T, Ida T, Sato T, Nakashima Y, Nakamura Y, Tsuji A, Kojima M (2009) Production of n-octanoyl-modified ghrelin in cultured cells requires prohormone processing protease and ghrelin O-acyltransferase, as well as n-octanoic acid. J Biochem 146:675–82

Tanaka M, Hayashida Y, Iguchi T, Nakao N, Nakai N, Nakashima K (2001) Organization of the mouse ghrelin gene and promoter: occurrence of a short noncoding first exon. Endocrinology 142:3697–700

Tena-Sempere M (2008) Ghrelin and reproduction: ghrelin as novel regulator of the gonadotropic axis. Vitam Horm 77:285–300

Theander-Carrillo C, Wiedmer P, Cettour-Rose P, Nogueiras R, Perez-Tilve D, Pfluger P, Castaneda TR, Muzzin P, Schurmann A, Szanto I, Tschop MH, Rohner-Jeanrenaud F (2006) Ghrelin action in the brain controls adipocyte metabolism. J Clin Invest 116:1983–93

Thompson NM, Gill DA, Davies R, Loveridge N, Houston PA, Robinson IC, Wells T (2004) Ghrelin and des-octanoyl ghrelin promote adipogenesis directly in vivo by a mechanism independent of the type 1a growth hormone secretagogue receptor. Endocrinology 145:234–42

Toshinai K, Yamaguchi H, Sun Y, Smith RG, Yamanaka A, Sakurai T, Date Y, Mondal MS, Shimbara T, Kawagoe T, Murakami N, Miyazato M, Kangawa K, Nakazato M (2006) Des-acyl ghrelin induces food intake by a mechanism independent of the growth hormone secretagogue receptor. Endocrinology 147:2306–14

Tschop M, Smiley DL, Heiman ML (2000) Ghrelin induces adiposity in rodents. Nature 407:908–13

Tschop M, Wawarta R, Riepl RL, Friedrich S, Bidlingmaier M, Landgraf R, Folwaczny C (2001a) Post-prandial decrease of circulating human ghrelin levels. J Endocrinol Invest 24:19–21

Tschop M, Weyer C, Tataranni PA, Devanarayan V, Ravussin E, Heiman ML (2001b) Circulating ghrelin levels are decreased in human obesity. Diabetes 50:707–9

Ukkola O, Ravussin E, Jacobson P, Snyder EE, Chagnon M, Sjostrom L, Bouchard C (2001) Mutations in the preproghrelin/ghrelin gene associated with obesity in humans. J Clin Endocrinol Metab 86:3996–9

Ukkola O, Ravussin E, Jacobson P, Perusse L, Rankinen T, Tschop M, Heiman ML, Leon AS, Rao DC, Skinner JS, Wilmore JH, Sjostrom L, Bouchard C (2002) Role of ghrelin polymorphisms in obesity based on three different studies. Obes Res 10:782–91

van der Lely AJ, Tschop M, Heiman ML, Ghigo E (2004) Biological, physiological, pathophysiological, and pharmacological aspects of ghrelin. Endocr Rev 25:426–57

Venkova K, Fraser G, Hoveyda HR, Greenwood-Van Meerveld B (2007) Prokinetic effects of a new ghrelin receptor agonist TZP-101 in a rat model of postoperative ileus. Dig Dis Sci 52:2241–8

Wo JM, Ejskjaer N, Hellstrom PM, Malik RA, Pezzullo JC, Shaughnessy L, Charlton P, Kosutic G, McCallum RW (2011) Randomised clinical trial: ghrelin agonist TZP-101 relieves gastroparesis associated with severe nausea and vomiting–randomised clinical study subset data. Aliment Pharmacol Ther 33:679–88

Wortley KE, Anderson KD, Garcia K, Murray JD, Malinova L, Liu R, Moncrieffe M, Thabet K, Cox HJ, Yancopoulos GD, Wiegand SJ, Sleeman MW (2004) Genetic deletion of ghrelin does not decrease food intake but influences metabolic fuel preference. Proc Natl Acad Sci USA 101:8227–32

Wortley KE, del Rincon JP, Murray JD, Garcia K, Iida K, Thorner MO, Sleeman MW (2005) Absence of ghrelin protects against early-onset obesity. J Clin Invest 115:3573–8

Wren AM, Seal LJ, Cohen MA, Brynes AE, Frost GS, Murphy KG, Dhillo WS, Ghatei MA, Bloom SR (2001) Ghrelin enhances appetite and increases food intake in humans. J Clin Endocrinol Metab 86:5992

Xin Z, Serby MD, Zhao H, Kosogof C, Szczepankiewicz BG, Liu M, Liu B, Hutchins CW, Sarris KA, Hoff ED, Falls HD, Lin CW, Ogiela CA, Collins CA, Brune ME, Bush EN, Droz BA, Fey TA, Knourek-Segel VE, Shapiro R, Jacobson PB, Beno DW, Turner TM, Sham HL, Liu G (2006) Discovery and pharmacological evaluation of growth hormone secretagogue receptor antagonists. J Med Chem 49:4459–69

Yang J, Brown MS, Liang G, Grishin NV, Goldstein JL (2008a) Identification of the acyltransferase that octanoylates ghrelin, an appetite-stimulating peptide hormone. Cell 132:387–96

Yang J, Zhao TJ, Goldstein JL, Brown MS (2008b) Inhibition of ghrelin O-acyltransferase (GOAT) by octanoylated pentapeptides. Proc Natl Acad Sci USA 105:10750–5

Zhang JV, Ren PG, Avsian-Kretchmer O, Luo CW, Rauch R, Klein C, Hsueh AJ (2005) Obestatin, a peptide encoded by the ghrelin gene, opposes ghrelin's effects on food intake. Science 310:996–9

Zhang W, Chai B, Li JY, Wang H, Mulholland MW (2008) Effect of des-acyl ghrelin on adiposity and glucose metabolism. Endocrinology 149:4710–6

Zhao H, Liu G (2006) Growth hormone secretagogue receptor antagonists as anti-obesity therapies? Still an open question. Curr Opin Drug Discov Devel 9:509–15

Zhu X, Cao Y, Voogd K, Steiner DF (2006) On the processing of proghrelin to ghrelin. J Biol Chem 281:38867–70

Zigman JM, Nakano Y, Coppari R, Balthasar N, Marcus JN, Lee CE, Jones JE, Deysher AE, Waxman AR, White RD, Williams TD, Lachey JL, Seeley RJ, Lowell BB, Elmquist JK (2005) Mice lacking ghrelin receptors resist the development of diet-induced obesity. J Clin Invest 115:3564–72

Zizzari P, Hassouna R, Grouselle D, Epelbaum J, Tolle V (2011) Physiological roles of preproghrelin-derived peptides in GH secretion and feeding. Peptides. doi:10.1016/j.peptides.2011.04.014

Zorrilla EP, Iwasaki S, Moss JA, Chang J, Otsuji J, Inoue K, Meijler MM, Janda KD (2006) Vaccination against weight gain. Proc Natl Acad Sci USA 103:13226–31

Anorexigenic Effects of GLP-1 and Its Analogues

Baptist Gallwitz

Contents

Abstract GLP-1 receptors are expressed in the brain, especially in the regions responsible for the regulation of food intake, and intracerebroventricular injection of GLP-1 results in inhibition of food intake. Peripheral administration of GLP-1 dose-dependently enhances satiety and reduces food intake in normal and obese subjects as well as in type 2 diabetic patients. So far, the mechanisms by which GLP-1 exerts its effects are not completely clear. Interactions with neurons in the gastrointestinal tract or possibly direct access to the brain through the blood–brain barrier as observed in rats are possible and discussed in this chapter as well as a novel hypothesis based on the finding that GLP-1 is also expressed in taste cells. Finally, the role of GLP-1 receptor agonists as a possible treatment option in obesity is discussed as well as the role of GLP-1 in the effects of bariatric surgery on adiposity and glucose homeostasis.

B. Gallwitz (✉)
Medizinische Klinik IV, Otfried-Müller-Str. 10, 72076 Tübingen, Germany
e-mail: baptist.gallwitz@med.uni-tuebingen.de

H.-G. Joost (ed.), *Appetite Control*, Handbook of Experimental Pharmacology 209,
DOI 10.1007/978-3-642-24716-3_8, © Springer-Verlag Berlin Heidelberg 2012

Keywords Bariatric surgery • GLP-1 • GLP-1 receptor agonists • Gut-brain axis • Incretins • Obesity

1 Incretin Hormones

Upon stimulation by nutrients after a meal, the intestinal mucosa secretes the gastrointestinal hormones GLP-1 and GIP (*g*astric *i*nhibitory *p*olypeptide or *g*lucose-dependent *i*nsulinotropic *p*olypeptide) from the endocrine L and K cells, respectively (Wellendorph et al. 2009). Both hormones are responsible for approximately 60% of the postprandial insulin secretion and contribute to the so-called incretin effect. This effect describes the phenomenon that orally ingested glucose leads to a much larger insulin response than an isoglycemic, intravenous glucose load (Creutzfeldt 1979; Nauck et al. 1986).

Exogenous GLP-1 application either by subcutaneous or intravenous injection resulting in supraphysiological GLP-1 plasma concentrations restores the incretin effect with an adequate insulin response under hyperglycemic conditions (Nauck et al. 1993).

2 Biosynthesis of GLP-1 and Overview on Its Pleiotropic Actions

The glucagon gene encodes a large peptide sequence of 158 amino acids that contains not only the sequence of glucagon but also that of other peptides that are formed posttranslationally by organ-specific and cell-specific processing (Bell et al. 1983; Holst et al. 2007; Ørskov et al. 1987). In the neuroendocrine L cells of the intestinal mucosa and in the central nervous system, preproglucagon is cleaved mainly to generate GLP-1. In the pancreatic alpha cells of the islets, glucagon is the major biologically active peptide generated from preproglucagon. Figure 1 shows the schematic tissue-specific posttranslational processing of GLP-1.

GLP-1 binds to highly specific GLP-1 receptors that belong to the G protein coupled seven-transmembrane-spanning (7TM) receptors (Drucker and Nauck 2006). After binding to its receptor, adenylate cyclase is activated, and GLP-1 effects are mediated mainly via the cAMP and protein kinase A pathways (Gromada et al. 1996; Reimer 2006). GLP-1 shows numerous physiological actions in various tissues and a broad therapeutic potential (see Fig. 2 for details).

The two most important metabolic actions of GLP-1 are that it stimulates insulin secretion of the pancreatic beta cells and additionally inhibits glucagon secretion from the alpha cells. These two actions on the islet occur in a strictly glucose-dependent manner and lead to a normalization of hyperglycemia. Under hypoglycemic conditions, the counter-regulation by glucagon is not affected, and insulin secretion is not stimulated. GLP-1 is therefore not able to elicit hypoglycemia by

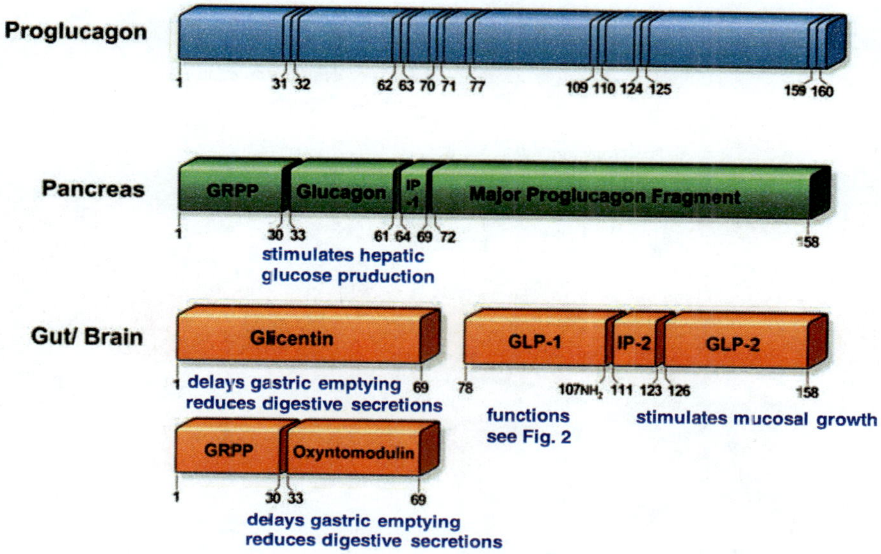

Fig. 1 *Posttranslational processing of proglucagon in different tissues*. Differential posttranslational processing of proglucagon in the pancreas and in the gut and brain. The *numbers* indicate amino acid positions in the 160-amino acid proglucagon sequence. The *vertical lines* indicate positions of basic amino acid residues, typical cleavage sites. *GRPP* glicentin-related pancreatic polypeptide, IP-1 *intervening peptide-1*, IP-2 *intervening peptide-2*. The major known biological actions of the peptides resulting from proglucagon processing are also shown (adapted from Holst 2007 and Pellissier et al. 2004)

itself. These physiological properties of GLP-1 later translated into the pharmacotherapy of type 2 diabetes with incretin-based therapies (Drucker and Nauck 2006).

Like other gastrointestinal regulatory peptides, GLP-1 has multiple further actions. In the gastrointestinal tract, GLP-1 slows gastric emptying after a meal. This effect also contributes to a normalization of postprandial hyperglycemia, promotes a feeling of fullness and possibly a secondary reduction of appetite (Meier et al. 2003a; Holst et al. 2008). In addition, GLP-1 binds to its receptor on hypothalamic neurons and stimulates satiety by direct actions described in detail in this chapter. These two satiety-promoting effects explain that long-term treatment with GLP-1 receptor agonists leads to long-term weight loss that persists as long as GLP-1 is given (Drucker and Nauck 2006).

Animal studies in different rodent species and studies in isolated human islets showed beneficial long-term actions of GLP-1: insulin synthesis is stimulated, and beta-cell mass is restored in rodent models of type 2 diabetes (Brubaker and Drucker 2004; Drucker and Nauck 2006; Fehmann and Habener 1992). Presently, it is not known whether these findings reflect an additional benefit in type 2 diabetes therapy in that GLP-1 slows or even stops disease progression. Long-term study data from clinical studies or clinical use of GLP-1 receptor agonists (GLP-1 RA) in type 2 diabetes with a sufficient observation time are still not yet available.

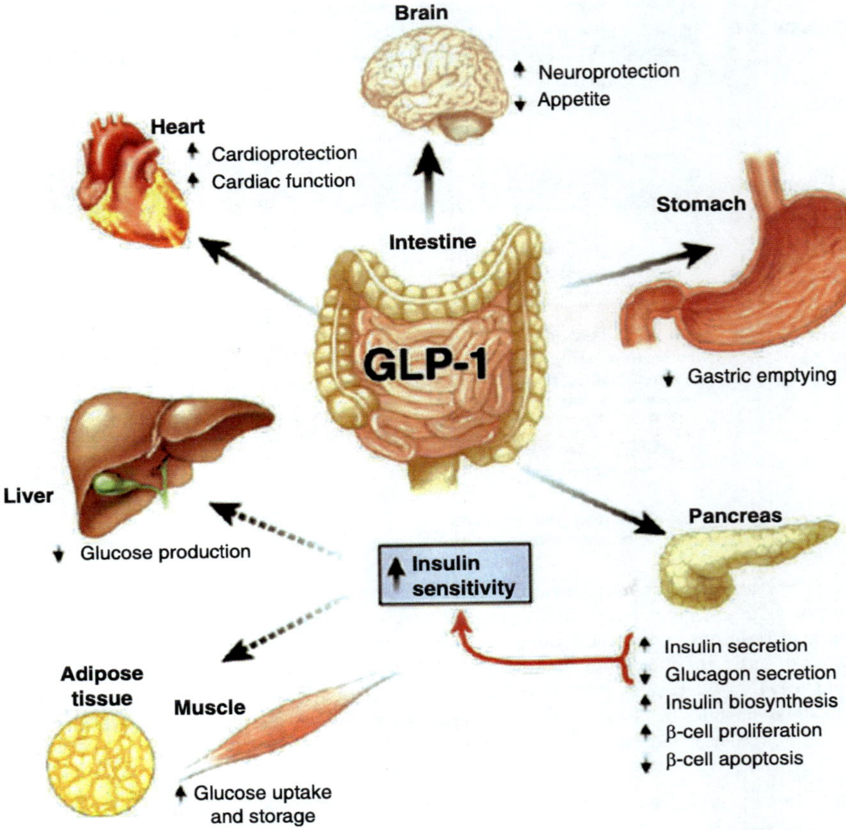

Fig. 2 *GLP-1 actions in peripheral tissues*. The majority of the effects of GLP-1 are mediated by direct interaction with GLP-1Rs on specific tissues. However, the actions of GLP-1 in liver, fat, and muscle most likely occur through indirect mechanisms (adapted from Baggio and Drucker 2007)

Furthermore, there are presently no reliable, validated methods to quantify beta cell mass in humans in a clinical setting.

Additionally, recent studies demonstrated that therapeutic application of GLP-1 or GLP-1 RA improved cardiovascular parameters. Systolic blood pressure is lowered by GLP-1 RA treatment for type 2 diabetes, and beneficial effects of GLP-1 on myocardial ischemia were observed in animal models as well as positive effects on left ventricular function in heart failure. These promising effects may also have important clinical implications for type 2 diabetes therapy with GLP-1 RA (Courreges et al. 2008; Klonoff et al. 2008; Sokos et al. 2006).

GLP-1 receptors are also expressed in the brain, especially in the regions responsible for the regulation of food intake (Göke et al. 1995), and intracerebroventricular injection of GLP-1 results in inhibition of food intake (Tang-Christensen et al. 1996; Turton et al. 1996). Peripheral administration of GLP-1 dose-dependently enhances satiety and reduces food intake in normal subjects (Flint et al. 1998; Verdich et al. 2001), obese subjects (Näslund et al. 1999), and type 2 diabetic patients (Gutzwiller et al. 1999; Zander et al. 2002). The mechanisms by which GLP-1 exerts its effects are not completely clear yet. Interactions with neurons in the gastrointestinal tract or possibly direct access to the brain through as observed in rats (Ørskov et al. 1996) are possible and discussed in this chapter as well as other novel hypothesis based on the finding that GLP-1 is also expressed in taste cells. Finally, the role of GLP-1 receptor agonists as possible treatment options in obesity is discussed as well as the role of GLP-1 in the weight losing and metabolic effects after various methods of bariatric surgery.

3 Central Effects of GLP-1

3.1 Direct Dentral Effects of GLP-1 in the Hypothalamus

As early as in 1988, it was shown that GLP-1 is also synthesized in the CNS in the caudal part of the nucleus of the solitary tract (Jin et al. 1988) in addition to its peripheral synthesis in the intestinal L cell. Receptors for GLP-1 are expressed throughout the brain widely, with highest levels in the paraventricular nucleus (Larsen et al. 1997b; Van Dijk et al. 1996; Turton et al. 1996). The presence of both, the peptide GLP-1 and the GLP-1 receptor, in the CNS points toward important physiological actions of GLP-1 in the CNS in addition to its actions on the peripheral system.

The first report that GLP-1 exerted effects in the central nervous system (CNS) came from Turton and colleagues. This group gave intracerebroventricular (icv) injections of GLP-1 to rats. The injections reduced the food intake of the animals compared to saline-treated control rats (Turton et al. 1996). The group also demonstrated the presence of GLP-1-containing neurons in the rat brain in hypothalamic areas that are known to be responsible for regulating satiety and food intake.

Since then, there has been a great interest in understanding the role of GLP-1 in the regulation of food intake and satiety. In the rat, other groups confirmed the findings of Turton by using either GLP-1 or exendin-4, a naturally occurring GLP-1 RA (Meeran et al. 1999; Turton et al. 1996; Tang-Christensen et al. 1996). In humans, subcutaneous or intravenous application of GLP-1 also reduced hunger and food intake, prandial injections in obese subjects led to a reduction in food intake (Näslund et al. 2004). Blocking GLP-1 action in the CNS by using the GLP-1 receptor antagonist exendin(9–39) increased food intake in rats that had had icv injections with exendin(9–39), and additionally facilitated weight gain in these animals after long-term administration (Turton et al. 1996).

Icv injections of GLP-1 receptor agonists inhibit food intake in rodents (Turton et al. 1996; Meeran et al. 1999). Repeated icv administration of GLP-1 in rats leads to weight loss (see Fig. 3). Conversely, icv injection of the GLP-1 receptor antagonist exendin(9–39) promoted weight gain in the animals, and exendin(9–39) administered simultaneously with the central orexigenic agent neuropeptide Y (NPY) resulted in an increased food intake and weight gain compared with that observed with neuropeptide Y alone (Meeran et al. 1999; Abu-Hamdah et al. 2009). In this respect, it is important that the intestinal L cells cosecrete GLP-1 and peptide YY (PYY). Immunohistological studies demonstrated that these peptides are

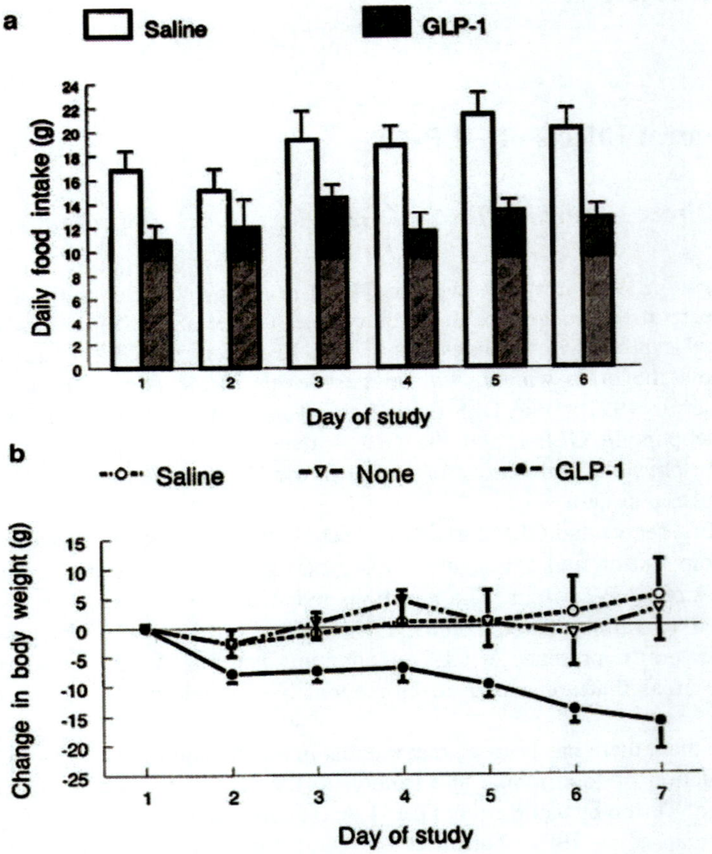

Fig. 3 *Effect of multiple icv injections of GLP-1 on food intake and body weight.* Food intake (**a**) and body weight (**b**) after daily icv injection of GLP-1 or saline as control are shown. The *hatched bars* and *filled circles* represent animals given 3 nmol GLP-1, and the *open bars* and *open circles* represent control animals that received saline. Food intake and body weight were significantly decreased through the study period in animals receiving GLP-1 ($P < 0.05$ for both groups). Body weight was similar in noninjected controls (*open triangles*) and those given icv normal saline (from Meeran et al. 1999 with permission)

colocalized and coreleased from these cells. The truncated PYY(3–36), comprising the major circulating form of PYY, has been reported to be a potent anorexigenic agent in rats as well as in man (Batterham et al. 2002, 2003). Although these effects have not been reproduced by others (Tschöp et al. 2004; Boggiano et al. 2005), it was suggested that the corelease of GLP-1 and PYY has an important roles in the mediation of satiety (Abu-Hamdah et al. 2009).

It has further been demonstrated that GLP-1 may play a role in the regulation of the hypothalamic pituitary axis via effects on CRH, LH, TSH, oxytocin, and vasopressin secretion (Beak et al. 1996, 1998). The available evidence suggests that taste and/or food aversion induced by GLP-1 is mediated by different CNS pathways (Kinzig et al. 2002; Seeley et al. 2000; Tang-Christensen et al. 1996).

The GLP-1 receptor is widely expressed in the rodent brain in the hypothalamic arcuate nucleus (ARC), the paraventricular nucleus (PVN), and supraoptic nuclei (Shughrue et al. 1996). Furthermore, GLP-1 neurons of the solitary tract predominantly project into the PVN (Larsen et al. 1997a). The food-intake decreasing effect of GLP-1 in rodents is associated with an increase in c-Fos expression in the ARC (Larsen et al. 1997b). In rats treated with monosodium glutamate, the inhibitory effect of GLP-1 on hunger-induced feeding was completely abolished (Tang-Christensen et al. 1998). Additionally, GLP-1 stimulates the electrical activity of proopiomelanocortin (POMC) neurons via the protein kinase A pathway and a consecutive increase of L-type calcium currents (Ma et al. 2007). These findings suggest that the hypothalamic ARC may play a role in GLP-1-induced inhibition of food intake.

The regulation of energy balance involves the interaction of numerous regulatory peptides and neurotransmitters in the hypothalamus. The orexigenic peptides neuropeptide Y (NPY) and agouti-related peptide (AgRP) as well as the anorexigenic peptides POMC and cocaine- and amphetamine-related transcripts (CART) are produced in the ARC of the hypothalamus and play an important role in the regulation of energy intake and energy expenditure (Schwartz et al. 2000). The hypothalamic neurons are sensitive to satiety and hunger signals such as cholecystokinin (CCK) and ghrelin. These hypothalamic neurons are also sensitive to signals of long-term energy stores such as insulin and leptin (Schwartz et al. 2000). However, the effect of GLP-1 on the expression of these hypothalamic mediators is not completely known. NPY and AgRP expression as measured by mRNA concentrations is increased during fasting. These hunger-induced increases are significantly diminished by icv injections of GLP-1. Conversely, the expressions of POMC and CART are decreased during fasting, and again, these changes are attenuated by the icv injection of GLP-1. Additionally, when determining mRNA concentrations of AMP-activated kinase (AMPK), a stimulation of hypothalamic AMPKα2 could be observed during fasting that was also inhibited by GLP-1 application. In summary, these findings suggest that the decreased food intake mediated by GLP-1 is facilitated by the above mentioned changes of orexigenic and anorexigenic hypothalamic neurotransmitter expression changes (Seo et al. 2008).

Another regulatory system most likely to be involved is a GLP-1-mediated activation of the hypothalamo–pituitary–adrenocortical (HPA) axis (Larsen et al. 1997a; Larsen et al. 1997b). This mechanism primarily involves the stimulation of corticotropin-releasing factor (CRF) neurons by GLP-1, and this activation may also be responsible for the inhibition of feeding behavior. There seem to be species differences regarding this regulatory mechanism: in rats, plasma concentrations of corticosterone were rapidly increased after central administration of GLP-1, whereas icv injections of GLP-1 did not alter plasma corticosterone concentrations in the neonatal chick (Furuse et al. 1997).

In neonatal chicks, a noradrenergic mechanism was shown to contribute to the anorexigenic effect of GLP-1 (Bungo et al. 2001a). Icv administration of norepinephrine (NE) suppressed food intake and produced narcolepsy comparable to the effect of GLP-1 in chicks. Although dopamine (DA) did not alter food intake, the coadministration of inhibitors of dopamine-b-hydroxylase (DBH) or fusaric acid (FA) attenuated the suppressive effect of GLP-1 on feeding behavior. Thus, it is suggested that there may be interactive relationships between GLP-1 and noradrenergic regulatory systems in chicks (Bungo et al. 2001b), and that additionally, there may be species differences in GLP-1-mediated appetite control.

3.2 GLP-1 in Taste Cells

GLP-1 and PYY are secreted not only from L cells in the small intestine but also from mammalian taste cells. Both cell types, human duodenal L cells and taste cells of the tongue, express the sweet taste receptor G protein gustducin that may additionally be involved in the regulation of GLP-1 release (Jang et al. 2007). In many L cells, GLP-1, gustducin, and PYY are colocalized (Jang et al. 2007). Furthermore, GLP-1 is produced in two subsets of mammalian taste cells (type 2 and type 3).The corresponding GLP-1 receptors are present on adjacent intragemmal afferent nerve fibers (Shin et al. 2008). It is therefore hypothesized that GLP-1 (and PYY) activates anorexigenic CNS events prior to stimulating islet hormones (Egan and Margolskee 2008).

3.3 Central Effects of GLP-1 Influencing Gastrointestinal Functions

It has also been demonstrated already in 1997 that icv injections of GLP-1 cause a retardation of liquid gastric emptying (Imeryüz et al. 1997). Nakade and his group showed that the peripheral sympathetic nervous system and the central CRF receptors are involved in the central GLP-1-mediated delay of solid gastric emptying in rats (Nakade et al. 2006).

4 Peripheral Effects of GLP-1 Affecting Satiety

GLP-1 exerts potent and important inhibitory effects on gastric emptying and gastric acid secretion. It is primarily responsible for the "ileal break," a tightly regulated process under neural and hormonal control that regulates the passage of nutrients through the digestive tract. GLP-1 enhances satiety and reduces food intake (Pitomobo 2008). GLP-1 inhibits these proximal events of the gastrointestinal tract in a negative feedback manner (Ahren 2004). Nauck and colleagues were able to inhibit gastric emptying after a liquid meal by icv administration of GLP-1 in healthy, normoglycemic volunteers (Nauck et al. 1997). The observed effect of GLP-1 on gastric emptying was dose dependent and highly significant with physiological GLP-1 plasma concentrations (Nauck et al. 1997; Meier et al. 2002, 2003a) (see Fig. 4). Another study in healthy volunteers investigated the effect of two different doses of GLP-1 (0.125-nmol/kg or 0.25-nmol/kg body weight) administered subcutaneously 5 min prior to a mixed test meal (Schirra et al. 1997). The pattern of gastric emptying of the mixed meal as well as pancreatic secretion, antroduodenal motility, and the glycemic response and the release of insulin, C-peptide, and glucagon were quantified. The lag period or the time to reach maximal velocity of gastric emptying was dose-dependently prolonged in response to the subcutaneous application of GLP-1. However, the maximal emptying velocity, the total emptying rate, and the exponential emptying rate were unaltered (Schirra et al. 1997). The subcutaneous infusion of GLP-1 resulted in a dose-dependent inhibition of antral and duodenal motility, and both doses of GLP-1 led to coordinated antroduodenal contractions. GLP-1 initially reduced and then transiently stimulated the secretion of pancreatic enzymes. Both doses of GLP-1 delayed the postprandial insulin peak and enhanced total insulin release. The postprandial response of pancreatic polypeptide and glucagon was diminished (Schirra et al. 1997).

In another study, the same group investigated the antropyloroduodenal motility in humans and the actions of endogenously released GLP-1 on endocrine pancreas secretion (Schirra et al. 2006). In this study, the GLP-1 receptor antagonist exendin (9–39) was used to test whether GLP-1 acts as an incretin and/or as an enterogastrone in humans. The endogenously secreted GLP-1 significantly enhanced postprandial insulin secretion and suppressed the secretion of glucagon (Schirra et al. 2006). During the fasting and postprandial state, antroduodenal motility was inhibited by GLP-1, which qualifies GLP-1 as an enterogastrone. The stimulation of pyloric motility that is induced by intestinal glucose was mediated by GLP-1. The presence of nutrients in the small intestine stimulates the L cells to release GLP-1 into the circulation. The rise in GLP-1 concentrations not only stimulates the beta cells to produce insulin but also slows gastric emptying and may lead to a decrease in appetite and a sensation of fullness (Näslund et al. 1999; Flint et al. 2001; Meier et al. 2003b; Silvestre et al. 2003; Ling et al. 2001; Nagai et al. 2004).

The mechanisms by which GLP-1 inhibits gastric emptying appear to be complex and to involve communication with the central and peripheral nervous systems

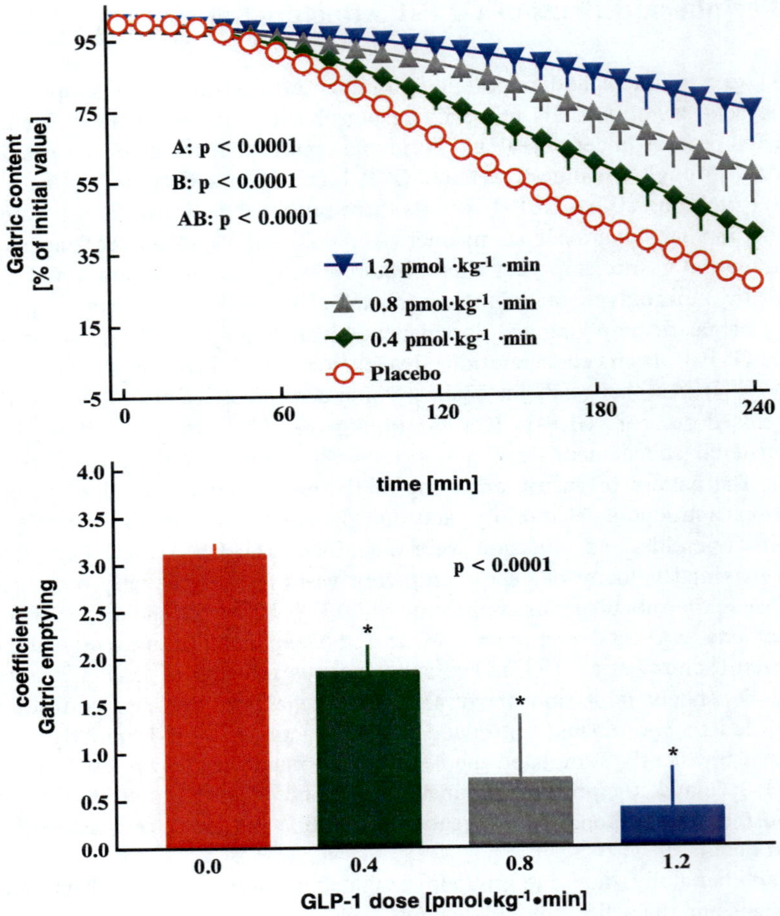

Fig. 4 *Time and dose dependency of GLP-1 on gastric emptying in humans*. Panel A: Time pattern of gastric emptying of a solid meal (250 kcal) during icv administration of different doses of GLP-1 (0.4, 0.8, and 1.2 pmol/kg/min; *filled symbols*) or placebo (*open symbols*) in patients with type 2 diabetes ($n = 12$). Gastric emptying was determined from the measurement of $^{13}CO_2$ in breath samples collected after the ingestion of the test meal labeled with [^{13}C] octanoic acid using infrared absorptiometry. Data are expressed as the mean + SE. *P* values were calculated using repeated measures ANOVA and denote: A, differences between the doses tested; B, differences over time; and AB, differences due to the interaction of experiment and time. Panel B: Gastric emptying coefficients. Data are expressed as the mean + SE. Asterisks indicate significant differences ($P < 0.05$) versus placebo. (from Meier et al. 2003a with permission)

(D'Alessio 2008; Drucker 2006). Gastric distension increases the expression of c-Fos in brain stem neurons that produce GLP-1 (Vrang et al. 2003). Furthermore, icv administration of GLP-1 resulted in reduction of food intake (Kinzig et al. 2002), which is accompanied with increased expression of c-Fos in the brain stem of the rat (Larsen et al. 1997b; Dakin et al. 2004). The denervation of afferent vagal fibers abolishes the effects of GLP-1 on gastric emptying in the rat (Imeryüz et al. 1997). The stimulation to the CNS is most likely responsible for the reduction in food intake, inhibition of gastric emptying, as well as inhibitory action on gastric motor function (Kinzig et al. 2002; Imeryüz et al. 1997). These actions are most likely mediated by increased action potential and calcium influx in neurons of the nodose ganglion (Kakei et al. 2002).

Although small peptides such as GLP-1 and exendin-4 are capable of rapidly crossing the blood–brain barrier and directly accessing the CNS, GLP-1 receptor agonists with a larger molecule size and a higher molecular weight, such as albumin-bound GLP-1, that do not cross the blood–brain barrier, are still capable of inhibiting gastric emptying and food intake (Baggio et al. 2004). These findings underline the importance of ascending vagal afferents for the GLP-1 receptor-dependent control of gastrointestinal motility. Interestingly, studies by Meier and his group (Meier et al. 2005) showed that antagonizing the delaying effects of GLP-1 on gastric emptying by a prokinetic agent such as erythromycin resulted in an augmentation of the insulin secretory response after meal ingestion. GLP-1 receptors are also directly expressed in the stomach on gastric parietal cells, where GLP-1 may directly regulate gastric acid secretion (Schmidtler et al. 1994). However, the effects of GLP-1 on gastric acid secretion were found to be absent in vagotomized human subjects (Wettergren et al. 1997). Hence, considerable evidence supports the importance of vagal innervation for GLP-1 regulation of gastric secretion and motility.

These observed effects of delayed gastric emptying have been generally demonstrated with at least physiological or, as in most studies, supraphysiological doses of exogenously administered GLP-1 (Delgado-Aros et al. 2002; Schirra et al. 1996). Therefore, it remains unclear whether endogenously released GLP-1 has a significant effect on gastric emptying. Studies in healthy baboons have shown that with intragastric infusion of glucose and D-xylose (a marker for rate of emptying of glucose from stomach), plasma levels of D-xylose were similar when the effects of GLP-1 were blocked with exendin(9–36) amide or with a specific monoclonal antibody to GLP-1 (D'Alessio et al. 1996; D'Alessio 2008). These findings suggest that gastric emptying is not increased when the effects of GLP-1 are blocked, at least in the baboon. The use of a DPP-4 inhibitor, which increases plasma concentrations of endogenous GLP-1, might be expected to delay gastric emptying, but a study in patients with type 2 diabetes and DPP-4 inhibitor treatment did not reveal any changes in the gastric emptying of a solid meal (Vella et al. 2007). Most recently, an iv-oral hyperglycemic clamp study in humans was reported during which 75-g glucose-containing D-xylose was ingested. During the entire clamp, plasma glucose levels were held at a steady level despite the ingestion of glucose. Two studies were conducted, with blockade of GLP-1 receptor in one. The rate of appearance of ingested D-xylose was not different between the two studies, indicating that endogenously released GLP-1 has at best only a modest effect on gastric emptying (Salehi et al. 2008).

In a variety of endocrine regulatory systems, a negative feedback mechanism regulates the secretion of the hormone, e.g., the reproductive hormone regulation by the hypothalamus. Exogenous infusion of a hormone may also exert negative feedback regulation of the endogenously released hormone. An example of this is the documented suppression of C-peptide plasma concentrations when insulin is infused (Elahi et al. 1982). In this context, it is presently still not known whether exogenously administered GLP-1 really has a significant impact on the regulation of endogenously released GLP-1.

5 Effects of GLP-1 Receptor Agonists on Body Weight in Humans

Native GLP-1 cannot be used in a feasible way for treatment of type 2 diabetes or obesity due to its very short biological half-life of 1–2 min. For this reason, long-acting GLP-1 receptor agonists were developed. In 1992 exendin-4, a reptilian peptide isolated from the lizard *Heloderma suspectum* was identified as a long-acting GLP-1 receptor agonist (Raufman et al. 1992; Göke et al. 1993). Exendin-4 has a 53% sequence homology with GLP-1 and has a biological half-life of approximately 3.5 h. The synthetic form of exendin-4, exenatide, was the first GLP-1 receptor agonist that was approved for type 2 diabetes therapy in patients not sufficiently controlled on a therapy with metformin or sulfonylureas or a combination of both (Klonoff et al. 2008). Exenatide (Byetta®, Eli Lilly Pharmaceuticals, Indianapolis, USA and Amylin Pharmaceuticals, San Diego, USA) is given subcutaneously twice daily.

Liraglutide (Victoza®, Novo Nordisk Pharmaceuticals, Copenhagen, Denmark) was the first human GLP-1 analogue that was developed for once daily subcutaneous application. Liraglutide has a half-life of 13.5 h. It has a fatty acid side chain that allows heptamer formation of the molecule that prevents direct DPP-4 action as well as fast dissociation from the subcutaneous tissue into the circulation. Furthermore, the fatty acid side chain allows albumin binding that further protracts degradation and prolongs biological availability (Garber et al. 2009; Madsbad et al. 2011).

Presently, even longer-acting GLP-1 receptor analogues are being developed in order to reduce the frequency of injections, to reduce fasting glucose more efficiently, and to reduce the gastrointestinal side effects (mainly fullness and nausea) that are associated with the fluctuation of GLP-1 receptor agonist plasma concentrations. Compounds with an exendin-4/exenatide backbone are exenatide once weekly (Bydureon®, Eli Lilly Pharmaceuticals, Indianapolis, USA and Amylin Pharmaceuticals, San Diego, USA) that has just received a positive opinion by the regulatory agencies in the United States and Europe and is expected to be marketed later in 2011 (Madsbad et al. 2011).

Lixisenatide (Sanofi-Aventis Pharmaceuticals, Paris, France) is an exendin-4 analogue for once daily application, presently in phase III of the clinical study program (Christensen et al. 2011; Madsbad et al. 2011).

Albiglutide, an albumin-bound-human GLP-1 fusion protein (GlaxoSmithKline Pharmaceuticals, London, UK) is feasible for once weekly dosing and is also presently in phase III of the clinical study program (Madsbad et al. 2011; St Onge and Miller 2011).

Another human GLP-1 receptor agonist in earlier development for once weekly dosing is LY2189265 (Dulaglutide, Eli Lilly Pharmaceuticals, Indianapolis, USA) (Glaesner et al. 2010; Madsbad et al. 2011). Table 1 gives an overview on the GLP-1 receptor agonists available and in development and their respective characteristics.

Chronic peripheral administration of GLP-1 RA agonists (exendin-4/exenatide and liraglutide) has consistently been associated with reductions in food intake and weight loss in animal studies and in humans (Szayna et al. 2000; Young et al. 1999; Schnabel et al. 2006; Garber et al. 2009). Also, weight loss was documented in a pivotal study in which a continuous 6-week infusion of GLP-1 was given to obese type 2 diabetic patients (Zander et al. 2002). Conversely, a continuous subcutaneous administration study of a lower dose of GLP-1 (1.5 pmol·kg^{-1}·min^{-1}) for 12 weeks

Table 1 *GLP-1 receptor agonists and their characteristics.* The available GLP-1 receptor agonists as well as substances advanced in clinical development are shown

Substance	Chemical backbone	Dosing interval	Approval/ developmental status
Exenatide (Byetta®, Eli Lilly and Amylin)	Exendin-4	Twice daily	Approved 2005
Liraglutide (Victoza®, NovoNordisk)	Human GLP-1 with two amino acid exchanges and a c-16 fatty acid side chain	Once daily	Approved 2009
Exenatide QW (Bydureon®, Eli Lilly and Amylin)	Exendin-4, incorporated into a matrix of poly(D,L-lactide-co-glycolide) (PLG) to prolong action	Once weekly	Approved 2011
Lixisenatide (Sanofi-Aventis)	Exendin-4, 44 amino acids with C-terminal extension (C-terminal with six Lys residues and one Pro deleted)	Once daily	Phase III clinical investigation
Albiglutide (Syncrea®, GlaxoSmithKline)	Human GLP-1 receptor agonist consisting of two copies of a 30-amino acid sequence of a dipeptyl peptidase-4-resistant human GLP-1 (as a tandem repeat) coupled to serum human albumin	Once weekly (planned)	Phase III clinical investigation
LY2189265 (Dulaglutide®, Eli Lilly)	DPP-4-protected GLP-1 analogue is fused to a modified immunoglobulin G4 (IgG4) Fc fragment	Once weekly (planned)	Phase III clinical investigation

produced no significant weight loss (Meneilly et al. 2003). This study most likely demonstrates that weight loss with exogenous GLP-1 administration is only possible, when much larger doses are given.

In most phase 3 studies with exenatide and liraglutide, the weight loss was in the range of 2–3 kg after 26 weeks of treatment compared with placebo, and greatest when added to metformin (Madsbad 2009). In studies with a longer duration, a plateau of the weight loss is seen in the same range (Garber et al. 2009, 2011).

Presently, long-acting GLP-1 receptor agonists that require a single weekly dose only, or other long-range dosing regimens, are in clinical development (Madsbad et al. 2011). Summarizing the known effects on body weight, there does not seem to be a difference between the short-acting, established GLP-1 RA and the novel long-acting ones. The weight loss in a study comparing the novel, once weekly GLP-1 RA albiglutide with exenatide did not reveal a significant difference in the body weight development. The weight loss amounted from −1.1 to −1.7 kg in the albiglutide groups compared with −0.7 kg in the placebo group and −2.4 kg in the exenatide group (Rosenstock et al. 2009).

Exenatide once weekly (exenatide QW) is a new dosage form of the active drug exenatide. Exenatide QW's microsphere technology enables very slow release due to the use of a slowly biodegradable polymer as the exenatide carrier. This changes both the effect and adverse effect profiles versus the short-acting receptor stimulation produced by the established, unretarded exenatide for twice daily application. Long-term stimulation of GLP-1 receptors results in superior lowering of fasting blood glucose levels and HbA1c (Kim et al. 2007). In subjects on first-line treatment with metformin, exenatide QW produced a superior HbA1c reduction. Because of the weaker inhibition of gastric emptying, gastrointestinal side effects are reduced by about one-third (20% with exenatide QW versus 35% with exenatide) (Linnebjerg et al. 2008; Wang et al. 2008). In addition, a relevant loss in mean weight (ranging from 2.3 kg to 3.7 kg) was seen in all studies (Drucker et al. 2008; Buse et al. 2010; Bergenstal et al. 2010; Diamant et al. 2010; Kim et al. 2007).

Another GLP-1 RA in development is CJC-1134-PC (ConjuChem, Montreal, Quebec, Canada), which consists of an exendin-4 molecule covalently linked to human recombinant albumin. Its half-life of approximately 8 days corresponds to that of circulating albumin (Thibaudeau et al. 2006; Baggio et al. 2008). At present, it is unclear whether the modest effect on body weight is explained by a reduced efficacy in engaging the central nervous system regions regulating appetite and body weight, because large proteins like albumin are not expected to cross the blood–brain barrier (Chuang et al. 2002). Alternatively, the compound can still regulate feeding and body weight via the vagus nerve (Abbott et al. 2005; Imeryüz et al. 1997).

With respect to weight control, no clinically significant differences seem to exist within the entire group of GLP-1 receptor agonists, although it remains possible that the CJC-1134-PC is less effective (Chuang et al. 2002; Wang et al. 2009).

With the short-acting GLP-1 RA liraglutide, a clinical study was performed in obese, nondiabetic subjects to investigate the efficacy and safety as a weight-loss-promoting drug. In a placebo-controlled 20-week trial, with an open-label

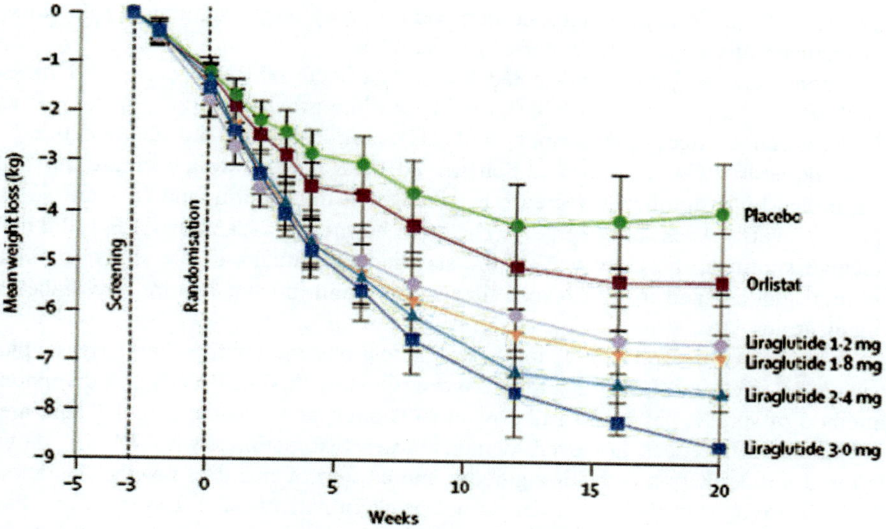

Fig. 5 *Dose dependent effects of the GLP-1 RA liraglutide on body weight in obese subjects*. Five hundred sixty-four individuals (age range 18–65 years, BMI 30–40 kg/m) were randomized to one of four liraglutide doses (1·2 mg, 1·8 mg, 2·4 mg, or 3·0 mg, $n = 90–95$) or to placebo ($n = 98$) administered once a day subcutaneously, or orlistat (120 mg, $n = 95$) three times a day orally. Weight change was analyzed by intention to treat. Data are mean (95% CI) (ANCOVA estimate) for the intention-to-treat population with the last observation carried forward (from Astrup et al. 2009 with permission)

orlistat (120 mg t.i.d.) comparator arm, study participants were assigned to arms of four liraglutide doses (1.2 mg/d, 1.8 mg/d, 2.4 mg/d, or 3.0 mg/d) or to placebo. Participants on liraglutide lost significantly more weight than did those on placebo and orlistat. The mean weight loss caused by liraglutide was dose dependent and amounted to 4.8 kg, 5.5 kg, 6.3 kg, and 7.2 kg for the respective liraglutide doses (1.2 mg/d, 1.8 mg/d, 2.4 mg/d, or 3.0 mg/d) and was 2.1 kg greater than that in the placebo group. The weight loss with placebo amounted to 2.8 kg and with orlistat 4.1 kg (Astrup et al. 2009) (see Fig. 5).

6 GLP-1 and Bariatric Surgery

The mechanisms involved in weight loss and metabolic improvements after bariatric surgery are dependent on the type of surgery performed and are not yet completely understood. They are presently under thorough investigation. Changes in the size of the gastric pouch and the length and parts of the intestinal surfaces after bariatric surgery determine the contact of food with enteroendocrine cells and consequently also the gut hormonal response. Each type of bariatric surgery has a different effect on hormonal secretion and thus may play a significant role in the mechanism of weight loss (Vetter et al. 2009).

Rubino and colleagues evaluated the early effects of Roux-Y gastric bypass (RYGB) on glucose, insulin, glucagon, insulin-like growth factor-1, GIP, GLP-1, CCK, adrenocorticotropic hormone (ACTH), corticosterone, and neuropeptide Y (Rubino et al. 2004; Thomas and Schauer 2010). RYGB led to a decrease in BMI paralleled by a significant decrease in glucose, insulin, leptin, and an increase in ACTH levels 3 weeks after surgery. The other hormones, especially GLP-1, did not change significantly. However, of the six diabetic patients in the study, all had normal glucose and insulin levels after surgery and did not require any diabetic medications.

It has been observed that the initial weight loss observed with either laparoscopic sleeve gastrectomy (LSG) as a restrictive surgical method or RYGB as a diversion method of surgery leads to similar results (Thomas and Schauer 2010). There are metabolic differences however, demonstrating that patients with RYGB show a rapid normalization of fasting glucose and an improvement of insulin clearance and sensitivity, but these changes do not occur in patients with LSG. Testing the different cohorts of patients after RYGB or LSG with a mixed meal tolerance test, the dramatic increase in insulin secretion and an increase in GLP-1 are only observed in the RYGB group (Thomas and Schauer 2010).

In obese patients with coexisting type 2 diabetes, both types of bariatric procedures were associated with an improvement of hyperglycemia. RYGB was associated with an insulinotropic response with an oral mixed meal but not with intravenous glucose, consistent with an incretin effect. These data suggest a different effect of the two procedures on pancreatic beta-cell function. The improvement may be due to insulin sensitivity and therefore a reduced insulin response following gastric restriction only (Thomas and Schauer 2010). It is still not known, however, whether the portion that is bypassed causes this effect or whether this effect is due to nutrients that rapidly reach the distal ileum and release insulinotropic and beta-cell-enhancing hormones (Kashyap et al. 2010). The limitations of the study are the small sample size and the sample being a nonrandomized convenience sample.

To evaluate the potential role of the exclusion of the proximal small intestine in the improvement of diabetes mellitus after gastric bypass surgery, Peterli and his group conducted a prospective, randomized, controlled trial comparing RYGB and LSG. Patients were evaluated 1 week and 3 months after surgery before and after a standard test meal. At 3 months, body weight and BMI decreased significantly and comparably in both groups with markedly increased postprandial plasma insulin and GLP-1 levels (Dirksen et al. 2010). RYGB patients had increased insulin responses as early as 1 week after the surgery; however, no significant differences were seen at 3 months in insulin or GLP-1 levels. Thus, both procedures improved glucose homeostasis, insulin, and GLP-1 and PYY levels (Peterli et al. 2009).

Two different hypotheses have been proposed to explain these conflicting data. One offered is the "hindgut explanation," suggesting that the rapid transit of nutrients to the distal intestine improves glucose metabolism by stimulating secretion of GLP-1 and other appetite-suppressing gut peptides such as PYY. Insulin secretion is increased and glucose tolerance improves, affecting body weight and food intake (Cummings et al. 2007; Patrita et al. 2007). On the other hand, Rubino and his group proposed the "foregut hypothesis." They propose that there is a yet-unknown factor that promotes insulin resistance and type 2 diabetes. When food bypasses the duodenum and proximal jejunum after bariatric surgery, this so-called anti-incretin or decretin factor is inhibited, and thus insulin resistance is decreased and glucose tolerance improves. Other factors may help to explain the differences seen among the various types of procedures (Vetter et al. 2009; Cummings et al. 2008). Likewise, because GLP-1, PYY, and GIP are secreted by the small intestine, differences in the length of the roux limb may contribute to the secretion of these gut hormones and thus the results seen. Finally, as this is a relatively new, evolving field of research, there are most likely unknown factors to be considered (Thomas and Schauer 2010).

References

Abbott CR, Monteiro M, Small CJ, Sajedi A, Smith KL, Parkinson JR, Ghatei MA, Bloom SR (2005) The inhibitory effects of peripheral administration of peptide YY(3–36) and glucagon-like peptide-1 on food intake are attenuated by ablation of the vagal-brainstem-hypothalamic pathway. Brain Res 1044:127–131

Abu-Hamdah R, Rabiee A, Meneilly GS, Shannon RP, Andersen DK, Elahi D (2009) Clinical review: the extrapancreatic effects of glucagon-like peptide-1 and related peptides. J Clin Endocrinol Metab 94:1843–1852

Ahren B (2004) GLP-1 and extra-islet effects. Horm Metab Res 36:842–845

Astrup A, Rössner S, Van Gaal L, Rissanen A, Niskanen L, Al Hakim M, Madsen J, Rasmussen MF, Lean ME, NN8022-1807 Study Group (2009) Effects of liraglutide in the treatment of obesity: a randomised, double-blind, placebo-controlled study. Lancet 374:1606–1616

Baggio LL, Drucker DJ (2007) Biology of incretins: GLP-1 and GIP. Gastroenterology 132:2131–2157

Baggio LL, Huang Q, Brown TJ, Drucker DJ (2004) A recombinant human glucagon-like peptide (GLP)-1-albumin protein (albugon) mimics peptidergic activation of GLP-1R-dependent pathways coupled with satiety, gastrointestinal motility, and glucose homeostasis. Diabetes 53:2492–2500

Baggio LL, Huang Q, Cao X, Drucker DJ (2008) An albumin-exendin-4 conjugate engages central and peripheral circuits regulating murine energy and glucose homeostasis. Gastroenterology 134:1137–1147

Batterham RL, Cowley MA, Small CJ, Herzog H, Cohen MA, Dakin CL, Wren AM, Brynes AE, Low MJ, Ghatei MA, Cone RD, Bloom SR (2002) Guthormone PYY(3–36) physiologically inhibits food intake. Nature 418:650–654

Batterham RL, Cohen MA, Ellis SM, Le Roux CW, Withers DJ, Frost GS, Ghatei MA, Bloom SR (2003) Inhibition of food intake in obese subjects by peptide YY3-36. N Engl J Med 349:941–948

Beak SA, Small CJ, Ilovaiskaia I, Hurley JD, Ghatei MA, Bloom SR, Smith DM (1996) Glucagon-like peptide-1 (GLP-1) releases thyrotropin (TSH): characterization of binding sites for GLP-1 on α-TSH cells. Endocrinology 137:4130–4138

Beak SA, Heath MM, Small CJ, Morgan DG, Ghatei MA, Taylor AD, Buckingham JC, Bloom SR, Smith DM (1998) Glucagon-like peptide-1 stimulates luteinizing hormone-releasing hormone secretion in a rodent hypothalamic neuronal cell line. J Clin Invest 101:1334–1341

Bell GI, Sanchez-Pescador R, Laybourn PJ, Najarian RC (1983) Exon duplication and divergence in the human preproglucagon gene. Nature 304:368–371
Bergenstal RM, Wysham C, Macconell L, Malloy J, Walsh B, Yan P, Wilhelm K, Malone J, Porter LE (2010) Efficacy and safety of exenatide once weekly versus sitagliptin or pioglitazone as an adjunct to metformin for treatment of type 2 diabetes (DURATION-2): a randomised trial. Lancet 376:431–439
Boggiano MM, Chandler PC, Oswald KD, Rodgers RJ, Blundell JE, Ishii Y, Beattie AH, Holch P, Allison DB, Schindler M, Arndt K, Rudolf K, Mark M, Schoelch C, Joost HG, Klaus S, Thone-Reineke C, Benoit SC, Seeley RJ, Beck-Sickinger AG, Koglin N, Raun K, Madsen K, Wulff BS, Stidsen CE, Birringer M, Kreuzer OJ, Deng XY, Whitcomb DC, Halem H, Taylor J, Dong J, Datta R, Culler M, Ortmann S, Castaneda TR, Tschop M (2005) PYY3-36 as an anti-obesity drug target. Obesity Rev 6:307–322
Brubaker PL, Drucker DJ (2004) Minireview: glucagon-like peptides regulate cell proliferation and apoptosis in the pancreas, gut, and central nervous system. Endocrinology 145:2653–2659
Bungo T, Ando R, Kawakami S-I, Ohgushi A, Furuse M (2001a) The role of central catecholaminergic systems in regulation of food intake of chicks. J Poult Sci 38:35–40
Bungo T, Kawakami S-I, Ohgushi A, Sashihara K, Saito N, Sugahara K, Hasegawa S, Denbow DM, Furuse M (2001b) Intracerebroventricular injection of fusaric acid attenuates the anorexia by glucagon-like peptide-1 in the neonatal chick. Pharmacol Biochem Behav 70:251–255
Buse JB, Drucker DJ, Taylor KL, Kim T, Walsh B, Hu H, Wilhelm K, Trautmann M, Shen LZ, Porter LE (2010) DURATION-1: exenatide once weekly produces sustained glycemic control and weight loss over 52 weeks. Diabetes Care 33:1255–1261
Christensen M, Knop FK, Vilsbøll T, Holst JJ (2011) Lixisenatide for type 2 diabetes mellitus. Expert Opin Investig Drugs 20:549–557
Chuang VT, Kragh-Hansen U, Otagiri M (2002) Pharmaceutical strategies utilizing recombinant human serum albumin. Pharm Res 19:569–577
Courreges JP, Vilsboll T, Zdravkovic M, Le-Thi T, Krarup T, Schmitz O, Verhoeven R, Buganova I, Madsbad S (2008) Beneficial effects of once-daily liraglutide, a human glucagon-like peptide-1 analogue, on cardiovascular risk biomarkers in patients with Type 2 diabetes. Diabet Med 25:1129–1131
Creutzfeldt W (1979) The incretin concept today. Diabetologia 16:75–85
Cummings D, Overduin J, Foster-Schubert K, Carlson M (2007) Role of the bypassed proximal intestine in the anti-diabetic effects of bariatric surgery. Surg Obes Relat Dis 3:109–115
Cummings D, Foster-Schubert K, Carlson M, Shannon M, Overduin J (2008) Possible hormonal mechanisms mediating the effects of bariatric surgery. In: Pitomobo C, Jones K, Higa K, Pareja J (eds) Obesity surgery: principles and practice. McGraw-Hill, New York, NY, pp 137–147
D'Alessio D (2008) Intestinal hormones and regulation of satiety: the case for CCK, GLP-1, PYY, and Apo A-IV. JPEN J Parenter Enteral Nutr 32:567–568
D'Alessio DA, Vogel R, Prigeon R, Laschansky E, Koerker D, Eng J, Ensinck JW (1996) Elimination of the action of glucagon-like peptide 1 causes an impairment of glucose tolerance after nutrient ingestion by healthy baboons. J Clin Invest 97:133–138
Dakin CL, Small CJ, Batterham RL, Neary NM, Cohen MA, Patterson M, Ghatei MA, Bloom SR (2004) Peripheral oxyntomodulin reduces food intake and body weight gain in rats. Endocrinology 145:2687–2695
Delgado-Aros S, Kim DY, Burton DD, Thomforde GM, Stephens D, Brinkmann BH, Vella A, Camilleri M (2002) Effect of GLP-1 on gastric volume, emptying, maximum volume ingested, and postprandial symptoms in humans. Am J Physiol Gastrointest Liver Physiol 282: G424–G431
Diamant M, Van GL, Stranks S, Northrup J, Cao D, Taylor K, Trautmann M (2010) Once weekly exenatide compared with insulin glargine titrated to target in patients with type 2 diabetes (DURATION-3): an open-label randomised trial. Lancet 375:2234–2243
Dirksen C, Hansen DL, Madsbad S, Hvolris LE, Naver LS, Holst JJ, Worm D (2010) Postprandial diabetic glucose tolerance is normalized by gastric bypass feeding as opposed to gastric

feeding and is associated with exaggerated GLP-1 secretion: a case report. Diabetes Care 33:375–377
Drucker DJ (2006) The biology of incretin hormones. Cell Metab 3:153–165
Drucker DJ, Nauck MA (2006) The incretin system: glucagon-like peptide-1 receptor agonists and dipeptidyl peptidase-4 inhibitors in type 2 diabetes. Lancet 368:1696–1705
Drucker DJ, Buse JB, Taylor K, Kendall DM, Trautmann M, Zhuang D, Porter L (2008) Exenatide once weekly versus twice daily for the treatment of type 2 diabetes: a randomised, open-label, non-inferiority study. Lancet 372:1240–1250
Egan JM, Margolskee RF (2008) Taste cells of the gut and gastrointestinal chemosensation. Mol Interv 8:78–81
Elahi D, Nagulesparan M, Hershcopf RJ, Muller DC, Tobin JD, Blix PM, Rubenstein AH, Unger RH, Andres R (1982) Feedback inhibition of insulin secretion by insulin: relation to the hyperinsulinemia of obesity. N Engl J Med 306:1196–1202
Fehmann HC, Habener JF (1992) Insulinotropic hormone glucagon-like peptide-I(7–37) stimulation of proinsulin gene expression and proinsulin biosynthesis in insulinoma beta TC-1 cells. Endocrinology 130:159–166
Flint A, Raben A, Astrup A, Holst JJ (1998) Glucagon-like peptide 1 promotes satiety and suppresses energy intake in humans. J Clin Invest 101:515–520
Flint A, Raben A, Ersbøll AK, Holst JJ, Astrup A (2001) The effect of physiological levels of glucagon-like peptide-1 on appetite, gastric emptying, energy and substrate metabolism in obesity. Int J Obes Relat Metab Disord 25:781–792
Furuse M, Matsumoto M, Saito N, Sugahara K, Hasegawa S (1997) The central corticotropin-releasing factor and glucagon-like peptide-1 in food intake of the neonatal chick. Eur J Pharmacol 339:211–213
Garber A, Henry R, Ratner R, Garcia-Hernandez PA, Rodriguez-Pattzi H, Olvera-Alvarez I, Hale PM, Zdravkovic M, Bode B (2009) Liraglutide versus glimepiride monotherapy for type 2 diabetes (LEAD-3 Mono): a randomised,52-week, phase III, double-blind, parallel-treatment trial. Lancet 373:473–481
Garber A, Henry RR, Ratner R, Hale P, Chang CT, Bode B, LEAD-3 (Mono) Study Group (2011) Liraglutide, a once-daily human glucagon-like peptide 1 analogue, provides sustained improvements in glycaemic control and weight for 2 years as monotherapy compared with glimepiride in patients with type 2 diabetes. Diabetes Obes Metab 13:348–356
Glaesner W, Vick AM, Millican R, Ellis B, Tschang SH, Tian Y, Bokvist K, Brenner M, Koester A, Porksen N, Etgen G, Bumol T (2010) Engineering and characterization of the long-acting glucagon-like peptide-1 analogue LY2189265, an Fc fusion protein. Diabetes Metab Res Rev 26:287–296
Göke R, Fehmann HC, Linn T, Schmidt H, Krause M, Eng J, Göke B (1993) Exendin-4 is a high potency agonist and truncated exendin-(9–39)-amide an antagonist at the glucagon-like peptide 1-(7–36)-amide receptor of insulin-secreting beta-cells. J Biol Chem 268:19650–19655
Göke R, Larsen PJ, Mikkelsen JD, Sheikh SP (1995) Distribution of GLP-1 binding sites in the rat brain: evidence that exendin-4 is a ligand of brain GLP-1 binding sites. Eur J Neurosci 7:2294–2300
Gromada J, Rorsman P, Dissing S, Wulff BS (1996) Stimulation of cloned human glucagon-like peptide 1 receptor expressed in HEK 293 cells induces cAMP-dependent activation of calcium-induced calcium release. FEBS Lett 373:182–186
Gutzwiller JP, Drewe J, Goke B, Schmidt H, Rohrer B, Lareida J, Beglinger C (1999) Glucagon-like peptide-1 promotes satiety and reduces food intake in patients with diabetes mellitus type 2. Am J Physiol 276:R1541–R1544
Holst JJ (2007) The physiology of glucagon-like peptide 1. Physiol Rev 87:1409–1439
Holst JJ, Deacon CF, Vilsbøll T, Krarup T, Madsbad S (2008) Glucagon-like peptide-1, glucose homeostasis and diabetes. Trends Mol Med 14:161–168
Imeryüz N, Yeden BC, Bozkurt A, Copkun T, Villanueva-Peñacarrillo ML, Ulusoy NB (1997) Glucagon-like peptide-1 inhibits gastric emptying via vagal afferent-mediated central mechanisms. Am J Physiol 273:G920–G927

Jang HJ, Kokrashvili Z, Theodorakis MJ, Carlson OD, Kim BJ, Zhou J, KimHH XuX, Chan SL, Juhaszova M, Bernier M, Mosinger B, Margolskee RF, Egan JM (2007) Gut-expressed gustducin and taste receptors regulate secretion of glucagon-like peptide-1. Proc Natl Acad Sci USA 104:15069–15074

Jin SL, Han VK, Simmons JG, Towle AC, Lauder JM, Lund PK (1988) Distribution of glucagonlike peptide I (GLP-I), glucagon, and glicentin in the rat brain: an immunocytochemical study. J Comp Neurol 271:519–532

Kakei M, Yada T, Nakagawa A, Nakabayashi H (2002) Glucagon-like peptide-1 evokes action potentials and increases cytosolic Ca2+ in rat nodose ganglion neurons. Auton Neurosci 102:39–44

Kashyap SR, Daud S, Kelly KR, Gastaldelli A, Win H, Brethauer S, Kirwan JP, Schauer PR (2010) Acute effects of gastric bypass versus gastric restrictive surgery on beta-cell function and insulinotropic hormones in severely obese patients with type 2 diabetes. Int J Obes (Lond) 34:462–471

Kim D, Macconell L, Zhuang D, Kothare PA, Trautmann M, Fineman M, Taylor K (2007) Effects of once-weekly dosing of a long-acting release formulation of exenatide on glucose control and body weight in subjects with type 2 diabetes. Diabetes Care 30:1487–1493

Kinzig KP, D'Alessio DA, Seeley RJ (2002) The diverse roles of specific GLP-1Rs in the control of food intake and the response to visceral illness. J Neurosci 22:10470–10476

Klonoff DC, Buse JB, Nielsen LL, Guan X, Bowlus CL, Holcombe JH, Wintle ME, Maggs DG (2008) Exenatide effects on diabetes, obesity, cardiovascular risk factors and hepatic biomarkers in patients with type 2 diabetes treated for at least 3 years. Curr Med Res Opin 24:275–286

Larsen PJ, Tang-Christensen M, Holst JJ, Ørskov C (1997a) Distribution of glucagon-like peptide-1 and other preproglucagon-derived peptides in the rat hypothalamus and brainstem. Neuroscience 77:257–270

Larsen PJ, Tang-Christensen M, Jessop DS (1997b) Central administration of glucagon-like peptide-1 activates hypothalamic neuroendocrine neurons in the rat. Endocrinology 138:4445–4455

Ling Z, Wu D, Zambre Y, Flamez D, Drucker DJ, Pipeleers DG, Schuit FC (2001) Glucagon-like peptide 1 receptor signaling influences topography of islet cells in mice. Virchows Arch 438:382–387

Linnebjerg H, Park S, Kothare PA, Trautmann ME, Mace K, Fineman M, Wilding I, Nauck M, Horowitz M (2008) Effect of exenatide on gastric emptying and relationship to postprandial glycemia in type 2 diabetes. Regul Pept 151:123–129

Ma X, Bruning J, Ashcroft FM (2007) Glucagon-like peptide 1 stimulates hypothalamic proopiomelanocortin neurons. J Neurosci 27:7125–7129

Madsbad S (2009) Exenatide and liraglutide: different approaches to develop GLP-1 receptor agonists (incretin mimetics) - preclinical and clinical results. Best Pract Res Clin Endocrinol Metab 23:463–477

Madsbad S, Kielgast U, Asmar M, Deacon C, Torekov SS, Holst JJ (2011) An overview of once-weekly GLP-1 receptor agonists - available efficacy and safety data and perspectives for the future. Diabetes Obes Metab 13:394–407

Meeran K, O'Shea D, Edwards CM, Turton MD, Heath MM, Gunn I, Abusnana S, Rossi M, Small CJ, Goldstone AP, Taylor GM, Sunter D, SteereJ CSJ, Ghatei MA, Bloom SR (1999) Repeated intracerebroventricular administration of glucagon-like peptide-1-(7–36) amide or exendin-(9–39) alters body weight in the rat. Endocrinology 140:244–250

Meier JJ, Gallwitz B, Schmidt WE, Nauck MA (2002) Glucagon-like peptide 1 as a regulator of food intake and body weight: therapeutic perspectives. Eur J Pharmacol 440:269–279

Meier JJ, Gallwitz B, Salmen S, Goetze O, Holst JJ, Schmidt WE, Nauck MA (2003a) Normalization of glucose concentrations and deceleration of gastric emptying after solid meals during intravenous glucagon-like peptide 1 in patients with type 2 diabetes. J Clin Endocrinol Metab 88:2719–2725

Meier S, Hücking K, Ritzel R, Holst JJ, Schmiegel WH, Nauck MA (2003b) Absence of a memory effect for the insulinotropic action of glucagon-like peptide 1 (GLP-1) in healthy volunteers. Horm Metab Res 35:551–556

Meier JJ, Kemmeries G, Holst JJ, Nauck MA (2005) Erythromycin antagonizes the deceleration of gastric emptying by glucagon-like peptide 1 and unmasks its insulinotropic effect in healthy subjects. Diabetes 54:2212–2218

Meneilly GS, Greig N, Tildesley H, Habener JF, Egan JM, Elahi D (2003) Effects of 3 months of continuous subcutaneous administration of glucagon-like peptide 1 in elderly patients with type 2 diabetes. Diabetes Care 26:2835–2841

Nagai K, Tsuchiya K, Ezaki T, Tsuchiya M, Ohgawara H (2004) Effect of GLP-1 (glucagon-like peptide 1:7–36 amide) on porcine pancreatic endocrine cell proliferation and insulin secretion. Pancreas 28:138–145

Nakade Y, Tsukamoto K, Pappas TN, Takahashi T (2006) Central glucagon like peptide-1 delays solid gastric emptying via central CRF and peripheral sympathetic pathway in rats. Brain Res 1111:117–121

Näslund E, Barkeling B, King N, Gutniak M, Blundell JE, Holst JJ, Rössner S, Hellström PM (1999) Energy intake and appetite are suppressed by glucagon-like peptide-1 (GLP-1) in obese men. Int J Obes Relat Metab Disord 23:304–311

Näslund E, King N, Mansten S, Adner N, Holst JJ, Gutniak M, Hellström PM (2004) Prandial subcutaneous injections of glucagon-like peptide-1 cause weight loss in obese human subjects. Br J Nutr 91:439–446

Nauck M, Stöckmann F, Ebert R, Creutzfeldt W (1986) Reduced incretin effect in type 2 (non-insulin-dependent) diabetes. Diabetologia 29:46–52

Nauck MA, Kleine N, Ørskov C, Holst JJ, Willms B, Creutzfeldt W (1993) Normalization of fasting hyperglycaemia by exogenous glucagon-like peptide 1 (7–36 amide) in type 2 (non-insulin-dependent) diabetic patients. Diabetologia 36:741–744

Nauck MA, Niedereichholz U, Ettler R, Holst JJ, Ørskov C, Ritzel R, Schmiegel WH (1997) Glucagon-like peptide 1 inhibition of gastric emptying outweighs its insulinotropic effects in healthy humans. Am J Physiol 273:E981–E988

Ørskov C, Holst JJ, Poulsen SS, Kirkegaard P (1987) Pancreatic and intestinal processing of proglucagon in man. Diabetologia 30:874–881

Ørskov C, Poulsen SS, Moller M, Holst JJ (1996) Glucagon-like peptide I receptors in the subfornical organ and the area postrema are accessible to circulating glucagon-like peptide I. Diabetes 45:832–835

Patrita A, Aisa M, Annetti C, Sidoni A, Galli F, Ferri I, Gullà N, Donini A (2007) How the hindgut can cure type 2 diabetes. Ileal transposition improves glucose metabolism and β-cell function in Goto-kakizaki rats through enhanced proglucagon gene expression and L-cell number. Surgery 142:74–85

Pellissier S, Sasaki K, Le-Nguyen D, Bataille D, Jarrousse C (2004) Oxyntomodulin and glicentin are potent inhibitors of the fed motility pattern in small intestine. Neurogastroenterol Motil 16:455–463

Peterli R, Wölnerhanssen B, Peters T, Devaux N, Kern B, Christoffel-Courtin C, Drewe J, von Flüe M, Beglinger C (2009) Improvement in glucose metabolism after bariatric surgery: comparison of laparoscopic Roux-en-Y gastric bypass and laparoscopic sleeve gastrectomy: a prospective randomized trial. Ann Surg 250:234–241

Pitomobo C (2008) Central nervous system regulation and hormonal signaling. In: Pitomobo C, Jones K, Higa K, Pareja C (eds) Obesity surgery: principles and practice. McGraw-Hill, New York, NY

Raufman JP, Singh L, Singh G, Eng J (1992) Truncated glucagon-like peptide-1 interacts with exendin receptors on dispersed acini from guinea pig pancreas. Identification of a mammalian analogue of the reptilian peptide exendin-4. J Biol Chem 267:21432–21437

Reimer RA (2006) Meat hydrolysate and essential amino acid-induced glucagon-like peptide-1 secretion, in the human NCI-H716 enteroendocrine cell line, is regulated by extracellular

signal-regulated kinase1/2 and p38 mitogen-activated protein kinases. J Endocrinol 191:159–170
Rosenstock J, Reusch J, Bush M, Yang F, Stewart M (2009) Potential of albiglutide, a long-acting GLP-1 receptor agonist, in type 2 diabetes: a randomized controlled trial exploring weekly, biweekly, and monthly dosing. Diabetes Care 32:1880–1886
Rubino F, Gagner M, Gentileschi P, Kini S, Fukuyama S, Feng J, Diamond E (2004) The early effect of the Roux-en-Y gastric bypass on hormones involved in body weight regulation and glucose metabolism. Ann Surg 240:236–242
Salehi M, Vahl TP, D'Alessio DA (2008) Regulation of islet hormone release and gastric emptying by endogenous GLP-1 following glucose ingestion. J Clin Endocrinol Metab 93:4909–4916
Schirra J, Katschinski M, Weidmann C, Schäfer T, Wank U, Arnold R, Göke B (1996) Gastric emptying and release of incretin hormones after glucose ingestion in humans. J Clin Invest 97:92–103
Schirra J, Kuwert P, Wank U, Leicht P, Arnold R, Göke B, Katschinski M (1997) Differential effects of subcutaneous GLP-1 on gastric emptying, antroduodenal motility, and pancreatic function in men. Proc Assoc Am Physicians 109:84–97
Schirra J, Nicolaus M, Roggel R, Katschinski M, Storr M, Woerle HJ, Göke B (2006) Endogenous glucagon-like peptide 1 controls endocrine pancreatic secretion and antro-pyloro-duodenal motility in humans. Gut 55:243–251
Schmidtler J, Dehne K, Allescher HD, Schusdziarra V, Classen M, Holst JJ, Polack A, Schepp W (1994) Rat parietal cell receptors for GLP-1-(7–36) amide: Northern blot, cross-linking, and radioligand binding. Am J Physiol 267:G423–G432
Schnabel CA, Wintle M, Kolterman O (2006) Metabolic effects of the incretin mimetic exenatide in the treatment of type 2 diabetes. Vasc Health Risk Manag 2:69–77
Schwartz MW, Woods SC, Porte D Jr, Seeley RJ, Baskin DG (2000) Central nervous system control of food intake. Nature 404:661–671
Seeley RJ, Blake K, Rushing PA, Benoit S, Eng J, Woods SC, D'Alessio D (2000) The role of CNS glucagon-like peptide-1 (7–36) amide receptors in mediating the visceral illness effects of lithium chloride. J Neurosci 20:1616–1621
Seo S, Ju S, Chung H, Lee D, Park S (2008) Acute effects of glucagon-like peptide-1 on hypothalamic neuropeptide and AMP activated kinase expression in fasted rats. Endocr J 55:867–874
Shin YK, Martin B, Golden E, Dotson CD, Maudsley S, Kim W, Jang HJ, Mattson MP, Drucker DJ, Egan JM, Munger SD (2008) Modulation of taste sensitivity by GLP-1 signaling. J Neurochem 106:455–463
Shughrue PJ, Lane MV, Merchenthaler I (1996) Glucagon-like peptide-1 receptor (GLP1-R) mRNA in the rat hypothalamus. Endocrinology 137:5159–5162
Silvestre RA, Rodríguez-Gallardo J, Egido EM, Marco J (2003) Interrelationship among insulin, glucagon and somatostatin secretory responses to exendin-4 in the perfused rat pancreas. Eur J Pharmacol 469:195–200
Sokos GG, Nikolaidis LA, Mankad S, Elahi D, Shannon RP (2006) Glucagon-like peptide-1 infusion improves left ventricular ejection fraction and functional status in patients with chronic heart failure. J Card Fail 12:694–699
St Onge EL, Miller SA (2011) Albiglutide: a new GLP-1 analog for the treatment of type 2 diabetes. Expert Opin Biol Ther 10:801–806
Szayna M, Doyle ME, Betkey JA, Holloway HW, Spencer RG, Greig NH, Egan JM (2000) Exendin-4 decelerates food intake, weight gain, and fat deposition in Zucker rats. Endocrinology 141:1936–1941
Tang-Christensen M, Larsen PJ, Göke R, Fink-Jensen A, Jessop DS, Møller M, Sheikh SP (1996) Central administration of GLP-1-(7–36) amide inhibits food and water intake in rats. Am J Physiol 271:R848–R856
Tang-Christensen M, Vrang N, Larsen PJ (1998) Glucagon-like peptide 1(7–36) amide's central inhibition of feeding and peripheral inhibition of drinking are abolished by neonatal monosodium glutamate treatment. Diabetes 47:530–537

Thibaudeau K, Robitaille M, Wen S et al (2006) CJC-1134-PC: an exendin-4 conjugate with extended pharmacodynamic profiles in rodents. Diabetes 55(suppl 1):A103

Thomas S, Schauer P (2010) Bariatric surgery and the gut hormone response. Nutr Clin Pract 25:175–182

Tschöp M, Castaneda T, Joost HG, Thöne-Reineke C, Ortmann S, Klaus S, Hagan HH, Chandler PC, Oswald KD, Benoit SC, Seeley RJ, Kinzig KP, Moran TH, Beck-Sickinger AG, Koglin N, Rodgers RJ, Blundell JE, Ishii Y, Beattie AH, Holch P, Allison DB, Raun K, Madsen K, Wulff BS, Stidsen CE, Birringer M, Kreuzer O, Schindler M, Arndt K, Rudolf K, Mark M, Deng XY, Withcomb DC, Halem H, Taylor J, Dong J, Datta R, Culler M, Craney S, Flora D, Smiley D, Heiman ML (2004) Does gut hormone PYY3-36 decrease food intake in rodents? Nature 430:165–166

Turton MD, O'Shea D, Gunn I, Beak SA, Edwards CM, Meeran K, Choi SJ, Taylor GM, Heath MM, Lambert PD, Wilding JP, Smith DM, Ghatei MA, Herbert J, Bloom SR (1996) A role for glucagon-like peptide-1 in the central regulation of feeding. Nature 379:69–72

Van Dijk G, Thiele TE, Donahey JC, Campfield LA, Smith FJ, Burn P, Bernstein IL, Woods SC, Seeley RJ (1996) Central infusions of leptin and GLP-1-(7–36) amide differentially stimulate c-FLI in the rat brain. Am J Physiol 271:R1096–R1100

Vella A, Bock G, Giesler PD, Burton DB, Serra DB, Saylan ML, Dunning BE, Foley JE, Rizza RA, Camilleri M (2007) Effects of dipeptidyl peptidase-4 inhibition on gastrointestinal function, meal appearance, and glucose metabolism in type 2 diabetes. Diabetes 56:1475–1480

Verdich C, Flint A, Gutzwiller JP, Näslund E, Beglinger C, Hellström PM, Long SJ, Morgan LM, Holst JJ, Astrup A (2001) A meta-analysis of the effect of glucagon-like peptide-1 (7–36) amide on ad libitum energy intake in humans. J Clin Endocrinol Metab 86:4382–4389

Vetter ML, Cardillo S, Rickels MR, Iqbal N (2009) Narrative review: effect of bariatric surgery on type 2 diabetes mellitus. Ann Intern Med 150:94–103

Vrang N, Phifer CB, Corkern MM, Berthoud HR (2003) Gastric distension induces c-Fos in medullary GLP-1/2-containing neurons. Am J Physiol Regul Integr Comp Physiol 285: R470–R478

Wang GJ, Tomasi D, Backus W, Wang R, Telang F, Geliebter A, Korner J, Bauman A, Fowler JS, Thanos PK, Volkow ND (2008) Gastric distention activates satiety circuitry in the human brain. Neuroimage 39:1824–1831

Wang M, Matheson S, Picard J, Pezzullo J, Ulich T (2009) PC-DAC (TM): exendin-4 (CJC-1134-PC) significantly reduces HbA1c and body weight as an adjunct therapy to metformin: two randomized, double-blind, placebo-controlled, 12 week, phase II studies in patients with type 2 diabetes mellitus. Diabetes 58(suppl 1):A148

Wellendorph P, Johansen LD, Brauner-Osborne H (2009) Molecular pharmacology of promiscuous seven transmembrane receptors sensing organic nutrients. Mol Pharmacol 76:453–465

Wettergren A, Wøjdemann M, Meisner S, Stadil F, Holst JJ (1997) The inhibitory effect of glucagon-like peptide-1 (GLP-1) 7–36 amide on gastric acid secretion in humans depends on an intact vagal innervation. Gut 40:597–601

Young AA, Gedulin BR, Bhavsar S, Bodkin N, Jodka C, Hansen B, Denaro M (1999) Glucose-lowering and insulin-sensitizing actions of exendin-4: studies in obese diabetic (ob/ob, db/db) mice, diabetic fatty Zucker rats, and diabetic rhesus monkeys (Macacamulatta). Diabetes 48:1026–1034

Zander M, Madsbad S, Madsen JL, Holst JJ (2002) Effect of 6-week course of glucagon-like peptide 1 on glycaemic control, insulin sensitivity, and β-cell function in type 2 diabetes: a parallel-group study. Lancet 359:824–830

CCK, PYY and PP: The Control of Energy Balance

K. Simpson, J. Parker, J. Plumer, and S. Bloom

Contents

Abstract The control of food intake consists of neural and hormonal signals between the gut and central nervous system (CNS). Gut hormones such as CCK, PYY and PP signal to important areas in the CNS involved in appetite regulation to terminate a meal. These hormones can act directly via the circulation and activate their respective receptors in the hypothalamus and brainstem. In addition, gut vagal afferents also exist, providing an alternative pathway through which gut hormones can communicate with higher centres through the brainstem. Animal and human studies have demonstrated that peripheral administration of certain gut hormones reduces food intake and leads to weight loss. Gut hormones are therefore potential

K. Simpson • J. Parker • J. Plumer • S. Bloom (✉)
Department of Investigative Medicine, Imperial College London, London W12 0HS, UK
e-mail: s.bloom@imperial.ac.uk

H.-G. Joost (ed.), *Appetite Control*, Handbook of Experimental Pharmacology 209,
DOI 10.1007/978-3-642-24716-3_9, © Springer-Verlag Berlin Heidelberg 2012

targets in the development of novel treatments for obesity and analogue therapies are currently under investigation.

Keywords Brainstem • Cholecystokinin • Gut • Hypothalamus • Obesity • Pancreatic polypeptide • Peptide YY • Vagal afferents

1 Introduction

The World Health Organisation estimates that over one billion adults and over forty-three million children under the age of five are overweight (World Health 2011; Tunstall-Pedoe 2005). When homeostatic mechanisms controlling food intake are unable to compensate for excess energy intake and decreased exercise, the end result is weight gain. The control of food intake consists of neural and hormonal signals between the gut and central nervous system (CNS). The brain is responsible for the interpretation of these signals from the periphery and modifying feeding behaviour depending on energy requirements and the "wanting" of food. Even prior to eating, gut hormones are released and begin the process of meal termination and feelings of satiation. Gut hormones act within key brain areas such as the hypothalamus and brainstem, which contain intricate neuronal networks and connections related to energy homeostasis, regulation of food intake and glucose control.

1.1 Gut–Brain Axis

Anticipation of a meal, mechanical stimulation due to the presence of food in the stomach and gut nutrient content, stimulate secretion of gut hormones which activate signalling pathways from the gut to the brainstem and hypothalamus (the gut–brain axis) to terminate food consumption. Such "anorectic" hormones include peptide tyrosine tyrosine (PYY), pancreatic polypeptide (PP), cholecystokinin (CCK), oxyntomodulin (OXM) and glucagon-like peptide 1 (GLP-1). Circulating levels of these hormones rise following a meal and are proportional to the caloric intake and composition of a meal (Adrian et al. 1985; Jorde and Burhol 1984; Vahl et al. 2010; Ghatei et al. 1983). In contrast, ghrelin is an "orexigenic" hormone and initiates hunger prior to a meal.

Nuclei within the hypothalamus are integrated within networks that signal to other areas in the brain involved in appetite control. These areas include the brainstem, reward centres such as the amygdala and nucleus accumbens, and the prefrontal cortex, which is involved in conditioned taste aversion. Some gut hormones can bind and activate their respective receptors within the hypothalamus and brainstem via the blood–brain barrier. However, evidence suggests that alternative pathways also exist via vagal afferents from the gut.

The vagus nerve is the major neuroanatomical link between the gastrointestinal tract and the brain. Transection of all gut sensory vagal fibres results in increased meal size and duration (Schwartz 2000; 1999). Cell bodies of afferent fibres of the abdominal vagus nerve are located in the nodose ganglia, which project onto the brainstem. Here, the dorsal vagal complex (DVC), consisting of the dorsal motor nucleus of the vagus nerve (DMV), the area postrema (AP) and the sensory nucleus of the tractus solitarius (NTS), interfaces with hypothalamic and higher centres (Ter Horst et al. 1989; Chaudhri et al. 2008). In addition, the parabrachial nucleus (PBN) receives information related to feeding including taste, gastric distension and hepatic vagal afferents (Andrews 1986; Yuan and Barber 1991; Hajnal et al. 1999; Baird et al. 2001). Areas within the brainstem contain extensive reciprocal neuronal projections to hypothalamic feeding circuits and provide an alternative pathway through which circulating satiety factors can communicate with the hypothalamus (Ter Horst et al. 1989; Ter Horst et al. 1984). These signalling pathways are summarised in Fig. 1. This chapter will focus on three gut hormones

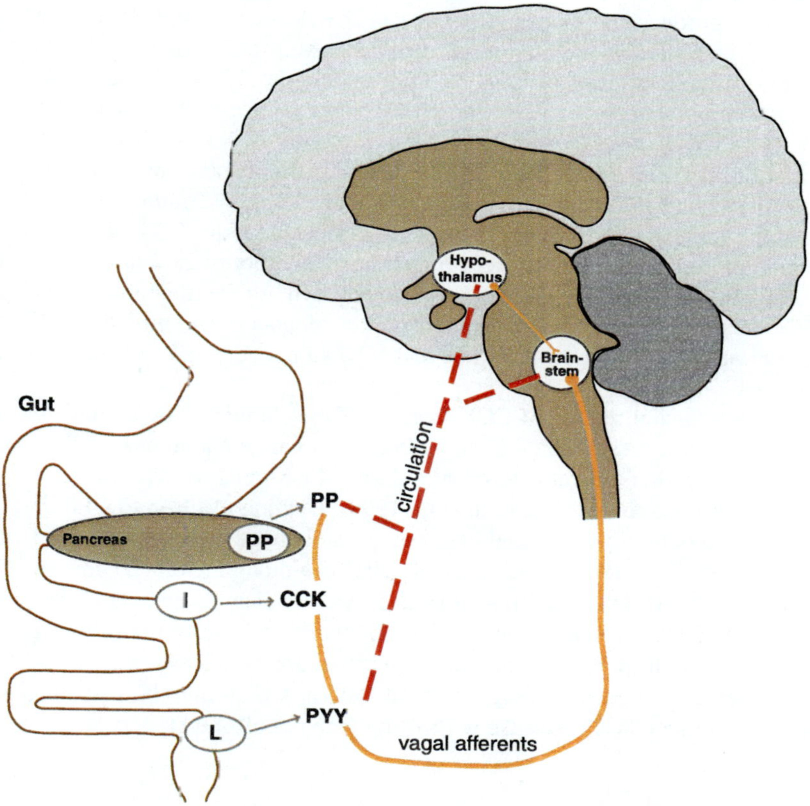

Fig. 1 Pathways from the gut through which gut hormones CCK, PYY and PP signal to the CNS to modify food intake and energy expenditure. Circulating PYY and PP can act directly in the hypothalamus and brainstem via the circulation and activate their respective receptors. Vagotomy and lesioning studies also suggest that CCK, PYY and PP may act via gut vagal afferents that signal to the dorsal vagal complex within the brainstem. Abbreviations: *CCK* cholecystokinin; *PYY* peptide tyrosine tyrosine; *PP* pancreatic polypeptide

that are implicated in the control of food intake and energy expenditure: CCK, PYY and PP.

2 Cholecystokinin

In 1973, CCK was the first gut hormone found to be capable of regulating appetite after it was shown to reduce meal size in rats (Gibbs et al. 1973). Prior to this, CCK was best characterised for its role in digestion, including stimulation of gall bladder contraction and pancreatic exocrine secretion. In addition, CCK inhibits gastric emptying. The anorectic effect of CCK is in keeping with a pro-digestive hormone: limiting the volume of nutrients entering the small intestine retains an optimal ratio of bile and digestive enzymes to nutrients for digestion.

CCK is produced in endocrine I-cells in the duodenum and ileum. Although originally identified as a 33 amino acid protein (Mutt and Jorpes 1971), there are multiple circulating forms of CCK of different lengths. The predominant forms of CCK found in the upper small intestine vary between species. In humans, CCK-58, CCK-39, CCK-33 and CCK-8 have been identified (Eberlein et al. 1988). The circulating half-life varies between the different fragments; however, one study determined a half-life in humans of approximately 2.5 min (Thompson et al. 1975). The active sequence of CCK is the C-terminal octapeptide (Ondetti et al. 1970), which is present in all the different fragments of the hormone found in the small intestine. In addition to the gut, CCK is also released as a neuropeptide in the CNS. CCK expressing neurons are found in the cortex, hippocampus, amygdala, olfactory bulb, preoptic nucleus, hypothalamus and NTS (Innis et al. 1979; Vanderhaeghen et al. 1980).

The postprandial release of CCK from I-cells is stimulated by amino acids and free fatty acids (FFAs) (Liddle et al. 1985). FFAs with a chain length of eleven or more carbon atoms are capable of stimulating CCK release, whereas those with shorter chains are not (McLaughlin et al. 1999). Binding of FFAs to GPR40, a cell surface receptor highly expressed in I-cells, is likely to be the major mechanism for FFA-induced CCK release (Liou et al. 2010), although GPR120 may also be implicated (Tanaka et al. 2008). Amino acids such as phenylalanine and tryptophan also stimulate CCK release from I-cells, although less readily than FFAs (Meyer et al. 1976). Both phenylalanine and tryptophan are agonists at the extracellular calcium sensing receptor (Conigrave et al. 2000), which may be responsible for amino acid-induced CCK release in the gut (Wang et al. 2011).

2.1 CCK Receptors

Two subtypes of the CCK receptor exist, CCK-1R and CCK-2R, both of which belong to the class 1 G protein coupled receptor family. CCK-1R and CCK-2R are

429 and 452 amino acids in length, respectively. Binding of CCK to both receptors leads to activation of phospholipase C and subsequent release of intracellular Ca^{2+} (Wank 1995). There is also some activation of adenylate cyclase and resultant cyclic adenosine monophosphate (cAMP) production as a result of CCK binding to the CCK-1R, although this has no known physiological relevance (Marino et al. 1993).

CCK-1R contains four sites for NH_2-linked glycosylation: three on the N-terminal tail and one on an extracellular loop, while CCK-2R has three sites, all on the N-terminal tail. Glycosylation has no reported effects on the properties of these receptors, which are fully functional even when not glycosylated (Hadac et al. 1996). CCK-1R has three sites for protein kinase C-mediated phosphorylation on the third intracellular loop and one on the C-terminal tail. Agonist-stimulated phosphorylation of the CCK-1R occurs at two sites on the third intracellular loop and less frequently on the C-terminal tail (Ozcelebi 1995). Phosphorylation at one site on the third intracellular loop has been shown to be involved in agonist-stimulated receptor desensitisation, although this process does occur at a reduced rate without phosphorylation (Rao et al. 1997). The effect of CCK-2R phosphorylation is less clear but may involve receptor internalisation (Pohl et al. 1997).

Although CCK-1R and CCK-2R are 48% homologous in sequence (Wank et al. 1992), CCK-1R has greater affinity for CCK than the related peptide, gastrin, while CCK-2R binds with equal affinity to both peptides. This is because CCK-2R recognises only the N-terminal tetrapeptide, which is identical in CCK and gastrin, whereas CCK-1R recognises the N-terminal heptapeptide, including a sulphated tyrosine residue (Miller and Gao 2008). These differences have enabled the development of specific agonists and antagonists to the two receptor types and investigation into their specific roles.

Studies have shown that CCK-1R is the predominant receptor involved with food intake and satiety. The CCK-1R exists in two different affinity states, a high-affinity, low-capacity state and a low-affinity, high-capacity state (Sankaran et al. 1982). Interestingly, activation of the low-affinity receptors appears to be important for CCK-induced satiety in rats, whereas activation of both forms of the CCK receptor can cause satiety in mice (Weatherford et al. 1993).

CCK-1Rs are predominantly expressed in the pancreas, gall bladder, stomach, kidney and lung (Regard et al. 2008), in addition to the vagus nerve (Zarbin et al. 1981). Although once thought to be expressed only in the periphery, CCK-1Rs are also expressed in the CNS. In the brainstem, CCK-1Rs have been identified in the AP, NTS and DMV. In the hypothalamus, CCK-1Rs are expressed in the supraoptic nucleus (SON), paraventricular nucleus (PVN) and dorsomedial nucleus (DMN). CCK-1Rs are also present in the substantia nigra, ventral tegmental area and nucleus accumbens (Moran et al. 1986; Hill et al. 1987, 1990; Mercer and Beart 1997). CCK-2Rs are widely expressed in the CNS, particularly in the olfactory bulb, cerebral cortex, hippocampus, striatum, hypothalamus and brainstem, whereas peripheral expression is limited to the stomach and uterus (Regard et al. 2008).

2.2 *Mechanisms of CCK-Induced Satiety*

CCK-induced satiety occurs predominantly through mechanisms involving the CCK-1R. CCK-1R selective agonists inhibit food intake (Simmons et al. 1998) whereas antagonists increase food intake (Corwin et al. 1991). The CCK-2R may also have a minor role since CCK-2R knockout mice are hyperphagic and 28% heavier than wild types (Clerc et al. 2007). However, satiety in response to exogenously administered CCK is retained in CCK-2 knockout mice (Kopin et al. 1999). Furthermore, agonists and antagonists at the CCK-2R show no effect on food intake (Corwin et al. 1991; Parrott 1993).

The anorectic effect of CCK was originally thought to be due to reduced gastric emptying (Moran and McHugh 1982). However, peripherally administered CCK inhibits feeding in sham-fed animals in whom an open gastric fistula prevents accumulation of food in the stomach, thereby removing any effect of gastric emptying rate on food intake (Bado et al. 1988). A peripheral site of action for CCK is supported by the finding that it is unable to cross the blood–brain barrier from the peripheral circulation into cerebrospinal fluid (Passaro et al. 1982). In addition, vagotomy abolishes the satiety effect of peripherally administered CCK (Smith et al. 1981). CCK-1Rs have also been identified on abdominal vagal afferents (Moran et al. 1987) and vagal fibres originating in the gastric mucosa have been shown to be excited by CCK-8 (Blackshaw and Grundy 1990).

The precise mechanism by which CCK reduces food intake via the vagus nerve is unclear; however, CCK has been shown to alter gene expression within vagal neurons. CCK activation of CCK-1Rs suppresses vagal expression of the cannabinoid receptor CB1, melanin-concentrating hormone (MCH) and its receptor MCHR-1 (Burdyga et al. 2004, 2006). CCK also increases the expression of the anorectic neurotransmitter cocaine- and amphetamine-regulated transcript (CART) (De Lartigue et al. 2007) and the Y2 receptor (Y2R), a receptor for the anorectic hormone $PYY_{3\text{-}36}$ (Burdyga et al. 2008). There is also evidence that CCK and leptin have a synergistic action in food intake inhibition which is thought to occur at a peripheral site (Barrachina et al. 1997). The leptin receptor is co-expressed with the CCK-1R in the rat nodose ganglion (Li et al. 2011), and a subset of gastric vagal afferents, which are unresponsive to leptin administration alone, become leptin sensitive after prior treatment with CCK (Wang et al. 1997). Therefore, CCK appears to reduce the capacity of vagal afferent neurons to respond to orexigenic signals and increases their ability to respond to anorectic signals.

Peripherally administered CCK causes neuronal activation, as assessed by c-fos immunohistochemistry, in the NTS and AP in the brainstem, in addition to the hypothalamic SON, PVN and central nucleus of the amygdala (CeA) (Chen et al. 1993; Day et al. 1994). Induction of this c-fos pattern by CCK is abolished by vagotomy, again highlighting the importance of vagal-mediated pathways (Li and Rowland 1995; Sayegh and Ritter 2000). Although the AP is activated by CCK, lesions of this area do not reduce CCK-induced satiety (Edwards et al. 1986). In contrast, ascending projections from the NTS appear to be important for

CCK-induced satiety as midbrain transection and lesions of the NTS abolish the anorectic effect (Crawley and Schwaber 1984; Crawley et al. 1984). In addition, lesions in the PVN abolish the effect of CCK on feeding, suggesting that projections from the NTS to the PVN may be responsible for CCK-induced satiety (Crawley and Kiss 1985). The central pathways involved are thought to involve 5-HT-releasing neurons. CCK-induced satiety is attenuated when co-administered with the 5-HT_3 receptor antagonist, ondansetron, and reduced neuronal activation in the NTS, and AP is also seen (Daughters et al. 2001).

In addition to its role in the periphery, there is some evidence that CCK acts within the CNS as a neurotransmitter to modulate food intake. Centrally administered CCK reduces food intake in mice, an effect which is blocked by centrally administered CCK-1R antagonists (Hirosue et al. 1993). Selective CCK-1R antagonists can also stimulate food intake, even in vagotomised animals (Reidelberger 1992). Furthermore, injections of CCK into the NTS and fourth ventricle in the brainstem and several hypothalamic nuclei, including the DMN, potently reduces food intake in fasted animals (Blevins et al. 2000). The ability of centrally administered CCK to reduce food intake and CCK-1R antagonists to increase food intake does not necessarily imply that CCK released from I-cells in the gut acts directly at central sites. Instead, CCK may act as a neurotransmitter which is released in response to peripheral signals and acts in areas within the brain to control food intake.

2.3 *Physiological Relevance of CCK-Induced Satiety*

If an endogenous hormone, such as CCK, inhibits appetite in a physiological setting, it might be expected that blocking the action of this hormone using receptor antagonists or removal of the gene coding for the hormone or its receptor by genetic manipulation would cause the animal to become hyperphagic. The Otsuka Long-Evans Tokushima Fatty (OLETF) rat lacks CCK-1Rs. Centrally or peripherally administered CCK does not induce satiety in the OLETF rat. In addition, OLETF rats consume 80% more in a single meal and are 30–40% heavier than controls (Bi and Moran 2002).

In contrast, it appears that CCK-1Rs are not involved in the physiological regulation of satiety in mice. CCK-1R knockout mice display no difference in food intake or body weight compared to wild types (Bi et al. 2004). However, CCK-1R knockout mice do eat larger meals, although less frequently, thereby explaining the apparent lack of effect of this mutation on overall food intake and body weight (Bi et al. 2004). This apparent difference in phenotype between CCK-1R null mice and rats is unknown; however, a species difference in CCK-1R expression in the DMN has been suggested. Immunohistochemical studies in rats have revealed that CCK-1Rs and neuropeptide Y (NPY) are co-localised in DMN neurons, and administration of CCK into the DMN downregulates NPY gene expression and inhibits food intake (Bi et al. 2004). Furthermore, OLETF rats demonstrate hyperphagia and increased NPY mRNA expression in the rat DMN (Moran 2008; Funakoshi 2000). In contrast, mice do not express the CCK-1R in the DMN (Bi et al. 2004), and CCK-1R knockout mice show no increase in NPY in the DMN. However, a lack of phenotype in CCK-1R knockout mice does not necessarily mean CCK is not involved in the physiological regulation of appetite and body weight in this species. CCK is one of many gut hormones involved in satiety, and removal of just one of these in a knockout model is easily compensated for by upregulation of another satiety signal in order to prevent disruption of energy homeostasis.

2.4 CCK and Human Obesity

In humans, it is unclear whether disruption of CCK-1R expression has detrimental effects on energy homeostasis. As yet no gene mutations of the CCK-1R have been conclusively linked to obesity however, a polymorphism in the CCK-1R receptor gene promoter is associated with increased percentage body fat (Funakoshi 2000). Early human studies showed that peripheral administration of CCK affects food intake. CCK-8 infusion in lean volunteers decreases meal size (Kissileff et al. 1981), an effect which is maintained in obese humans (Pi-Sunyer et al. 1982). As such, CCK-1R agonists have been developed with the intention to reduce appetite and thus aid weight loss in obese patients. Although highly specific agonists have been developed, the effect on food intake has been disappointing. In one study, the agonist GI181771 demonstrated no increased weight loss compared to placebo controls and resulted in undesirable gastrointestinal side effects (Jordan et al. 2008). It is conceivable that chronic use of CCK-1R agonists may lead to receptor desensitisation or compensation by other satiety signals. It has been suggested that the development of allosteric modulators at the CCK-1R might circumvent this problem as stimulation of the CCK-1R would only occur in the presence of endogenous CCK, and the modulator would merely enhance this response (Cawston and Miller 2010). This could induce less desensitisation of the receptors than that which may occur in the presence of longer-lasting orthosteric agonists. An allosteric binding site has been identified in the CCK-1R (Hadac et al. 2006; Gao et al. 2008) but as yet no agonists binding to this site have been tested for efficacy in vivo.

3 Peptide Tyrosine Tyrosine

PYY is a member of the PP-fold family, which also includes NPY and PP. All three peptides play a role in appetite regulation and share common structural features including a characteristic hairpin-shaped PP-fold, comprising an α-helix and polyproline helix linked with a β-turn (Berglund et al. 2003). PP-fold peptides mediate their effects on appetite regulation via five different G-protein coupled "Y-receptors": Y1, Y2, Y4, Y5 and Y6, although Y6 is inactivated in primates.

Table 1 (below) describes each of these receptors and the endogenous PP-fold peptides acting on them, in addition to agonists and antagonists that have been developed to regulate food intake and body weight. These receptors are linked to adenylyl cyclase, working to inhibit the formation of secondary messenger cAMP. However, the receptors differ in their tissue distribution, function and their selectivity for binding NPY, PP and PYY, thereby presenting a unique example of a multi-ligand/multi-receptor system. It must be noted that while there is hormone/receptor selectivity, there is also the potential for considerable promiscuity of these hormones, especially when administered at supraphysiological doses. All three PP-fold peptides have established roles in energy homeostasis. NPY has a fundamental orexigenic role in the brain, and PYY and PP released peripherally regulate energy homeostasis through both homeostatic and hedonic circuits. PYY signals to circuits in the CNS to inhibit food intake in addition to regulating gastric emptying, long-term energy homeostasis and insulin sensitivity.

3.1 PYY and the Y2 Receptor

PYY is a 36 amino acid peptide released from enteroendocrine L-cells in the distal gut following a meal. PYY immunoreactivity is predominantly expressed in the rectum with lower levels found in the jejunum and duodenum (Adrian et al. 1985). Initially, the major role of PYY was understood to be in the regulation of gastric motility and gastric secretion. Acting as a mediator of the "ileal brake," it slows the transit of food from the stomach into the small intestine in response to nutrient sensing. However, in the past decade it has become apparent that $PYY_{3\text{-}36}$ also plays a role in central appetite regulation (Savage et al. 1987).

Following a meal, an initial postprandial rise in circulating PYY is seen in less than 15 min, reaching a peak at around 1–2 h followed by a plateau period of several hours (Adrian et al. 1985). Interestingly, after 15 min, nutrients have not yet had time to reach the distal gut, suggesting that other factors contribute to the release of PYY from L-cells in the distal gut. Indeed, vasoactive intestinal peptide, CCK and vagal impulses have all been suggested to regulate PYY secretion (Ballantyne et al. 1993; McFadden et al. 1992; Zhang et al. 1993). Furthermore, the caloric load, consistency and nutrient composition of food all affect the resultant circulating PYY levels (Adrian et al. 1985; Batterham et al. 2006; Helou et al. 2008).

Two endogenous forms of PYY have been identified in the circulation, full length $PYY_{1\text{-}36}$ and the truncated $PYY_{3\text{-}36}$, the latter representing the majority of total circulating PYY. The membrane-bound enzyme, dipeptidyl peptidase IV (DPPIV), is responsible for the removal of N-terminal tyrosine-proline residues of $PYY_{1\text{-}36}$ to produce $PYY_{3\text{-}36}$ (Mentlein et al. 1993). Whereas full length PYY binds to Y1, Y2 and Y5 receptors, $PYY_{3\text{-}36}$ selectively binds and activates the Y2

Table 1 Summary of agonists and antagonists of each of the PP-fold family receptors, showing endogenous peptides in addition to synthetic peptide analogues and small molecule therapeutics

Receptor	Agonists		Antagonists	Physiological effect of receptor activation
	Endogenous	Synthetic		
Y1	NPY	Leu^{31}, Pro^{34} NPY	BIBP 3226 (Rudolf et al. 1994)	NPY: Increased peripheral vascular resistance, anxiety and depression. Increased food intake and insulin resistance
			LY357897 (Hipskind et al. 1997)	
	PYY	Leu^{31}, Pro^{34} PYY	J-104870 (Kanatani et al. 1999)	
			BMS-193885 (Antal-Zimanyi et al. 2008)	
			BIBO 3304 (Wieland et al. 1998)	
Y2 pre-synaptic	NPY	PYY 3-36 nasal spray (Gantz et al. 2007)	BIIE0246 (Doods et al. 1999)	PYY: Decreased body weight and food intake, nausea, decreased gastric emptying, increased intestinal transit time
	NPY (3-36)	Pegylated PYY (25-36) (Lumb et al. 2007)		
	NPY (13-36)			
	PYY	Obinepitide (7TM Pharma 2007)		
	PYY (3-36)			
Y4	PP	Leu^{31}, Pro^{34} NPY		PP: Decreased food intake, body weight and adiposity, decreased pancreatic secretions and gallbladder motility. Delayed gastric emptying. Increased oxygen consumption and discharge rate of sympathetic efferent nerves to adrenal glands and brown adipose tissue
		Leu^{31}, Pro^{34} PYY		
	NPY	Obinepitide (7TM Pharma 2007)		
		PP1420 (Gibb 2010)		
		TM30339 (7TM Pharma 2011)		
Y5	NPY	Leu^{31}, Pro^{34} NPY	CGP71683A (Rueeger et al. 2000)	NPY: increased food intake
	NPY (2-36)	Leu^{31}, Pro^{34} PYY	GW438014A (Daniels et al. 2002)	
	NPY (3-36)		FMS586 (Kakui et al. 2006)	
	NPY (13-36)		Benzimidazole compounds (Pizzi et al. 2010)	
	PYY			
	PYY (3-36)			
	PYY (13-36)			

receptor (Y2R) to bring about its effects on energy homeostasis (Grandt et al. 1992; Browning and Travagli 2009).

The Y2R is found predominantly in the CNS, and when activated, causes inhibition of cAMP production (Browning and Travagli 2009). Calcium signalling is also reduced which presumably interferes with exocytosis of neurotransmitter vesicles (Toth et al. 1993; Wiley et al. 1993). Y2R mRNA has been observed by in situ hybridisation in the hippocampus, amygdala, hypothalamus and brainstem in addition to peripheral sympathetic, parasympathetic and sensory neurons (Gustafson et al. 1997). In the hypothalamus, evidence of Y2R mRNA is seen most strongly in the SON and medial ARC (Parker and Herzog 1999).

3.2 *Mechanisms of PYY-Induced Satiety*

PYY knockout mice are hyperphagic with increased body mass and adiposity, a phenotype which can be abrogated with exogenous administration of $PYY_{3\text{-}36}$ (Batterham et al. 2002). Peripheral administration of $PYY_{3\text{-}36}$ inhibits food intake in rodents and lean and obese humans (Batterham et al. 2002, 2003). Some published studies have failed to find a significant effect of $PYY_{3\text{-}36}$ on food intake, but this may be due to an anorectic stress response masking any effect (Tschop et al. 2004; Boggiano et al. 2005). With sufficient acclimatisation and handling, the acute anorectic effect of $PYY_{3\text{-}36}$ has been witnessed by a large number of different groups. $PYY_{3\text{-}36}$ has also been implicated in the long-term regulation of body weight. Chronic administration of $PYY_{3\text{-}36}$ via either an osmotic mini-pump or continuous intravenous infusion significantly reduces body weight and adiposity in normal and obese rodents (Vrang et al. 2006). Furthermore, it has been suggested that $PYY_{3\text{-}36}$ increases energy expenditure through an increase in postprandial thermogenesis, resting metabolic rate and 24-h respiratory quotient (Guo et al. 2006; Sloth et al. 2007).

The exact mechanisms underlying the anorectic effects of $PYY_{3\text{-}36}$ are unclear. $PYY_{3\text{-}36}$ may act directly via an incomplete blood–brain barrier in the median eminence of the hypothalamus and act in the ARC. Indeed, peripheral administration of $PYY_{3\text{-}36}$ induces the expression of c-fos in the ARC (Batterham et al. 2002). Moreover, intra-arcuate injection of the Y2R antagonist BIIE0246 abolishes the anorectic effects of both endogenous and exogenous $PYY_{3\text{-}36}$ (Abbott et al. 2005a). The medial ARC is populated with orexigenic NPY/AgRP neurons, and there is a high degree of Y2R/NPY co-expression in these neurons (Broberger et al. 1997). $PYY_{3\text{-}36}$ may reduce food intake by inhibiting NPY release via autoinhibitory Y2Rs. $PYY_{3\text{-}36}$ decreases NPY release and increases α-MSH release in vitro from hypothalamic explants (Batterham et al. 2002). Furthermore, electrophysiology studies have shown that $PYY_{3\text{-}36}$ directly inhibits activity of ARC NPY neurons, thereby secondarily disinhibiting anorectic POMC neurons (Acuna-Goycolea and van den Pol 2005). However, further studies suggest that the anorectic effects of $PYY_{3\text{-}36}$ are more complex than a simple action on the ARC, since both POMC-null and MC4-R null mice respond normally to the anorectic effects of PYY_{3-36} (Challis et al. 2004; Halatchev et al. 2004).

An alternative mechanism for the anorectic effect of PYY may be via gut vagal afferents that signal to the brainstem, similar to that of CCK as discussed above. Transectioning of brainstem–hypothalamic pathways attenuates the effects of PYY_{3-36} on food intake inhibition (Abbott et al. 2005b), although others have failed to replicate this (Halatchev and Cone 2005; Koda et al. 2005; Talsania et al. 2005). In addition, Y2R mRNA is expressed in the NTS and nodose ganglion of the vagus nerve (Gustafson et al. 1997; Koda et al. 2005), and c-fos immunoreactivity is detected in the NTS and AP following intravenous infusion of PYY_{3-36} (Halatchev and Cone 2005; Koda et al. 2005). These findings have led to the proposal that PYY_{3-36} may regulate ARC neuronal activity indirectly via vagal–brainstem pathways. Alternatively, PYY_{3-36} may act directly in the brainstem since areas of incomplete blood–brain barrier are also present at the AP.

3.3 *PYY and Human Obesity*

PYY_{3-36} has also been shown to regulate food intake in humans. High levels of circulating PYY_{3-36} are found in patients suffering from anorexia nervosa whilst frequently low levels are measured in obese subjects (Pfluger et al. 2007). Obese people also have a blunted rise in PYY_{3-36} after a meal, possibly resulting in impaired satiety and hence greater food intake (Ashby and Bloom 2007). In addition, circulating levels of PYY_{3-36} following bariatric surgery are raised and coincide with reduced appetite in patients (Korner et al. 2006).

Peripheral administration of PYY_{3-36} reduces food intake in human volunteers (Batterham et al. 2002) and increases neuronal activity in brain regions known to be involved in appetite regulation (Batterham et al. 2007). As such, PYY_{3-36} offers a potentially valuable therapeutic tool in the treatment of obesity. However, endogenous PYY_{3-36} has a short half-life and frequently causes nausea following administration in humans (Batterham et al. 2007; le Roux et al. 2008). The development of long-acting analogues may circumvent these problems. Several trials have been undertaken using analogues of PYY_{3-36}, including a nasal spray which reached phase II clinical trials and resulted in weight loss over a 6-day period (MDRNA 2008). A dual analogue combining both PYY_{3-36} and PP, called obinepitide, has also been developed and has been shown to reduce food intake in human volunteers (7TM Pharma 2007). Similarly, the combination of GLP-1 and PYY_{3-36} significantly reduces food intake (Emisphere 2009).

3.4 *Additional Roles of PYY*

Peripherally administered PYY_{3-36} has been shown to stimulate the release of gonadotropins (Fernandez-Fernandez et al. 2005) and to inhibit the release of thyroid stimulating hormone from the pituitary gland in rats (Pinilla et al. 2007). However, there is, to date, no evidence that endogenous PYY_{3-36} modulates the release of these hormones. A role for PYY in bone remodelling was postulated after observations of an osteoporotic phenotype in PYY null mice (Wortley et al. 2007) and circulating PYY levels in patients with anorexia nervosa appear to be the major determinant of bone density in the spine (Utz et al. 2008). Colonic motility appears to be inhibited by peripherally administered PYY_{3-36} acting via the Y2 and Y1 receptors (Wang et al. 2010; Tough et al. 2011).

4 Pancreatic Polypeptide

PP is an amidated 36 amino acid PP-fold peptide. It is thought to have been formed through duplication of the PYY gene (Hort et al. 1995) and acts at the Y4 receptor (Y4R) to reduce food intake and delay gastric emptying and motility.

PP is released from PP cells, also known as F cells, in the pancreatic islets of Langerhans. Release of PP in response to meal ingestion is proportional to caloric intake and levels remain elevated for 6 h (Adrian et al. 1976). While PP secretion is dependent on a vagal cholinergic mechanism (Taylor et al. 1978), PP release is also stimulated by CCK, ghrelin, motilin, secretin and adrenergic activation during hypoglycaemia or exercise (Kojima et al. 2007). PP levels are also influenced by circadian rhythm with greater postprandial circulating PP levels seen following a meal in the evening compared to an identical meal ingested in the morning (Johns et al. 2006).

PP exerts a range of regulatory functions including the inhibition of pancreatic secretions and gallbladder motility. In addition, PP delays gastric emptying (Kojima et al. 2007) and has a role in energy homeostasis. Obese subjects have a blunted postprandial PP response, while subjects with anorexia nervosa show an exaggerated response (Lassmann et al. 1980; Uhe et al. 1992). PP may also contribute to the pathogenesis of Prader–Willi syndrome, a disease characterised by hyperphagia, obesity, short stature, intellectual impairment and infertility (Zipf et al. 1983).

Peripheral administration of PP reduces food intake in both humans and mice (Asakawa et al. 1999). Furthermore, over-expression of PP in the pancreatic islets of mice produce a hypophagic and thin phenotype, suggesting that chronic exposure to PP does not lead to desensitisation or attenuation of the anorectic effect (Ueno et al. 1999). In addition, PP has a role in regulating energy expenditure. Repeated PP administration in mice increases oxygen consumption and the discharge rate of sympathetic efferent nerves innervating the adrenal glands and brown adipose tissue (Asakawa et al. 2003).

PP has high affinity for the Y4R and is thought to act specifically through this receptor to inhibit food intake. Y4R knockout mice exhibit no change in food intake following PP administration compared to wild-type mice (Lin et al. 2009). Y4Rs have been shown by in situ hybridisation and autoradiography to be located in the AP, NTS, DMV, ARC and PVN (Parker and Herzog 1999; Whitcomb et al. 1990).

The AP and ARC are located close to regions with an incomplete blood–brain barrier. Therefore, the actions of PP could be mediated through direct actions here via the circulation. PP decreases orexigenic NPY, orexin and ghrelin expression while increasing the expression of anorexigenic urocortin and brain-derived neurotrophic factor (BDNF) in the hypothalamus (Asakawa et al. 2003; Sainsbury et al. 2010). This indicates the importance of hypothalamic circuits in regulating the effects of PP. Similarly, manganese-enhanced magnetic resonance imaging (MEMRI) investigations have shown that subcutaneous administration of PP results in a significant reduction in signal intensity in the ARC, VMN and PVN of the hypothalamus in fasted mice (Hankir et al. 2011). However, the vagus nerve may also be involved since vagotomy abolishes the anorectic effects of PP in rats (Asakawa et al. 2003). It is currently unknown whether the effect of PP on appetite is mediated directly via the circulation at the brainstem or hypothalamus, through vagal afferents or through a combination of both.

PP has been proposed as a potentially important drug target for the treatment of obesity. However, the endogenous form of PP has a short half-life of approximately 7 min due to degradation by enzymes, including DPPIV and neprilysin. A long-acting, degradation-resistant PP analogue, PP-1420, has been recently developed and is undergoing investigation in phase 1 clinical trials (Gibb 2010). A PP analogue has also been conjugated to a lipid moiety allowing increased plasma half-life and sustained food intake inhibition (Bellmann-Sickert et al. 2011).

5 Conclusion

Gut hormones convey important information to the CNS in order to control food intake and energy expenditure. Key areas in the CNS involved with appetite control include the hypothalamus and brainstem, and hormones such as PYY and PP can act directly in these areas via the circulation. In addition, some gut hormones such as CCK have actions at the brainstem via gut vagal afferents. Animal and human studies have demonstrated that peripheral administration of these gut hormones can reduce food intake and cause weight loss. These gut hormones are therefore potential targets in the development of novel treatments for obesity, and analogue therapies are currently under investigation.

References

7TM Pharma (2007) Results from a Phase I/II clinical study with the drug candidate Obinepitide for the treatment of obesity. Press release 28th Nov 2007. http://www.7tm.com/News/News_Archive.aspx?M=News&PID=45&NewsID=9

7TM Pharma (2011) TM30339. Metabolic disorders http://www.7tm.com/R-D/Metabolic_Disorders/TM30339.aspx

Abbott CR, Small CJ, Kennedy AR, Neary NM, Sajedi A, Ghatei MA et al (2005a) Blockade of the neuropeptide Y Y2 receptor with the specific antagonist BIIE0246 attenuates the effect of endogenous and exogenous peptide YY(3-36) on food intake. Brain Res 1043(1–2):139–144

Abbott CR, Monteiro M, Small CJ, Sajedi A, Smith KL, Parkinson JR et al (2005b) The inhibitory effects of peripheral administration of peptide YY(3-36) and glucagon-like peptide-1 on food intake are attenuated by ablation of the vagal-brainstem-hypothalamic pathway. Brain Res 1044(1):127–131

Acuna-Goycolea C, van den Pol AN (2005) Peptide YY(3-36) inhibits both anorexigenic proopiomelanocortin and orexigenic neuropeptide Y neurons: implications for hypothalamic regulation of energy homeostasis. J Neurosci 25(45):10510–10519

Adrian TE, Bloom SR, Bryant MG, Polak JM, Heitz PH, Barnes AJ (1976) Distribution and release of human pancreatic polypeptide. Gut 17(12):940–944

Adrian TE, Ferri GL, Bacarese-Hamilton AJ, Fuessl HS, Polak JM, Bloom SR (1985) Human distribution and release of a putative new gut hormone, peptide YY. Gastroenterology 89 (5):1070–1077

Andrews PL (1986) Vagal afferent innervation of the gastrointestinal tract. Prog Brain Res 67:65–86

Antal-Zimanyi I, Bruce MA, Leboulluec KL, Iben LG, Mattson GK, McGovern RT et al (2008) Pharmacological characterization and appetite suppressive properties of BMS-193885, a novel and selective neuropeptide Y(1) receptor antagonist. Eur J Pharmacol 590(1–3):224–232

Asakawa A, Inui A, Ueno N, Fujimiya M, Fujino MA, Kasuga M (1999) Mouse pancreatic polypeptide modulates food intake, while not influencing anxiety in mice. Peptides 20 (12):1445–1448

Asakawa A, Inui A, Yuzuriha H, Ueno N, Katsuura G, Fujimiya M et al (2003) Characterization of the effects of pancreatic polypeptide in the regulation of energy balance. Gastroenterology 124 (5):1325–1336

Ashby D, Bloom SR (2007) Recent progress in PYY research – an update report for 8th NPY meeting. Peptides 28(2):198–202

Bado A, Rodriguez M, Lewin MJ, Martinez J, Dubrasquet M (1988) Cholecystokinin suppresses food intake in cats: structure-activity characterization. Pharmacol Biochem Behav 31 (2):297–303

Baird JP, Travers JB, Travers SP (2001) Parametric analysis of gastric distension responses in the parabrachial nucleus. Am J Physiol Regul Integr Comp Physiol 281(5):R1568–R1580

Ballantyne GH, Goldenring JR, Savoca PE, Kranz HK, Adrian TE, Bilchik AJ et al (1993) Cyclic AMP-mediated release of peptide YY (PYY) from the isolated perfused rabbit distal colon. Regul Pept 47(2):117–126

Barrachina MD, Martinez V, Wang L, Wei JY, Tache Y (1997) Synergistic interaction between leptin and cholecystokinin to reduce short-term food intake in lean mice. Proc Natl Acad Sci U S A 94(19):10455–10460

Batterham RL, Cowley MA, Small CJ, Herzog H, Cohen MA, Dakin CL et al (2002) Gut hormone PYY(3-36) physiologically inhibits food intake. Nature 418(6898):650–654

Batterham RL, Cohen MA, Ellis SM, le Roux CW, Withers DJ, Frost GS et al (2003) Inhibition of food intake in obese subjects by peptide YY3-36. N Engl J Med 349(10):941–948

Batterham RL, Heffron H, Kapoor S, Chivers JE, Chandarana K, Herzog H et al (2006) Critical role for peptide YY in protein-mediated satiation and body-weight regulation. Cell Metab 4 (3):223–233

Batterham RL, ffytche DH, Rosenthal JM, Zelaya FO, Barker GJ, Withers DJ et al (2007) PYY modulation of cortical and hypothalamic brain areas predicts feeding behaviour in humans. Nature 450(7166):106–109

Bellmann-Sickert K, Elling CE, Madsen AN, Little PB, Lundgren K, Gerlach LO et al (2011) Long-acting lipidated analog of human pancreatic polypeptide is slowly released into circulation. J Med Chem 54(8):2658–2667

Berglund MM, Hipskind PA, Gehlert DR (2003) Recent developments in our understanding of the physiological role of PP-fold peptide receptor subtypes. Exp Biol Med (Maywood) 228 (3):217–244

Bi S, Moran TH (2002) Actions of CCK in the controls of food intake and body weight: lessons from the CCK-A receptor deficient OLETF rat. Neuropeptides 36(2–3):171–181
Bi S, Scott KA, Kopin AS, Moran TH (2004) Differential roles for cholecystokinin a receptors in energy balance in rats and mice. Endocrinology 145(8):3873–3880
Blackshaw LA, Grundy D (1990) Effects of cholecystokinin (CCK-8) on two classes of gastroduodenal vagal afferent fibre. J Auton Nerv Syst 31(3):191–201
Blevins JE, Stanley BG, Reidelberger RD (2000) Brain regions where cholecystokinin suppresses feeding in rats. Brain Res 860(1–2):1–10
Boggiano MM, Chandler PC, Oswald KD, Rodgers RJ, Blundell JE, Ishii Y et al (2005) PYY3-36 as an anti-obesity drug target. Obes Rev 6(4):307–322
Broberger C, Landry M, Wong H, Walsh JN, Hokfelt T (1997) Subtypes Y1 and Y2 of the neuropeptide Y receptor are respectively expressed in pro-opiomelanocortin- and neuropeptide-Y-containing neurons of the rat hypothalamic arcuate nucleus. Neuroendocrinology 66 (6):393–408
Browning KN, Travagli RA (2009) Modulation of inhibitory neurotransmission in brainstem vagal circuits by NPY and PYY is controlled by cAMP levels. Neurogastroenterol Motil 21 (12):1309–e126
Burdyga G, Lal S, Varro A, Dimaline R, Thompson DG, Dockray GJ (2004) Expression of cannabinoid CB1 receptors by vagal afferent neurons is inhibited by cholecystokinin. J Neurosci 24(11):2708–2715
Burdyga G, Varro A, Dimaline R, Thompson DG, Dockray GJ (2006) Feeding-dependent depression of melanin-concentrating hormone and melanin-concentrating hormone receptor-1 expression in vagal afferent neurones. Neuroscience 137(4):1405–1415
Burdyga G, de LG, Raybould HE, Morris R, Dimaline R, Varro A et al (2008) Cholecystokinin regulates expression of Y2 receptors in vagal afferent neurons serving the stomach. J Neurosci 28(45):11583–11592
Cawston EE, Miller LJ (2010) Therapeutic potential for novel drugs targeting the type 1 cholecystokinin receptor. Br J Pharmacol 159(5):1009–1021
Challis BG, Coll AP, Yeo GS, Pinnock SB, Dickson SL, Thresher RR et al (2004) Mice lacking pro-opiomelanocortin are sensitive to high-fat feeding but respond normally to the acute anorectic effects of peptide-YY(3-36). Proc Natl Acad Sci U S A 101(13):4695–4700
Chaudhri OB, Wynne K, Bloom SR (2008) Can gut hormones control appetite and prevent obesity? Diabetes Care 31(Suppl 2):S284–S289
Chen DY, Deutsch JA, Gonzalez MF, Gu Y (1993) The induction and suppression of c-fos expression in the rat brain by cholecystokinin and its antagonist L364,718. Neurosci Lett 149(1):91–94
Clerc P, Coll Constans MG, Lulka H, Broussaud S, Guigne C, Leung-Theung-Long S et al (2007) Involvement of cholecystokinin 2 receptor in food intake regulation: hyperphagia and increased fat deposition in cholecystokinin 2 receptor-deficient mice. Endocrinology 148 (3):1039–1049
Conigrave AD, Quinn SJ, Brown EM (2000) L-Amino acid sensing by the extracellular Ca^{2+}-sensing receptor. Proc Natl Acad Sci U S A 97(9):4814–4819
Corwin RL, Gibbs J, Smith GP (1991) Increased food intake after type A but not type B cholecystokinin receptor blockade. Physiol Behav 50(1):255–258
Crawley JN, Kiss JZ (1985) Paraventricular nucleus lesions abolish the inhibition of feeding induced by systemic cholecystokinin. Peptides 6(5):927–935
Crawley JN, Schwaber JS (1984) Abolition of the behavioral effects of cholecystokinin following bilateral radiofrequency lesions of the parvocellular subdivision of the nucleus tractus solitarius. Brain Res 295(2):289–299
Crawley JN, Kiss JZ, Mezey E (1984) Bilateral midbrain transections block the behavioral effects of cholecystokinin on feeding and exploration in rats. Brain Res 322(2):316–321
Daniels AJ, Grizzle MK, Wiard RP, Matthews JE, Heyer D (2002) Food intake inhibition and reduction in body weight gain in lean and obese rodents treated with GW438014A, a potent and selective NPY-Y5 receptor antagonist. Regul Pept 106(1–3):47–54

Daughters RS, Hofbauer RD, Grossman AW, Marshall AM, Brown EM, Hartman BK et al (2001) Ondansetron attenuates CCK induced satiety and c-fos labeling in the dorsal medulla. Peptides 22(8):1331–1338

Day HE, McKnight AT, Poat JA, Hughes J (1994) Evidence that cholecystokinin induces immediate early gene expression in the brainstem, hypothalamus and amygdala of the rat by a CCKA receptor mechanism. Neuropharmacology 33(6):719–727

De Lartigue G, Dimaline R, Varro A, Dockray GJ (2007) Cocaine- and amphetamine-regulated transcript: stimulation of expression in rat vagal afferent neurons by cholecystokinin and suppression by ghrelin. J Neurosci 27(11):2876–2882

Doods H, Gaida W, Wieland HA, Dollinger H, Schnorrenberg G, Esser F et al (1999) BIIE0246: a selective and high affinity neuropeptide Y Y(2) receptor antagonist. Eur J Pharmacol 384 (2–3):R3–R5

Eberlein GA, Eysselein VE, Goebell H (1988) Cholecystokinin-58 is the major molecular form in man, dog and cat but not in pig, beef and rat intestine. Peptides 9(5):993–998

Edwards GL, Ladenheim EE, Ritter RC (1986) Dorsomedial hindbrain participation in cholecystokinin-induced satiety. Am J Physiol 251(5 Pt 2):R971–R977

Emisphere Technologies (2009) Emisphere technologies reports encouraging data from independent clinical study assessing the effects of oral GLP-1 and PYY3-36, combined with Eligen(R) technology on appetite suppression. Press release 26th May 2009. http://ir.emisphere.com/releasedetail.cfm?ReleaseID=385806

Fernandez-Fernandez R, Aguilar E, Tena-Sempere M, Pinilla L (2005) Effects of polypeptide YY (3-36) upon luteinizing hormone-releasing hormone and gonadotropin secretion in prepubertal rats: in vivo and in vitro studies. Endocrinology 146(3):1403–1410

Funakoshi A (2000) Gene structure of human cholecystokinin (CCK) type-A receptor: body fat content is related to CCK type-A receptor gene promoter polymorphism. FEBS Lett 466 (2–3):264

Gantz I, Erondu N, Mallick M, Musser B, Krishna R, Tanaka WK et al (2007) Efficacy and safety of intranasal peptide YY3-36 for weight reduction in obese adults. J Clin Endocrinol Metab 92 (5):1754–1757

Gao F, Sexton PM, Christopoulos A, Miller LJ (2008) Benzodiazepine ligands can act as allosteric modulators of the Type 1 cholecystokinin receptor. Bioorg Med Chem Lett 18(15):4401–4404

Ghatei MA, Uttenthal LO, Christofides ND, Bryant MG, Bloom SR (1983) Molecular forms of human enteroglucagon in tissue and plasma: plasma responses to nutrient stimuli in health and in disorders of the upper gastrointestinal tract. J Clin Endocrinol Metab 57(3):488–495

Gibb BJ (2010) Wellcome Trust. PP 1420: a breakthrough in appetite suppression. 6th Oct 2010. http://wellcometrust.wordpress.com/2010/10/06/pp1420/

Gibbs J, Young RC, Smith GP (1973) Cholecystokinin decreases food intake in rats. J Comp Physiol Psychol 84(3):488–495

Grandt D, Teyssen S, Schimiczek M, Reeve JR Jr, Feth F, Rascher W et al (1992) Novel generation of hormone receptor specificity by amino terminal processing of peptide YY. Biochem Biophys Res Commun 186(3):1299–1306

Guo Y, Ma L, Enriori PJ, Koska J, Franks PW, Brookshire T et al (2006) Physiological evidence for the involvement of peptide YY in the regulation of energy homeostasis in humans. Obesity (Silver Spring) 14(9):1562–1570

Gustafson EL, Smith KE, Durkin MM, Walker MW, Gerald C, Weinshank R et al (1997) Distribution of the neuropeptide Y Y2 receptor mRNA in rat central nervous system. Brain Res Mol Brain Res 46(1–2):223–235

Hadac EM, Ghanekar DV, Holicky EL, Pinon DI, Dougherty RW, Miller LJ (1996) Relationship between native and recombinant cholecystokinin receptors: role of differential glycosylation. Pancreas 13(2):130–139

Hadac EM, Dawson ES, Darrow JW, Sugg EE, Lybrand TP, Miller LJ (2006) Novel benzodiazepine photoaffinity probe stereoselectively labels a site deep within the membrane-spanning domain of the cholecystokinin receptor. J Med Chem 49(3):850–863

Hajnal A, Takenouchi K, Norgren R (1999) Effect of intraduodenal lipid on parabrachial gustatory coding in awake rats. J Neurosci 19(16):7182–7190

Halatchev IG, Cone RD (2005) Peripheral administration of PYY(3-36) produces conditioned taste aversion in mice. Cell Metab 1(3):159–168

Halatchev IG, Ellacott KL, Fan W, Cone RD (2004) Peptide YY3-36 inhibits food intake in mice through a melanocortin-4 receptor-independent mechanism. Endocrinology 145(6):2585–2590

Hankir MK, Parkinson JR, Minnion JS, Addison M, Bloom SR, Bell JD (2011) PYY(3-36) and pancreatic polypeptide differentially regulate hypothalamic neuronal activity in mice in vivo as measured by manganese enhanced MRI. J Neuroendocrinol 23(4):371–380

Helou N, Obeid O, Azar ST, Hwalla N (2008) Variation of postprandial PYY 3-36 response following ingestion of differing macronutrient meals in obese females. Ann Nutr Metab 52 (3):188–195

Hill DR, Campbell NJ, Shaw TM, Woodruff GN (1987) Autoradiographic localization and biochemical characterization of peripheral type CCK receptors in rat CNS using highly selective nonpeptide CCK antagonists. J Neurosci 7(9):2967–2976

Hill DR, Shaw TM, Graham W, Woodruff GN (1990) Autoradiographical detection of cholecystokinin-A receptors in primate brain using 125I-Bolton Hunter CCK-8 and 3H-MK-329. J Neurosci 10(4):1070–1081

Hipskind PA, Lobb KL, Nixon JA, Britton TC, Bruns RF, Catlow J et al (1997) Potent and selective 1,2,3-trisubstituted indole NPY Y-1 antagonists. J Med Chem 40(23):3712–3714

Hirosue Y, Inui A, Teranishi A, Miura M, Nakajima M, Okita M et al (1993) Cholecystokinin octapeptide analogues suppress food intake via central CCK-A receptors in mice. Am J Physiol 265(3 Pt 2):R481–R486

Hort Y, Baker E, Sutherland GR, Shine J, Herzog H (1995) Gene duplication of the human peptide YY gene (PYY) generated the pancreatic polypeptide gene (PPY) on chromosome 17q21.1. Genomics 26(1):77–83

Innis RB, Correa FM, Uhl GR, Schneider B, Snyder SH (1979) Cholecystokinin octapeptide-like immunoreactivity: histochemical localization in rat brain. Proc Natl Acad Sci U S A 76 (1):521–525

Johns CE, Newton JL, Westley BR, May FE (2006) Human pancreatic polypeptide has a marked diurnal rhythm that is affected by ageing and is associated with the gastric TFF2 circadian rhythm. Peptides 27(6):1341–1348

Jordan J, Greenway FL, Leiter LA, Li Z, Jacobson P, Murphy K et al (2008) Stimulation of cholecystokinin-A receptors with GI181771X does not cause weight loss in overweight or obese patients. Clin Pharmacol Ther 83(2):281–287

Jorde R, Burhol PG (1984) Fasting and postprandial plasma pancreatic polypeptide (PP) levels in obesity. Int J Obes 8(5):393–397

Kakui N, Tanaka J, Tabata Y, Asai K, Masuda N, Miyara T et al (2006) Pharmacological characterization and feeding-suppressive property of FMS586 [3-(5,6,7,8-tetrahydro-9-isopropyl-carbazol-3-yl)-1-methyl-1-(2-pyridin-4-yl-ethy l)-urea hydrochloride], a novel, selective, and orally active antagonist for neuropeptide Y Y5 receptor. J Pharmacol Exp Ther 317 (2):562–570

Kanatani A, Kanno T, Ishihara A, Hata M, Sakuraba A, Tanaka T et al (1999) The novel neuropeptide Y Y(1) receptor antagonist J-104870: a potent feeding suppressant with oral bioavailability. Biochem Biophys Res Commun 266(1):88–91

Kissileff HR, Pi-Sunyer FX, Thornton J, Smith GP (1981) C-terminal octapeptide of cholecystokinin decreases food intake in man. Am J Clin Nutr 34(2):154–160

Koda S, Date Y, Murakami N, Shimbara T, Hanada T, Toshinai K et al (2005) The role of the vagal nerve in peripheral PYY3-36-induced feeding reduction in rats. Endocrinology 146 (5):2369–2375

Kojima S, Ueno N, Asakawa A, Sagiyama K, Naruo T, Mizuno S et al (2007) A role for pancreatic polypeptide in feeding and body weight regulation. Peptides 28(2):459–463

Kopin AS, Mathes WF, McBride EW, Nguyen M, Al-Haider W, Schmitz F et al (1999) The cholecystokinin-A receptor mediates inhibition of food intake yet is not essential for the maintenance of body weight. J Clin Invest 103(3):383–391

Korner J, Inabnet W, Conwell IM, Taveras C, Daud A, Olivero-Rivera L et al (2006) Differential effects of gastric bypass and banding on circulating gut hormone and leptin levels. Obesity (Silver Spring) 14(9):1553–1561

Lassmann V, Vague P, Vialettes B, Simon MC (1980) Low plasma levels of pancreatic polypeptide in obesity. Diabetes 29(6):428–430

le Roux CW, Borg CM, Murphy KG, Vincent RP, Ghatei MA, Bloom SR (2008) Supraphysiological doses of intravenous PYY3-36 cause nausea, but no additional reduction in food intake. Ann Clin Biochem 45(Pt 1):93–95

Li BH, Rowland NE (1995) Effects of vagotomy on cholecystokinin- and dexfenfluramine-induced Fos-like immunoreactivity in the rat brain. Brain Res Bull 37(6):589–593

Li Y, Wu X, Zhou S, Owyang C (2011) Low-affinity CCK-A receptors are coexpressed with leptin receptors in rat nodose ganglia: implications for leptin as a regulator of short-term satiety. Am J Physiol Gastrointest Liver Physiol 300(2):G217–G227

Liddle RA, Goldfine ID, Rosen MS, Taplitz RA, Williams JA (1985) Cholecystokinin bioactivity in human plasma. Molecular forms, responses to feeding, and relationship to gallbladder contraction. J Clin Invest 75(4):1144–1152

Lin S, Shi YC, Yulyaningsih E, Aljanova A, Zhang L, Macia L et al (2009) Critical role of arcuate Y4 receptors and the melanocortin system in pancreatic polypeptide-induced reduction in food intake in mice. PLoS One 4(12):e8488

Liou AP, Lu X, Sei Y, Zhao X, Pechhold S, Carrero RJ et al (2010) The G-protein-coupled receptor GPR40 directly mediates long-chain fatty acid-induced secretion of cholecystokinin. Gastroenterology 140(3):903–912

Lumb KJ, DeCarr LB, Milardo LF, Mays MR, Buckholz TM, Fisk SE et al (2007) Novel selective neuropeptide Y2 receptor PEGylated peptide agonists reduce food intake and body weight in mice. J Med Chem 50(9):2264–2268

Marino CR, Leach SD, Schaefer JF, Miller LJ, Gorelick FS (1993) Characterization of cAMP-dependent protein kinase activation by CCK in rat pancreas. FEBS Lett 316(1):48–52

McFadden DW, Rudnicki M, Kuvshinoff B, Fischer JE (1992) Postprandial peptide YY release is mediated by cholecystokinin. Surg Gynecol Obstet 175(2):145–150

McLaughlin J, Grazia LM, Jones MN, D'Amato M, Dockray GJ, Thompson DG (1999) Fatty acid chain length determines cholecystokinin secretion and effect on human gastric motility. Gastroenterology 116(1):46–53

MDRNA (2008) MDRNA announces phase 2 trial of PYY(3–36) does not meet weight loss endpoint. MDRNA Press release 31st July 2008. http://phx.corporate-ir.net/phoenix.zhtml?c=83674&p=irol-newsArticle&ID=1182214&highlight=PYY%20nasal%20spray

Mentlein R, Dahms P, Grandt D, Kruger R (1993) Proteolytic processing of neuropeptide Y and peptide YY by dipeptidyl peptidase IV. Regul Pept 49(2):133–144

Mercer LD, Beart PM (1997) Histochemistry in rat brain and spinal cord with an antibody directed at the cholecystokininA receptor. Neurosci Lett 225(2):97–100

Meyer JH, Kelly GA, Spingola LJ, Jones RS (1976) Canine gut receptors mediating pancreatic responses to luminal L-amino acids. Am J Physiol 231(3):669–677

Miller LJ, Gao F (2008) Structural basis of cholecystokinin receptor binding and regulation. Pharmacol Ther 119(1):83–95

Moran TH (2008) Unraveling the obesity of OLETF rats. Physiol Behav 94(1):71–78

Moran TH, McHugh PR (1982) Cholecystokinin suppresses food intake by inhibiting gastric emptying. Am J Physiol 242(5):R491–R497

Moran TH, Robinson PH, Goldrich MS, McHugh PR (1986) Two brain cholecystokinin receptors: implications for behavioral actions. Brain Res 362(1):175–179

Moran TH, Smith GP, Hostetler AM, McHugh PR (1987) Transport of cholecystokinin (CCK) binding sites in subdiaphragmatic vagal branches. Brain Res 415(1):149–152

Mutt V, Jorpes E (1971) Hormonal polypeptides of the upper intestine. Biochem J 125(3):57P–58P
Ondetti MA, Pluscec J, Sabo EF, Sheehan JT, Williams N (1970) Synthesis of cholecystokinin-pancreozymin. I. The C-terminal dodecapeptide. J Am Chem Soc 92(1):195–199
Ozcelebi F (1995) Phosphopeptide mapping of cholecystokinin receptors on agonist-stimulated native pancreatic acinar cells. J Biol Chem 270(7):3435
Parker RM, Herzog H (1999) Regional distribution of Y-receptor subtype mRNAs in rat brain. Eur J Neurosci 11(4):1431–1448
Parrott RF (1993) Peripheral and central effects of CCK receptor agonists on operant feeding in pigs. Physiol Behav 53(2):367–372
Passaro E Jr, Debas H, Oldendorf W, Yamada T (1982) Rapid appearance of intraventricularly administered neuropeptides in the peripheral circulation. Brain Res 241(2):335–340
Pfluger PT, Kampe J, Castaneda TR, Vahl T, D'Alessio DA, Kruthaupt T et al (2007) Effect of human body weight changes on circulating levels of peptide YY and peptide YY3-36. J Clin Endocrinol Metab 92(2):583–588
Pinilla L, Fernandez-Fernandez R, Roa J, Castellano JM, Tena-Sempere M, Aguilar E (2007) Selective role of neuropeptide Y receptor subtype Y2 in the control of gonadotropin secretion in the rat. Am J Physiol Endocrinol Metab 293(5):E1385–E1392
Pi-Sunyer X, Kissileff HR, Thornton J, Smith GP (1982) C-terminal octapeptide of cholecystokinin decreases food intake in obese men. Physiol Behav 29(4):627–630
Pizzi DA, Leslie CP, Mazzali A, Seri C, Biagetti M, Bentley J et al (2010) Design, synthesis and SAR of a novel series of benzimidazoles as potent NPY Y5 antagonists. Bioorg Med Chem Lett 20(23):7120–7123
Pohl M, Silvente-Poirot S, Pisegna JR, Tarasova NI, Wank SA (1997) Ligand-induced internalization of cholecystokinin receptors. Demonstration of the importance of the carboxyl terminus for ligand-induced internalization of the rat cholecystokinin type B receptor but not the type A receptor. J Biol Chem 272(29):18179–18184
Rao RV, Roettger BF, Hadac EM, Miller LJ (1997) Roles of cholecystokinin receptor phosphorylation in agonist-stimulated desensitization of pancreatic acinar cells and receptor-bearing Chinese hamster ovary cholecystokinin receptor cells. Mol Pharmacol 51(2):185–192
Regard JB, Sato IT, Coughlin SR (2008) Anatomical profiling of G protein-coupled receptor expression. Cell 135(3):561–571
Reidelberger RD (1992) Abdominal vagal mediation of the satiety effects of exogenous and endogenous cholecystokinin in rats. Am J Physiol 263(6 Pt 2):R1354–R1358
Rudolf K, Eberlein W, Engel W, Wieland HA, Willim KD, Entzeroth M et al (1994) The first highly potent and selective non-peptide neuropeptide Y Y1 receptor antagonist: BIBP3226. Eur J Pharmacol 271(2–3):R11–R13
Rueeger H, Rigollier P, Yamaguchi Y, Schmidlin T, Schilling W, Criscione L et al (2000) Design, synthesis and SAR of a series of 2-substituted 4-amino-quinazoline neuropeptide Y Y5 receptor antagonists. Bioorg Med Chem Lett 10(11):1175–1179
Sainsbury A, Shi YC, Zhang L, Aljanova A, Lin Z, Nguyen AD et al (2010) Y4 receptors and pancreatic polypeptide regulate food intake via hypothalamic orexin and brain-derived neurotropic factor dependent pathways. Neuropeptides 44(3):261–268
Sankaran H, Goldfine ID, Bailey A, Licko V, Williams JA (1982) Relationship of cholecystokinin receptor binding to regulation of biological functions in pancreatic acini. Am J Physiol 242(3): G250–G257
Savage AP, Adrian TE, Carolan G, Chatterjee VK, Bloom SR (1987) Effects of peptide YY (PYY) on mouth to caecum intestinal transit time and on the rate of gastric emptying in healthy volunteers. Gut 28(2):166–170
Sayegh AI, Ritter RC (2000) Vagus nerve participates in CCK-induced Fos expression in hindbrain but not myenteric plexus. Brain Res 878(1–2):155–162
Schwartz GJ (2000) The role of gastrointestinal vagal afferents in the control of food intake: current prospects. Nutrition 16(10):866–873
Schwartz GJ, Salorio CF, Skoglund C, Moran TH (1999) Gut vagal afferent lesions increase meal size but do not block gastric preload-induced feeding suppression. Am J Physiol 276(6 Pt 2): R1623–R1629

Simmons RD, Kaiser FC, Pierson ME, Rosamond JR (1998) ARL 15849: a selective CCK-A agonist with anorectic activity in the rat and dog. Pharmacol Biochem Behav 59(2):439–444
Sloth B, Holst JJ, Flint A, Gregersen NT, Astrup A (2007) Effects of PYY1-36 and PYY3-36 on appetite, energy intake, energy expenditure, glucose and fat metabolism in obese and lean subjects. Am J Physiol Endocrinol Metab 292(4):E1062–E1068
Smith GP, Jerome C, Cushin BJ, Eterno R, Simansky KJ (1981) Abdominal vagotomy blocks the satiety effect of cholecystokinin in the rat. Science 213(4511):1036–1037
Talsania T, Anini Y, Siu S, Drucker DJ, Brubaker PL (2005) Peripheral exendin-4 and peptide YY (3-36) synergistically reduce food intake through different mechanisms in mice. Endocrinology 146(9):3748–3756
Tanaka T, Katsuma S, Adachi T, Koshimizu TA, Hirasawa A, Tsujimoto G (2008) Free fatty acids induce cholecystokinin secretion through GPR120. Naunyn Schmiedebergs Arch Pharmacol 377(4–6):523–527
Taylor IL, Impicciatore M, Carter DC, Walsh JH (1978) Effect of atropine and vagotomy on pancreatic polypeptide response to a meal in dogs. Am J Physiol 235(4):E443–E447
Ter Horst GJ, Luiten PG, Kuipers F (1984) Descending pathways from hypothalamus to dorsal motor vagus and ambiguus nuclei in the rat. J Auton Nerv Syst 11(1):59–75
Ter Horst GJ, de Boer P, Luiten PG, van Willigen JD (1989) Ascending projections from the solitary tract nucleus to the hypothalamus. A *Phaseolus vulgaris* lectin tracing study in the rat. Neuroscience 31(3):785–797
Thompson JC, Fender HR, Ramus NI, Villar HV, Rayford PL (1975) Cholecystokinin metabolism in man and dogs. Ann Surg 182(4):496–504
Toth PT, Bindokas VP, Bleakman D, Colmers WF, Miller RJ (1993) Mechanism of presynaptic inhibition by neuropeptide Y at sympathetic nerve terminals. Nature 364(6438):635–639
Tough IR, Forbes S, Tolhurst R, Ellis M, Herzog H, Bornstein JC et al (2011) Endogenous peptide YY and neuropeptide Y inhibit colonic ion transport, contractility and transit differentially via Y(1) and Y(2) receptors. Br J Pharmacol 164(2b):471–484
Tschop M, Castaneda TR, Joost HG, Thone-Reineke C, Ortmann S, Klaus S et al (2004) Physiology: does gut hormone PYY3-36 decrease food intake in rodents? Nature 430(6996):1
Tunstall-Pedoe H (2005) Preventing chronic diseases. A vital investment. WHO Global Report, Geneva
Ueno N, Inui A, Iwamoto M, Kaga T, Asakawa A, Okita M et al (1999) Decreased food intake and body weight in pancreatic polypeptide-overexpressing mice. Gastroenterology 117 (6):1427–1432
Uhe AM, Szmukler GI, Collier GR, Hansky J, O'Dea K, Young GP (1992) Potential regulators of feeding behavior in anorexia nervosa. Am J Clin Nutr 55(1):28–32
Utz AL, Lawson EA, Misra M, Mickley D, Gleysteen S, Herzog DB et al (2008) Peptide YY (PYY) levels and bone mineral density (BMD) in women with anorexia nervosa. Bone 43 (1):135–139
Vahl TP, Drazen DL, Seeley RJ, D'Alessio DA, Woods SC (2010) Meal-anticipatory glucagon-like peptide-1 secretion in rats. Endocrinology 151(2):569–575
Vanderhaeghen JJ, Lotstra F, De MJ, Gilles C (1980) Immunohistochemical localization of cholecystokinin- and gastrin-like peptides in the brain and hypophysis of the rat. Proc Natl Acad Sci U S A 77(2):1190–1194
Vrang N, Madsen AN, Tang-Christensen M, Hansen G, Larsen PJ (2006) PYY(3-36) reduces food intake and body weight and improves insulin sensitivity in rodent models of diet-induced obesity. Am J Physiol Regul Integr Comp Physiol 291(2):R367–R375
Wang YH, Tache Y, Sheibel AB, Go VL, Wei JY (1997) Two types of leptin-responsive gastric vagal afferent terminals: an in vitro single-unit study in rats. Am J Physiol 273(2 Pt 2): R833–R837
Wang L, Gourcerol G, Yuan PQ, Wu SV, Million M, Larauche M et al (2010) Peripheral peptide YY inhibits propulsive colonic motor function through Y2 receptor in conscious mice. Am J Physiol Gastrointest Liver Physiol 298(1):G45–G56

Wang Y, Chandra R, Samsa LA, Gooch B, Fee BE, Cook JM et al (2011) Amino acids stimulate cholecystokinin release through the calcium-sensing receptor. Am J Physiol Gastrointest Liver Physiol 300(4):G528–G537

Wank SA (1995) Cholecystokinin receptors. Am J Physiol 269(5 Pt 1):G628–G646

Wank SA, Pisegna JR, de Weerth A (1992) Brain and gastrointestinal cholecystokinin receptor family: structure and functional expression. Proc Natl Acad Sci U S A 89(18):8691–8695

Weatherford SC, Laughton WB, Salabarria J, Danho W, Tilley JW, Netterville LA et al (1993) CCK satiety is differentially mediated by high- and low-affinity CCK receptors in mice and rats. Am J Physiol 264(2 Pt 2):R244–R249

Whitcomb DC, Taylor IL, Vigna SR (1990) Characterization of saturable binding sites for circulating pancreatic polypeptide in rat brain. Am J Physiol 259(4 Pt 1):G687–G691

Wieland HA, Engel W, Eberlein W, Rudolf K, Doods HN (1998) Subtype selectivity of the novel nonpeptide neuropeptide Y Y1 receptor antagonist BIBO 3304 and its effect on feeding in rodents. Br J Pharmacol 125(3):549–555

Wiley JW, Gross RA, MacDonald RL (1993) Agonists for neuropeptide Y receptor subtypes NPY-1 and NPY-2 have opposite actions on rat nodose neuron calcium currents. J Neurophysiol 70 (1):324–330

World Health Organisation (2011) World Health Organisation report factsheet. No. 311. Obesity and Overweight

Wortley KE, Garcia K, Okamoto H, Thabet K, Anderson KD, Shen V et al (2007) Peptide YY regulates bone turnover in rodents. Gastroenterology 133(5):1534–1543

Yuan CS, Barber WD (1991) Parabrachial nucleus: neuronal evoked responses to gastric vagal and greater splanchnic nerve stimulation. Brain Res Bull 27(6):797–803

Zarbin MA, Wamsley JK, Innis RB, Kuhar MJ (1981) Cholecystokinin receptors: presence and axonal flow in the rat vagus nerve. Life Sci 29(7):697–705

Zhang T, Uchida T, Gomez G, Lluis F, Thompson JC, Greeley GH Jr (1993) Neural regulation of peptide YY secretion. Regul Pept 48(3):321–328

Zipf WB, O'Dorisio TM, Cataland S, Dixon K (1983) Pancreatic polypeptide responses to protein meal challenges in obese but otherwise normal children and obese children with Prader–Willi syndrome. J Clin Endocrinol Metab 57(5):1074–1080

Effects of Amylin on Eating and Adiposity

Thomas Alexander Lutz

Contents

Abstract Amylin's best investigated function is to reduce eating via a meal size effect by promoting meal-ending satiation. This effect seems to depend on an

T.A. Lutz (✉)
Institute of Veterinary Physiology and Zurich Center for Integrative Human Physiology, University of Zurich, Winterthurerstrasse 260, 8057 Zurich, Switzerland
e-mail: tomlutz@vetphys.uzh.ch

H.-G. Joost (ed.), *Appetite Control*, Handbook of Experimental Pharmacology 209,
DOI 10.1007/978-3-642-24716-3_10, © Springer-Verlag Berlin Heidelberg 2012

activation of specific area postrema neurons. Brain areas that convey the neural signal to the forebrain include the nucleus of the solitary tract and the lateral parabrachial nucleus. Acute application of amylin modulates the activity of hypothalamic areas involved in the control of eating, namely, the lateral hypothalamic area and possibly the ventromedial hypothalamic nucleus. Amylin also interacts with other satiating signals, such as cholecystokinin, presumably in the brainstem. Interestingly, amylin also exhibits characteristics of adiposity signals; plasma levels of amylin are higher in obese individuals, chronic infusion of amylin into the brain reduces body weight gain and adiposity, and infusion of amylin antagonists increases adiposity. Furthermore, amylin maintains energy expenditure at higher levels than would be expected considering its body weight-lowering effect. However, much less is known (e.g., site of action, signaling pathways, differential activation of brain sites, and, most importantly, physiological relevance) with respect to its role as adiposity signal and regulator of energy expenditure than about its satiating action. Notwithstanding, and perhaps because amylin resistance does not seem to be a general and prohibitive concomitant of obesity, animal data and recent clinical data in humans indicate that amylin is a very promising candidate for the treatment of obesity. Amylin seems to be particularly effective when combined with other hormones such as leptin.

Keywords Amylin • Satiation • Area postrema • Adiposity signal • Energy expenditure

1 Introduction

Pancreatic beta cells are the major, though not the only, source of circulating amylin. Changes in circulating amylin levels are thought to directly reflect changes in beta-cell secretion, and the contribution of other amylin-secreting cells to circulating amylin levels is considered minor. It is generally believed that these fluctuations in beta-cell secretion (e.g., postprandial increase) are the physiological basis for amylin's effect on eating and energy homeostasis (Young 2005b). Whether centrally synthesized amylin also contributes to this eating control is still a matter of debate, and it is in fact still not clear whether amylin is synthesized in the mammalian brain. A recent study suggested that at least in female rats, central amylin production may contribute to the control of maternal regulations because amylin was specifically upregulated in the preoptic area of the hypothalamus in the early postpartum period (Dobolyi 2009); however, the role of central amylin in the control of eating in this particular metabolic situation has not been tested.

1.1 Amylin and the Control of Nutrient Fluxes

The best investigated function of amylin, and the main focus of this chapter, is its role as a control of eating. However, the anorectic action of amylin appears to be only one of several important factors in amylin's overall role to control the influx of nutrients into the circulation. Amylin's actions to reduce gastric acid secretion, to limit the rate of gastric emptying, and to diminish pancreatic glucagon and digestive enzyme secretion are other factors that serve the same purpose. All these effects seem to be mediated by a direct action in the central nervous system, specifically via area postrema (AP) neurons. Hence, amylin is considered a necessary and complementary factor to insulin in the control of nutrient flux by regulating nutrient appearance and the postprandial glucose concentration. These actions are also the basis for the use of the amylin analogue pramlintide as adjunct treatment in type 1 and type 2 diabetes mellitus (Young and Denaro 1998; Weyer et al. 2001a; summarized in Young 2005b).

1.2 *Amylin's Effects on Eating*

It seems well accepted that peripheral amylin acts directly in the brain to affect eating and that amylin influences well-defined brain areas. The multiple signals in the control of eating interact in the brain (e.g., cholecystokinin (CCK) and amylin; Bhavsar et al. 1998; Mollet et al. 2003a); according to a generally agreed concept, the various controls of eating can be classified as adiposity, or tonic, signals that enhance the effect of satiation, or episodic/phasic, signals (Barrachina et al. 1997; Riedy et al. 1995; Woods 2004, 2005). Amylin may perhaps play a particular role in the control of eating because it has properties of both adiposity and satiating signals (Lutz 2006; Woods et al. 2006). This, of course, questions the clear distinction between the two classes of signals. Whether the idea of complementary actions of satiating and adiposity signals (Woods 2004, 2005) also holds up in the case of amylin is at present unknown. In other words, it is not known how tonic levels of amylin may affect the satiating effect of phasic, meal-associated changes in amylin levels (see also paragraph 3.2). Amylin-deficient mice may also provide some insight in this respect; these mice show a discrete phenotype of unaltered total food intake (measured as total food consumption in a cohort of group-housed animals), combined with a slightly higher rate of body weight gain compared to wild-type controls (Devine and Young 1998; Mollet et al. 2003a; reviewed in Lutz 2006). Importantly, adult amylin-deficient mice eat less after acute peripheral amylin injection, and the magnitude of this effect on eating is similar in amylin-deficient mice and their controls. Hence, at least the acute eating-inhibitory effect of amylin does not require a "background" level of endogenous amylin.

1.3 *Amylin Agonists and Antagonists*

Amylin is structurally and functionally related to the calcitonin family of peptides that apart from amylin includes calcitonin (CT), calcitonin gene-related peptide

(CGRP), adrenomedullin (ADM), and others (Cooper 1994; Lutz 2006, 2010; Young 2005a). Amylin, CGRP, and CTs reduce food intake in a variety of species. Interestingly, the anorectic effect of salmon CT (sCT) is far more potent than that of mammalian CT in rats. Hence, sCT has been used in a number of studies as potent and long-lasting amylin agonist (Lutz et al. 2000). Another commonly used agonist is the amylin analogue pramlintide which is derived from human amylin. Native human amylin, like feline amylin, has a strong propensity to precipitate in vitro and to form amyloid deposits in vivo (Lutz and Rand 1993). This propensity seems to be related to slight differences in the amino acid structure of human (or feline) amylin versus rat amylin (which does not precipitate) in the 20–29 regions of the molecule; in particular, the amino acids in positions 25, 28, and 29 seem to be critically important in this respect (Cooper 1994). These three amino acids have been replaced in pramlintide by proline; notably, rat amylin also contains proline residues in these positions. Pramlintide is an approved drug (SymlinR) for the treatment of type 1 and type 2 diabetics (Weyer et al. 2001a; Young 2005a; see paragraphs 1.1 and 7).

Several peptide antagonists have been used to antagonize the action of endogenous or exogenous amylin (Cooper 1994; Young 2005a). Because the first seven amino acids seem to be necessary for biological action of amylin, CTs, or CGRP, typical antagonists are fragments derived from the native peptides. The most commonly used amylin antagonist is AC187 which is a derivative of sCT (30N32Y[8–32]sCT); other antagonists are CGRP(8–37) or amylin(8–37) (for review, see Young 2005a). The amino acid structures of rat and human amylin, of the amylin agonists sCT and pramlintide, and of AC187 are shown in Table 1.

Table 1 Amino acid sequence of amylin (rat; human), the amylin analogue pramlintide (25P28P29P human amylin), the amylin agonist salmon calcitonin (sCT), and the amylin antagonist AC187 (30N32Y[8-32]sCT)

Amino acid position			
1	11	21	31
Amylin rat (37 amino acids)			
K C N T A T C A T Q	R L A N F L V **R** S S	N N L G **P** **V** L **P** **P** T	N V G S N T Y-NH2
Amylin human (37 amino acids)			
K C N T A T C A T Q	R L A N F L V **H** S S	N N **F** G **A** **I** L **S** **S** T	N V G S N T Y-NH2
Pramlintide (37 amino acids; 25P28P29P human amylin)			
K C N T A T C A T Q	R L A N F L V H S S	N N F G **P** I L **P** **P** T	N V G S N T Y-NH2
sCT (32 amino acids)			
C S N L S T C V L G	K L S Q E L H K L Q	T Y P R T N T G S G	T P-NH2
AC187 (25 amino acids; 30N32Y[8-32]sCT)			
V L G	K L S Q E L H K L Q	T Y P R T N T G S **N**	T **Y**-NH2

Amino acids in bold differ between rat and human amylin, between human amylin and pramlintide, or between sCT and AC187, respectively. Amino acids are denoted by the one-letter abbreviation

2 Amylin Has Typical Properties of a Satiation Signal

Amylin has properties of a physiological control of meal size (Geary 2004; Lutz et al. 1995; Lutz 2006; Young and Denaro 1998); the meal-induced changes in endogenous plasma amylin levels are rapid, and exogenous amylin given at the onset of spontaneous meals decreases eating in rats within few minutes (Lutz et al. 1995). Importantly, even when given chronically via osmotic minipumps, amylin's overall inhibitory effect on eating relies mainly on an amylin-induced reduction in meal size without a compensatory increase in meal frequency (Lutz et al. 1995, 2001a; Mack et al. 2007). Administration of the amylin receptor antagonist AC187 stimulates eating by increasing meal size, presumably by a blockade of endogenous amylin action (Mollet et al. 2004; Reidelberger et al. 2004). All in all, these results support the idea of a physiological role of amylin as satiation signal.

2.1 *Amylin Site of Action at the Brainstem*

The satiating effect of peripheral amylin is mediated by direct action on AP neurons; only specific AP lesions, but not interruption of afferent nerve signaling from the periphery to the brain, in particular to the nucleus of the solitary tract (NTS), reduced the peripheral effect of amylin on eating (for review: Potes and Lutz 2010). Further, local injection of amylin into the AP also inhibited eating by a meal size effect, and injections of the amylin receptor antagonist AC187 increased meal size. Finally, a specific chemical lesion of noradrenergic AP neurons was sufficient to block amylin's effect on eating, indicating that these neurons play a necessary part in mediating amylin's anorectic action (Potes et al. 2010c). The in vivo behavioral data are consistent with electrophysiological and immunohistochemical studies that confirmed a direct influence of amylin on the AP and an activation of AP neurons (Lutz et al. 1998; Lutz 2006; Mollet et al. 2004; Riediger et al. 2001, 2004).

Amylin displays strong binding to the AP (Sexton et al. 1994), and all identified components of the specific amylin receptor [the calcitonin receptor core (CTR) and receptor-activity-modifying proteins (RAMPs)] are expressed in the AP. The typical amylin receptor is a heterodimer of the type A or type B CTR and RAMP1 or 3. RAMPs alter CTR's specificity for amylin (Christopoulos et al. 1999; Muff et al. 1999), i.e., they alter CTR pharmacology from calcitonin-preferring to amylin-preferring receptors; RAMPs regulate the transport of the core receptor to the cell surface and their glycosylation state which eventually determines ligand specificity (Fischer et al. 2002; McLatchie et al. 1998). The CTR is present in the AP (Becskei et al. 2004), and RAMP1 and RAMP3 mRNA are highly expressed in the mouse AP (Ueda et al. 2001). Finally, amylin-induced c-Fos mRNA and CTR and RAMP3 mRNA expressions all colocalize in the rat AP (Barth et al. 2004), and most AP neurons in which systemic amylin specifically induces cyclic guanosine monophosphate (cGMP) formation carry the CTR (Becskei et al. 2004; Riediger et al. 2001). Of note, it still needs to be shown that the CTR and RAMP1 or RAMP3 colocalize in the same amylin-sensitive AP neurons.

2.2 Modulation of Amylin Sensitivity of AP Neurons by Nutrients

Amylin and glucose coactivate AP neurons (Riediger et al. 2002), and a certain (blood) glucose level may be necessary for full amylin action; this coactivation seems to have a functional correlate for amylin's effect on gastric emptying which is not present under hypoglycemic conditions (Gedulin and Young 1998). Whether the anorectic effect of amylin is also affected by hypoglycemia, i.e., whether amylin reduces eating under euglycemic and hyperglycemic, but not under hypoglycemic conditions, is currently unknown.

Recent findings indicate that the action of amylin on AP neurons may also be affected by protein. Initial experiments showed that a low dose of peripheral amylin induced a strong c-Fos expression in the AP and NTS of 24-h-fasted rats, but not in rats fed ad libitum (Michel et al. 2007). Based on subsequent experiments using nutrient-deficient noncaloric mash diets (NCM) that were selectively supplemented with protein, glucose, or fat (lard) (Michel et al. 2007), we concluded that protein—or perhaps single amino acids—attenuated the amylin-induced c-Fos response in the AP. Consistent with this view, peripheral injection of an amino acid mixture also significantly attenuated the amylin-induced c-Fos expression in the AP in fasted rats (Riediger et al. 2009); further, amylin's anorectic action was stronger in rats fed with a 1% protein diet (1% weight/weight) compared to its action in rats fed with an isocaloric 8% or 18% protein diet (Riediger et al. 2009). The underlying mechanisms are still not clear; it needs, e.g., to be tested whether specific amino acids reduce the effect of amylin to activate AP neurons or whether the effect is indirect, e.g., by a protein-induced specific release of some hormone that in part counteracts amylin action in the AP. Further, the functional implications of these findings, and in particular their clinical relevance for the use of the amylin analogue pramlintide in the treatment of obesity, need to be defined.

2.3 Differences Between Amylin and CCK as Satiation Signals

Amylin produces acute satiating effects like CCK, but an important difference is worth to mention. Chronic CCK seems unable to reduce body weight under most experimental conditions. First, rats infused continuously with CCK do not show a sustained reduction in eating and in body weight; second, CCK infused intraperitoneally prior to spontaneous meals only reduces food intake initially because the decrease in meal size is soon compensated by an increase in meal frequency (Crawley and Beinfeld 1983; West et al. 1984; reviewed in Smith and Gibbs 1985). The mechanisms of this compensatory increase in meal frequency are unknown. Importantly, such mechanisms—at least to the same extent as for CCK—do not seem to occur with amylin, because chronic amylin produces a sustained reduction in meal size, total food intake, and ultimately body weight (Arnelo et al. 1996, 1998; Lutz et al. 2001a).

2.4 Central Processes of Amylin Signaling

The amylin-induced activation of the AP is generally assumed to be synaptically transmitted to the forebrain via the NTS and the lateral parabrachial nucleus (lPBN) (Riediger et al. 2004). Lesions of the AP, the NTS, or the lPBN [but not of the central nucleus of the amygdala (CeA)] blocked the eating-inhibitory effect of peripheral amylin, and these lesions also blocked amylin-induced c-Fos expression in areas *rostral* to the lesion, i.e., NTS, lPBN, and CeA in AP-lesioned rats, and in the CeA in lPBN-lesioned rats. Hence, the AP seems to be amylin's primary target, with the NTS and lPBN as direct projection areas (Becskei et al. 2007; Potes et al. 2010a; Riediger et al. 2004; Rowland and Richmond 1999).

The lPBN also appears to act as an important relay station between the hindbrain and the lateral hypothalamic area (LHA), where amylin reduces the fasting-induced c-Fos expression (Potes et al. 2010a; Riediger et al. 2004). Tracing studies revealed ascending projections from the lPBN to other hypothalamic nuclei, and in particular the ventromedial hypothalamic nucleus (Potes et al. 2010a; see also Mollet et al. 2003b); the role of these hypothalamic projection areas in the inhibitory effect of amylin on eating is currently under investigation.

Central pathways involved in the signaling cascade of amylin and other eating controls have often been characterized by use of c-Fos protein expression as a marker of neuronal activation; however, the functional implications of c-Fos expression for hormonal action on eating behavior are unknown (see Lutz 2010 for a discussion on the limitations using c-Fos immunohistochemistry). Another marker of neuronal activation that has been used more recently is the phosphorylated form of the extracellular signal-regulated kinase 1/2 (pERK). The kinetics but also the cellular effects of this pathway seem to parallel much better the characteristics of the eating-inhibitory effects of short-acting signals such as amylin or CCK. Activation of the ERK pathway leads to a rapid ERK1/2 phosphorylation; this may lead to acute neuronal responses such as activation or inhibition of ion channels that directly affect neuronal excitability (Nishimoto and Nishida 2006; Yuan et al. 2002). Hence, it is plausible that the ERK pathway may be functionally involved in fast eating-inhibitory effects, not only for CCK (Sutton et al. 2004), but potentially also for amylin. We have recently shown that amylin increases pERK formation time and dose dependently (Potes et al. 2010b); most interestingly, the peak of pERK detected by immunocytochemistry was observed at a time when amylin exerts its action to end a meal, i.e., within about 10–15 min after administration (Lutz et al. 1995). Further, blockade of ERK phosphorylation in the AP by a local infusion of the MEK inhibitor U0126 into the forth ventricle reduced amylin's potency to inhibit eating (unpublished). These studies provide good evidence for the use of ERK phosphorylation as a neuronal correlate for fast behavioral effects in general, and for an important functional role of this pathway in amylin's satiating effect.

2.5 *Future Areas of Research*

Despite the increasing knowledge on the mechanisms that are involved in amylin's anorectic action, there are still many open questions that need to be answered to obtain a full picture on how the brain processes amylin signaling. These include questions on specific aspects of amylin physiology and pathophysiology in the AP (e.g., is amylin signaling different in obese versus lean individuals), and on the temporal sequence of activation or inhibition of other brain areas that may be involved in amylin signaling. Further, it will be important to study efferent pathways that are triggered by amylin because no information is available concerning these specific circuits involved in amylin's effect on eating.

3 Amylin Has Typical Properties of an Adiposity Signal

Amylin also shares characteristics of adiposity signals, like leptin or insulin (Hillebrand and Geary 2010; Schwartz et al. 2003; Woods 2004, 2005). The basal plasma levels of amylin are generally higher in obese individuals (Enoki et al. 1992; Hanabusa et al. 1992; Leckström et al. 1999; Martin et al. 2010; Pieber et al. 1994), indicating an association between body adiposity and plasma amylin, but it is not yet clear whether changes in body adiposity result directly and with the same temporal pattern in changes of amylin levels (Gloy et al. 2010). Further, chronic peripheral (Mack et al. 2007; Roth et al. 2006) or central (Rushing et al. 2000) amylin infusion decreases body weight gain specifically by an effect on fat mass, and central administration of the amylin antagonist AC187 increases body adiposity (Rushing et al. 2001). Recent experiments provided further support for the idea to consider amylin a potential adiposity signal (Wielinga et al. 2008, 2010). In these experiments, the body weight of rats was manipulated by prior overfeeding or food restriction. Subsequent central amylin infusion resulted in a body weight trajectory that within few days was indistinguishable from that of rats infused with amylin and fed ad libitum; however, body weight was markedly lower than in respective saline-treated controls. In other words, the central amylin level appeared to be an important determinant for the body weight to be reached (Wielinga et al. 2010). Hence, the results are similar to what has been reported for leptin or insulin (Chavez et al. 1995; Woods 2005); amylin may encode the achieved level of body weight and contribute to the relative constancy of body weight throughout adult life.

3.1 *Resistance to Leptin and Insulin in Obesity*

Most cases of obesity in humans and in animal models are typically associated with the development of resistance to the central effects of leptin and insulin. This results in a reduced eating-inhibitory response to exogenous administration of these hormones; in other words, higher doses of leptin and insulin are necessary to exert the same magnitude of effect as observed in lean individuals. The causes of this resistance phenomenon are most likely numerous; in addition to reduced leptin and insulin transfer via the blood–brain barrier (Banks and Kastin 1998; Banks 2008, 2010), the direct action of leptin and insulin at their most important hypothalamic target sites is also altered in obesity (for review: Münzberg 2010).

3.2 Sensitivity to Amylin in Obesity

Until recently, only few studies directly investigated the influence of obesity on the sensitivity of rats or mice to anorectic doses of amylin or its agonists, e.g., salmon calcitonin (sCT).

Most studies indicated that amylin retains at least partial efficiency to reduce eating and body weight gain, despite prevailing obesity. It was, e.g., reported that *ob/ob* mice, melanocortin-4 receptor knockout mice, and obese Zucker *fa/fa* rats all respond to amylin agonism (Eiden et al. 2002; Grabler and Lutz 2004; Morley et al. 1994), and antagonism to endogenous amylin with peripheral AC187 also increased eating in Zucker rats (Grabler and Lutz 2004). Further, at least using high doses, acute administration of the amylin analogue pramlintide decreased the size of test meals by about 20% in obese humans (Chapman et al. 2005; Hollander et al. 2004).

Recent studies in our own laboratory tested the effect of specific obesity-related phenomena on the anorectic and body weight-lowering properties of amylin (Boyle et al. 2010, 2011). In the first study, rats maintained on a high-fat diet (HF; 60% fat by calories) showed an unaltered response to acute amylin for about 3 months on this diet. Only after that, rats appeared to be less sensitive to amylin. We concluded that the HF diet alone does not seem to alter amylin sensitivity but that the metabolic changes that occur following prolonged HF intake may modify amylin's efficacy. It is unclear at present which parameters associated with the development of obesity in HF fed animals may be responsible for this phenomenon.

To rule out that chronic hyperamylinemia, which typically parallels the development of obesity (Hanabusa et al. 1992; Leckström et al. 1999; Pieber et al. 1994; Reinehr et al. 2007), may cause a reduction in amylin sensitivity, we tested whether an elevation of baseline plasma amylin levels independent of obesity would render rats less sensitive to acute amylin injections. In contrast to hyperleptinemia or high circulating levels of CCK, which can induce leptin resistance (Knight et al. 2010) or reduced sensitivity to CCK in rats (Covasa et al. 2001), respectively, no such phenomenon was observed in the case of amylin. Lean rats that were chronically infused with amylin, resulting in baseline amylin concentrations typically observed in obesity, demonstrated a dose-dependent decrease in food intake after acute amylin administration, regardless of circulating amylin levels (Boyle et al. 2011). We therefore concluded that amylin sensitivity seems unaltered by chronically elevated amylin; hence, a downregulation of amylin receptors or a decrease in postreceptor signaling appears to be unlikely.

Overall, we observed reduced amylin sensitivity under some but not all experimental conditions that would typically be associated with leptin resistance; in all cases, at least high doses of amylin still caused a significant effect on eating. Why amylin sensitivity may be somewhat lower in some cases is not yet clear. Because the AP has an open blood–brain barrier (reviewed in Potes and Lutz 2010), resistance at the level of amylin access to the brain and its target neurons seems an unlikely explanation for reduced amylin sensitivity in obesity. This clearly distinguishes amylin from leptin because decreased leptin transport across the blood–brain barrier appears to play an important role in leptin resistance (Banks et al. 1999). Further, at least when gauged by c-Fos expression, the amylin-induced activation of its target neurons in the AP was similar between rats fed standard chow or HF. Thus, amylin insensitivity is probably also not caused by direct changes in amylin receptor function. Potential alterations in intracellular postreceptor effects (e.g., activation of the cGMP or the pERK systems (Riediger et al. 2001; Potes et al. 2010b)) have not yet been tested.

3.3 *Amylin Secretion in Obesity*

An alternative possibility for reduced action of the (endogenous) amylin system in obesity could be deficient amylin release in response to a meal. We therefore recently tested whether rats exposed to a HF diet exhibit an altered meal-induced amylin release; we compared hepatic portal vein concentrations of amylin and insulin in rats maintained on a standard diet to those on a HF diet. Part of the HF fed rats developed diet-induced obesity. We observed that rats maintained on the HF diet exhibited an earlier meal-induced increase in plasma amylin levels, independent of the level of obesity. The meal-induced insulin release appeared unchanged in HF versus chow-fed rats. This can potentially be explained by a selective effect of elevated fatty acids to enhance mRNA expression of amylin, but not of insulin (Qi et al. 2009). More importantly, these data do not provide evidence for a reduction in meal-induced amylin release. Hence, it seems unlikely that overeating in obesity may be due to an insufficiency of endogenous amylin release.

4 How Do the Two Roles of Amylin as Satiation and Adiposity Signals Combine?

As discussed in the previous paragraphs, amylin has characteristics of satiation and of adiposity signals. An important difference between amylin and the "classical" satiation signal CCK is that continuous infusion of amylin in rats (Arnelo et al.

1996; Lutz et al. 2001a), but not of CCK (Crawley and Beinfeld 1983), results in a sustained reduction in food intake and body weight; in contrast to CCK (West et al. 1984), the amylin-induced decrease in meal size is not fully compensated by an increase in meal frequency (Arnelo et al. 1996; Lutz et al. 2001a).

An important question that may come up is how we can differentiate mechanistically between these two "roles" of amylin. This question is in fact not limited to amylin and could also be asked in respect to other signals involved in the control of eating, such as insulin or ghrelin (Surina-Baumgartner et al. 1995; Williams and Cummings 2005). One way to answer this question would be to investigate whether the roles of amylin as satiating and as adiposity signals are processed by different parts of the central nervous system; one could imagine processing of the signals in different brain sites or by different neuronal populations in a given brain site. At present, these questions remain unanswered.

5 Amylin's Effect to Increase Energy Expenditure

Amylin, like other signals involved in the control of eating, seems to increase energy expenditure in rats and mice. One of the first experiments to study this phenomenon indicated that the amylin receptor agonist sCT reduced body weight and body fat in fasted rats compared to saline-treated fasted controls (Lutz et al. 2001b). Further, body fat loss in rats centrally infused with amylin was more pronounced than in pair-fed controls (Wielinga et al. 2008, 2010); this again indicated that amylin increased energy expenditure in addition to reducing food intake.

Over recent years, several reports consistently indicated that both acute and chronic amylin receptor activation seems to increase energy expenditure, or at least to prevent a decrease in energy expenditure that would typically result from the amylin-induced reduction in eating and body weight (Isaksson et al. 2005; Mack et al. 2007; Osaka et al. 2008; Roth et al. 2006). It has been suggested that this effect on energy expenditure may in part be secondary to the amylin-induced reduction in adiposity, and to a relative increase in the metabolically more active lean body mass (Roth et al. 2006). Acute peripheral injection of an anorectic dose of amylin failed to increase energy expenditure in rats, whereas the long-acting agonist sCT increased it (Wielinga et al. 2007). Thus, we presume that the lack of effect of peripheral amylin may be due to its short half-life in the peripheral circulation. When administered centrally, both sCT and amylin acutely increased energy expenditure by about 25%. Because central amylin specifically lowers body adiposity (Rushing et al. 2000), we presume that amylin increases fat oxidation as indicated by a lower respiratory quotient.

Despite the clear effects of exogenous amylin (or sCT) on energy expenditure, the physiological relevance of these effects, or whether these effects are purely pharmacological, is still an open question. In other words, the role of endogenous amylin in the control of energy expenditure has not yet been tested. Keeping this reservation in mind, we believe that the data reported above are consistent with the idea that amylin affects energy balance in part by increasing energy expenditure, but final proof is missing. It is important to note that this effect has to be seen in light of amylin's body weight-lowering effect. Hence, we consider the effect of amylin to prevent the compensatory decrease in energy expenditure that is typically seen in weight-reduced animals or fasted humans (Weyer et al. 2001b) as indicative of a physiologically relevant role in energy balance.

Another open question is what neural targets mediate the effect of exogenous amylin on energy expenditure. At present, the (presumably) central mediators and the exact effector mechanisms which may underlie amylin's action on energy expenditure are unknown. Increased physical activity is unlikely to play an important role (Wielinga et al. 2007), and the same may be true for an effect of amylin on the expression of uncoupling protein (Roth et al. 2006).

6 Basic Research on the Interaction Between Amylin and Leptin

The system controlling energy balance is very complex; due to the multitude of signals involved in this system, it is not surprising that many of these signals interact, but detailed research on the potential interactions among these signals is limited. In recent years, the interactions between amylin and leptin have been a specific focus of research; the results of basic research but also of clinical studies are encouraging because they point to a potential use of this combination therapy in antiobesity treatment. This interaction is most likely not based on a direct effect of amylin at leptin receptors or vice versa and may involve secondary projection areas where the amylin signal (most likely originating in the AP) and leptin signal (possibly originating in the ARC) converge.

6.1 Interaction of Amylin and Leptin in Rats and Mice

The first study in this context showed that acute central administration of leptin increased the acute eating-inhibitory effect of peripheral amylin in rats (Osto et al. 2007). Subsequent studies, most of them involving chronic administrations, yielded similar and consistent results. Roth and colleagues (Roth et al. 2008) described the effects on eating and body weight of combined 2-week peripheral infusions of amylin and leptin in rats. In obese rats, exogenous leptin alone had no effect on food intake or on the development of body weight, while the same dose of leptin was effective in lean animals. Amylin alone was still effective in the obese rats and reduced body weight by about 5%; this is consistent with the findings discussed above that obesity does not eliminate amylin sensitivity in rats (Boyle et al. 2011).

When leptin was combined with amylin, its effect on food intake and body weight in obese rats was enhanced by a factor of 2. In addition, coadministration of leptin and amylin produced a greater effect on adiposity than amylin alone or than leptin combined with pair feeding to amylin-treated rats. Furthermore, dark-phase energy expenditure was highest in the rats that received both amylin and leptin. Overall, these data indicate that amylin reversed the leptin resistance of obese rats (Roth et al. 2008).

A series of subsequent studies confirmed that amylin and leptin have synergistic effects on eating, body weight, and body adiposity (Trevaskis et al. 2008; Turek et al. 2010). The major effect on body weight of the combination of amylin and leptin seemed to be due to the reduction in food intake; however, similar to our recent studies with central amylin infusion (Wielinga et al. 2008, 2010), the decrease in energy expenditure that might be expected due to lower food intake and body weight was prevented by amylin and leptin. Hence, maintenance of energy expenditure probably contributed to the overall effect of amylin and leptin on body weight.

The combined treatment with amylin plus leptin seemed to preferentially favor fat oxidation because the respiratory quotient was maintained at a low level throughout the study period; subsequently, loss of body fat was more pronounced in rats that received amylin and leptin than in rats pair-fed to the treated rats (Roth et al. 2008; Trevaskis et al. 2008). This metabolic effect was consistent with gene expression profiles in the amylin- and leptin-treated rats; the expression of genes for hepatic lipogenesis was reduced, and the expression of genes for lipid utilization was increased (Trevaskis et al. 2008).

While these studies provide evidence for the pharmacological interaction between amylin and leptin, it is less clear whether this interaction is also of physiological relevance. At the behavioral level, this would indicate that endogenous amylin and leptin interact to decrease eating and body weight. In fact, reduction of body weight and adiposity by leptin was lower in amylin-deficient mice than in wild-type controls (Turek et al. 2010). Conversely, the amylin agonist sCT seemed to be less effective in reducing eating in leptin-deficient *ob/ob* mice (Eiden et al. 2002), indicating that endogenous leptin may be required for a full action of amylin. Finally, amylin-deficient mice have a reduction in the leptin-induced pSTAT3 formation in the ARC and VMH, and in leptin receptor expression in the mediobasal hypothalamus (Turek et al. 2010). Overall, these experiments indicate that the functional interaction between amylin and leptin may be physiologically relevant, but more studies are warranted to fully investigate the relevance of this interaction under physiological conditions.

6.2 Mechanism(s) of Interaction

The mechanisms involved in the interaction between amylin and leptin seem to involve secondary projection areas where the amylin and leptin signals converge. Recent studies indicate that amylin mainly seems to influence the central processing of the leptin signal (Roth et al. 2008; Turek et al. 2010) and to affect leptin sensitivity of rats; hence, amylin reverses leptin resistance in obese rats (Roth et al. 2008) and increases leptin sensitivity in lean rats (Turek et al. 2010). The studies also suggest that the AP does not seem to be the major converging site for this interaction because leptin did not enhance the amylin-induced activation of AP neurons when assessed by c-Fos immunocytochemistry (Turek et al. 2010).

The amylin–leptin interaction appears to reside in the hypothalamus, possibly after polysynaptic input from the AP (Lutz et al. 1998, 2001a; Turek et al. 2010). Amylin increased the effect of leptin to induce pSTAT3 formation in the ARC (Turek et al. 2010), and at least high doses of leptin resulted in increased pSTAT3 signaling in the ventromedial hypothalamus (VMH) in obese, amylin-pretreated rats (Roth et al. 2008); in other words, amylin restored the leptin-induced pSTAT3 signal in the VMH of obese rats to a level of leptin-treated lean rats.

These results are consistent with reduced leptin receptor expression in the mediobasal hypothalamus in amylin-deficient mice mentioned above (Turek et al. 2010). Because acute amylin treatment upregulated leptin receptor expression in the hypothalamus and because it also increased leptin binding in the VMH (and some other distinct hypothalamic nuclei), it is plausible that these effects of amylin on pSTAT3 expression, leptin receptor expression, and on leptin binding are all causally linked and correlate with the increase in leptin sensitivity and with the stronger effect of amylin plus leptin on eating and body weight.

7 Clinical Studies on the Potential of Amylin in Antiobesity Therapy

The positive outcome of rodent studies on the interaction between amylin and leptin encouraged translational research for potential application in humans. Preclinical tests in humans showed that the coadministration of the amylin and leptin analogues pramlintide and metreleptin, respectively, in overweight humans lowered body weight in a clinically relevant fashion (Ravussin et al. 2009; Roth et al. 2008). A decrease in body weight of more than 12% was seen in patients receiving the combination of pramlintide and metreleptin, and body weight did not yet stabilize at the end of the observation period; this means that body weight continued to decrease even after half year of treatment.

However, similar to many other chronic diseases (e.g., type 2 diabetes), antiobesity therapy and maintenance of lower body weight may require continuous and perhaps lifelong treatment. At least in diet-induced obese rats, body weight seems to increase again on the cessation of amylin and leptin therapy; the body weight loss was only maintained in rats that received continued treatment (Trevaskis et al. 2010).

8 Summary

To summarize, amylin is a physiological satiating signal that controls meal size by acting at amylin receptors in the AP. Several data indicate that amylin may also act as an adiposity signal. Importantly, chronic amylin treatment lowers body weight gain in rats and mice, and the same effect has also been observed in humans. The mechanisms of amylin action have been explored at least in part, and it remains to be studied whether adjunct modulation of amylin signaling pathways may enhance the anorectic effect triggered by amylin alone. In recent years, animal and human studies on amylin and leptin or their analogues suggest that this combination therapy may be an effective and promising treatment strategy for obesity.

9 Perspectives

Based on the promising preclinical findings to treat human obesity with a combination of amylin and leptin analogues, further research is warranted to investigate the long-term usefulness and potential late side effects. An interesting question is whether lower doses of amylin or leptin analogues can be used to maintain lower body weight once marked body weight loss has been achieved. Further, amylin-based pharmacotherapy is also interesting in respect to recent findings in a rat model of Roux-en-Y gastric bypass (RYGB) surgery where it was reported that postprandial amylin levels, similar to GLP-1 and PYY, were higher in RYGB than sham-operated rats (Bueter et al. 2010; Shin et al. 2010). Hence, it may be possible in the future to mimic the effects of RYGB surgery by application of a well-defined mixture of hormones that lower eating and increase energy expenditure similar to RYGB, but without its only partly explored undesired side effects such as massive loss of bone mineral density (Lutz et al. 2010).

Acknowledgement Our work has been financially supported by the Swiss National Science Foundation. The support of the Zurich Center of Integrative Human Physiology, the Novartis Foundation, the Olga Mayenfisch Foundation, the Vontobel Foundation, and the Ciba-Geigy Jubilee Foundation are also gratefully acknowledged.

References

Arnelo U, Permert J, Adrian TE, Larsson J, Westermark P, Reidelberger RD (1996) Chronic infusion of IAPP causes anorexia in rats. Am J Physiol 271:R1654–R1659

Arnelo U, Reidelberger R, Adrian TE, Larsson J, Permert J (1998) Sufficiency of postprandial plasma levels of islet amyloid polypeptide for suppression of feeding in rats. Am J Physiol 275: R1537–R1542

Banks WA (2008) The blood-brain barrier: connecting the gut and the brain. Regul Pept 149:11–14
Banks WA (2010) Blood-brain barrier as a regulatory interface. Forum Nutr 63:102–110
Banks WA, Kastin AJ (1998) Differential permeability of the blood-brain barrier to two pancreatic peptides: insulin and amylin. Peptides 19:883–889
Banks WA, DiPalma CR, Farrell CL (1999) Impaired transport of leptin across the blood-brain barrier in obesity. Peptides 20:1341–1345
Barrachina MD, Martinez V, Wang LX, Wei JY, Tachè Y (1997) Synergistic interaction between leptin and cholecystokinin to reduce short-term food intake in lean mice. Proc Natl Acad Sci USA 94:10455–10460
Barth SW, Riediger T, Lutz TA, Rechkemmer G (2004) Peripheral amylin activates circumventricular organs expressing calcitonin receptor a/b subtypes and receptor-activity modifying proteins in the rat. Brain Res 997:97–102
Becskei C, Riediger T, Zund D, Wookey P, Lutz TA (2004) Immunohistochemical mapping of calcitonin receptors in the adult rat brain. Brain Res 1030:221–233
Becskei C, Grabler V, Edwards GL, Riediger T, Lutz TA (2007) Lesion of the lateral parabrachial nucleus attenuates the anorectic effect of peripheral amylin and CCK. Brain Res 1162:76–84
Bhavsar S, Watkins J, Young A (1998) Synergy between amylin and cholecystokinin for inhibition of food intake in mice. Physiol Behav 64:557–561
Boyle CN, Munz M, Wielinga PY, Stöcker D, Lutz TA (2010) Short-term, but not extended, access to palatable diet diminishes amylin responsiveness in rat. Appetite 54:636
Boyle CN, Rossier MM, Lutz TA (2011) Influence of high-fat feeding, diet-induced obesity, and hyperamylinemia on the sensitivity to acute amylin. Physiol Behav 104(1):20–28
Bueter M, Löwenstein C, Olbers T, Wang M, Cluny NL, Bloom SR, Sharkey KA, Lutz T, Le Roux CW (2010) Gastric bypass increases energy expenditure in rats. Gastroenterology 138:1845–1853
Chapman I, Parker B, Doran S, Feinle-Bisset C, Wishart J, Strobel S, Wang Y, Burns C, Lush C, Weyer C, Horowitz M (2005) Effect of pramlintide on satiety and food intake in obese subjects. Diabetologia 48:838–848
Chavez M, Kaiyala K, Madden LJ, Schwartz MW, Woods SC (1995) Intraventricular insulin and the level of maintained body weight in rats. Behav Neurosci 109:528–531
Christopoulos G, Perry KJ, Morfis M, Tilakaratne N, Gao Y, Fraser NJ, Main MJ, Foord SM, Sexton PM (1999) Multiple amylin receptors arise from receptor activity-modifying protein interaction with the calcitonin receptor gene product. Mol Pharmacol 56:235–242
Cooper GJ (1994) Amylin compared with calcitonin gene-related peptide: structure, biology, and relevance to metabolic disease. Endocr Rev 15:163–201
Covasa M, Marcuson JK, Ritter RC (2001) Diminished satiation in rats exposed to elevated levels of endogenous or exogenous cholecystokinin. Am J Physiol Regul Integr Comp Physiol 280: R331–R337
Crawley JN, Beinfeld MC (1983) Rapid development of tolerance to the behavioural actions of cholecystokinin. Nature 302:703–706
Devine E, Young AA (1998) Weight gain in male and female mice with amylin knockout. Diabetes 47:A317
Dobolyi A (2009) Central amylin expression and its induction in rat dams. J Neurochem 111:1490–1500
Eiden S, Daniel C, Steinbrück A, Schmidt I, Simon E (2002) Salmon calcitonin—a potent inhibitor of food intake in states of impaired leptin signaling in laboratory rodents. J Physiol 541:1041–1048
Enoki S, Mitsukawa T, Takemura J, Nakazato M, Aburaya J, Toshimori H, Matsukara S (1992) Plasma islet amyloid polypeptide levels in obesity, impaired glucose tolerance and non-insulin-dependent diabetes mellitus. Diabetes Res Clin Pract 15:97–102
Fischer JA, Muff R, Born W (2002) Functional relevance of G-protein-coupled-receptor-associated proteins, exemplified by receptor-activity-modifying proteins (RAMPs). Biochem Soc Trans 30:455–460

Geary N (2004) Endocrine controls of eating: CCK, leptin, and ghrelin. Physiol Behav 81:719–733

Gedulin BR, Young AA (1998) Hypoglycemia overrides amylin-mediated regulation of gastric emptying in rats. Diabetes 47:93–97

Gloy VL, Lutz TA, Langhans W, Geary N, Hillebrand JJ (2010) Basal plasma levels of insulin, leptin, ghrelin and amylin do not signal adiposity in rats recovering from forced overweight. Endocrinology 151:4280–4288

Grabler V, Lutz TA (2004) Chronic infusion of the amylin antagonist AC 187 increases feeding in Zucker fa/fa rats but not in lean controls. Physiol Behav 81:481–488

Hanabusa T, Kubo K, Oki C, Nakano Y, Okai K, Sanke T, Nanjo K (1992) Islet amyloid polypeptide (IAPP) secretion from islet cells and its plasma concentration in patients with non-insulin-dependent diabetes mellitus. Diabetes Res Clin Pract 15:89–96

Hillebrand JJ, Geary N (2010) Do leptin and insulin signal adiposity? Forum Nutr 63:111–122

Hollander P, Maggs DG, Ruggles JA, Fineman M, Shen L, Kolterman OG, Weyer C (2004) Effect of pramlintide on weight in overweight and obese insulin-treated type 2 diabetes patients. Obesity Res 12:661–668

Isaksson B, Wang F, Permert J, Olsson M, Fruin B, Herrington MK, Enochsson L, Erlanson-Albertsson C, Arnelo U (2005) Chronically administered islet amyloid polypeptide in rats serves as an adiposity inhibitor and regulates energy homeostasis. Pancreatology 5:29–36

Knight ZA, Hannan KS, Greenberg ML, Friedman JM (2010) Hyperleptinemia is required for the development of leptin resistance. PLoS One 5:e11376

Leckström A, Lundquist I, Ma Z, Westermark P (1999) Islet amyloid polypeptide and insulin relationship in a longitudinal study of the genetically obese (ob/ob) mouse. Pancreas 18:266–273

Lutz TA (2006) Amylinergic control of feeding. Physiol Behav 89:465–471

Lutz TA (2010) The role of amylin in the control of energy homeostasis. Am J Physiol 298: R1475–R1484

Lutz TA, Rand JS (1993) A review of new developments in type 2 diabetes in human beings and cats. Br Vet J 149:527–536

Lutz TA, Geary N, Szabady MM, Del Prete E, Scharrer E (1995) Amylin decreases meal size in rats. Physiol Behav 58:1197–1202

Lutz TA, Senn M, Althaus J, Del Prete E, Ehrensperger F, Scharrer E (1998) Lesion of the area postrema/nucleus of the solitary tract (AP/NTS) attenuates the anorectic effects of amylin and calcitonin gene-related peptide (CGRP) in rats. Peptides 19:309–317

Lutz TA, Tschudy S, Rushing PA, Scharrer E (2000) Amylin receptors mediate the anorectic action of salmon calcitonin (sCT). Peptides 21:233–238

Lutz TA, Mollet A, Rushing PA, Riediger T, Scharrer E (2001a) The anorectic effect of a chronic peripheral infusion of amylin is abolished in area postrema/nucleus of the solitary tract (AP/NTS) lesioned rats. Int J Obes Relat Metab Disord 25:1005–1011

Lutz TA, Riediger T, Rushing PA, Scharrer E (2001b) The anorectic effect of amylin and its interaction with other anorectic signals. Proceedings, Neuropeptide interactions in the control of ingestive behavior, International Conference on the Physiology of Food and Fluid Intake

Lutz TA, Bueter M, Hillebrand JJ, Liesegang A, LeRoux CW (2010) Roux-en-Y gastric bypass reduces bone mineral density independent of body weight in rats. Appetite 54:660

Mack C, Wilson J, Athanacio J, Reynolds J, Laugero K, Guss S, Vu C, Roth J, Parkes D (2007) Pharmacological actions of the peptide hormone amylin in the long-term regulation of food intake, food preference, and body weight. Am J Physiol 293:R1855–R1863

Martin L, Siliart B, Lutz T, Biourge V, Nguyen P, Dumon H (2010) Postprandial response of plasma insulin, amylin and acylated ghrelin to various test meals in lean and obese cats. Br J Nutr 26:1–10

McLatchie LM, Fraser NJ, Main MJ, Wise A, Brown J, Thompson N, Solari R, Lee MG, Foord SM (1998) RAMPs regulate the transport and ligand specificity of the calcitonin-receptor-like receptor. Nature 393:333–339

Michel S, Becskei C, Erguven E, Lutz TA, Riediger T (2007) Diet-derived nutrients modulate the effects of amylin on c-Fos expression in the area postrema and on food intake. Neuroendocrinol 86:124–135
Mollet A, Meier S, Grabler V, Gilg S, Scharrer E, Lutz TA (2003a) Endogenous amylin contributes to the anorectic effects of cholecystokinin and bombesin. Peptides 24:91–98
Mollet A, Meier S, Riediger T, Lutz TA (2003b) Histamine H1 receptors in the ventromedial hypothalamus mediate the anorectic action of the pancreatic hormone amylin. Peptides 24:155–158
Mollet A, Gilg S, Riediger T, Lutz TA (2004) Infusion of the amylin antagonist AC 187 into the area postrema increases food intake in rats. Physiol Behav 81:149–155
Morley JE, Flood JF, Horowitz M, Morley PM, Walter MJ (1994) Modulation of food intake by peripherally administered amylin. Am J Physiol 67:R178–R184
Muff R, Buhlmann N, Fischer JA, Born W (1999) An amylin receptor is revealed following co-transfection of a calcitonin receptor with receptor activity modifying proteins-1 or -3. Endocrinology 140:2924–2927
Münzberg H (2010) Leptin-signaling pathways and leptin resistance. Forum Nutr 63:123–132
Nishimoto S, Nishida E (2006) MAPK signalling: ERK5 versus ERK1/2. EMBO Rep 7:782–786
Osaka T, Tsukamoto A, Koyama Y, Inoue S (2008) Central and peripheral administration of amylin induces energy expenditure in anesthetized rats. Peptides 29:1028–1035
Osto M, Wielinga PY, Alder B, Walser N, Lutz TA (2007) Modulation of the satiating effect of amylin by central ghrelin, leptin and insulin. Physiol Behav 91:566–572
Pieber TR, Roitelman J, Lee Y, Luskey KL, Stein DT (1994) Direct plasma radioimmunoassay for rat amylin-(1-37): concentrations with acquired and genetic obesity. Am J Physiol 267: E156–E164
Potes CS, Lutz TA (2010) Brainstem mechanisms of amylin induced anorexia. Physiol Behav 100:511–518
Potes CS, Lutz TA, Riediger T (2010a) Identification of central projections from amylin-activated neurons to the lateral hypothalamus. Brain Res 1334:31–44
Potes CS, Riediger T, Lutz TA (2010b) Amylin induces ERK 1/2 phosphorylation in structures of the AP/NTS-LPB-Ce-BSTL axis. Appetite 54:670
Potes C, Turek V, Cole R, Vu C, Roland B, Roth J, Riediger T, Lutz TA (2010c) Noradrenergic neurons of the area postrema mediate amylin's hypophagic action. Am J Physiol 299: R623–R631
Qi D, Cai K, Wang O, Li Z, Chen J, Deng B, Qian L, Le Y (2009) Fatty acids induce amylin expression and secretion by pancreatic beta-cells. Am J Physiol Endocrinol Metab 298: E99–E107
Ravussin E, Smith SR, Mitchell JA, Shringarpure R, Shan K, Maier H, Koda JE, Weyer C (2009) Enhanced weight loss with pramlintide/metreleptin: an integrated neurohormonal approach to obesity pharmacotherapy. Obesity 17:1736–1743
Reidelberger RD, Haver AC, Arnelo U, Smith DD, Schaffert CS, Permert J (2004) Amylin receptor blockade stimulates food intake in rats. Am J Physiol 287:R568–R574
Reinehr T, de Sousa G, Niklowitz P, Roth CL (2007) Amylin and its relation to insulin and lipids in obese children before and after weight loss. Obesity (Silver Spring) 15:2006–2011
Riediger T, Schmid HA, Lutz T, Simon E (2001) Amylin potently activates AP neurons possibly via formation of the excitatory second messenger cGMP. Am J Physiol 281:R1833–R1843
Riediger T, Schmid HA, Lutz TA, Simon E (2002) Amylin and glucose co-activate area postrema neurons of the rat. Neurosci Lett 328:121–124
Riediger T, Zünd D, Becskei C, Lutz TA (2004) The anorectic hormone amylin contributes to feeding-related changes of neuronal activity in key structures of the gut-brain axis. Am J Physiol 286:R114–R122
Riediger T, Michel S, Forster K, Lutz TA (2009) The ability of amylin to reduce eating depends on the protein content of the diet. Appetite 52:854
Riedy CA, Chavez M, Figlewicz DP, Woods SC (1995) Central insulin enhances sensitivity to cholecystokinin. Physiol Behav 58:755–760

Roth JD, Hughes H, Kendall E, Baron AD, Anderson CM (2006) Antiobesity effects of the beta-cell hormone amylin in diet-induced obese rats: effects on food intake, body weight, composition, energy expenditure, and gene expression. Endocrinology 147:5855–5864

Roth JD, Roland BL, Cole RL, Trevaskis JL, Weyer C, Koda JE, Anderson CM, Parkes DG, Baron AD (2008) Leptin responsiveness restored by amylin agonism in diet-induced obesity: evidence from nonclinical and clinical studies. Proc Natl Acad Sci USA 105:7257–7262

Rowland NE, Richmond RM (1999) Area postrema and the anorectic actions of dexfenfluramine and amylin. Brain Res 820:86–91

Rushing PA, Hagan MM, Seeley RJ, Lutz TA, Woods SC (2000) Amylin: a novel action in the brain to reduce body weight. Endocrinology 141:850–853

Rushing PA, Hagan MM, Seeley RJ, Lutz TA, D'Alessio DA, Air EL, Woods SC (2001) Inhibition of central amylin signaling increases food intake and body adiposity in rats. Endocrinology 142:5035–5038

Schwartz MW, Woods SC, Seeley RJ, Barsh GS, Baskin DG, Leibel RL (2003) Is the energy homeostasis system inherently biased toward weight gain? Diabetes 52:232–238

Sexton PM, Paxinos G, Kenney MA, Wookey PJ, Beaumont K (1994) In vitro autoradiographic localization of amylin binding sites in rat brain. Neurosci 62:553–567

Shin AC, Zheng H, Townsend RL, Sigalet DL, Berthoud H-R (2010) Meal-induced hormone responses in a rat model of Roux-en-Y gastric bypass surgery. Endocrinology 151:1588–1597

Smith GP, Gibbs J (1985) The satiety effect of cholecystokinin. Recent progress and current problems. Ann N Y Acad Sci 448:417–423

Surina-Baumgartner DM, Langhans W, Geary N (1995) Hepatic portal insulin antibody infusion increases, but insulin does not alter, spontaneous meal size in rats. Am J Physiol 269: R978–R982

Sutton GM, Patterson LM, Berthoud HR (2004) Extracellular signal-regulated kinase 1/2 signaling pathway in solitary nucleus mediates cholecystokinin-induced suppression of food intake in rats. J Neurosci 24:10240–10247

Trevaskis JL, Coffey T, Cole R, Lei C, Wittmer C, Walsh B, Weyer C, Koda J, Baron AD, Parkes DG, Roth JD (2008) Amylin-mediated restoration of leptin responsiveness in diet-induced obesity: magnitude and mechanisms. Endocrinology 149:5679–5687

Trevaskis JL, Lei C, Koda JE, Weyer C, Parkes DG, Roth JD (2010) Interaction of leptin and amylin in the long-term maintenance of weight loss in diet-induced obese rats. Obesity 18:21–26

Turek VF, Trevaskis JL, Levin BE, Dunn-Meynell AA, Irani B, Gu G, Wittmer C, Griffin PS, Vu C, Parkes DG, Roth J (2010) Mechanisms of amylin/leptin synergy in rodent models. Endocrinology 151:143–152

Ueda T, Ugawa S, Saishin Y, Shimada S (2001) Expression of receptor-activity modifying protein (RAMP) mRNAs in the mouse brain. Brain Res Mol Brain Res 93:36–45

West DB, Fey D, Woods SC (1984) Cholecystokinin persistently suppresses meal size but not food intake in free-feeding rats. Am J Physiol 246:R776–R787

Weyer C, Maggs DG, Young AA, Kolterman OG (2001a) Amylin replacement with pramlintide as an adjunct to insulin therapy in type 1 and type 2 diabetes mellitus: a physiological approach toward improved metabolic control. Curr Pharm Des 7:1353–1373

Weyer C, Vozarova B, Ravussin E, Tataranni PA (2001b) Changes in energy metabolism in response to 48 h of overfeeding and fasting in Caucasians and Pima Indians. Int J Obes 25:593–600

Wielinga PY, Alder B, Lutz TA (2007) The acute effect of amylin and salmon calcitonin on energy expenditure. Physiol Behav 91:212–217

Wielinga PY, Muff S, Alder B, Woods SC, Lutz TA (2008) Amylin levels in the brain influence the level of body weight maintenance implying that amylin acts as adiposity signal. Int J Obes 32:S51

Wielinga PY, Löwenstein C, Muff S, Munz M, Woods SC, Lutz TA (2010) Central amylin acts as an adiposity signal to control body weight and energy expenditure. Physiol Behav 101:45–52

Williams DL, Cummings DE (2005) Regulation of ghrelin in physiologic and pathophysiologic states. J Nutr 135:1320–1325

Woods SC (2004) Gastrointestinal satiety signals. I. An overview of gastrointestinal signals that influence food intake. Am J Physiol 286:G7–G13

Woods SC (2005) Signals that influence food intake and body weight. Physiol Behav 86:709–716

Woods SC, Lutz TA, Geary N, Langhans W (2006) Pancreatic signals controlling food intake; insulin, glucagon and amylin. Philos Trans R Soc Lond B Biol Sci 361:1219–1235

Young A (2005a) Amylin: physiology and pharmacology. Adv Pharmacol 52:1–66

Young A (2005b) Amylin and the integrated control of nutrient flux. Adv Pharmacol 52:67–77

Young A, Denaro M (1998) Roles of amylin in diabetes and in regulation of nutrient load. Nutrition 14:524–527

Yuan LL, Adams JP, Swank M, Sweatt JD, Johnston D (2002) Protein kinase modulation of dendritic K^+ channels in hippocampus involves a mitogen-activated protein kinase pathway. J Neurosci 22:4860–4868

Intestinal Microbiota and Obesity

Michael Blaut and Susanne Klaus

Contents

Abstract The human gut harbors a highly diverse microbial ecosystem of approximately 400 different species, which is characterized by a high interindividual variability. The intestinal microbiota has recently been suggested to contribute to the development of obesity and the metabolic syndrome. Transplantation of gut microbiota from obese mice to nonobese, germ-free mice resulted in transfer of

M. Blaut (✉)
Department of Gastrointestinal Microbiology, German Institute of Human Nutrition Potsdam-Rehbruecke, Arthur-Scheunert-Allee 114-116, 14558 Nuthetal, Germany
e-mail: blaut@dife.de

S. Klaus
Research Group Physiology of Energy Metabolism, German Institute of Human Nutrition Potsdam-Rehbruecke, Arthur-Scheunert-Allee 114-116, 14558 Nuthetal, Germany
e-mail: klaus@dife.de

H.-G. Joost (ed.), *Appetite Control*, Handbook of Experimental Pharmacology 209,
DOI 10.1007/978-3-642-24716-3_11,

metabolic syndrome–associated features from the donor to the recipient. Proposed mechanisms for the role of gut microbiota include the provision of additional energy by the conversion of dietary fiber to short-chain fatty acids, effects on gut-hormone production, and increased intestinal permeability causing elevated systemic levels of lipopolysaccharides (LPS). This metabolic endotoxemia is suggested to contribute to low-grade inflammation, a characteristic trait of obesity and the metabolic syndrome. Finally, activation of the endocannabinoid system by LPS and/or high-fat diets is discussed as another causal factor. In conclusion, there is ample evidence for a role of gut microbiota in the development of obesity in rodents. However, the magnitude of its contribution to human obesity is still unknown.

Keywords Bifidobacteria • Diet • Endotoxemia • Energy harvest • Intestinal microbiota • Low-grade inflammation • Metabolic syndrome • Obesity

1 Introduction

The worldwide epidemic in obesity and its associated metabolic disorders, such as type-2 diabetes, fatty liver, and cardiovascular disease, has triggered an intense search for genetic and environmental factors that contribute to this development. The high prevalence of the metabolic syndrome in affluent societies has been attributed to a predominantly sedentary lifestyle and unhealthy eating behavior (Kushner & Choi 2010). Excessive consumption of high-energy diets, lack of physical activity, and genetic susceptibility are the main factors for obesity. However, the sequence of events leading to the loss of energy homeostasis in obese subjects is largely unknown. A number of functional parameters involved in appetite control and metabolic regulation are altered in obese subjects, but it is unclear whether these changes are the cause or rather the consequence of the ensuing metabolic disorder. In 2004 and the following years, studies of the group led by Jeffrey Gordon in St. Louis proposed the intestinal microbiota as one intestinal factor that may contribute to the development of obesity (Backhed et al. 2004; Ley et al. 2005, 2006a; Turnbaugh et al. 2006). This has triggered intense research in this area. Even though the exact role of gut bacteria in obesity development is still obscure, it has been proposed that changes in gut microbiota composition in response to high-fat diets affect lipogenesis (Backhed et al. 2007), gut permeability for lipopolysaccharides (LPS), and inflammatory status (Cani et al. 2007a) as well as the endocannabinoid system tone (Muccioli et al. 2010) (Fig. 1).

2 The Intestinal Microbiota

The microbial community resident in the human digestive tract represents a highly diverse microbial ecosystem as indicated by the presence of ~400 species-level phylotypes based on 16S rRNA gene analysis (Ley et al. 2006b; Eckburg et al.

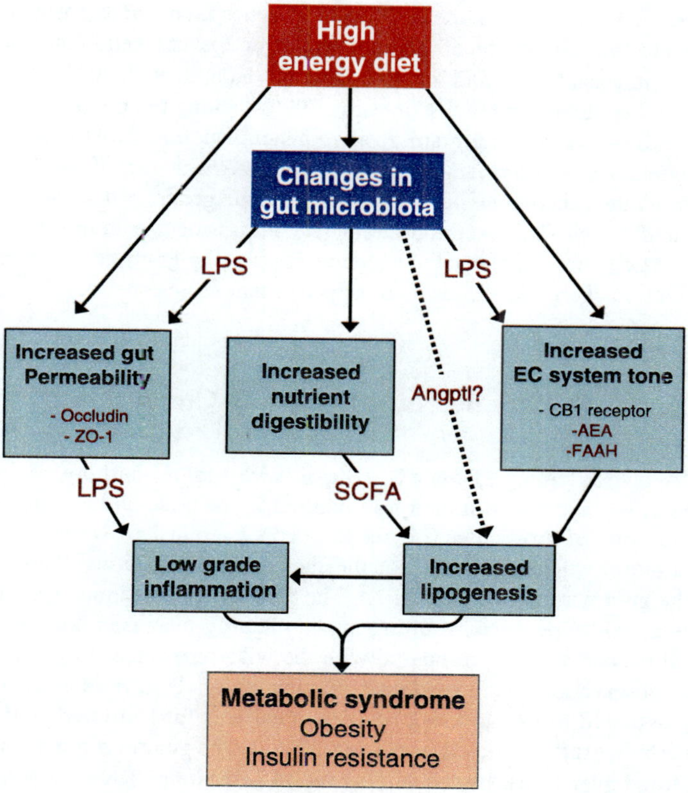

Fig. 1 Hypothetical scheme depicting how intestinal microbiota may influence the development of diet-induced obesity and metabolic syndrome. For detailed description, see text. *AEA* anandamide, *Angptl4* angiopoietin-like protein 4, *CB1* endocannabinoid receptor 1, *EC system* Endocannabinoid system, *FAAH* fatty acid amid hydrolase, *LPS* Lipopolysaccharide, *SCFA* short-chain fatty acids, *ZO-1* tight-junction protein zonola occludens-1

2005). The microbial cell concentration in the gastrointestinal tract increases considerably from the stomach ($<10^3$/ml) to the colon (10^{12}/g) (Finegold et al. 1983). The high interindividual variability in gut microbiota composition makes it difficult to define a core microbiome (all members of a microbial ecosystem) that is shared by everyone (Qin et al. 2010). In spite of these differences in species composition, many functions exerted by the gut microbiota are apparently shared among individuals. Known functions of the gut microbiota include the conversion of nondigestible carbohydrates (dietary fiber) to short-chain fatty acids, transformation of bile acids, the provision of a barrier against pathogenic bacteria, and modulation of the innate and the adaptive immune systems.

More recent work indicates that the gut microbiota affects host physiology to a larger extent than previously assumed. Association of germ-free mice with *Bacteroides thetaiotaomicron*, a dominant species in the gut microbiota of mice and humans, led to changes in the expression of genes involved in postnatal maturation of the intestine, absorption of nutrients, angiogenesis, and xenobiotic metabolism as well as in mucosal barrier and immune function (Hooper et al. 2001). The majority of bacteria in the human and mouse intestine (99%) belong to only four major phyla: the Firmicutes (encompassing gram-positive genera, such as *Clostridium*, *Eubacterium*, *Ruminococcus*, *Butyrivibrio*, *Anaerostipes*, *Roseburia*, *Faecalibacterium*), the Bacteroidetes (encompassing the gram-negative genera *Bacteroides*, *Porphyromonas*, and *Prevotella*), Actinobacteria (encompassing the gram-positive genus *Bifidobacterium*), and Proteobacteria (encompassing the gram-negative Enterobacteriaceae with *Escherichia coli* as its most prominent representative).

3 Impact of Obesity on Gut Microbiota Composition

Genetically obese, leptin-deficient C57BL/6J mice (*ob*/*ob*) harbored a 50% lower proportion of Bacteroidetes in their gut than wild-type mice and a correspondingly higher proportion of Firmicutes (Ley et al. 2005). Both groups were fed a standard chow ad libitum, suggesting that not the diet but the nutritional status (obesity) affected the gut microbiota composition. The *ob*/*ob* mice consumed 4% more chow than the lean wild-type mice, resulting in significantly increased body weights and epididymal fat-pad weights compared with the wild-type mice (Ley et al. 2005).

Similar observations were made in a human study. Twelve obese people were randomly assigned to two low-calorie diets: One was fat-restricted, and the other one was carbohydrate-restricted (Ley et al. 2006a). The gut microbiota composition was monitored over 1 year by sequencing the 16S ribosomal RNA (rRNA) genes in fecal DNA. Before starting the weight-loss diets, the obese study participants harbored fewer Bacteroidetes in their gut microbiota than the lean control subjects (3% versus 25%). The Firmicutes were correspondingly higher in the obese than in the normal-weight individuals (89% versus 73%). The proportion of Bacteroidetes in the obese subjects increased within 1 year from 3 to 18% in response to the restriction of energy intake.

These findings are in disagreement with two other, more recent human studies (Duncan et al. 2008; Schwiertz et al. 2010). Duncan and coworkers compared the gut microbiota composition of 23 obese (body mass index, BMI > 30 kg/m^2) and 14 nonobese subjects (BMI < 30 kg/m^2) by 16S rRNA-targeted fluorescence in situ hybridization (FISH) and quantitative real-time PCR (qRT-PCR) (Duncan et al. 2008). The obese and nonobese subjects did not differ in the proportion of fecal Bacteroides, and there was also no correlation between BMI and the proportion of Bacteroides over a BMI range of 20–44 kg/m^2. The obese study participants underwent a dietary intervention for the purpose of body-weight reduction. They consumed two energy-restricted diets offered in a crossover design for 4 weeks each. One of the diets was high in protein, low in carbohydrate, and ketogenic, while the other diet was high in protein, moderate in carbohydrate, and nonketogenic. There was no significant change in the proportion of fecal Bacteroides in the obese subjects in response to any of the low-calorie diets (Duncan et al. 2008).

The study by Schwiertz et al. (2010) included 30 normal-weight (BMI = 18.5–24.9 kg/m^2), 35 overweight (BMI = 25–30 kg/m^2), and 33 obese subjects (BMI >30 kg/m^2). As determined by qRT-PCR, the proportion of Bacteroidetes was found to be 23% for the normal-weight subjects, 47% for the overweight subjects, and 45% for the obese study participants, respectively (Schwiertz et al. 2010). This result is in contradiction to the results by Ley et al. (2006a).

Several reasons could be responsible for the observed differences. First, the detection methods used for quantification of the intestinal bacterial groups differed. While Duncan et al. (2008) and Schwiertz et al. (2010) used FISH and qRT-PCR, respectively, Ley et al. (2006a) employed 16S rRNA gene sequencing. For both FISH and qRT-PCR, appropriate oligonucleotide probes and primers, respectively, have to be selected. This selection determines which bacteria are targeted. In contrast, 16S rRNA gene sequencing is expected to essentially detect any bacterial strain whose 16S rRNA gene contains the conserved site to which the primers can bind. It may thus be possible that changes within the bacterial groups of interest, the Bacteroidetes and the Firmicutes, remained undetected by FISH and qRT-PCR because the probes and primers, respectively, failed to target all members of these phyla. Hence, changes within these phyla might have occurred but were not detected because the selected probes or primers missed those bacterial groups undergoing changes in their relative proportion. Second, the participants in the three studies and also the study designs are not really comparable. It thus remains unclear whether obesity inevitably leads to changes in the human gut microbiota, in particular to a decrease in the Bacteroidetes.

4 Influence of High-Fat Diets on Gut Microbiota Composition

A more recent investigation in mice supports the notion that diet rather than obesity affects the intestinal microbiota composition (Hildebrandt et al. 2009). Both wild-type 129Svev/C57BL/6 mice and RELMβ knockout (KO) mice were fed a standard chow or a high-fat diet to study the influence of the genotype and the resulting phenotype on the composition of gut microbiota using 16S rRNA gene sequencing of fecal DNA. RELMβ is a colonic goblet cell–specific gene, whose expression depends on the presence of the gut microbiota. On the standard chow, both strains remained lean with a body weight of 20 g for both strains. In contrast, on the high-fat diet (45% energy), the wild-type mice became obese (45 g body weight, 11 g body fat), while the RELMβ KO mice remained relatively lean (32 g body weight, 6 g body fat). The 16S rRNA gene–sequence analysis revealed a high similarity in the composition of the bacterial communities from both wild-type and RELMβ KO mice when the standard chow was fed. However, the fecal microbiota changed in both mice strains after being fed the high-fat diet for 3 months. These changes were characterized by a decrease of the Bacteroidetes from 55% (standard chow) to 13% (high-fat diet) and increases of Firmicutes from 10 to 30%, Proteobacteria from 29 to 45%, and Actinobacteria from 3 to 8%. Since these diet-dependent changes occurred both in the wild-type strain, which is susceptible to diet-induced obesity, and in the relatively obesity-resistant RELMβ KO strain, the authors concluded that the high-fat diet rather than obesity determines the composition of the gut microbiota (Hildebrandt et al. 2009). This notion is supported by another recent study in which 7-week-old wild-type mice were switched from a low-fat to a high-fat diet for 8 weeks. The changes in their gut microbiota composition were compared with those of wild-type and genetically obese mice (leptin-deficient *ob/ob* mice) fed a low-fat diet throughout the same time period (Murphy et al. 2010). The microbial composition of fresh fecal samples obtained at 7, 11, and 15 weeks of age was determined by pyrosequencing of 16S rRNA. Microbiota composition at week 7 was similar in all three groups, the main phyla detected being Firmicutes (56–57%), Bacteroidetes (24–25%), and Actinobacteria (almost exclusively represented by *Bifidobacterium*, 16%). There were no significant changes in microbiota composition in wild-type mice fed the low-fat diet over time, whereas in mice fed the high-fat diet, the proportion of Firmicutes increased from 57 to 70%. The authors conclude that changes in gut microbiota composition are a feature of high-fat feeding rather than related to genetically induced obesity (Murphy et al. 2010).

It is noteworthy that the actual proportions of the dominant microbial phyla reported in the two reports above (Hildebrandt et al. 2009; Murphy et al. 2010) differ considerably, also from those reported by others (Ley et al. 2005; Turnbaugh et al. 2006, 2008), even though all studies used 16S rRNA–based molecular population analysis. Major differences between the studies were reported for all phyla. The reasons for these differences are unclear, but it may be surmised that different environmental and housing condition in the breeding facilities from which the animals were obtained may have affected their initial colonization (Hildebrandt et al. 2009). Discrepant findings may also result from the use of different diets. Furthermore, they could be related to technical aspects such as efficiency of DNA extraction and choice of primers for 16S rRNA gene amplification. (Murphy et al. 2010) speculated that the high levels of Actinobacteria and the associated genus *Bifidobacterium* detected in their study could be related to the particular DNA-extraction protocol employed and to the use of primers predicted to bind to 95% of all 16S rRNA genes (Murphy et al. 2010).

5 Does the Gut Microbiota Contribute to Obesity?

Since the availability of dietary substrates is a major determinant for establishment and persistence of microorganisms in the intestinal tract, it is not surprising that nutrition influences the activity and composition of gut microbiota. Diets devoid of dietary fiber do not provide fermentable carbohydrates to the gut microbiota. In this situation, growth of the gut microbes depends on the utilization of endogenous substrates such as mucins, digestive proteins, and desquamated epithelial cells. In accordance with this notion, the number of fecal anaerobes per gram of dry matter is

1.3 $\log_{10}$ higher in rats fed a high-fiber diet than in those fed a low-fiber diet (Maczulak et al. 1993). Along the same line, the type of fermentable fiber influences the composition of gut microbiota because bacteria capable of taking advantage of a given dietary fiber will grow faster than those unable to do so. For example, oligofructose has repeatedly been demonstrated to stimulate the growth of intestinal bifidobacteria (Gibson et al. 1995; Kleessen et al. 2001). Therefore, changes in gut microbiota composition after shifting from a standard chow to a semisynthetic high-fat diet are mainly due to a lower proportion of fermentable dietary fiber in the latter diet (Fleissner et al. 2010). It has to be emphasized that fat itself cannot be metabolized under anaerobic conditions and therefore cannot serve as energy source for strictly anaerobic bacteria.

Are the reported dietary effects on the gut microbiota of any relevance for the host or merely a side effect with little or no impact? Turnbaugh et al. (2006) demonstrated that the transfer of microbiota from obese *ob*/*ob* mice to germ-free wild-type mice within 2 weeks resulted in a significantly greater relative increase in body fat (47%) than microbiota transfer from normal-weight wild-type mice to germ-free wild-type mice (26%). The groups did not significantly differ in chow consumption, initial body fat, and initial body weight of the recipients. The donor microbiota of the *ob*/*ob* mice and the microbiota of the recipient wild-type mice inoculated with the "*ob*/*ob* microbiota" contained higher proportions of Firmicutes (69 and 63%, respectively) than the microbiota of the lean donor mice (55%) or the recipients of their microbiota (49%). The relative abundance of Bacteroidetes in the "*ob*/*ob* microbiota" of both donor and recipient mice was accordingly lower (28 and 32%, respectively) than in the "lean microbiota" of donor and recipient mice (44 and 48%, respectively). These findings were interpreted to indicate that the microbiota from the obese mice has a higher capacity to harvest energy from the diet. This notion is supported by a metagenomic analysis which revealed that genes encoding enzymes involved in the degradation of indigestible polysaccharides were enriched in the microbiome of the obese *ob*/*ob* mice as compared to the lean wild-type mice. Genes encoding transport proteins for uptake of the cleavage products and enzymes involved in the formation of acetate and butyrate were also enriched in the microbiome of the *ob*/*ob* mice. In agreement with these findings, the authors detected approximately 20% higher acetate and almost twofold higher butyrate concentrations in the cecal contents of *ob*/*ob* mice than of wild-type mice (Turnbaugh et al. 2006). Both acetate and butyrate can be absorbed by the host and oxidized, thereby providing additional energy.

In order to further support the notion that the microbiota from obese mice is more efficient in energy extraction from indigestible polysaccharides, and in order to exclude that this finding was restricted to genetically obese mice, the microbiota-transplantation experiments were extended to mice with a diet-induced obesity (Turnbaugh et al. 2008). To exclude differences in microbiota composition, previously germ-free mice on a low-fat and polysaccharide-rich diet (LFD) were colonized with the microbiota from a conventional mouse. Subsequently, five of the ten mice were kept on this diet, while the remaining mice were switched to a high-fat high-sugar diet (HFD: energy from fat, 41%; sucrose, 18%; (w/w) maltodextrin, 12%; and corn starch, 16%). Eight weeks later, the mice on the HFD had gained 5.3 g body weight, whereas the mice on the LFD had only gained 1.5 g body weight. The cecal microbiota of the mice fed the HFD had significantly higher proportions of Firmicutes and lower proportions of Bacteroidetes than the mice fed the LFD. Interestingly, the increase of the Firmicutes in the HFD-fed mice was restricted to one bacterial family within this phylum, namely, the Erysipelotrichaceae, which are referred to as Mollicutes by Turnbaugh et al. (Turnbaugh et al. 2008). The Mollicutes sequences represented approximately 65% of all 16S rRNA sequences analyzed. Such a shift in the Erysipelotrichaceae in response to an adipogenic diet has also been observed in a more recent mouse study (Fleissner et al. 2010) but not in *ob/ob* mice, in which the high relative abundance of Firmicutes extended across the whole phylum (Turnbaugh et al. 2006). Transplantation of the microbiota from HFD-fed conventional mice to LFD-fed germ-free recipients led to a greater relative increase in body fat (43%) as compared to the transplantation of microbiota from the LFD-fed mice (25%). There was no difference in food consumption or initial weight between the recipients of the HFD microbiota and the LFD microbiota. A metagenomic comparison of the HFD and the LFD cecal microbiomes revealed an enrichment of genes involved in uptake and utilization of dietary carbohydrates and host glycans (mucins and other glycoproteins). In particular, genes encoding components of the phosphotransferase system, the main sugar uptake system for a large proportion of intestinal bacteria, were enriched in the HFD microbiome, while genes involved in motility were depleted. The simultaneous enrichment of Erysipelotrichaceae and the above genes in the HFD microbiome may indicate that these genes and their encoded functions are contributed by these bacteria enabling a more efficient extraction of energy from nondigestible dietary components (Turnbaugh et al. 2008).

A more recent metagenomic study in mice associated with a human intestinal microbiota supports these findings: It demonstrated that switching these humanized mice from an LFD to an HFD not only changed the composition of their gut microbiota within 1 day but also made them obese (Turnbaugh et al. 2009). Metagenomic analysis of the cecal microbiome revealed changes in the representation of genes encoding metabolic enzymes in response to dietary shifts. These changes were accompanied by an altered gene expression. Microbiota transfers between various combinations of donor and recipient mice fed with different diets showed that colonization history affects the initial microbiota composition but that diet overrides these effects within a certain time period (Turnbaugh et al. 2009).

6 Are Germ-Free Mice Protected Against Obesity?

The above described studies suggested that microbial communities in the gut of mice or humans may differ in their capacity to extract energy from indigestible dietary components by fermentation. In line with this observation, germ-free mice were reported to be protected from diet-induced obesity (Backhed et al. 2004;

Backhed et al. 2007). Germ-free and conventional C57BL/6 mice were fed a high-fat diet ad libitum. After 2 weeks on this diet, the conventional mice had higher total body-fat contents (13% versus 8%) and also higher epididymal fat-pad weights than the germ-free mice (0.15 g versus 0.10 g), despite a higher chow consumption by the germ-free mice than by the conventional mice (4.3 g/day versus 3.3 g/day). Interestingly, there was no difference in total body weight between the two mouse groups. Germ-free mice exhibited a 27% lower metabolic rate in spite of the lower total and epididymal body fat as compared with conventional mice. The conventional mice also displayed higher serum levels of leptin (2.2 ng/ml versus 0.7 ng/ml), insulin (1.2 ng/ml versus 0.3 ng/ml), and glucose (10 mM versus 7.5 mM) (Backhed et al. 2004). Higher levels of serum leptin in the conventional animals are in accordance with the observed lower chow consumption by these mice because leptin reduces appetite (Friedman and Halaas 1998). The conventional animals also showed signs of impaired glucose homeostasis: Glucose tolerance and insulin sensitivity were reduced, liver triglyceride content increased 2.3-fold, and expression of genes (mRNA) encoding acetyl-CoA carboxylase (Acc1) and fatty acid synthase (Fas), key enzymes of fatty acid biosynthesis, was elevated 2.4- to 2.6-fold in the liver compared to germ-free mice. The mRNA level of the carbohydrate responsive element binding protein (ChREBP), which regulates the expression of Acc1 and Fas, was twofold higher in the liver of conventional mice. These observations indicated that the conventional mice had an increased hepatic lipogenesis (Backhed et al. 2004). The authors concluded that the gut microbiota stimulated triglyceride formation in the liver of the mice.

In another recent investigation, germ-free C3H mice fed a high-fat diet had the same or an even higher increase in body fat than conventional C3H mice fed the same diet (Fleissner et al. 2010). This high-fat diet (HFD) was almost identical in energy content and carbohydrate–protein–fat ratio (41:16:43, 21.4 kJ/g) to the diet used by Backhed et al. (2004), referred to as Western diet (WD 41:19:41, 21.5 kJ/g), but differed in the type of carbohydrates and fats. The WD contains considerably more sucrose than the HFD (183 g/kg versus 50 g/kg). In addition, the WD contains 10% vegetable shortening and 10% beef tallow, while the HFD contains 18% coconut oil as the main fat component. Interestingly, feeding the WD to these mice reproduced the results reported by Backhed et al. (2004). This led to the conclusion that germ-free mice are not generally protected against obesity but that this effect is diet-dependent (Fleissner et al. 2010). The exact reasons for this diet-dependent effect have remained elusive.

7 Does the Gut Microbiota Modulate Energy Harvest from Diet?

As already pointed out, the intestinal microbiota converts nondigestible dietary carbohydrates (dietary fiber) into short-chain fatty acids (SCFA), which in turn can be oxidized by the host and thus provide additional energy. It has been estimated that up to 10% of human energy needs may be covered by fermentation of fiber in the colon (McNeil 1984). In mice, the addition of fermentable fiber to a high-fat diet leads to significantly increased fermentation rates resulting in increased dietary energy extraction and increased obesity compared to mice receiving the same diet supplemented with nonfermentable fiber (Isken et al. 2010). It is conceivable that gut microbiota composition affects the efficiency of energy extraction from the diet, which in turn might lead to the development of obesity. Indeed, it has been shown that gnotobiotic mice colonized with *B. thetaiotaomicron* and *Methanobrevibacter smithii* display increased SCFA concentrations in the cecum compared to mice colonized with either one of the two species, and that they also show increased white fat and liver triglycerides compared to germ-free mice (Samuel & Gordon 2006). Aside from this direct effect of bacterial SCFA production on energy extraction, indirect effects of SCFA on intestinal motility and intestinal hormone production could also play a role. The formation of SCFA has been associated with increased expression and production of glucagon-like peptide-1 (GLP-1) and peptide YY (PYY) in the intestine (Zhou et al. 2008). Both are gut-secreted peptide hormones implicated in intestinal function and appetite regulation (Wren and Bloom 2007). The effect of SCFA on gut hormone production could be mediated by G protein-coupled receptor (Gpr)41, which has been demonstrated to play a role in improved energy harvest (Samuel et al. 2008). Gpr41 and Gpr43 are expressed in adipocytes, in the distal small intestine, and in the large intestine. They are activated by their ligands acetate, propionate, and butyrate. In the intestine, they are expressed in enteroendocrine cells. Activation of Gpr41 leads to secretion of leptin from adipocytes (Xiong et al. 2004) and formation of PYY in enteroendocrine cells (Tazoe et al. 2008). Both peptide hormones exert an anorexigenic effect. In addition, PYY also regulates gut motility. Mice deficient in Gpr41 (*Gpr41*$^{-/-}$) did not differ in weight gain and fat-pad weight as long as they were germ-free (Samuel et al. 2008). However, *Gpr41*$^{-/-}$ mice colonized with a conventional microbiota had a 30% lower fat-pad weight, 30% less weight gain, and 25% less body fat than the corresponding wild-type mice (*Gpr41*$^{+/+}$). The serum PYY levels of germ-free *Gpr41*$^{-/-}$ mice and *Gpr41*$^{+/+}$ mice did not differ from each other, but they were two- to fourfold lower than those of mice that were colonized with *B. thetaiotaomicron* and *Methanobrevibacter smithii*, indicating that intestinal bacteria induce the formation of PYY. In the disassociated state, *Gpr41*$^{-/-}$ mice had 40% lower serum PYY levels than *Gpr41*$^{+/+}$ mice. There were no differences in chow consumption, but the intestinal transit in the *Gpr41*$^{-/-}$ mice was faster than in the *Gpr41*$^{+/+}$ mice, suggesting that the main target of PYY produced upon activation of Gpr41 by SCFA was gut transit time. Indeed, a shorter transit time in the *Gpr41*$^{-/-}$ mice compared to the *Gpr41*$^{+/+}$ mice coincided with 30% higher levels of unabsorbed fecal SCFA and 25% higher fecal energy content, despite the same energy intake. In agreement with these findings, disassociated *Gpr41*$^{+/+}$ mice had a greater hepatic lipogenesis than the corresponding *Gpr41*$^{-/-}$ mice. Hence, the SCFA produced by the gut microbiota activate Gpr41, and the resulting increase in PYY slows intestinal transit, which in turn leads to a more complete absorption of intestinal nutrients including SCFA.

It should be pointed out that the above described observations were made in mice fed polysaccharide-rich chow diets, which presumably contain a high but undefined amount of fiber. Digestibility of semisynthetic high-fat diets with low-fiber content, as assessed by analysis of diet and feces energy content, was found to be the same in germ-free and conventional mice, and not related to differential weight gain of mice on different high-fat diets (Fleissner et al. 2010). Murphy et al. (2010) also found no association between changes in gut microbiota composition and markers of energy harvest such as fecal SCFA and energy content in mice fed low- or high-fat diets. They concluded that gut microbiota contributes little to energy extraction (Murphy et al. 2010).

Taken together, the data imply that the gut microbiota may affect energy harvest by producing SCFA from dietary fiber. However, this contribution to the energy demand of the host is highly dependent on the amount and type of dietary fiber consumed. So far, it is unknown whether this mechanism is relevant to humans in Western societies, who on average have a rather low dietary fiber intake.

8 The Role of Fasting-Induced Adipose Factor/ Angiopoietin-Like Protein 4 (Fiaf/Angptl4)

The inhibition of fasting-induced adipose factor (Fiaf), also known as angiopoietin-like protein 4 (Angptl4), by gut microbiota has been suggested to play a causal role in microbiota-mediated increase of fat storage (Backhed et al. 2007). Fiaf/Angptl4 is a circulating inhibitor of lipoprotein lipase (LPL)-mediated lipolysis of plasma triglyceride-rich lipoproteins and thus increases plasma triglycerides levels and decreases uptake of fatty acids into tissues (Lichtenstein and Kersten 2010). LPL is considered a main gatekeeper for fatty acid entry into adipose tissue (Voshol et al. 2009). Consequently, Fiaf/Angptl4-overexpressing mice have 50% decreased white fat stores (Mandard et al. 2006), whereas Fiaf/Angptl4-null mice showed slightly increased white fat weight (in males only, (Kim et al. 2010)). Fiaf/Angptl4 is a secreted protein member of the angiopoietin family, which in 2000 was independently identified by three different groups as a novel target of peroxisome proliferator–activated receptor (PPAR) (Yoon et al. 2000; Kersten et al. 2000; Kim et al. 2000; Lichtenstein and Kersten 2010). Fiaf/Angptl4 mRNA expression is highest in white and brown adipose tissue and in liver but also detectable at lower levels in other tissues such as kidney, lung, ovaries, hypothalamus, and small intestine (Yoon et al. 2000; Kersten et al. 2000; Kim et al. 2000). Fiaf/Angptl4 is abundant in plasma, where it can easily be detected by Western blotting. It has a structure similar to other angiopoietin-like proteins consisting of an N-terminal coiled-coil domain and a C-terminal fibrinogen-like domain. Fiaf/Angptl4 forms oligomers inside the cell and is cleaved upon secretion into an N-terminal portion (nAngptl4) and a C-terminal portion (cAngptl4) (Lichtenstein and Kersten 2010). A short amino acid consensus motive close to the N-terminus of nAngptl4 was found to be responsible for the interaction with and inhibition of LPL by blocking its dimerization (Yau et al. 2009).

As early as in 2001 did the group of Jeffrey Gordon identify Fiaf/Angptl4 as a factor whose gene expression in small intestine was largely repressed in conventionalized compared to germ-free mice (Hooper et al. 2001). The same group showed later that conventionalization of germ-free mice with cecal microbiota from normal mice resulted in an almost 60% increase in body fat within 2 weeks (Backhed et al. 2004). Furthermore, they reported that germ-free mice were protected against diet-induced obesity (Backhed et al. 2007). The increased intestinal expression of Fiaf/Angptl4 in germ-free mice as compared to conventional mice was suggested to be—at least partially—responsible for the resistance to diet-induced obesity in germ-free mice based on the following evidence: (1) LPL activity in white fat and heart was higher in conventionalized mice than in germ-free mice, (2) germ-free Fiaf/Angptl4-null (*fiaf*$^{-/-}$) mice had the same amount of body fat as age-matched conventional wild-type mice, with only a slight increase after conventionalization, and (iii) when fed a high-fat diet, germ-free *fiaf*$^{-/-}$ mice gained substantially more body weight and white fat than *fiaf*$^{+/+}$ mice (Backhed et al. 2004, 2007). However, a crucial point of evidence is missing in this reasoning, namely, that intestinal secretion of Fiaf/Angptl4 contributes substantially to Fiaf/Angptl4 plasma levels. This would be a prerequisite for its effects on adipose tissue LPL activity and, in turn, on body-fat accumulation. Furthermore, it was recently shown that hypothalamic Fiaf/Angptl4 is a regulator of food intake and body weight (Kim et al. 2010), which makes it difficult to interpret the above-reported findings in *fiaf*$^{-/-}$ mice.

As mentioned above, the work by Fleissner et al. (2010) suggests that the absence of a gut microbiota does not generally protect mice from diet-induced obesity. Rather, the data indicate that body-fat accumulation in germ-free mice very much depends on the composition of the high-fat diet. Fleissner et al. (2010) confirmed that the levels of Fiaf/Angptl4 mRNA in intestinal mucosa were higher in germ-free mice than in conventional mice as reported by Backhed et al. (2004). However, these higher Fiaf/Angptl4 mRNA levels in intestinal mucosa were independent of the composition of the high-fat diet and also independent of body-fat accumulation in the germ-free mice. Moreover, Western blot analysis of plasma Fiaf/Angptl4 did not show increased levels of any of the detectable Fiaf/Angptl4 isoforms in serum of germ-free mice. Conversely, levels were slightly higher in conventional mice compared to germ-free mice (Fleissner et al. 2010). A similar increase in Fiaf/Angptl4 was seen after monocolonization of germ-free mice with *Lactobacillus paracasei* (Aronsson et al. 2010). Furthermore, Fiaf/Angptl4 could not be detected in intestinal mucosa with an antibody against the N-terminus of the protein (Fleissner et al. 2010). Preliminary data suggest that the cAngptl4 form rather than the nAngptl4 form (which contains the LPL-inhibiting sequences) is present in intestinal mucosa (Fleissner et al., unpublished data). These findings suggest that the intestinal mucosa is not a major contributor to circulating Fiaf/Angptl4 levels and argue against a role of the intestinal Fiaf/Angptl4 as an inhibitor of LPL in peripheral tissues of germ-free mice.

In addition to its role in lipid metabolism, Fiaf/Angptl4 is a key player in angiogenesis, exhibiting pro- and antiangiogenic activities (Le Jan et al. 2003; Cazes et al. 2006). Since intestinal bacteria modulate the density of the capillary network in the intestine (Stappenbeck et al. 2002), it may be speculated that intestinal Fiaf/Angptl4 is involved in this process. The role of Fiaf/Angptl4 in intestinal integrity and function and its modulation by microbiota are thus far from clear and certainly deserve further attention. Fiaf/Angptl4-ablated mice were reported to display intestinal pathologies and decreased survival on a high-fat diet (Desai et al. 2007). This suggests that differences between *fiaf*$^{-/-}$ mice and wild-type mice in intestinal microbiota-mediated effects on energy metabolism are due to general intestinal pathologies of *fiaf*$^{-/-}$ mice rather than to the specific lack of Fiaf/Angptl4. Taken together, the currently available data do not support the hypothesis that the intestinal microbiota affects body-fat accretion by influencing intestinal Fiaf/Angptl4 gene expression.

9 Role of Gut Microbiota in Low-Grade Inflammation

Tumor necrosis factor-alpha (TNF-alpha) not only plays a role in innate immunity but also has catabolic effects (Hotamisligil et al. 1993), indicating that several links exist between immune responses and metabolic control. Inflammatory signaling pathways are induced by metabolic stress resulting from nutrient overload and endoplasmic reticulum stress (Hotamisligil and Erbay 2008). Therefore, metabolic signaling pathways may affect the immune response. Along the same line, metabolic hormones, including adipokines such as adiponectin, leptin, and resistin, are also immunologically active. Hence, there are several links between metabolism, insulin action, and inflammation. However, the underlying mechanisms are not yet understood.

It has recently been proposed that the gut microbiota plays an important role in the development of symptoms of the metabolic syndrome by contributing to the low-grade inflammation observed under metabolic stress caused by diet-induced obesity (Cani and Delzenne 2009). Lipopolysaccharide (LPS), a characteristic cell-wall component of gram-negative bacteria, is continuously released in the intestinal tract as a consequence of bacterial cell lysis. Serum LPS was shown to be 76% higher in type 2 diabetic subjects compared to control subjects. This was accompanied by a higher expression in abdominal adipose cells of proteins involved in the innate immune response, such as Toll-like receptor 2 (TLR2) and nuclear factor kappa B (NFκB) (Creely et al. 2007). Moreover, fasting serum insulin was correlated with serum LPS levels in these subjects. In another human study, the consumption of a high-fat meal resulted in 50% higher endotoxin levels (Erridge et al. 2007). Feeding mice a high-fat diet for 4 weeks led to a two- to threefold increase in plasma LPS concentrations (Cani et al. 2007a), indicating that such a diet facilitates the transfer of LPS from the intestine into the bloodstream. Chylomicrons synthesized in response to a high-fat diet were proposed to play a prominent role in this LPS transfer. Conversely, fasting led to decreased serum LPS levels in mice. The increase in serum LPS in response to high-fat diet is referred to as metabolic endotoxemia (Cani et al. 2007a). The authors of this study also reported differences in the gut microbiota of mice fed a high-fat diet as compared to mice fed a control diet. Mice on the high-fat diet harbored lower concentrations of gram-positive bifidobacteria and bacteria of the *Eubacterium rectale–Clostridium coccoides* cluster than mice on the control diet. Interestingly, concentrations of none of the gram-negative bacteria, including Enterobacteriaceae and Bacteroides increased in response to the high-fat diet. However, since the concentration of the gram-positive bacteria decreased to a larger extent than those of the gram-negative bacteria, the relative proportion of LPS-containing intestinal bacteria increased. To mimic the higher LPS transfer from the intestine to the blood as observed for mice on a high-fat diet, the authors induced metabolic endotoxemia in mice by infusing LPS subcutaneously for 4 weeks. As a result of this treatment, body weight, body fat, liver weight, fasted serum glucose and insulin levels, liver triglycerides, as well as inflammatory cytokines increased in the same way as observed for mice on a high-fat diet (Cani et al. 2007a). This was accompanied by signs of hepatic insulin resistance in the LPS-treated mice. Mutant mice devoid of the cluster of differentiation 14 (CD14) did not show these signs of metabolic disease upon treatment with LPS or in response to a high-fat diet. Since CD14 in conjunction with Toll-like receptor 4 (TLR4) acts as a coreceptor for LPS (Kitchens 2000), LPS from gut bacteria was proposed to play a key role in the development of metabolic endotoxemia, which is accompanied by symptoms of the metabolic syndrome (Cani et al. 2007a). A role of gut microbiota in the development of obesity and diabetes is also supported by the observation that treatment of obese and diabetic mice with antibiotics (ampicillin and neomycin) for 4 weeks led to a reduction of metabolic endotoxemia, body weight, and body fat (Cani et al. 2008). In addition, glucose intolerance and insulin resistance improved in response to the antibiotic treatment. Moreover, following antibiotic treatment, markers of inflammation and oxidative stress as well as macrophage infiltration decreased in adipose tissue of mice fed a high-fat diet.

A second link between immunity and metabolic syndrome has been observed recently. Mice lacking TLR5 ($TLR5^{-/-}$) displayed an altered gut microbiota composition and symptoms of the metabolic syndrome compared to corresponding wild-type mice (Vijay-Kumar et al. 2010). TLR5 is a transmembrane protein expressed in the gut mucosa and a component of the innate immune system. It recognizes bacterial flagellin and contributes to the host response against bacterial infections. $TLR5^{-/-}$ mice had a 20% higher body weight than wild-type mice, 1.5- to 2.5-fold higher fat-pad weights, 20% higher serum triglycerides, 60% higher serum cholesterol, increased visceral fat, elevated systolic and diastolic blood pressure, as well as increased glucose and insulin levels. Treatment of the $TLR5^{-/-}$ mice with a broad-spectrum antibiotic for 12 weeks reduced the bacterial cell concentration in the intestine to 10% of its original value. This was accompanied by an improvement of symptoms of the metabolic syndrome. The $TLR5^{-/-}$ mice did not differ from the wild-type mice in the proportion of the major phyla, but they differed in species composition. Transplantation of the gut microbiota from *TLR5*$^{-/-}$ mice to germ-free wild-type mice resulted in the transfer of many *TLR5*$^{-/-}$ mouse-specific phenotypic features to the recipients, such as hyperphagia, obesity, hyperglycemia, insulin resistance, and increased levels of proinflammatory cytokines. The authors speculated that lack of TLR5 led to changes in the gut microbiota which in turn contributed to development of low-grade inflammation (Vijay-Kumar et al. 2010). However, the exact mechanism how the observed changes in the gut microbiota contribute to the development of the metabolic syndrome remains elusive.

10 Effect of Bifidobacteria on Serum LPS Levels

Supplementation of newborn Balb/c mice with *Bifidobacterium infantis* and *Bifidobacterium bifidum* resulted in consistently lower intestinal LPS levels in ileocecal filtrates at 1, 2, 3, and 4 weeks after delivery as compared to unsupplemented mice (Griffiths et al. 2004), possibly by improving the gut barrier function (Wang et al. 2006). Cani et al. therefore investigated whether bifidobacteria are capable of improving the symptoms of metabolic endotoxemia (Cani et al. 2007b). They used oligofructose to stimulate the growth of endogenous intestinal bifidobacteria. Oligofructose had previously been shown to stimulate growth of intestinal bifidobacteria in both humans (Gibson et al. 1995) and rodents (Kleessen et al. 2001) because it escapes digestion and absorption in the small intestine. Cani et al. (2007b) fed mice either a high-fat diet or a high-fat diet supplemented with oligofructose. While the high-fat diet led to a reduction in the concentration of most bacterial groups, including the bifidobacteria, mice fed the high-fat diet supplemented with oligofructose had normal bifidobacteria concentrations just like mice fed a control diet. In conjunction with these effects on the gut microbiota, oligofructose supplementation lowered the endotoxin levels and improved glucose tolerance and glucose-induced insulin secretion. Cell numbers of bifidobacteria correlated negatively with various parameters of endotoxemia and positively with improved glucose tolerance and insulin secretion. Therefore, the authors of this study proposed that stimulating the growth of intestinal bifidobacteria helps to alleviate the pathophysiological consequences of endotoxemia and thereby prevent the occurrence of diabetes and obesity (Cani et al. 2007b).

The observed increase in plasma LPS in response to high-fat diets was suggested to result from increased gut permeability. Indeed, oral application of fluorescein isothiocyanate-labeled dextran (molecular weight 4,000, DX-4,000-FITC) resulted in a higher recovery of DX-4000-FITC in obese mice as compared to control mice, indicating a higher permeability of the gut wall in obese mice fed a high-fat diet (Cani et al. 2008). Interestingly, antibiotic treatment reduced the gut permeability in these mice. Changes in gut permeability correlated with the expression of the tight-junction protein zonola occludens-1 (ZO-1), whose expression was lower in mice fed a high-fat diet than in mice fed a normal diet. Antibiotic treatment of the high-fat diet-fed mice normalized the expression of ZO-1, suggesting that the gut microbiota modifies the permeability of the gut epithelial layer. It is plausible that an increased permeability of the gut epithelium may result in higher serum LPS concentrations and that the decimation of intestinal bacteria with antibiotics decreases the concentration of LPS in the intestine. However, it remains unclear which bacterial components or products prompt the host to modify the expression of tight junction proteins and how this is regulated.

In this context, it is interesting to note that oral administration of a *Bifidobacterium infantis* cell-culture supernatant enhanced the barrier function of the epithelial cell layer (Ewaschuk et al. 2008). Specifically, it reduced the permeability of the colonic epithelium in mice and also attenuated inflammation in interleukin 10 (IL-10)-deficient mice. Addition of *B. infantis* cell-culture supernatant to T84 human epithelial cells increased the transepithelial resistance (TER) and enhanced the expression of the tight junction proteins ZO-1 and occludin, while that of claudin decreased. The supernatant of the *B. infantis* culture also prevented the decrease in TER, which occurred in response to the addition of TNF-α and interferon-γ (IFN-γ). The factor(s) responsible for these effects have not yet been identified.

11 Effect of Oligofructose on Symptoms of the Metabolic Syndrome

Since the feeding of oligofructose to obese mice not only stimulated the growth of endogenous intestinal bifidobacteria but also reduced endotoxemia and improved a number of symptoms associated with the metabolic syndrome, it was concluded that the beneficial effects observed in response to supplementing a high-fat diet with oligofructose were mediated by bifidobacteria (Cani et al. 2009). However, it cannot be ruled out that oligofructose exerted these effects by influencing other intestinal factors. It also cannot be excluded that the replacement of cellulose by oligofructose, which differs in its physicochemical properties, lowered mucosal permeability for LPS, and thereby alleviated endotoxemia and the symptoms of the metabolic syndrome. Nevertheless, the observed lowering of serum LPS levels in response to oral supplementation of newborn Balb/c mice with bifidobacteria (Griffiths et al. 2004) argues in favor of the concept that the effects of oligofructose on symptoms of the metabolic syndrome were indeed mediated by bifidobacteria, even though the cytokine production (IL-6, TNF-α, INF-γ) in Peyer's patches of these mice did not change.

The effect of direct oral administration of bifidobacteria was recently investigated in mice fed a high-fat diet (Kondo et al. 2010). After 8 weeks on the high-fat diet, two mouse groups receiving either 10^8 or 10^9 colony-forming units (cfu)/d of *Bifidobacterium breve* B3 and an unsupplemented control group did not differ significantly in the amount of chow consumed. However, they differed in body weight gain: While the control mice gained 12.8 g of body weight, the mice supplemented with the lower or higher dose of *B. breve* only gained 11.3 g or 10.5 g body weight. The mean body weights of the *B. breve*-supplemented mouse groups were 12 and 18% lower than the mean body weight of the control group. These differences were similar for the epididymal fat pads of the mice. In addition, the oral application of *B. breve* B3 at a dose of 10^9 cfu/d reduced serum glucose, insulin, and cholesterol levels by 17, 14, and 58%, respectively, compared to the control mice. Interestingly, mRNA levels of proglucagon in colonic tissue and of adiponectin in epididymal fat pads were elevated up to 35 and 39%, respectively, in the mice supplemented with *B. breve* B3 as compared with the control mice. These effects were not observed with heat-treated bacteria. The authors of this study therefore speculated that the observed improvement of a number of metabolic parameters may be related to the reported ability of some bifidobacterial strains to convert linoleic acid to conjugated linoleic acid (CLA) (Kondo et al. 2010). They refer to a crossover study involving 55 diabetic postmenopausal women who had a significant reduction in BMI in response to the consumption of CLA-containing oil for 16 weeks (6.4 g CLA/d), while the consumption of a CLA-free oil did not affect BMI (Norris et al. 2009).

Taken together, there are some good indications that the beneficial effects observed in response to the consumption of oligofructose are at least to some extent mediated by bifidobacteria because this oligosaccharide stimulates the growth of bifidobacteria, and a number of studies support a role of these bacteria in the alleviation of endotoxemia and ensuing metabolic disorders. However, the available evidence is not as clear-cut as it would be desirable. Moreover, knowledge of the underlying molecular mechanism could help to increase the credibility of this concept. For obtaining experimental proof, it is necessary to identify the bacterial factors that mediate the observed reduction in gut permeability as well as the host components targeted by such factors. A direct effect of oligofructose on symptoms of the metabolic syndrome cannot be excluded at present.

12 Effects of the Gut Microbiota on the Endocannabinoid System

Recently, a third system involved in energy homeostasis was proposed to be affected by the gut microbiota, namely, the endocannabinoid system (Muccioli et al. 2010). Elements of this system include the endocannabinoid receptors CB1 and CB2, which belong to the family of G protein-coupled receptors (Scherer and Buettner 2009). A role in the regulation of energy metabolism has only been defined for CB1. This receptor can be found in the central nervous system, liver, muscle, and white adipose tissue. The CB1 receptor has two different functions in energy homeostasis: In the brain, CB1 signaling contributes to appetite regulation, and it has been proposed that there is cross talk between endocannabinoid and leptin signaling (Di Marzo et al. 2001). Physiological ligands of the CB1 receptor are lipid signals derived from polyunsaturated fatty acids, which are called endocannabinoids

(ECs). The two major ECs are anandamide (AEA) and 2-arachidonoyl glycerol (2-AG) – both derivatives of arachidonic acid, an n-6 essential fatty acid (Maccarrone et al. 2010). The CB1 receptor also affects functions that are independent of food intake. This may be concluded from the observation that CB1 receptor knockout mice in contrast to the corresponding wildtype mice do not develop diet-induced obesity, even though their chow consumption did not differ from that of the wild-type mice (Ravinet Trillou et al. 2004). Plasma levels of 2-AG in obese human subjects are higher than in lean control subjects (Bluher et al. 2006). Adipocytes and macrophages have the ability to synthesize ECs (Scherer and Buettner 2009). Various mechanisms regulating the EC tone have been proposed, including enzymes involved in the degradation of ECs such as fatty acid amide hydrolase (FAAH), which degrades AEA, and monoacylglycerol lipase (MAGL), which degrades 2-AG. The FAAH mRNA level in abdominal fat of obese subjects was lower than in normal weight subjects (Bluher et al. 2006). Insulin treatment of adipocytes led to a decrease of intracellular EC concentrations and simultaneously to an increase of mRNA expression levels of FAAH and MAGL, suggesting that insulin-resistant adipocytes are no longer capable of maintaining the EC tone in a physiological range (D'Eon et al. 2008).

The gut microbiota is the main source of LPS, which not only triggers a proinflammatory immune response but also affects the EC system. For example, in mouse macrophages, LPS leads to increased AEA levels by activation of the AEA biosynthetic enzymes (Liu et al. 2003). To further elucidate the role of the gut microbiota in the EC system, Muccioli et al. studied the CB1 receptor mRNA expression in colonic and jejunal tissue of various mouse models differing in their microbial status and/or the type of diet (Muccioli et al. 2010): Germ-free mice had a twofold higher CB1 receptor mRNA expression in colonic tissue than conventional mice, and conventional mice treated with ampicillin and neomycin displayed a 60% lower CB1 receptor mRNA expression. Obese *ob/ob* mice treated with oligofructose had a 25% lower CB1 receptor mRNA level than such mice on a control diet. Subcutaneous adipose tissue of obese *ob/ob* mice also had more than twofold higher mRNA levels of CB1 receptor and of *N*-acylphosphatidylethanolamine-phospholipase D, which plays a major role in AEA synthesis, than their lean littermates (*Lean-ob*). The mRNA of the FAAH was eightfold higher in the lean than in the obese mice. These differences are in accordance with the 20% higher AEA levels observed in the *ob/ob* mice compared with their lean littermates, demonstrating that the EC system tone in adipose tissues of *ob/ob* mice differs from that of their lean littermates (Muccioli et al. 2010).

Mice on a high-fat diet had a 2.5-fold higher colon CB1 receptor mRNA expression than mice on a control diet (Muccioli et al. 2010). These results indicate that genotype, microbial status, and diet influence the expression of the CB1 receptor and thereby the EC tone. The gut microbiota not only influenced the expression of the CB1 receptor in colonic tissue but also that of the EC-degrading enzymes FAAH and MGL (Muccioli et al. 2010). The authors hypothesized that the elevated plasma LPS levels and the increased gut permeability as observed in obese mice are linked by the intestinal EC system. The intestinal AEA concentration in colonic tissue of *ob/ob* mice fed an oligofructose-containing diet was 25% lower than that of *ob/ob* mice fed a control diet. In accordance with this finding, the AEA-degrading FAAH was 25% higher in the oligofructose-fed mice than in the mice on the control diet. The 2-AG concentrations in colonic tissue did not differ between these groups. *Ob/ob* mice treated with the CB1 antagonist SR141716A displayed 50% lower plasma LPS levels than untreated *ob/ob* mice. Treatment of wild-type mice with the CB1 receptor agonist HU-210 led to a 2.5-fold increase in plasma LPS levels and to a threefold increase in gut permeability as measured with DX-4000-FITC. It was concluded that the EC system plays an important role in the regulation of gut permeability, which in turn influences the systemic LPS levels (Muccioli et al. 2010). This view is supported by in vitro investigations in epithelial Caco-2 cell monolayers, which demonstrated that LPS leads to a 15% reduction in the mRNA levels of the tight junction proteins occludin and ZO-1. A 30 to 35% decrease in the mRNA levels of these tight junction proteins was observed when LPS was added together with the CB1 receptor agonist HU-210. The addition of the CB1 receptor antagonist SR141716A on top of LPS and HU-210 completely abolished the reduced mRNA expression of occludin and ZO-1.

Genetically obese *ob/ob* mice supplemented with oligofructose had a 10% lower adiposity index, 50% lower CB1 receptor mRNA levels in spite of an increased gene expression of markers of lipogenesis, such as sterol responsive element binding protein (SREBP)-1c, FAS, and ACC, as well as of markers of adipocyte differentiation, such as PPAR-γ and fatty acid-binding protein (FABP). The change in these markers in response to oligofructose supplementation coincided with a decreased fat mass. Interestingly, blocking of the CB1 receptor had the same effect as the oral application of oligofructose. Since oligofructose has previously been shown to alter the gut microbiota, these results were taken as an indication for an existing link between the gut microbiota, LPS, the EC system, gut permeability, and symptoms of the metabolic syndrome. However, it is noteworthy that activation of the EC system by HU-210 under physiological conditions leads to an increase in the expression of adipocyte differentiation and adipogenesis markers without changing the CB1 mRNA levels (Muccioli et al. 2010).

It has been speculated that the changes in the composition of gut microbiota observed in obesity lead to increased plasma LPS levels, which in turn result in low-grade inflammation and a greater EC system tone. These changes are thought to lead to the dysregulation of adipogenesis (Muccioli et al. 2010). The dietary fatty acid intake has been shown to influence endocannabinoid levels in different tissues including the intestine (Artmann et al. 2008). It can therefore be speculated that the consumption of high-fat diets rich in n-6 polyunsaturated fatty acids (as typical in Western societies) in conjunction with microbiota-mediated effects leads to deterioration of the EC system tone in the human metabolic syndrome.

13 Conclusions

There is no doubt that diet affects gut microbiota composition. This is not really surprising because diet is the primary source of substrates for intestinal bacteria. However, it is an astonishing finding that transplantation of a gut microbiota from obese mice to nonobese germ-free mice results in a transfer of metabolic features from the donor animal to the recipients. This has been demonstrated for genetically obese mice such as *ob*/*ob* mice and *TLR5*$^{-/-}$ mice as well as for mice with diet-induced obesity. These observations suggest that the gut microbiota affects host energy metabolism by improving energy harvest from the diet. This may be of advantage to the host when nutrients are scarce, but it may be of disadvantage to the host when the nutrient supply is ample, a situation typical of modern societies, in which we observe an increase in the proportion of people afflicted by the metabolic syndrome. Possible mechanisms underlying the improved energy harvest include a more effective conversion of dietary fiber to SCFA, which provides additional energy to the host. In addition, SCFA produced by the gut microbiota extends the intestinal transit time by activating Gpr41 and influencing gut hormone secretion, which leads to a more complete absorption of intestinal nutrients. The gut microbiota may also influence energy harvest and fat storage by modulating the expression of proteins that play a role in energy homeostasis, such as hepatic ChREBP and SREBP-1, which are involved in liver triglyceride accumulation, and Angptl4/Fiaf, a circulating inhibitor of adipose tissue fatty acid uptake. However, regarding the latter, the available data are still controversial. Furthermore, the gut microbiota and obesity are linked by low-grade inflammation and elevated serum LPS levels (endotoxemia). Components of the innate immune system respond to endotoxemia with symptoms characteristic of the metabolic syndrome. A recent study indicates that this response involves the EC system. The improvement of endotoxemia by oligofructose has been linked to stimulation of intestinal bifidobacteria, which are proposed to reduce gut permeability by changes in the expression of tight junction proteins. With all these links, there are several important questions that have not yet been answered (1) How much does any of these factors contribute to the development of obesity and the metabolic syndrome? (2) Which bacteria do actually modify energy balance of the host? (3) Which bacterial molecules are involved in the improvement of metabolic endotoxemia observed in response to oligofructose supplementation?

References

Aronsson L, Huang Y, Parini P et al (2010) Decreased fat storage by *Lactobacillus paracasei* is associated with increased levels of angiopoietin-like 4 protein (ANGPTL4). PLoS One 5: e13087. doi:doi:10.1371/journal.pone.0013087

Artmann A, Petersen G, Hellgren LI et al (2008) Influence of dietary fatty acids on endocannabinoid and N-acylethanolamine levels in rat brain, liver and small intestine. Biochim Biophys Acta 1781:200–212

Backhed F, Ding H, Wang T et al (2004) The gut microbiota as an environmental factor that regulates fat storage. Proc Natl Acad Sci USA 101:15718–15723

Backhed F, Manchester JK, Semenkovich CF et al (2007) Mechanisms underlying the resistance to diet-induced obesity in germ-free mice. Proc Natl Acad Sci USA 104:979–984

Bluher M, Engeli S, Kloting N et al (2006) Dysregulation of the peripheral and adipose tissue endocannabinoid system in human abdominal obesity. Diabetes 55:3053–3060

Cani PD, Delzenne NM (2009) The role of the gut microbiota in energy metabolism and metabolic disease. Curr Pharm Des 15:1546–1558

Cani PD, Amar J, Iglesias MA et al (2007a) Metabolic endotoxemia initiates obesity and insulin resistance. Diabetes 56:1761–1772

Cani PD, Neyrinck AM, Fava F et al (2007b) Selective increases of bifidobacteria in gut microflora improve high-fat-diet-induced diabetes in mice through a mechanism associated with endotoxaemia. Diabetologia 50:2374–2383

Cani PD, Bibiloni R, Knauf C et al (2008) Changes in gut microbiota control metabolic endotoxemia-induced inflammation in high-fat diet-induced obesity and diabetes in mice. Diabetes 57:1470–1481

Cani PD, Possemiers S, Van de Wiele T et al (2009) Changes in gut microbiota control inflammation in obese mice through a mechanism involving GLP-2-driven improvement of gut permeability. Gut 58:1091–1103

Cazes A, Galaup A, Chomel C et al (2006) Extracellular matrix-bound angiopoietin-like 4 inhibits endothelial cell adhesion, migration, and sprouting and alters actin cytoskeleton. Circ Res 99:1207–1215

Creely SJ, McTernan PG, Kusminski CM et al (2007) Lipopolysaccharide activates an innate immune system response in human adipose tissue in obesity and type 2 diabetes. Am J Physiol Endocrinol Metab 292:E740–747

D'Eon TM, Pierce KA, Roix JJ et al (2008) The role of adipocyte insulin resistance in the pathogenesis of obesity-related elevations in endocannabinoids. Diabetes 57:1262–1268

Desai U, Lee EC, Chung K et al (2007) Lipid-lowering effects of anti-angiopoietin-like 4 antibody recapitulate the lipid phenotype found in angiopoietin-like 4 knockout mice. Proc Natl Acad Sci USA 104:11766–11771

Di Marzo V, Goparaju SK, Wang L et al (2001) Leptin-regulated endocannabinoids are involved in maintaining food intake. Nature 410:822–825

Duncan SH, Lobley GE, Holtrop G et al (2008) Human colonic microbiota associated with diet, obesity and weight loss. Int J Obes (Lond) 32:1720–1724

Eckburg PB, Bik EM, Bernstein CN et al (2005) Diversity of the human intestinal microbial flora. Science 308:1635–1638

Erridge C, Attina T, Spickett CM et al (2007) A high-fat meal induces low-grade endotoxemia: evidence of a novel mechanism of postprandial inflammation. Am J Clin Nutr 86:1286–1292

Ewaschuk JB, Diaz H, Meddings L et al (2008) Secreted bioactive factors from Bifidobacterium infantis enhance epithelial cell barrier function. Am J Physiol Gastrointest Liver Physiol 295: G1025–1034

Finegold SM, Sutter VL, Mathisen GE (1983) Normal indigenous intestinal flora. In: Hentges DJ (ed) Human intestinal microflora in health and disease. Academic, New York/London, pp 3–31

Fleissner CK, Huebel N, Abd El-Bary MM et al (2010) Absence of intestinal microbiota does not protect mice from diet-induced obesity. Br J Nutr 104:919–929

Friedman JM, Halaas JL (1998) Leptin and the regulation of body weight in mammals. Nature 395:763–770

Gibson GR, Beatty ER, Wang X et al (1995) Selective stimulation of bifidobacteria in the human colon by oligofructose and inulin. Gastroenterology 108:975–982

Griffiths EA, Duffy LC, Schanbacher FL et al (2004) In vivo effects of bifidobacteria and lactoferrin on gut endotoxin concentration and mucosal immunity in Balb/c mice. Dig Dis Sci 49:579–589

Hildebrandt MA, Hoffmann C, Sherrill-Mix SA et al (2009) High-fat diet determines the composition of the murine gut microbiome independently of obesity. Gastroenterology 137:1716–1724

Hooper LV, Wong MH, Thelin A et al (2001) Molecular analysis of commensal host-microbial relationships in the intestine. Science 291:881–884

Hotamisligil GS, Erbay E (2008) Nutrient sensing and inflammation in metabolic diseases. Nat Rev Immunol 8:923–934

Hotamisligil GS, Shargill NS, Spiegelman BM (1993) Adipose expression of tumor necrosis factor-alpha: direct role in obesity-linked insulin resistance. Science 259:87–91

Isken F, Klaus S, Osterhoff M et al (2010) Effects of long-term soluble vs. insoluble dietary fiber intake on high-fat diet-induced obesity in C57BL/6J mice. J Nutr Biochem 21:278–284

Kersten S, Mandard S, Tan NS et al (2000) Characterization of the fasting-induced adipose factor FIAF, a novel peroxisome proliferator-activated receptor target gene. J Biol Chem 275:28488–28493

Kim I, Kim HG, Kim H et al (2000) Hepatic expression, synthesis and secretion of a novel fibrinogen/angiopoietin-related protein that prevents endothelial-cell apoptosis. Biochem J 346 (Pt 3):603–610

Kim HK, Youn BS, Shin MS et al (2010) Hypothalamic Angptl4/Fiaf is a novel regulator of food intake and body weight. Diabetes 59:2772–2780

Kitchens RL (2000) Role of CD14 in cellular recognition of bacterial lipopolysaccharides. Chem Immunol 74:61–82

Kleessen B, Hartmann L, Blaut M (2001) Oligofructose and long-chain inulin: influence on the gut microbial ecology of rats associated with a human faecal flora. Br J Nutr 86:291–300

Kondo S, Xiao JZ, Satoh T et al (2010) Antiobesity effects of *Bifidobacterium breve* strain B-3 supplementation in a mouse model with high-fat diet-induced obesity. Biosci Biotechnol Biochem 74:1656–1661

Kushner RF, Choi SW (2010) Prevalence of unhealthy lifestyle patterns among overweight and obese adults. Obesity 18:1160–1167

Le Jan S, Amy C, Cazes A et al (2003) Angiopoietin-like 4 is a proangiogenic factor produced during ischemia and in conventional renal cell carcinoma. Am J Pathol 162:1521–1528

Ley RE, Backhed F, Turnbaugh P et al (2005) Obesity alters gut microbial ecology. Proc Natl Acad Sci USA 102:11070–11075

Ley RE, Turnbaugh PJ, Klein S et al (2006a) Human gut microbes associated with obesity. Nature 444:1022–1023

Ley RE, Peterson DA, Gordon JI (2006b) Ecological and evolutionary forces shaping microbial diversity in the human intestine. Cell 124:837–848

Lichtenstein L, Kersten S (2010) Modulation of plasma TG lipolysis by Angiopoietin-like proteins and GPIHBP1. Biochim Biophys Acta 1801:415–420

Liu J, Batkai S, Pacher P et al (2003) Lipopolysaccharide induces anandamide synthesis in macrophages via CD14/MAPK/phosphoinositide 3-kinase/NF-kappaB independently of platelet-activating factor. J Biol Chem 278:45034–45039

Maccarrone M, Gasperi V, Catani MV et al (2010) The endocannabinoid system and its relevance for nutrition. Annu Rev Nutr 30:423–440

Maczulak AE, Wolin MJ, Miller TL (1993) Amounts of viable anaerobes, methanogens, and bacterial fermentation products in feces of rats fed high-fiber or fiber-free diets. Appl Environ Microbiol 59:657–662

Mandard S, Zandbergen F, van Straten E et al (2006) The fasting-induced adipose factor/angiopoietin-like protein 4 is physically associated with lipoproteins and governs plasma lipid levels and adiposity. J Biol Chem 281:934–944

McNeil NI (1984) The contribution of the large intestine to energy supplies in man. Am J Clin Nutr 39:338–342

Muccioli GG, Naslain D, Backhed F et al (2010) The endocannabinoid system links gut microbiota to adipogenesis. Mol Syst Biol 6:392

Murphy EF, Cotter PD, Healy S et al (2010) Composition and energy harvesting capacity of the gut microbiota: relationship to diet, obesity and time in mouse models. Gut 59:1635–1642

Norris LE, Collene AL, Asp ML et al (2009) Comparison of dietary conjugated linoleic acid with safflower oil on body composition in obese postmenopausal women with type 2 diabetes mellitus. Am J Clin Nutr 90:468–476

Qin J, Li R, Raes J et al (2010) A human gut microbial gene catalogue established by metagenomic sequencing. Nature 464:59–65

Ravinet Trillou C, Delgorge C, Menet C et al (2004) CB1 cannabinoid receptor knockout in mice leads to leanness, resistance to diet-induced obesity and enhanced leptin sensitivity. Int J Obes Relat Metab Disord 28:640–648

Samuel BS, Gordon JI (2006) A humanized gnotobiotic mouse model of host-archaeal-bacterial mutualism. Proc Natl Acad Sci USA 103:10011–10016

Samuel BS, Shaito A, Motoike T et al (2008) Effects of the gut microbiota on host adiposity are modulated by the short-chain fatty-acid binding G protein-coupled receptor, Gpr41. Proc Natl Acad Sci USA 105:16767–16772

Scherer T, Buettner C (2009) The dysregulation of the endocannabinoid system in diabesity-a tricky problem. J Mol Med 87:663–668

Schwiertz A, Taras D, Schafer K et al (2010) Microbiota and SCFA in lean and overweight healthy subjects. Obesity (Silver Spring) 18:190–195

Stappenbeck TS, Hooper LV, Gordon JI (2002) Developmental regulation of intestinal angiogenesis by indigenous microbes via Paneth cells. Proc Natl Acad Sci USA 99:15451–15455

Tazoe H, Otomo Y, Kaji I et al (2008) Roles of short-chain fatty acids receptors, GPR41 and GPR43 on colonic functions. J Physiol Pharmacol 59(Suppl 2):251–262

Turnbaugh PJ, Ley RE, Mahowald MA et al (2006) An obesity-associated gut microbiome with increased capacity for energy harvest. Nature 444:1027–1031

Turnbaugh PJ, Backhed F, Fulton L et al (2008) Diet-induced obesity is linked to marked but reversible alterations in the mouse distal gut microbiome. Cell Host Microbe 3:213–223

Turnbaugh PJ, Ridaura VK, Faith JJ, et al. (2009) The effect of diet on the human gut microbiome: a metagenomic analysis in humanized gnotobiotic mice. Sci Transl Med 1:6ra14. doi: 1/6/6ra14

Vijay-Kumar M, Aitken JD, Carvalho FA et al (2010) Metabolic syndrome and altered gut microbiota in mice lacking toll-like receptor 5. Science 328:228–231

Voshol PJ, Rensen PC, van Dijk KW et al (2009) Effect of plasma triglyceride metabolism on lipid storage in adipose tissue: studies using genetically engineered mouse models. Biochim Biophys Acta 1791:479–485

Wang Z, Xiao G, Yao Y et al (2006) The role of bifidobacteria in gut barrier function after thermal injury in rats. J Trauma 61:650–657

Wren AM, Bloom SR (2007) Gut hormones and appetite control. Gastroenterology 132:2116–2130

Xiong Y, Miyamoto N, Shibata K et al (2004) Short-chain fatty acids stimulate leptin production in adipocytes through the G protein-coupled receptor GPR41. Proc Natl Acad Sci USA 101:1045–1050

Yau MH, Wang Y, Lam KS et al (2009) A highly conserved motif within the NH2-terminal coiled-coil domain of angiopoietin-like protein 4 confers its inhibitory effects on lipoprotein lipase by disrupting the enzyme dimerization. J Biol Chem 284:11942–11952

Yoon JC, Chickering TW, Rosen ED et al (2000) Peroxisome proliferator-activated receptor gamma target gene encoding a novel angiopoietin-related protein associated with adipose differentiation. Mol Cell Biol 20:5343–5349

Zhou J, Martin RJ, Tulley RT et al (2008) Dietary resistant starch upregulates total GLP-1 and PYY in a sustained day-long manner through fermentation in rodents. Am J Physiol Endocrinol Metab 295:E1160–1166

Part III
Nutrient Sensing

Sensing of Glucose in the Brain

Bernard Thorens

Contents

Abstract The brain, and in particular the hypothalamus and brainstem, have been recognized for decades as important centers for the homeostatic control of feeding, energy expenditure, and glucose homeostasis. These structures contain neurons and neuronal circuits that may be directly or indirectly activated or inhibited by glucose, lipids, or amino acids. The detection by neurons of these nutrient cues may become deregulated, and possibly cause metabolic diseases such as obesity and diabetes. Thus, there is a major interest in identifying these neurons, how they respond to nutrients, the neuronal circuits they form, and the physiological function they control. Here I will review some aspects of glucose sensing by the brain. The

B. Thorens (✉)
Center for Integrative Genomics, University of Lausanne, Genopode Building, CH-1015 Lausanne, Switzerland
e-mail: Bernard.thorens@unil.ch

H.-G. Joost (ed.), *Appetite Control*, Handbook of Experimental Pharmacology 209,
DOI 10.1007/978-3-642-24716-3_12,

brain is responsive to both hyperglycemia and hypoglycemia, and the glucose sensing cells involved are distributed in several anatomical sites that are connected to each other. These eventually control the activity of the sympathetic or parasympathetic nervous system, which regulates the function of peripheral organs such as liver, white and brown fat, muscle, and pancreatic islets alpha and beta cells. There is now evidence for an extreme diversity in the sensing mechanisms used, and these will be reviewed.

Keywords Brainstem • Counterregulation • Food intake • Glucogen • Glucokinase • Glucose transporters • Glucose Sensing • Hypothalamus

1 Introduction

Since the initial observation by Claude Bernard that a puncture of the floor of the fourth ventricle of the dog induces diabetes (Bernard 1849), the brain has been recognized as an important regulator of glucose homeostasis. Subsequent studies have demonstrated that feeding behavior was also regulated by central glucose sensing, leading to the glucostatic hypothesis of feeding control by J. Mayer (1953). It was further demonstrated that distinct hypothalamic nuclei were involved in the regulation of feeding and fasting, since lesion of the lateral hypothalamus reduced feeding and body weight, whereas lesion of the ventromedial hypothalamus (VMH) induced hyperphagia and hyperinsulinemia (Bray 1985; Hoebel 1965; King 2006). A widely used animal model of obesity was also established when it was shown that administration of gold thioglucose, which causes the destruction of VMH neurons, induces obesity (Marshall and Mayer 1956; Mayer and Thomas 1967). The toxic effect of gold thioglucose is not duplicated when gold is conjugated with other metabolites or nutrients, suggesting a specific effect on glucose-sensitive neurons. Intracerebroventricular injection of the glucose antimetabolite 2-deoxy-D-glucose, which inhibits glycolysis and creates a glucopenic state mimicking hypoglycemia, has been shown to induce feeding (Miselis and Epstein 1975) and glucagon secretion (Borg et al. 1995). In contrast, i.c.v. injection of glucose reduces feeding in fasted mice (Bady et al. 2006) and can prevent hypoinsulinemia-induced glucagon response (Biggers et al. 1989; Frizzell et al. 1993). A control of energy expenditure through glucose sensing has also been proven by i.c.v. 2-DG injection which induces a marked hypothermic response (Freinkel et al. 1972).

The above-described observations therefore indicated that both hypo- and hyperglycemia can be recognized by central glucose sensing cells to control feeding, energy expenditure, and counterregulation. It has been established for ~50 years that these glucose sensing responses depend on the firing activity of glucose-excited (GE) or glucose-inhibited (GI) neurons that is triggered by, respectively, rises or falls in glucose concentrations (Anand et al. 1964; Oomura and Yoshimatsu 1984; Routh 2002; Yang et al. 2004). Both types of neurons are widely distributed in the hypothalamus and brainstem. In the hypothalamus, GE and GI neurons are present in the arcuate (AN), ventromedial (VMN), paraventricular (PVN), and lateral (LH) hypothalamic nuclei (Dunn-Meynell et al. 1998; Silver and Erecinska 1998; Wang et al. 2004). Both types of neurons are also found in the brainstem, in particular in the nucleus of the tractus solitarius (NTS), the area postrema (AP), and the dorsal motor nucleus of the vagus (DMNX) (Adachi et al. 1984; Dallaporta et al. 1999; Mizuno and Oomura 1984; Yettefti et al. 1997). Recently, it has been suggested that subpopulations of GE and GI neurons in AN are actually responsive to glucose over a high glucose concentration range (5–20 mM) and are referred to as HGE (high-glucose-excited) or HGI (high-glucose-inhibited) neurons, respectively (Fioramonti et al. 2004; Penicaud et al. 2006).

Studies over the last several years have started to yield a molecular picture of the mechanisms of glucose sensing by GE and GI neurons. This is, however, still far from being complete, and new studies reveal the extreme diversity of the molecular basis for glucose recognition in the control of neuronal firing, suggesting complex regulatory networks activated by glucose to control physiology.

2 Anatomical Organization of Glucose Sensing Nuclei

2.1 *The Melanocortin Pathway*

An important site for integration of hormonal, nutritional, and neuronal signals is the melanocortin pathway which consists of AN neurons expressing the anorexigenic peptides POMC and CART as well as neurons expressing the orexigenic peptides NPY and AgRP. AgRP is an antagonist of the melanocortin receptors (MCR) 3 and 4, whereas α-MSH, derived from the POMC prohormone, is an agonist of these receptors. The NPY and POMC neurons project to neurons in the PVN and LH that express the melanocortin 3 and 4 receptors (Gautron and Elmquist 2011; Schwartz et al. 2000). Neurons in the PVN produce the anorexigenic neuropeptides TRH and CRF, whereas neurons in the LH produce the orexigenic peptides MCH and orexin (Schwartz et al. 2000). Together, these neurons form the melanocortin pathway and regulate peripheral metabolism through regulation of the activity of both the sympathetic and parasympathetic branches of the autonomic nervous system; they are also connected to higher brain structures to control feeding behavior, arousal, and reward (Adamantidis and de Lecea 2008; Berthoud 2002; Sakurai 2007).

The neurons in the AN are regulated by several hormones including ghrelin, insulin, PYY3-36, and most importantly leptin. They are also regulated by nutrients including lipids, amino acids, and glucose (Cummings and Schwartz 2000; Gale et al. 2004; Schwartz 2000; Schwartz et al. 2000; Thorens and Larsen 2004; Woods et al. 1998).

Although the role of leptin to regulate this pathway is critical (Gautron and Elmquist), there is also strong evidence for its modulation by glucose. Forty percent of NPY neurons have been found to be glucose inhibited; POMC neurons are typical GE neurons, and orexin neurons in the LH are GI, whereas those expressing MCH are GE neurons.

2.1.1 The Ventromedial Hypothalamus

The VMH has afferent connections with many hypothalamic nuclei, including the medial and lateral hypothalamus, but also with brainstem structures, including the NTS (Canteras et al. 1994). The VMH has been associated with regulation of the counterregulatory response to hypoglycemia, inducing glucagon secretion in response to fall in blood glucose concentrations. Lesion, pharmacological, and genetic studies have demonstrated the role of VMH glucose sensing in counterregulation. For instance, glucagon secretion can be induced by direct injection of 2-DG in the VMH (Borg et al. 1995) or, in contrast, hypoglycemia-induced glucagon secretion can be suppressed by direct VMH injection of glucose (Borg et al. 1997). Interestingly, VMH neurons are predominantly glutamatergic and express the vesicular glutamate transporter vGLUT2. Because the nuclear hormone receptor SF-1 is expressed selectively in VMH neurons, SF-1-Cre mice have been generated that allow specific deletion of floxed genes in the VMH (Dhillon et al. 2006). Deletion of vGLUT2 in the VMH generated mice that had marked defect in glucagon secretion in response to fasting or hypoglycemia (Tong et al. 2007), suggesting that glutamatergic neurons of the VMH are required for the counterregulatory response.

2.1.2 Brainstem, The Dorsal Vagal Complex, and the Basolateral Medulla

The hindbrain structures involved in glucose-dependent regulation of feeding and glucose homeostasis include the dorsal vagal complex (DVC), which consists of the AP, the NTS, and the DMNX, as well as the basolateral region (BLM) that contains the A1/C1 catecholamine neurons. The role of the hindbrain in glucoregulation has been proven by intracerebroventricular (i.c.v.) injection of 2-DG, which stimulates feeding only if the cerebral aqueduct is open to allow access of the injected substance to the brainstem (Berthoud and Mogenson 1977; Ritter et al. 1981), and food uptake can be activated by direct injection of 5-thioglucose (5-TG) into the NTS, DMNX, or BLM (Ritter et al. 2000). The importance of the NTS neurons in glucose sensing is also demonstrated by their sensitivity to small variations of blood glucose concentrations as determined by extracellular recording of their firing activity (Yettefti et al. 1995). Neurons from the NTS project to the LH and PVN, whereas neurons from the BLM project to the AN. Destruction by immunotoxins of the BLM projections to the AN suppresses the effect of 2-DG on food intake and on regulated expression of AgRP and NPY, suggesting a highly functional interrelationship between glucose-sensitive neurons from the brainstem and hypothalamus in integrated control of feeding (Fraley and Ritter 2003; Ritter et al. 2001).

3 Mechanisms of Glucodetection by GE and GI Neurons

3.1 *Glucose-Excited Neurons: The Glut2/Glucokinase/K_{ATP} Channel Signaling Pathway*

The mechanism of glucose sensing by GE neurons is thought to be similar to that of the pancreatic beta cells (Fig. 1), which depends on glucose metabolism and production of coupling factors, mostly derived from mitochondrial metabolism, which induce depolarization of plasma membrane prior to Ca^{2+} entry and stimulated secretion. In the beta-cell signaling pathway, Glut2 is the major glucose transporter isoform that allows a fast equilibration of glucose between the extra- and intracellular compartments. Glucokinase then phosphorylates glucose, and this is the rate-controlling step in glucose utilization and production of the coupling factors, the major one being the increase in the ATP/ADP ratio, which induces the closure of ATP-dependent K^+ (K_{ATP}) channels. This channel closure depolarizes the plasma membrane and opens voltage-gated calcium channels, resulting in Ca^{2+} influx which triggers insulin secretion. The Glut2/GK/K_{ATP} channel signaling pathway is probably also active in hypothalamus and brainstem to control neuronal excitability and control of feeding, energy expenditure, and glucose homeostasis. However, so far there is no direct proof that the three components of the Glut2/GK/K_{ATP} channel signaling pathway are present together in any given neuron.

POMC neurons in arcuate nucleus are GE neurons that express the K_{ATP} channel subunits SUR1 and Kir6.2 (Ibrahim et al. 2003). The importance of this channel in glucose sensing and glycemic control has been shown in mice in which a mutated form of Kir6.2, which prevents channel closure in response to increased ATP/ADP ratio, is expressed selectively in POMC neurons. The neurons of these mice no longer respond to glucose when tested by electrophysiological recording, and this is associated with the presence of mild glucose intolerance (Parton et al. 2007). As for pancreatic beta cells, expression of the uncoupling protein UCP2 in mitochondria is thought to reduce the production of ATP and therefore reduces glucose-stimulated membrane depolarization and induced firing. In agreement with this hypothesis, inhibition of UCP2 in POMC neurons by genipin increases their glucose responsiveness (Parton et al. 2007). In beta cells, it is, however, still debated whether the effect of UCP2 on secretion is explained only by its effect on intracellular ATP levels or whether it acts as a regulator of reactive oxygen species (ROS) production (Pi et al. 2009; Pi et al. 2007; Zhang et al. 2001). Indeed, ROS are also intracellular signaling molecules (Rhee 2006) that can regulate the activity of voltage-gated K^+ channels (Archer et al. 2004; Pan et al. 2008) or Ca^{2+} influx (Kraft et al. 2004; Tabet et al. 2004; Todt et al. 2001). ROS may also be part of the mechanisms controlling glucose signaling in the hypothalamus. For instance,

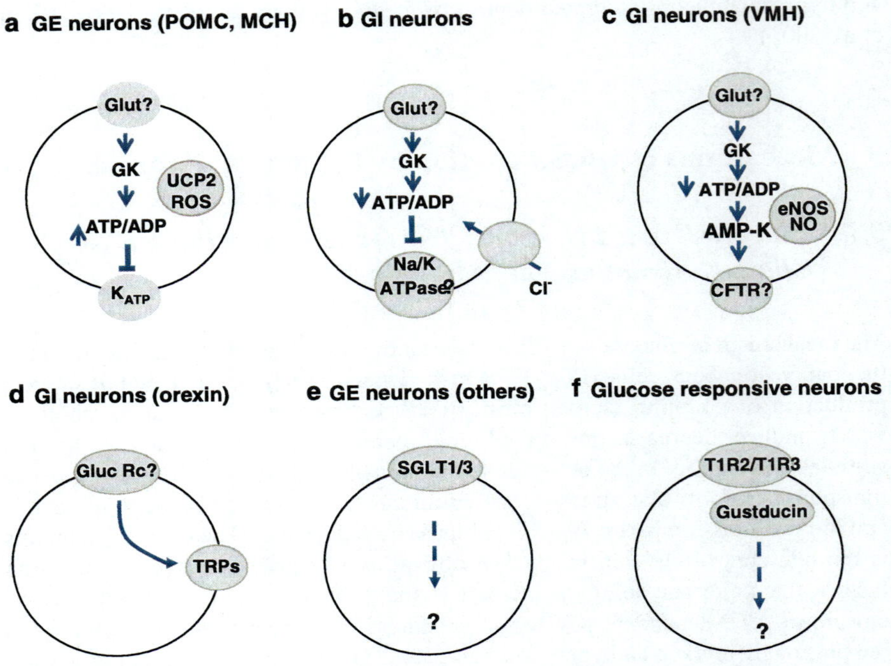

Fig. 1 Schematic representation of the glucose sensing mechanisms. (**a**) The classical model for GE neurons, found in POMC and MCH neurons, depends on glucose uptake and metabolism leading to increased ATP/ADP ratio and closure of K_{ATP} channels, and membrane depolarization induces influx of Ca^{2+} to induce firing activity. UCP2 and ROS can modulate this signaling pathway. (**b**) The initial description of glucose sensing by GI neurons suggested that decreases in intracellular ATP levels consequent to fall in extracellular glucose reduce the activity of the Na/K-ATPase. The resulting increase in intracellular Na^+ then closes a chloride conductance to induce nerve firing. (**c**) GI neurons of the VMH respond to hypoglycemia by activating AMPK, which can be further upregulated by an eNOS/NO/guanylate-cyclase-dependent mechanism; AMPK finally activates the chloride conductance of the CFTR. (**d**) The GI neurons of the LH orexin neurons are activated by low glucose in a glucose intake and metabolism-independent manner, possibly secondary to glucose interaction with a cell surface receptor that controls a K^+ conductance. (**e**) A large fraction of hypothalamic GE neurons, in dispersed neuronal populations, can be activated by the nonmetabolizable SGLT substrate α-MDG. This requires substrate and Na^+ uptake, which depolarizes the plasma membrane, and is independent of K_{ATP} channel activity. (**f**) The sweet receptors T1R3 and gustducin are present in neuronal populations. This receptor could contribute another glucose sensing mechanism

exposing hypothalamic slices to 20 mM glucose stimulates ROS generation. Also, intracarotid administration of antimycin or rotenone, which induces ROS formation, mimics the effect of glucose on activity of AN neurons and subsequent insulin release mediated by efferent neurons (Leloup et al. 2006).

A role for AMP-kinase has also been proposed for the regulation by glucose of POMC neuron activity. In mice with genetic inactivation of the α2 subunit of AMPK (and with only one allele of the α1 subunit), the POMC neurons no longer respond to extracellular glucose as assessed by electrophysiological recordings (Claret et al. 2007). How this kinase, which is activated only at low glucose concentration, can prevent the response to high glucose of these neurons is not clear.

In the lateral hypothalamus, the MCH neurons are also GE and probably share the same glucose signaling pathway as the POMC neurons. The same requirement for a functional K_{ATP} channel has been established, and knockout of UCP2 specifically in MCH neurons increases their glucose responsiveness (Kong et al. 2010). Genetic inhibition of the K_{ATP} channel also leads to glucose intolerance.

3.1.1 Other Glucose Sensing Mechanisms in GE Neurons

Variations from the Glut2/GK/K_{ATP} channel signaling pathway have been described in the stimulation by glucose of different GE neurons (Fig. 1). First, there is no evidence that Glut2 is expressed in POMC or MCH neurons, and the isoform of glucose transporter expressed by these neurons is not yet established, although genetic inactivation of Glut2 prevents the normal regulation of POMC expression in response to i.c.v. glucose or during the fast-to-refed transition. This suggests that the regulation by glucose of these neurons in physiological conditions may be indirect, through interaction with Glut2-expressing neurons (see discussion of Glut2 in central glucose sensing below). Second, there is evidence that glucose can excite neuronal activity through mechanisms that require glucose recognition by the Na^+-dependent glucose transporters SGLT1 or SGLT3. SGLT1 may play a role in central glucose sensing as suggested by the effect of i.c.v. injection of phlorizin, a specific inhibitor, which enhances food intake in rats (Tsujii and Bray 1990) and inhibits activation of GE neurons in the VMH (Yang et al. 1999). Strikingly, analysis of isolated hypothalamic neurons shows that a majority of GE neurons can be activated by α-MDG, a specific SGLT1 substrate. Furthermore, tolbutamide cannot increase the activity of these α-MDG-sensitive neurons, indicating that the K_{ATP} channel may not be involved in this signaling pathway (O'Malley et al. 2006).

SGLT3 is a member of the SGLT family, which has been reported to be a glucose sensor in cholinergic neurons present in the small intestine and at the neuromuscular junctions (Diez-Sampedro et al. 2003). In *Xenopus oocytes* expressing SGLT3, glucose produces a phlorizin-sensitive inward current that depolarizes the membrane potential by up to 50 mV (Diez-Sampedro et al. 2003). As SGLT3 mRNA is expressed in both cultured hypothalamic neurons and adult hypothalamus, this suggests that it may also be involved in central glucose sensing (O'Malley et al. 2006).

The G-protein-coupled taste receptors of the T1R family form heterodimers for sensing sweet taste (T1R2; T1R3) or amino acids (umami taste) (T1R1; T1R3). The sweet receptors are activated by a large number of artificial sweeteners but also by sucrose and glucose. These receptors are localized in the taste buds of the tongue, in the intestine where they may control secretion of the gluco-incretin hormone GLP-1

(Jang et al. 2007; Steinert et al. 2011), and in diverse brain areas, but in particular, in the hypothalamic PVN and AN (Ren et al. 2009). They are also found in the brainstem, in the NTS (Lemon and Margolskee 2009). Whether these receptors participate in the regulation of glucose homeostasis or of feeding behavior is not yet established.

3.2 Glucose-Inhibited Neurons

Glucose-inhibited neurons increase their firing activity when glycemic levels decrease. Several models have been proposed to account for the induction of membrane depolarization induced by hypoglycemia (Fig. 1). A first model proposed that a decrease in glucose uptake reduces ATP production, leading to a lower activity of the Na^+/K^+ ATPase and an increase in intracellular Na^+ that drives membrane depolarization through activation of a chloride conductance (Silver and Erecinska 1998). In recent years, other models have been suggested, with different mechanisms being proposed for GI neurons in the VMH and orexin neurons in the LH.

In VMH neurons, hypoglycemia induces firing by a glucose-metabolism-dependent signaling pathway. The glucose transporter involved in glucose uptake may be Glut1, Glut2, or Glut3, as different subpopulations of GI neurons express these transporters, as assessed by single-cell RT-PCR analysis (Kang et al. 2004). There is also evidence that glucose sensing by VMH neurons requires glucokinase expression (Kang et al. 2006). The following steps leading to neuronal firing in these neurons have been proposed by the group of V. Routh: A reduction in glucose metabolism leads to an increased intracellular AMP concentration. This activates AMPK which in turn triggers production of NO by eNOS. The activation of guanylate cyclase by NO further activates AMPK. The critical part is the subsequent regulation of the chloride conductance of the CFTR by AMPK which induces neuronal firing (Canabal et al. 2007; Fioramonti et al. 2010; Murphy et al. 2009).

Orexin neurons from the LH have been proposed by the group of Burdakov to function in a very different manner (Karnani and Burdakov 2011). Most strikingly, data published by this group indicate that the activation of these neurons can be triggered by the nonmetabolizable analogue 2-DG, that lactate cannot reproduce the glucose response, and glucokinase inhibitors did not prevent glucose activation (Gonzalez et al. 2008) in agreement with the reported absence of this enzyme from orexin neurons (Dunn-Meynell et al. 2002). This led to the suggestion that glucose activates a surface receptor that leads to regulation of channel activity. This activity was originally proposed as being controlled by tandem pore K^+ channels (TRPs) (Burdakov et al. 2006), but recent studies on TRP knockout mice failed to directly support this hypothesis (Gonzalez et al. 2009). Interestingly, these authors also showed that orexin GI neurons are sensitive to changes in ambient glucose concentrations rather than to absolute glycemic levels.

At the level of the brainstem, where both GE and GI neurons are detected, electrophysiological recordings indicate that GI neurons are activated in response to glucose removal by a signaling pathway that requires the presence of glucokinase and the regulation of a K^+ current (Balfour et al. 2006; Balfour and Trapp 2007).

In the arcuate nucleus, inactivation of AMPK is part of the response to leptin and insulin, whereas hypoglycemia or 2-DG activates AMPK. The activation by low glucose or neuroglucopenia of AMPK is observed only in the AN and PVN but not in the VMH, DMH, and LH nuclei (Minokoshi et al. 2004). Adenoviral delivery of constitutively active or dominant negative forms of AMPK in medial hypothalamic nuclei activates or, respectively, inhibits feeding (Minokoshi et al. 2004). How AMPK activity in hypothalamic neurons controls feeding is not fully understood. In neuronal cell lines and on ex vivo hypothalamic explants, low glucose concentrations and AICAR increase AMPK activity and AgRP expression (Lee et al. 2005). In accordance with these observations, the specific deletion of the α2-subunit of AMPK in POMC and AgRP neurons suppressed glucose sensing by these cells but preserved normal leptin or insulin action (Claret et al. 2007).

Together, the above-described data indicate that during evolution, the brain has developed several mechanisms for sensing hypoglycemia, either to induce counterregulatory hormone secretion or to induce a feeding response. This variety of mechanisms may be explained by the almost exclusive dependence of the brain on glucose as a source of metabolic energy. Fall of glucose below the normoglycemic concentrations dose-dependently impairs brain function, possibly leading to coma and death. Therefore, the multiplicity of mechanisms involved may reflect an adaptive process to ensure constant, optimal brain function and to maximize the chances of survival.

4 Indirect Control of Neuronal Activity by Glucose: Glial Cells and Tanycytes

4.1 Glial Cells

The utilization of glucose by neurons has been proposed to be mostly secondary to its initial uptake and metabolism by astrocytes that first produce lactate. Lactate is then transferred to neurons via specific monocarboxylate transporters, MCT1 present in astrocytes and MCT2 present in neurons, and utilized by neurons for ATP production (Magistretti et al. 1999; Pellerin et al. 2007). This metabolic coupling between astrocytes and neurons may also be used in some glucose sensing and glucoregulatory functions.

For instance, it has been shown that methyl sulfoximide, an astrocyte-specific inhibitor of glycolysis, blocks the increase in c-fos labeling in the AN induced by intracarotid or brainstem 2-DG injections (Guillod-Maximin et al. 2004; Young et al. 2000). It was also hypothesized that the release of lactate from neighboring glial cells is involved in glucose response of hypothalamic neurons (Ainscow et al. 2002; Lam et al. 2005). In the brainstem, the involvement of astrocyte-derived lactate in the control of glucose-sensitive neurons in the AP and NTS has been demonstrated by c-fos labeling studies, when monocarboxylate transporter is inhibited by α-cyano-4-hydroxycinnamate injected in the fourth ventricle. This treatment leads to elevations in blood glucose concentrations (Briski and Patil 2005; Patil and Briski 2005a, b).

4.2 *Tanycytes*

Tanycytes are glial cells lining the lateral lower part and the floor of the third ventricle. Their apical pole faces the ventricular lumen. They also have extended basal processes that reach regions of the median eminence devoid of blood–brain barrier and sometimes are in direct contact with microvessels present in the median eminence. These processes form extended contact with AN neurons, in particular NPY neurons (Akmayev and Fidelina 1974; Flament-Durand and Brion 1985; Kozlowski and Coates 1985). These cells express the glucose transporter Glut2 and glucokinase (Garcia Mde et al. 2003; Millan et al. 2010). Because of their strategic location, contacting both the cerebrospinal fluid and the general circulation, and because they express genes involved in glucose sensing, they may have a role in glucoregulation. A functional link between these cells and NPY neurons has been shown to rely on tanycytes expressing deiodinase II, and converting T4 into T3, thereby modulating glucose sensing in NPY cells by inducing UCP2 expression (Coppola et al. 2007). More studies are clearly needed to assess the potential role of tanycytes on glucoregulation, but the available information clearly suggests a potentially important function.

5 Glut2-Expressing Cells in Central Glucose Sensing

5.1 *Glut2-Expressing Cells in the Brain*

The glucose transporter Glut2 catalyzes the first step in the Glut2/GK/K_{ATP} signaling pathway that controls insulin secretion from beta cells. Glut2 is expressed in the mouse brain, in neurons, astrocytes, tanycytes, and endothelial cells (Arluison et al. 2004a; Arluison et al. 2004b; Marty et al. 2007a). However, because of the low level of Glut2 expression in the brain, its immunocytochemical distribution is relatively difficult to establish. As a result, there is no solid information about a colocalization of Glut2 with well-characterized GE or GI neurons of the melanocortin pathway or of other brain structures. In fact, the available evidence points to Glut2 not being present in NPY, POMC, orexin, or MCH neurons

(Mounien et al. 2010). By quantitative RT-PCR analysis, Glut2 expression has been found to be relatively low in the rat AN, VMH, PVN, and LH and at somewhat higher levels in brainstem nuclei, in particular nucleus 12 and inferior olive; it is also present in the AP and NTS (Li et al. 2003). In the VMH, single-cell RT-PCR analysis revealed expression of Glut2 in approximately one third of the GE, GI, and of non-glucose-sensitive neurons (Kang et al. 2004). Very good evidence demonstrates the expression of Glut2 in tanycytes and ependymal cells (Garcia Mde et al. 2003), as discussed above. In human brain, Glut2 is expressed at highest level in the hypothalamus and brainstem, where it is often colocalized with glucokinase (Roncero et al. 2004). Interestingly, in trout, Glut2 is expressed not only in the insulin secreting cells of the Brockmann body (which contain the insulin secreting cells) but also in the hypothalamus and hindbrain (Polakof et al. 2007). In the zebrafish, it is also present in the brain, although the exact localization has not yet been established (Castillo et al. 2009).

Collectively, the above-described information indicates that Glut2 is present in brain regions involved in glucoregulation, but not in clear association with the principal neurons of the melanocortin pathway, and that it is only present in a small subset of neurons in the VMH. In the brainstem, it is not possible to establish expression of Glut2, since glucose sensing cells in the DVC and the BLM cannot be identified by histological markers.

In an attempt to identify the Glut2-expressing cells, Mounien et al. (Mounien et al. 2010) generated transgenic mice expressing the Cre recombinase under the control of the Glut2 gene promoter (query: change correct? see changes). These transgenic mice were then crossed with Rosa26eYFP mice, and expression of the fluorescent reporter gene was used to identify sites of Glut2 expression. Expression of eYFP was found only in neurons. In the hypothalamus, the highest concentrations of eYFP cells were detected in the LH and the zona incerta; it was present in a few cells in the VMH, and no positive cells were detected in the AN. In this nucleus, however, numerous nerve endings were found associated with NPY and POMC neurons, suggesting synaptic contacts with Glut2-positive neurons located outside of the AN. In the brainstem, eYFP positive neurons were found in the NTS, the DMNX, the parasolitary tract, and in the A1/C1 region of the BLM. These eYFP neurons are glucose sensitive as demonstrated by their costaining with c-fos following i.p. glucose or 2-DG injection. In fact, at the brainstem level, the BLM eYFP neurons were activated following glucose but not 2-DG injections. In contrast, the eYFP neurons of the NTS and DMNX were activated by 2-DG but not glucose injections, suggesting that these are GI neurons. In LH, a similar fraction of neurons were activated by glucose or by 2-DG, suggesting that eYFP neurons in this structure are either GE or GI.

5.2 *Evidence for Glut2 in Central Glucose Sensing*

Studies of genetic inactivation of Glut2 in mice (with transgenic expression of glucose transporter in their beta cell to normalize glucose-stimulated secretion), were analyzed to assess the role of brain glucose sensing in the control of counterregulation, feeding, and thermoregulation (reviewed in (Marty et al. 2007b; Thorens 2003). The critical findings can be summarized as follows. In this mouse model, plasma glucagon levels were elevated in the fed state but could be normalized by ganglionic blockers, indicating that in the absence of Glut2, there was an abnormally high autonomic tone to the alpha cells stimulating glucagon secretion (Burcelin and Thorens 2001). In complementation experiments, transgenic reexpression of Glut2 in glial cells, but not in neurons, of the Glut2-null mice restored hypoglycemia-induced glucagon secretion. This was associated with a restoration of c-fos labeling in the dorsal vagal complex following i.p. 2-DG injections (Marty et al. 2005). This suggests that astrocyte–neuron coupling is required for normal hypoglycemia detection and counterregulatory response. In these experiments, however, c-fos labeling in the VMN induced by 2-DG injection was similar in the presence and absence of Glut2, suggesting that this transporter is not involved in neuroglucopenia activation of VMN neurons.

Absence of Glut2 was also associated with a defect in refeeding following a fast, and with hyperphagia in ad libitum–fed mice. These mutant mice also failed to respond to i.p. or i.c.v. injections of 2-DG (which normally stimulates feeding) or of glucose (which normally reduces feeding). This was further associated with a loss of regulated expression of NPY and POMC in the AN during the fast-to-refed transition, or following i.c.v. injections of glucose (Bady et al. 2006). A defect in thermogenesis was also described, with an impaired capacity of the Glut2-null mice to maintain their body temperature when exposed to 4°C, and their spontaneous entry into torpor when fasted overnight (Mounien et al. 2010). This was secondary to reduced activation of thermogenesis, as revealed by reduced UCP-1 and deiodinase II expression in the brown adipose tissue. Impaired activation of thermogenesis may be secondary to a defect in leptin action on AN neurons. Absence of Glut2 indeed led to a reduction in leptin signaling as assessed by phosphorylation of STAT3 in NPY and POMC neurons during the fast-to-refed transition or following i.p. injection of leptin.

Collectively, these results suggest that glucose sensing by Glut2-expressing cells is required for the normal sensitivity to leptin of NPY and POMC neurons. They also indicate that even though NPY and POMC neurons may be directly responsive to changes in glycemia, as assessed in hypothalamic slices, in physiological conditions their glucose responsiveness is also controlled by Glut2-expressing cells. These can be neighboring tanycytes or neurons located in other brain regions and which send projections to the AN. Finally, these data suggest that Glut2-expressing neurons may form a distinct class of GE and GI neurons that act as modulator of the more classical GE and GI neurons of the AN, LH, and VMH.

6 Conclusions

The studies reviewed here indicate a very high diversity in the mechanisms involved in detecting variations in blood glucose levels or glucose availability by the brain. The picture that is emerging is that there are multiple sites of glucose sensing located mostly in the hypothalamus and brainstem, regions involved in homeostatic regulation of feeding, energy expenditure, and glucose homeostasis. These regions are connected to peripheral sites of glucose sensing such as the gut and hepatoportal vein regions (Marty et al. 2007a), which monitor peripheral glycemic levels, and also to regions of the brain involved in control of feeding behavior and reward. It is puzzling to observe such diverse glucose sensing systems, and so far there is no real hypothesis for the importance of this diversity. It may be related to the fact that different GI neurons may be required for activation of counterregulatory hormone secretion, glucagon and catecholamines, at different hypoglycemic levels, in order to induce feeding or thermogenesis. These responses may be coordinated but still controlled differentially. Alternatively, different glucose sensing neurons may be recruited at different levels of hypoglycemia, in analogy to the activation of various TRP-expressing, temperature-sensitive neurons that are responsive to different temperature ranges (Voets et al. 2005).

Acknowledgements Work in the author laboratory has been supported by grants from the Swiss National Science Foundation No. 3100A0-113525 and by the National Competence Center in Research "Frontiers in Genetics."

References

Adachi A, Shimizu N, Oomura Y, Kobashi M (1984) Convergence of hepatoportal glucose-sensitive afferents signal to glucose sensitive units within the nucleus of the solitary tract. Neurosci Lett 46:215–218

Adamantidis A, de Lecea L (2008) Sleep and metabolism: shared circuits, new connections. Trends Endocrinol Metab 19:362–370

Ainscow EK, Mirshamsi S, Tang T, Ashford ML, Rutter GA (2002) Dynamic imaging of free cytosolic ATP concentration during fuel sensing by rat hypothalamic neurones: evidence for ATP-independent control of ATP-sensitive K(+) channels. J Physiol 544:429–445

Akmayev IG, Fidelina OV (1974) Morphological aspects of the hypothalamic-hypophyseal system. V. The tanycytes: their relation to the hypophyseal adrenocorticotrophic function. An enzyme-histochemical study. Cell Tissue Res 152:403–410

Anand BK, Chhina GS, Sharma KN, Dua S, Singh B (1964) Activity of single neurons in the hypothalamic feeding centers: effect of glucose. Am J Physiol 207:1146–1154

Archer SL, Wu XC, Thebaud B, Moudgil R, Hashimoto K, Michelakis ED (2004) O_2 sensing in the human ductus arteriosus: redox-sensitive K^+ channels are regulated by mitochondria-derived hydrogen peroxide. Biol Chem 385:205–216

Arluison M, Quignon M, Nguyen P, Thorens B, Leloup C, Penicaud L (2004a) Distribution and anatomical localization of the glucose transporter 2 (GLUT2) in the adult rat brain–an immunohistochemical study. J Chem Neuroanat 28:117–136

Arluison M, Quignon M, Thorens B, Leloup C, Penicaud L (2004b) Immunocytochemical localization of the glucose transporter 2 (GLUT2) in the adult rat brain. II. Electron microscopic study. J Chem Neuroanat 28:137–146

Bady I, Marty N, Dallaporta M, Emery M, Gyger J, Tarussio D, Foretz M, Thorens B (2006) Evidence from glut2-null mice that glucose is a critical physiological regulator of feeding. Diabetes 55:988–995

Balfour RH, Hansen AM, Trapp S (2006) Neuronal responses to transient hypoglycaemia in the dorsal vagal complex of the rat brainstem. J Physiol 570:469–484

Balfour RH, Trapp S (2007) Ionic currents underlying the response of rat dorsal vagal neurones to hypoglycaemia and chemical anoxia. J Physiol 579:691–702

Bernard C (1849) Chiens rendus diabétiques. CR Soc Biol 1:60

Berthoud H-R (2002) Multiple neural systems controlling food intake and body weight. Neurosci Biobehav Rev 26:393–428

Berthoud HR, Mogenson GJ (1977) Ingestive behavior after intracerebral and intracerebroventricular infusions of glucose and 2-deoxy-D-glucose. Am J Physiol 233:R127–R133

Biggers DW, Myers SR, Neal D, Stinson R, Cooper NB, Jaspan JB, Williams PE, Cherrington AD, Frizzell RT (1989) Role of brain in counterregulation of insulin-induced hypoglycemia in dogs. Diabetes 38:7–16

Borg MA, Sherwin RS, Borg WP, Tamborlane WV, Shulman GI (1997) Local ventromedial hypothalamus glucose perfusion blocks counterregulation during systemic hypoglycemia in awake rats. J Clin Invest 99:361–365

Borg WP, Sherwin RS, During MJ, Borg MA, Shulman GI (1995) Local ventromedial hypothalamus glucopenia triggers counterregulatory hormone release. Diabetes 44:180–184

Bray GA (1985) Autonomic and endocrine factors in the regulation of food intake. Brain Res Bull 14:505–510

Briski KP, Patil GD (2005) Induction of Fos immunoreactivity labeling in rat forebrain metabolic loci by caudal fourth ventricular infusion of the monocarboxylate transporter inhibitor, alpha-cyano-4-hydroxycinnamic acid. Neuroendocrinology 82:49–57

Burcelin R, Thorens B (2001) Evidence that extrapancreatic GLUT2-dependent glucose sensors control glucagon secretion. Diabetes 50:1282–1289

Burdakov D, Jensen LT, Alexopoulos H, Williams RH, Fearon IM, O'Kelly I, Gerasimenko O, Fugger L, Verkhratsky A (2006) Tandem-pore K^+ channels mediate inhibition of orexin neurons by glucose. Neuron 50:711–722

Canabal DD, Song Z, Potian JG, Beuve A, McArdle JJ, Routh VH (2007) Glucose, insulin, and leptin signaling pathways modulate nitric oxide synthesis in glucose-inhibited neurons in the ventromedial hypothalamus. Am J Physiol Regul Integr Comp Physiol 292:R1418–R1428

Canteras NS, Simerly RB, Swanson LW (1994) Organization of projections from the ventromedial nucleus of the hypothalamus: a *Phaseolus vulgaris*-leucoagglutinin study in the rat. J Comp Neurol 348:41–79

Castillo J, Crespo D, Capilla E, Diaz M, Chauvigne F, Cerda J, Planas JV (2009) Evolutionary structural and functional conservation of an ortholog of the GLUT2 glucose transporter gene (SLC2A2) in zebrafish. Am J Physiol Regul Integr Comp Physiol 297:R1570–R1581

Claret M, Smith MA, Batterham RL, Selman C, Choudhury AI, Fryer LG, Clements M, Al-Qassab H, Heffron H, Xu AW, Speakman JR, Barsh GS, Viollet B, Vaulont S, Ashford ML, Carling D, Withers DJ (2007) AMPK is essential for energy homeostasis regulation and glucose sensing by POMC and AgRP neurons. J Clin Invest 117:2325–2336

Coppola A, Liu ZW, Andrews ZB, Paradis E, Roy MC, Friedman JM, Ricquier D, Richard D, Horvath TL, Gao XB, Diano S (2007) A central thermogenic-like mechanism in feeding regulation: an interplay between arcuate nucleus T3 and UCP2. Cell Metab 5:21–33

Cummings DE, Schwartz MW (2000) Melanocortins and body weight: a tale of two receptors. Nat Genet 26:8–9

Dallaporta M, Himmi T, Perrin J, Orsini J-C (1999) A solitary tract nucleus sensitivity to moderate changes in glucose level. Neuroreport 10:1–4

Dhillon H, Zigman JM, Ye C, Lee CE, McGovern RA, Tang V, Kenny CD, Christiansen LM, White RD, Edelstein EA, Coppari R, Balthasar N, Cowley MA, Chua S Jr, Elmquist JK,

Lowell BB (2006) Leptin directly activates SF1 neurons in the VMH, and this action by leptin is required for normal body-weight homeostasis. Neuron 49:191–203
Diez-Sampedro A, Hirayama BA, Osswald C, Gorboulev V, Baumgarten K, Volk C, Wright EM, Koepsell H (2003) A glucose sensor hiding in a family of transporters. Proc Natl Acad Sci USA 100:11753–11758
Dunn-Meynell AA, Rawson NE, Levin BE (1998) Distribution and phenotype of neurons containing the ATP-sensitive K^+ channel in rat brain. Brain Res 814:41–54
Dunn-Meynell AA, Routh VH, Kang L, Gaspers L, Levin BE (2002) Glucokinase is the likely mediator of glucosensing in both glucose-excited and glucose-inhibited central neurons. Diabetes 51:2056–2065
Fioramonti X, Lorsignol A, Taupignon A, Penicaud L (2004) A new ATP-sensitive K^+ channel-independent mechanism is involved in glucose-excited neurons of mouse arcuate nucleus. Diabetes 53:2767–2775
Fioramonti X, Marsollier N, Song Z, Fakira KA, Patel RM, Brown S, Duparc T, Pica-Mendez A, Sanders NM, Knauf C, Valet P, McCrimmon RJ, Beuve A, Magnan C, Routh VH (2010) Ventromedial hypothalamic nitric oxide production is necessary for hypoglycemia detection and counterregulation. Diabetes 59:519–528
Flament-Durand J, Brion JP (1985) Tanycytes: morphology and functions: a review. Int Rev Cytol 96:121–155
Fraley GS, Ritter S (2003) Immunolesion of norepinephrine and epinephrine afferents to medial hypothalamus alters basal and 2-deoxy-D-glucose-induced neuropeptide Y and agouti-gene-related protein messenger ribonucleic acid expression in the arcuate nucleus. Endocrinology 411:75–83
Freinkel N, Metzger BE, Harris E, Robinson S, Mager M (1972) The hypothermia of hypoglycemia. Studies with 2-deoxy-D-glucose in normal human subjects and mice. N Engl J Med 287:841–845
Frizzell RT, Jones EM, Davis SN, Biggers DW, Myers SR, Connolly CC, Neal DW, Jaspan JB, Cherrington AD (1993) Counterregulation during hypoglycemia is directed by widespread brain regions. Diabetes 42:1253–1261
Gale SM, Castracane VD, Mantzoros CS (2004) Energy homeostasis, obesity and eating disorders: recent advances in endocrinology. J Nutr 134:295–298
Garcia Mde L, Millan C, Balmaceda-Aguilera C, Castro T, Pastor P, Montecinos H, Reinicke K, Zuniga F, Vera JC, Onate SA, Nualart F (2003) Hypothalamic ependymal-glial cells express the glucose transporter GLUT2, a protein involved in glucose sensing. J Neurochem 86:709–724
Gautron L, Elmquist JK (2011) Sixteen years and counting: an update on leptin in energy balance. J Clin Invest 121:2087–2093
Gonzalez JA, Jensen LT, Doyle SE, Miranda-Anaya M, Menaker M, Fugger L, Bayliss DA, Burdakov D (2009) Deletion of TASK1 and TASK3 channels disrupts intrinsic excitability but does not abolish glucose or pH responses of orexin/hypocretin neurons. Eur J Neurosci 30:57–64
Gonzalez JA, Jensen LT, Fugger L, Burdakov D (2008) Metabolism-independent sugar sensing in central orexin neurons. Diabetes 57:2569–2576
Guillod-Maximin E, Lorsignol A, Alquier T, Penicaud L (2004) Acute intracarotid glucose injection towards the brain induces specific c-fos activation in hypothalamic nuclei: involvement of astrocytes in cerebral glucose-sensing in rats. J Neuroendocrinol 16:464–471
Hoebel BG (1965) Hypothalamic lesions by electrocauterization: disinhibition of feeding and self-stimulation. Science 149:452–453
Ibrahim N, Bosch MA, Smart JL, Qiu J, Rubinstein M, Ronnekleiv OK, Low MJ, Kelly MJ (2003) Hypothalamic proopiomelanocortin neurons are glucose responsive and express K(ATP) channels. Endocrinology 144:1331–1340
Jang HJ, Kokrashvili Z, Theodorakis MJ, Carlson OD, Kim BJ, Zhou J, Kim HH, Xu X, Chan SL, Juhaszova M, Bernier M, Mosinger B, Margolskee RF, Egan JM (2007) Gut-expressed

gustducin and taste receptors regulate secretion of glucagon-like peptide-1. Proc Natl Acad Sci USA 104:15069–15074
Kang L, Dunn-Meynell AA, Routh VH, Gaspers LD, Nagata Y, Nishimura T, Eiki J, Zhang BB, Levin BE (2006) Glucokinase is a critical regulator of ventromedial hypothalamic neuronal glucosensing. Diabetes 55:412–420
Kang L, Routh VH, Kuzhikandathil EV, Gaspers LD, Levin BE (2004) Physiological and molecular characteristics of rat hypothalamic ventromedial nucleus glucosensing neurons. Diabetes 53:549–559
Karnani M, Burdakov D (2011) Multiple hypothalamic circuits sense and regulate glucose levels. Am J Physiol Regul Integr Comp Physiol 300:R47–R55
King BM (2006) The rise, fall, and resurrection of the ventromedial hypothalamus in the regulation of feeding behavior and body weight. Physiol Behav 87:221–244
Kong D, Vong L, Parton LE, Ye C, Tong Q, Hu X, Choi B, Bruning JC, Lowell BB (2010) Glucose stimulation of hypothalamic MCH neurons involves K(ATP) channels, is modulated by UCP2, and regulates peripheral glucose homeostasis. Cell Metab 12:545–552
Kozlowski GP, Coates PW (1985) Ependymoneuronal specializations between LHRH fibers and cells of the cerebroventricular system. Cell Tissue Res 242:301–311
Kraft R, Grimm C, Grosse K, Hoffmann A, Sauerbruch S, Kettenmann H, Schultz G, Harteneck C (2004) Hydrogen peroxide and ADP-ribose induce TRPM2-mediated calcium influx and cation currents in microglia. Am J Physiol Cell Physiol 286:C129–C137
Lam TK, Gutierrez-Juarez R, Pocai A, Rossetti L (2005) Regulation of blood glucose by hypothalamic pyruvate metabolism. Science 309:943–947
Lee K, Li B, Xi X, Suh Y, Martin RJ (2005) Role of neuronal energy status in the regulation of adenosine 5′-monophosphate-activated protein kinase, orexigenic neuropeptides expression, and feeding behavior. Endocrinology 146:3–10
Leloup C, Magnan C, Benani A, Bonnet E, Alquier T, Offer G, Carriere A, Periquet A, Fernandez Y, Ktorza A, Casteilla L, Penicaud L (2006) Mitochondrial reactive oxygen species are required for hypothalamic glucose sensing. Diabetes 55:2084–2090
Lemon CH, Margolskee RF (2009) Contribution of the T1r3 taste receptor to the response properties of central gustatory neurons. J Neurophysiol 101:2459–2471
Li B, Xi X, Roane DS, Ryan DH, Martin RJ (2003) Distribution of glucokinase, glucose transporter GLUT2, sulfonylurea receptor-1, glucagon-like peptide-1 receptor and neuropeptide Y messenger RNAs in rat brain by quantitative real time RT-PCR. Brain Res Mol Brain Res 113:139–142
Magistretti PJ, Pellerin L, Rothman DL, Shulman RG (1999) Energy on demand. Science 283:496–497
Marshall NB, Mayer J (1956) Specificity of gold thioglucose for ventromedial hypothalamic lesions and hyperphagia. Nature 178:1399–1400
Marty N, Dallaporta M, Foretz M, Emery M, Tarussio D, Bady I, Binnert C, Beermann F, Thorens B (2005) Regulation of glucagon secretion by glucose transporter type 2 (glut2) and astrocyte-dependent glucose sensors. J Clin Invest 115:3545–3553
Marty N, Dallaporta M, Thorens B (2007a) Brain glucose sensing, counterregulation and feeding behavior. Physiology (Bethesda) 22:241–251
Marty N, Dallaporta M, Thorens B (2007b) Brain glucose sensing, counterregulation, and energy homeostasis. Physiology (Bethesda) 22:241–251
Mayer J (1953) Glucostatic mechanism of regulation of food intake. N Engl J Med 249:13–16
Mayer J, Thomas DW (1967) Regulation of food intake and obesity. Science 156:328–337
Millan C, Martinez F, Cortes-Campos C, Lizama I, Yanez MJ, Llanos P, Reinicke K, Rodriguez F, Peruzzo B, Nualart F, Garcia MA (2010) Glial glucokinase expression in adult and post-natal development of the hypothalamic region. ASN Neuro 2:e00035
Minokoshi Y, Alquier T, Furukawa N, Kim YB, Lee A, Xue B, Mu J, Foufelle F, Ferre P, Birnbaum MJ, Stuck BJ, Kahn BB (2004) AMP-kinase regulates food intake by responding to hormonal and nutrient signals in the hypothalamus. Nature 428:569–574
Miselis RR, Epstein AN (1975) Feeding induced by intracerebroventricular 2-deoxy-D-glucose in the rat. Am J Physiol 229:1438–1447

Mizuno Y, Oomura Y (1984) Glucose responding neurons in the nucleus tractus solitarius of the rat: in vitro studies. Brain Res 307:109–116
Mounien L, Marty N, Tarussio D, Metref S, Genoux D, Preitner F, Foretz M, Thorens B (2010) Glut2-dependent glucose-sensing controls thermoregulation by enhancing the leptin sensitivity of NPY and POMC neurons. FASEB J 24:1747–1758
Murphy BA, Fakira KA, Song Z, Beuve A, Routh VH (2009) AMP-activated protein kinase and nitric oxide regulate the glucose sensitivity of ventromedial hypothalamic glucose-inhibited neurons. Am J Physiol Cell Physiol 297:C750–C758
O'Malley D, Reimann F, Simpson AK, Gribble FM (2006) Sodium-coupled glucose cotransporters contribute to hypothalamic glucose sensing. Diabetes 55:3381–3386
Oomura Y, Yoshimatsu H (1984) Neural network of glucose monitoring system. J Auton Nerv Syst 10:359–372
Pan Y, Weng J, Cao Y, Bhosle RC, Zhou M (2008) Functional coupling between the Kv1.1 channel and aldoketoreductase Kvbeta1. J Biol Chem 283:8634–8642
Parton LE, Ye CP, Coppari R, Enriori PJ, Choi B, Zhang CY, Xu C, Vianna CR, Balthasar N, Lee CE, Elmquist JK, Cowley MA, Lowell BB (2007) Glucose sensing by POMC neurons regulates glucose homeostasis and is impaired in obesity. Nature 449:228–232
Patil GD, Briski KP (2005a) Lactate is a critical "sensed" variable in caudal hindbrain monitoring of CNS metabolic stasis. Am J Physiol Regul Integr Comp Physiol 289:R1777–R1786
Patil GD, Briski KP (2005b) Transcriptional activation of nucleus tractus solitarii/area postrema catecholaminergic neurons by pharmacological inhibition of caudal hindbrain monocarboxylate transporter function. Neuroendocrinology 81:96–102
Pellerin L, Bouzier-Sore AK, Aubert A, Serres S, Merle M, Costalat R, Magistretti PJ (2007) Activity-dependent regulation of energy metabolism by astrocytes: an update. Glia 55:1251–1262
Penicaud L, Leloup C, Fioramonti X, Lorsignol A, Benani A (2006) Brain glucose sensing: a subtle mechanism. Curr Opin Clin Nutr Metab Care 9:458–462
Pi J, Bai Y, Daniel KW, Liu D, Lyght O, Edelstein D, Brownlee M, Corkey BE, Collins S (2009) Persistent oxidative stress due to absence of uncoupling protein 2 associated with impaired pancreatic beta-cell function. Endocrinology 150:3040–3048
Pi J, Bai Y, Zhang Q, Wong V, Floering LM, Daniel K, Reece JM, Deeney JT, Andersen ME, Corkey BE, Collins S (2007) Reactive oxygen species as a signal in glucose-stimulated insulin secretion. Diabetes 56:1783–1791
Polakof S, Miguez JM, Moon TW, Soengas JL (2007) Evidence for the presence of a glucosensor in hypothalamus, hindbrain, and Brockmann bodies of rainbow trout. Am J Physiol Regul Integr Comp Physiol 292:R1657–R1666
Ren X, Zhou L, Terwilliger R, Newton SS, de Araujo IE (2009) Sweet taste signaling functions as a hypothalamic glucose sensor. Front Integr Neurosci 3:12
Rhee SG (2006) Cell signaling. H_2O_2, a necessary evil for cell signaling. Science 312:1882–1883
Ritter RC, Slusser PG, Stone S (1981) Glucoreceptors controlling feeding and blood glucose: location in the hindbrain. Science 213:451–453
Ritter S, Bugarith K, Dinh TT (2001) Immunotoxic destruction of distinct catecholamine subgroups produces selective impairment of glucoregulatory responses and neuronal activation. J Comp Neurol 432:197–216
Ritter S, Dinh TT, Zhang Y (2000) Localization of hindbrain glucoreceptive sites controlling food intake and blood glucose. Brain Res 856:37–47
Roncero I, Alvarez E, Chowen JA, Sanz C, Rabano A, Vazquez P, Blazquez E (2004) Expression of glucose transporter isoform GLUT-2 and glucokinase genes in human brain. J Neurochem 88:1203–1210
Routh VH (2002) Glucose-sensing neurons: are they physiologically relevant ? Physiol Behav 76:403–413
Sakurai T (2007) The neural circuit of orexin (hypocretin): maintaining sleep and wakefulness. Nat Rev Neurosci 8:171–181

Schwartz GJ (2000) The role of gastrointestinal vagal afferents in the control of food intake: current prospects. Nutrition 16:866–873

Schwartz MW, Woods SC, Porte D, Seeley RJ, Baskin DG (2000) Central nervous system control of food intake. Nature 404:661–671

Silver IA, Erecinska M (1998) Glucose-induced intracellular ion changes in sugar-sensitive hypothalamic neurons. J Neurophysiol 79:1733–1745

Steinert RE, Gerspach AC, Gutmann H, Asarian L, Drewe J, Beglinger C (2011) The functional involvement of gut-expressed sweet taste receptors in glucose-stimulated secretion of glucagon-like peptide-1 (GLP-1) and peptide YY (PYY). Clin Nutr 30:524–532

Tabet F, Savoia C, Schiffrin EL, Touyz RM (2004) Differential calcium regulation by hydrogen peroxide and superoxide in vascular smooth muscle cells from spontaneously hypertensive rats. J Cardiovasc Pharmacol 44:200–208

Thorens B (2003) A gene knockout approach in mice to identify glucose sensors controlling glucose homeostasis. Pflügers Arch – Eur J Physiol 445:482–490

Thorens B, Larsen PJ (2004) Gut-derived signaling molecules and vagal afferents in the control of glucose and energy homeostasis. Curr Opin Clin Nutr Metab Care 7:471–478

Todt I, Ngezahayo A, Ernst A, Kolb HA (2001) Hydrogen peroxide inhibits gap junctional coupling and modulates intracellular free calcium in cochlear Hensen cells. J Membr Biol 181:107–114

Tong Q, Ye C, McCrimmon RJ, Dhillon H, Choi B, Kramer MD, Yu J, Yang Z, Christiansen LM, Lee CE, Choi CS, Zigman JM, Shulman GI, Sherwin RS, Elmquist JK, Lowell BB (2007) Synaptic glutamate release by ventromedial hypothalamic neurons is part of the neurocircuitry that prevents hypoglycemia. Cell Metab 5:383–393

Tsujii S, Bray GA (1990) Effects of glucose, 2-deoxyglucose, phlorizin, and insulin on food intake of lean and fatty rats. Am J Physiol 258:E476–E481

Voets T, Talavera K, Owsianik G, Nilius B (2005) Sensing with TRP channels. Nat Chem Biol 1:85–92

Wang R, Liu X, Hentges ST, Dunn-Meynell AA, Levin BE, Wang W, Routh VH (2004) The regulation of glucose-excited neurons in the hypothalamic arcuate nucleus by glucose and feeding-relevant peptides. Diabetes 53:1959–1965

Woods SC, Seeley RJ, Porte D, Schwartz MW (1998) Signals that regulate food intake and energy homeostasis. Science 280:1378–1383

Yang X-J, Kow L-M, Funabashi T, Mobbs CV (1999) Hypothalamic glucose sensor. Similarities to and differences from pancreatic b cell mechanisms. Diabetes 48:1763–1772

Yang XJ, Low LM, Pfaff DW, Mobbs CV (2004) Metabolic pathways that mediate inhibition of hypothalamic neurons by glucose. Diabetes 53:67–73

Yettefti K, Orsini J-C, Perrin J (1997) Characteristics of glycemia-sensitive neurons in the nucleus tractus solitarii: possible involvement in nutritional regulation. Physiol Behav 61:93–100

Yettefti K, Orsini JC, El Ouazzani T, Himmi T, Boyer A, Perrin J (1995) Sensitivity of nucleus tractus solitarius neurons to induced moderate hyperglycemia, with special reference to catecholaminergic regions. J Auton Nerv Syst 51:191–197

Young JK, Baker JH, Montes MI (2000) The brain response to 2-deoxy glucose is blocked by a glial drug. Pharmacol Biochem Behav 67:233–239

Zhang C-Y, Baffy G, Perret P, Krauss S, Peroni O, Grujic D, Hagen T, Vidal-Puig AJ, Boss O, Kim Y-B, Zheng XX, Wheeler MB, Shulman GI, Chan CB, Lowell BB (2001) Uncoupling protein-2 negatively regulates insulin secretion and is a major link between obesity, b cell dysfunction, and type 2 diabetes. Cell 105:745–755

Role of CD36 in Oral and Postoral Sensing of Lipids

M. Chevrot, C. Martin, P. Passilly-Degrace, and P. Besnard

Contents

Abstract Obesity and associated plethora of diseases constitute a major public health challenge worldwide. The conjunction of profound changes in our lifestyle and a thrifty genetic that evolved in an environment of food scarcity largely explains this epidemic situation. Food abundance promotes our specific appetite for the more palatable food generally rich in lipids. It is noteworthy that this attraction for fatty food is not specific to humans. Rats and mice also spontaneously prefer lipid-rich food in a free-choice situation. Detection of lipids in food requires the presence of specific sensors located in strategic places (e.g., oral cavity, small intestine, brain) whose activation results in a modulation of the eating behavior. Recent data strongly suggest that the glycoprotein CD36 plays a significant role in this sensing system.

Keywords Central nervous system • Dietary lipids • Eating behavior • Endocannabinoids • Health • Lipid receptors • Obesity risk • Sense of taste • Small intestine

---

M. Chevrot • C. Martin • P. Passilly-Degrace • P. Besnard (✉)
Physiologie de la Nutrition, UMR U866 INSERM/Université de Bourgogne, AgroSup Dijon, 1, Esplanade Erasme, 21000 Dijon, France
e-mail: pbesnard@u-bourgogne.fr

H.-G. Joost (ed.), *Appetite Control*, Handbook of Experimental Pharmacology 209,
DOI 10.1007/978-3-642-24716-3_13,

Abbreviations

BBB	Blood–brain barrier
BMI	Body mass index
CART	Cocaine-amphetamine-related peptide
CCK	Cholecystokinin
FATP	Fatty acid transport proteins
GLP-1	Glucagon-like peptide-1
GPCR	G protein-coupled receptors
HBMEC	Human brain microvessel endothelial cells
LCFA	Long-chain fatty acid
LPL	Lipoprotein lipase
MAPK	Mitogen-activated protein kinase
MTP	Microsomal triglyceride transfer protein
NPY	Neuropeptide Y
NST	Nucleus of solitary tract
OEA	Oleoylethanolamine
PTK	Protein tyrosine kinase
TBC	Taste bud cells
TG	Triglyceride
T1R	Taste 1 receptors
SSO	Sulfo-N-succinimidyl oleate ester

1 Introduction

Dietary fat consists mainly of long-chain triglycerides (TG). Because they cannot cross membranes of cells, TG must be hydrolyzed before utilization. This process takes place during digestion which releases long-chain fatty acids (LCFA, number of carbons $\geq$ 16) from TG. LCFA exert basic functions in the cell mainly as membrane components, metabolic fuel, and precursors of lipid mediators. They also modulate the expression of genes involved in the regulation of energy balance through the binding and activation of nuclear receptors (e.g., PPAR). In addition to this intracellular lipid sensing, the existence of plasma membrane lipid receptors responsible for the extracellular detection of LCFA was recently reported. Located in strategic places of the body (e.g., oral cavity, small intestine, central nervous system), these lipid sensors participate in the control of different sequences of the feeding behavior from the choice of food to eat to the satiety by providing real-time information about the lipid content of the diet and their subsequent metabolic use. The progressive deciphering of this oral and postoral lipid-sensing system is of a great interest since it might open new insights to limit the consumption of fat-rich food and decrease risk of obesity and associated diseases (non-insulin-dependent diabetes, atherosclerosis, hypertension, cancers). A growing body of evidence suggests that the plasma membrane lipid-binding protein CD36 (cluster of differentiation 36), which has been specifically identified in the gustatory papillae, enterocytes, and hypothalamic neurons, appears to be a good candidate for this function. The purpose of this minireview is to summarize and discuss the recent data in this new field of investigations.

2 Oral Sensing of Lipids

The first step of dietary fat detection takes place in the oral cavity. For a long time, it was thought that only textural and olfactory cues were responsible for the oro-sensory perception of lipids. Recent compelling evidences support that sense of taste also plays a role in this detection system in rodents (rats and mice) (Takeda et al. 2000, 2001; Fukuwatari et al. 2003) and probably also in humans (Chale-Rush et al. 2007). Oro-sensory perception of dietary lipids is dependent from LCFA (Tsuruta et al. 1999; Fukuwatari et al. 2003). As for other tastants, LCFA are detected by specific receptors located on the apical side of the taste bud cells (TBC) clustered in the taste buds of the gustatory papillae.

2.1 CD36 Displays Features of a Gustatory Lipid Sensor

The glycoprotein CD36 belongs to the scavenger receptor family. It is found in tissues involved in lipid absorption, storage, and utilization (e.g., small intestine, adipose tissue, skeletal and cardiac muscles, mammary glands) and in several hematopoietic cells (e.g., monocytes/macrophages, platelets). It is a plasma membrane protein which can bind a large number of ligands, which explains its multifunctional roles (for review see (Martin et al. 2011)). CD36 increases the uptake of LCFA by cardiomyocytes and adipocytes (Coburn et al. 2000; Hajri et al. 2001) and that of oxidized LDL by macrophages (Endemann et al. 1993), modifies platelet aggregation by binding to thrombospondin and collagen (Chen et al. 1997), facilitates the phagocytosis of apoptotic cells by macrophages (Ren et al. 1995), and increases the cytoadhesion of erythrocytes infected with *Plasmodium falciparum* (Oquendo et al. 1989; Febbraio et al. 2001). In addition, CD36 has also recently been shown to play a role in the taste reception of dietary lipids on the tongue (Laugerette et al. 2005; Gaillard et al. 2008). This last finding was surprising since the expression of taste receptors was previously thought to be restricted to TBC. Nevertheless, this dogma has been challenged recently by the identification of sweet and umami taste receptors (T1R) in various tissues including the small intestine, pancreas, liver, kidney, testis, or brain (Dyer et al. 2005; Bezencon et al. 2007; Mace et al. 2009; Nakagawa et al. 2009; Hass et al. 2010; Iwatsuki et al. 2010). Moreover, CD36 displays several features required to be a gustatory lipid sensor. First, it binds saturated and unsaturated LCFA with an affinity in the nanomolar range (Baillie et al. 1996). Second, CD36 expression appears to be strictly restricted to the apical side of some TBC lining the pore of gustatory papillae in the lingual epithelium of rodents (Fukuwatari et al. 1997; Laugerette et al. 2005) and humans (Simons et al. 2010). Third, it displays a receptor-like structure, with a large extracellular binding pocket (Rac et al. 2007) and a C-terminal cytoplasmic tail able to interact with cell signaling proteins of the Src protein-tyrosine kinase (PTK) family (Huang et al. 1991). This complex is capable of activating cell signaling in TBC in a lipid-dependent manner. Consistent with this assumption, experiments with CD36-positive TBC isolated by immuno-magnetism from mouse circumvallate papillae show that activation of CD36 by LCFA leads to the recruitment and activation of Src-PTK (El-Yassimi et al. 2008). This event triggered a prompt and huge rise in intracellular ionized calcium levels ($[Ca^{2+}]_i$) (Gaillard et al. 2008), leading to the release of serotonin and norepinephrine, neurotransmitters known to activate afferent gustatory nerve fibers (El-Yassimi et al. 2008). It is noteworthy that all steps of this cascade are strictly CD36 dependent since they do not occur in CD36-negative TBC or in CD36-positive TBC pretreated with the pharmacological CD36-binding inhibitor, sulfo-N-succinimidyl oleate ester (SSO) (El-Yassimi et al. 2008). The nucleus of the solitary tract (NST) is the first synaptic relay of gustatory pathway in the brain stem. We have shown that an LCFA deposition onto the tongue induces the neuronal activation of NST areas known to receive afferent fibers from gustatory nerves (i.e., chorda tympani and glossopharyngeal nerves). Interestingly, this activation appears to be CD36 dependent since it is not reproduced in CD36-null mice subjected to an oral LCFA stimulation (Gaillard et al. 2008).

2.2 *Physiological Consequences*

Collectively, these data are in agreement with the role of CD36 as a lipid sensor in the gustatory papillae. This unexpected finding is supported by the fact that the lack of CD36 in oral cavity affects the spontaneous preference for lipids and cephalic phase of the digestion.

2.2.1 Preference for Fat

The two-bottle preference test is used for studying the spontaneous preference of animals in a free-choice situation. The use of this simple behavioral paradigm has been at the origin of the first demonstration that the sense of taste is also involved in the oro-sensory detection of dietary lipids. Indeed, rats and mice maintained a strong preference for LCFA-enriched solutions even when olfactory (Takeda et al. 2001; Fukuwatari et al. 2003), somesthesic (Smith et al. 2000; Takeda et al. 2000), and postingestive signals (Tsuruta et al. 1999; Smith et al. 2000; McCormack et al. 2006) were simultaneously minimized. Interestingly, CD36 gene inactivation fully abolishes the spontaneous fat preference in mice subjected to long-term (i.e., 0.5- and 48-h) two-bottle preference tests (Laugerette et al. 2005). Similar data were reproduced when behavioral tests were performed in conditions minimizing postingestive influences (i.e., preference test during 5 min after the first lick using computer-controlled lickometers) and textural cues (i.e., use of mineral oil as vehicle, unpublished data).

2.2.2 Digestive Anticipation

Oral lipid detection might also provide a physiological advantage by preparing the digestive tract to the incoming dietary lipids. Consistent to this assumption, it has been shown that a fatty acid load in oral cavity is sufficient to enhance protein levels in the pancreatic juice in esophagostomized rats, avoiding postingestive influence (Hiraoka et al. 2003). This change, due to the release of digestive enzymes, appears to be tightly dependent on the type of fatty acids used. Indeed, it was only found with LCFA (Laugerette et al. 2005) known to bind CD36 with a high affinity (Baillie et al. 1996). While similar data were obtained in wild-type mice, this digestive effect was markedly blunted in CD36-null mice (Laugerette et al. 2005). This finding provided the first evidence for an involvement of lingual CD36 in the cephalic phase of digestion. This regulatory reflex loop operates through an activation of NST by the CD36-mediated sensing of dietary lipids in the oral cavity, subsequently stimulating the digestive secretion via the efferent vagus projections.

3 Postoral Sensing of Lipids

Collectively, these data demonstrate that CD36 is a lipid sensor involved in the regulation of eating behavior in the tongue. This finding raises the possibility that CD36 might also play a similar role in other places of the body, as small intestine and brain.

3.1 In the Small Intestine

In contrast to other lipid-utilizing cells, enterocytes are subjected to dramatic changes in the fat supply daily. However, in healthy humans, fecal lipid loss remains low even during high-fat challenges (Ross 1993). Specific adaptations explain the efficiency of intestinal fat absorption. The unstirred water layer lining the enterocytes constitutes a unique microclimate characterized by a low pH gradient which induces the protonation of LCFA, facilitating their subsequent membrane permeation. Indeed, neutral LCFA diffuse more easily through a phospholipid bilayer than their corresponding ionized species (Kamp et al. 1993, 1995).

Consistent with this assumption, pharmacological inhibition of proton channels found in the apical side of enterocytes induced a dose-dependent decrease of LCFA uptake in rabbit and rat jejunal sheets (Schoeller et al. 1995). Moreover, the small membrane curvature found in the top of microvilli is favorable for a fast LCFA flip-flop in the phospholipid bilayer (Kleinfeld and Storch 1993; Kleinfeld et al. 1997; Kampf et al. 2006). These specificities explain why passive diffusion plays a significant role in LCFA uptake in the small intestine (for review see (Niot et al. 2009)). This conclusion raises the question of the physiological role played by the plasma membrane lipid-binding proteins found in enterocytes like CD36.

3.1.1 CD36: Lipid Transporter or Lipid Sensor?

CD36 function in the small intestine remains a matter of debate. Because it is involved in the lipid transfer in adipose tissue and heart (for review see (Ibrahimi and Abumrad 2002)) and highly expressed in the duodeno-jejunum, known to be the major site of fat absorption (Poirier et al. 1996), CD36 is generally thought to be a lipid transporter in the gut. However, this conclusion is not consistent with several physiological observations. First, no intestinal malabsorption of LCFA is found in CD36-null mice (Nauli et al. 2006). Second, CD36 gene disruption does not affect LCFA uptake by in situ isolated jejunal loops, an in vivo technical approach known to respect intestinal environment (i.e., intact unstirred water layer, blood and lymph flow, and enteric nervous system) (Tran et al. 2011). Third, there is a rapid lipid-mediated disappearance of CD36 from the brush border membrane of enterocytes (Tran et al. 2011). This process is reminiscent of the progressive desensitization of receptors subjected to permanent ligand stimulation. Fourth, like numerous receptors (Miranda and Sorkin 2007), lipid-induced internalization of CD36 is followed by an ubiquitin/proteasome-mediated degradation in the enterocyte (Tran et al. 2011). Altogether, these observations seem to be more consistent with a role of CD36 as a lipid receptor, as reported in the taste buds, rather than as an efficient lipid transporter.

As a lipid sensor, CD36 might provide real-time information about the presence of lipids in the intestinal lumen, leading to local adaptation of absorption efficiency and providing central information contributing to the regulation of the eating behavior.

3.1.2 Physiological Consequences of the Intestinal Lipid Sensing

During the postprandial period, the presence of energy nutrients in the intestinal lumen triggers various regulatory responses optimizing their absorption and signaling their availability for utilization. It is a complex control not fully known which mobilizes both endocrine and nervous pathways, and requires the presence of specific nutrient receptors in the intestinal mucosa. It has been shown recently that the presence of glucose in the intestinal tube is detected by the heterodimer

T1R2/T1R3, known to be responsible for perception of the sweet taste in gustatory papillae (Mace et al. 2007; Margolskee et al. 2007). In the small intestine, this gluco-reception induces the release of the glucagon-like peptide-1 (GLP-1) by the entero-endocrine L cells, leading to the enhancement of the glucose uptake by the enterocytes and to the inhibition of the eating behavior (Brubaker and Anini 2003; Drucker 2006). It is likely that a similar scenario also exists for the dietary lipids.

Role of Intestinal CD36 in Fat Absorption

An adaptation of the intestinal absorption capacity to the lipid content of the diet has been recently demonstrated (Petit et al. 2007). Food rich in unsaturated fatty acids induces both an increase of intestinal absorptive area, due to a rise in cell proliferation, and a coordinated induction of genes involved in the lipoprotein synthesis. These changes facilitate the production of large chylomicrons rapidly cleared in the blood by the lipoprotein lipase (LPL). They improve the absorption of dietary lipids and their subsequent peripheral use. Such a regulation requires the existence of specific sensors able to detect the presence of lipids in the intestinal lumen. The fact that CD36 gene ablation affects the efficiency of lipoprotein synthesis by the small intestine in the mouse suggests that this lipid receptor plays a significant role in this phenomenon (Nassir et al. 2007). A cytoplasmic accumulation of TG and production of small chylomicrons have been reported during the postprandial period in the CD36-null mice (Drover et al. 2005; Nauli et al. 2006; Masuda et al. 2009). Since chylomicrons of small size are poorly hydrolyzed by the LPL, a blood accumulation of TG is found both in mice and humans deficient in CD36, increasing the atherogenic risk. The molecular mechanism linking luminal lipid sensing by CD36 to lipoprotein synthesis is progressively deciphered. We have recently reported that the lipid-dependent disappearance of CD36 from the brush border membrane of enterocytes is associated with a signaling cascade inducing the expression of genes involved in the lipidation of chylomicrons such as the microsomal triglyceride transfer protein (MTP) and ApoB48 (Tran et al. 2011).

CD36, OEA, and Eating Behavior

Intestinal CD36 is also thought to play a role in the postprandial regulation of eating behavior. Infusion of lipid emulsion into the intestinal lumen is associated with a decrease in the meal frequency (satiety) without change in the meal size (satiation) (Schwartz et al. 2008). Compelling evidence supports the conclusion that this anorexic effect is mediated by the oleoylethanolamine (OEA), an endocannabinoid-like molecule enzymatically released from the plasma membrane precursor phosphatidylethanolamine during the postprandial period (Serrano et al. 2011). In the proximal small intestine, OEA production by enterocytes is stimulated by dietary fat, whereas fasting exerts the opposite effect. Oral and parenteral OEA administration induces an inhibition of the food intake in a dose- and time-dependent manner (Rodriguez de Fonseca et al. 2001). Targeted overproduction of OEA in the small intestine by luminal infusion of an adenoviral vector driving the OEA synthesis leads to a prolongation of feeding latency (Fu et al. 2008). OEA appears to be a local satiety signal rather than a blood-borne hormone. Indeed, central administration of OEA does affect feeding behavior. By contrast, surgical or pharmacological ablation of vagal afferences in the gut abolishes the effect of OEA on food intake (Rodriguez de Fonseca et al. 2001). The OEA-mediated signal is next relayed by NST in the brain stem. Subsequently, it induces the production of the anorexigenic peptide CART (cocaine-amphetamine-related peptide) in the paraventricular nucleus of hypothalamus and enhances memory consolidation by stimulating neurons in the basolateral complex of amygdala (Serrano et al. 2011). Interestingly, CD36 plays a crucial role in the postprandial production of OEA by the small intestine since it is dramatically decreased in CD36-null mice (Schwartz et al. 2008; Guijarro et al. 2010). The molecular mechanism by which this regulation takes place remains elusive. A role of CD36 in uptake of the OEA precursor oleic acid by intestinal mucosa was suggested. However, it has recently been demonstrated by an in vivo experiment that CD36 gene ablation does not affect LCFA uptake (Tran et al. 2011). By contrast, LCFA trigger a CD36-dependent activation of the $MAPK^{erk}$ (mitogen-activated protein kinase) pathway in intestinal mucosa (Tran et al. 2011). These data suggest a role of intestinal CD36 in OEA-mediated satiety as a lipid sensor linking lipid ingestion to eating behavior, rather than as a lipid transporter. How the OEA satiety effect is modulated by the downstream signaling events induced by CD36 is not yet known.

3.2 *In the Brain*

When minute quantities of a lipid emulsion are offered to the brain by a direct infusion into the carotid flow, a decrease in food intake is observed in previously fasted mice (unpublished data). Since blood lipid levels remain unchanged in the experimental conditions used, this finding raises the possibility that a lipid-sensing system also exists in the brain. To directly modulate the feeding behavior, lipids must first cross the blood–brain barrier (BBB) before they are detected by specific lipid-sensitive neurons involved in control of the feeding behavior.

3.2.1 Role of the Blood–brain Barrier

The BBB is constituted by high-density capillary endothelial cells linked by tight junctions, pericytes, and astrocyte cell projections which serve of support. It is a selective barrier limiting the transfer of molecules from the bloodstream to the brain. How blood lipids cross the BBB is progressively deciphered. An endothelial lipase generating LCFA from blood, TG has been identified in the BBB (Sovic et al. 2005). Their subsequent transfer toward deeper regions of the brain might be ensured by a set of plasma membrane lipid-binding proteins including fatty acid transport proteins (FATP) 1 and 4 and CD36 (Mitchell et al. 2011). However, the molecular mechanism by which this vectorial transfer takes place remains elusive. FATP1 and FATP4 share several structural features with acyl-CoA synthetases (ACS) (i.e., high amino acid sequence identity and a comparable predicted structure mainly located at the intracellular side of the plasma membrane) and functional properties (i.e., fatty acid acylation activity), suggesting that these FATP are plasma membrane ACS. Since membranes are impermeable to fatty acyl-CoA, it has been proposed that FATP1 and 4 contribute to the net influx of LCFA into the cells by trapping them as CoA derivatives. If such an ACS function might also contribute to the net transfer of LCFA from blood to the brain through the BBB is not yet known. The fact that FATP1 and 4 gene ablation reduces LCFA transport across human brain microvessel endothelial cells (HBMEC) is consistent with this assumption (Mitchell et al. 2011). However, the precise location of these FATP in BBB remains to be established. CD36 gene is also weakly expressed in HBMEC. Specific siRNA knockdown of CD36 reduces transport of short-, medium-, long-, and very-long-chain fatty acids in HBMEC (Mitchell et al. 2011). These data are unexpected since the binding specificity of CD36 is known to be restricted to fatty acids with a number of carbons ≥ 16. It suggests a more global role of CD36 in the provision of lipids for the brain than what was reported for FATP. The mechanism by which this CD36-mediated effect takes place remains elusive.

3.2.2 CD36 in the Central Nervous System

Once they have crossed the BBB, fatty acids can act as a satiety signal. Indeed, a direct intracerebroventricular administration of an LCFA leads to a decrease in food intake and in expression of an orexigenic molecule like neuropeptide Y (NPY) in the rat (Obici et al. 2002). This physiological impact takes place through modifications in the firing rate of specific neurons located in hypothalamic areas like the arcuate nucleus (Wang et al. 2006), known to play a crucial role in regulation of eating behavior and energy balance. Electrophysiological approaches have showed that activity of LCFA-sensitive hypothalamic neurons was positively or negatively affected by LCFA (Oomura et al. 1975). One part of these effects appears to be dependent on CD36. Indeed, pharmacological inhibition of CD36-binding activity by SSO leads to a twofold reduction of the excitatory and inhibitory neuronal effects of LCFA in the ventromedial hypothalamic nucleus (Le Foll et al. 2009). How CD36 plays this role remains to be determined.

4 Conclusions and Future Directions

A growing body of evidences supports the conclusion that the multifunctional glycoprotein CD36 plays the role of a lipid sensor involved in the different aspects of eating behavior (i.e., selection of food to eat, memorization of food sources, regulation of food intake). Nevertheless, this oral and postoral lipid-sensing system appears to be more complex than what was initially thought. Indeed, other putative lipid sensors displaying an affinity for LCFA have recently been identified along the oro-intestinal tract. For instance, the members of the G protein-coupled receptors (GPCR) family, GPR40 and GPR120, raise new questions. Why are different lipid sensors expressed in the body? Do they play complementary or specific roles in the regulation of eating behavior?

The physiological role of this lipid-sensing system might be to build fat stores in times of nutritional abundance to survive periods of food scarcity. Paradoxically, this adaptation, presumably developed during evolution, appears to be especially maladaptive to environment of permanent food plethora and likely contributes to the present epidemic of obesity throughout the world. Alternatively, an inappropriate sensing of dietary lipids might also lead to obesity by altered feeding behavior. Consistent with this hypothesis, it has been reported that obese patients display a higher preference for fatty food than lean subjects (Drewnowski et al. 1984; Mela 1988). Recently, experiments performed in healthy humans have highlighted the relationship between oro-sensory lipid detection and body mass index (BMI). Hypersensitivity to lipids was associated with lower energy consumption, fat intake, and BMI (Stewart et al. 2010). Whether there is a threshold of sensitivity to dietary lipids in humans which is dependent on CD36 and/or GPCR remains to be determined.

Several functional convergences between cells from taste buds and intestinal mucosa have also been highlighted by this new field of investigations, namely, the existence of (1) a common detection system devoted to the sensing of energy nutrients (not only lipids as reported herein but also carbohydrates and amino acid receptors), (2) production of similar hormones (e.g., GLP-1, cholecystokinin (CCK), and others) and expression of their respective receptors, and (3) connection to afferent nerve fibers involved in feeding behavior (i.e., gustatory and vagus nerves). This observation suggests the existence of a "*functional continuum*" along the oro-intestinal tract responsible for the permanent analysis and regulation of ingestion, digestion, absorption, and metabolic fate of energy nutrients. Efficiency of such a detection system requires a local coordination between cells through paracrine and autocrine regulations associated with a permanent dialogue with the brain via neuroendocrine pathways. A continuum being *"a set of elements such that it is possible to pass from one to another continuously,"* we propose that fundamental knowledge derived, for instance, from the oro-sensory tract can be replicable to the small intestine and likely to the brain and reciprocally. There are limitations due to functional specificity of lingual gustatory epithelium, intestinal mucosa, and central nervous system. However, this concept opens new ways of investigations of the early molecular mechanisms responsible for the oro-sensory perception of lipids and their intestinal fate, with consequences for feeding behavior and health.

Acknowledgements This work was supported by the French Research Agency (ANR, sensoFAT project to P.B.) and the Burgundy Council and the Centre National Interprofessionnel de l'Economie Laitière (CNIEL) through the HumanFATaste program (to P.B.).

References

Baillie AG, Coburn CT, Abumrad NA (1996) Reversible binding of long-chain fatty acids to purified FAT, the adipose CD36 homolog. J Membr Biol 153:75–81

Bezencon C, le Coutre J, Damak S (2007) Taste-signaling proteins are coexpressed in solitary intestinal epithelial cells. Chem Senses 32:41–49

Brubaker PL, Anini Y (2003) Direct and indirect mechanisms regulating secretion of glucagon-like peptide-1 and glucagon-like peptide-2. Can J Physiol Pharmacol 81:1005–1012

Chale-Rush A, Burgess JR, Mattes RD (2007) Evidence for human orosensory (taste?) sensitivity to free fatty acids. Chem Senses 32:423–431

Chen CH, Cartwright J Jr, Li Z, Lou S, Nguyen HH, Gotto AM Jr, Henry PD (1997) Inhibitory effects of hypercholesterolemia and ox-LDL on angiogenesis-like endothelial growth in rabbit aortic explants Essential role of basic fibroblast growth factor. Arteriosclerosis Thrombosis Vasc Biol 17:1303–1312

Coburn CT, Knapp FF Jr, Febbraio M, Beets AL, Silverstein RL, Abumrad NA (2000) Defective uptake and utilization of long chain fatty acids in muscle and adipose tissues of CD36 knockout mice. J Biol Chem 275:32523–32529

Drewnowski A, Cohen AE, Faust IM, Grinker JA (1984) Meal-taking behavior is related to predisposition to dietary obesity in the rat. Physiol Behav 32:61–67

Drover VA, Ajmal M, Nassir F, Davidson NO, Nauli AM, Sahoo D, Tso P, Abumrad NA (2005) CD36 deficiency impairs intestinal lipid secretion and clearance of chylomicrons from the blood. J Clin Invest 115:1290–1297

Drucker DJ (2006) The biology of incretin hormones. Cell Metab 3:153–165

Dyer J, Salmon KS, Zibrik L, Shirazi-Beechey SP (2005) Expression of sweet taste receptors of the T1R family in the intestinal tract and enteroendocrine cells. Biochem Soc Trans 33:302–305

El-Yassimi A, Hichami A, Besnard P, Khan NA (2008) Linoleic acid induces calcium signaling, Src kinase phosphorylation, and neurotransmitter release in mouse CD36-positive gustatory cells. J Biol Chem 283:12949–12959

Endemann G, Stanton LW, Madden KS, Bryant CM, White RT, Protter AA (1993) CD36 is a receptor for oxidized low density lipoprotein. J Biol Chem 268:11811–11816

Febbraio M, Hajjar DP, Silverstein RL (2001) CD36: a class B scavenger receptor involved in angiogenesis, atherosclerosis, inflammation, and lipid metabolism. J Clin Invest 108:785–791

Fu J, Kim J, Oveisi F, Astarita G, Piomelli D (2008) Targeted enhancement of oleoylethanolamide production in proximal small intestine induces across-meal satiety in rats. Am J Physiol Regul Integr Comp Physiol 295:R45–50

Fukuwatari T, Kawada T, Tsuruta M, Hiraoka T, Iwanaga T, Sugimoto E, Fushiki T (1997) Expression of the putative membrane fatty acid transporter (FAT) in taste buds of the circumvallate papillae in rats. FEBS Lett 414:461–464

Fukuwatari T, Shibata K, Iguchi K, Saeki T, Iwata A, Tani K, Sugimoto E, Fushiki T (2003) Role of gustation in the recognition of oleate and triolein in anosmic rats. Physiol Behav 78:579–583

Gaillard D, Laugerette F, Darcel N, El-Yassimi A, Passilly-Degrace P, Hichami A, Khan NA, Montmayeur JP, Besnard P (2008) The gustatory pathway is involved in CD36-mediated orosensory perception of long-chain fatty acids in the mouse. FASEB J 22:1458–1468

Guijarro A, Fu J, Astarita G, Piomelli D (2010) CD36 gene deletion decreases oleoylethanolamide levels in small intestine of free-feeding mice. Pharmacol Res 61:27–33

Hajri T, Ibrahimi A, Coburn CT, Knapp FF Jr, Kurtz T, Pravenec M, Abumrad NA (2001) Defective fatty acid uptake in the spontaneously hypertensive rat is a primary determinant of altered glucose metabolism, hyperinsulinemia, and myocardial hypertrophy. J Biol Chem 276:23661–23666

Hass N, Schwarzenbacher K, Breer H (2010) T1R3 is expressed in brush cells and ghrelin-producing cells of murine stomach. Cell Tissue Res 339:493–504

Hiraoka T, Fukuwatari T, Imaizumi M, Fushiki T (2003) Effects of oral stimulation with fats on the cephalic phase of pancreatic enzyme secretion in esophagostomized rats. Physiol Behav 79:713–717

Huang MM, Bolen JB, Barnwell JW, Shattil SJ, Brugge JS (1991) Membrane glycoprotein IV (CD36) is physically associated with the Fyn, Lyn, and Yes protein-tyrosine kinases in human platelets. Proc Natl Acad Sci USA 88:7844–7848

Ibrahimi A, Abumrad NA (2002) Role of CD36 in membrane transport of long-chain fatty acids. Curr Opin Clin Nutr Metab Care 5:139–145

Iwatsuki K, Nomura M, Shibata A, Ichikawa R, Enciso PL, Wang L, Takayanagi R, Torii K, Uneyama H (2010) Generation and characterization of T1R2-LacZ knock-in mouse. Biochem Biophys Res Commun 402:495–499

Kamp F, Westterhoff HV, Hamilton JA (1993) Movement of fatty acids, fatty acid analogues and bile acids across phospholipid bilayers. Biochemistry 32:11074–11086

Kamp F, Zakim D, Zhang F, Noy N, Hamilton JA (1995) Fatty acid flip-flop in phospholipid bilayers is extremely fast. Biochemistry 34:11928–11937

Kampf JP, Cupp D, Kleinfeld AM (2006) Different mechanisms of free fatty acid flip-flop and dissociation revealed by temperature and molecular species dependence of transport across lipid vesicles. J Biol Chem 281:21566–21574

Kleinfeld AM, Chu P, Romero C (1997) Transport of long-chain native fatty acids across lipid bilayer membranes indicates that transbilayer flip-flop is rate limiting. Biochemistry 36:14146–14158

Kleinfeld AM, Storch J (1993) Transfer of long-chain fluorescent fatty acids between small and large unilamellar vesicles. Biochemistry 32:2053–2061

Laugerette F, Passilly-Degrace P, Patris B, Niot I, Febbraio M, Montmayeur JP, Besnard P (2005) CD36 involvement in orosensory detection of dietary lipids, spontaneous fat preference, and digestive secretions. J Clin Invest 115:3177–3184

Le Foll C, Irani BG, Magnan C, Dunn-Meynell AA, Levin BE (2009) Characteristics and mechanisms of hypothalamic neuronal fatty acid sensing. Am J Physiol Regul Integr Comp Physiol 297:R655–664

Mace OJ, Affleck J, Patel N, Kellett GL (2007) Sweet taste receptors in rat small intestine stimulate glucose absorption through apical GLUT2. J Physiol 582:379–392

Mace OJ, Lister N, Morgan E, Shepherd E, Affleck J, Helliwell P, Bronk JR, Kellett GL, Meredith D, Boyd R, Pieri M, Bailey PD, Pettcrew R, Foley D (2009) An energy supply network of nutrient absorption coordinated by calcium and T1R taste receptors in rat small intestine. J Physiol 587:195–210

Margolskee RF, Dyer J, Kokrashvili Z, Salmon KS, Ilegems E, Daly K, Maillet EL, Ninomiya Y, Mosinger B, Shirazi-Beechey SP (2007) T1R3 and gustducin in gut sense sugars to regulate expression of Na + −glucose cotransporter 1. Proc Natl Acad Sci USA 104:15075–15080

Martin C, Chevrot M, Poirier H, Passilly-Degrace P, Niot I, Besnard P (2011) CD36 as a lipid sensor. Physiol Behav 105(1):36–42

Masuda D, Hirano K, Oku H, Sandoval JC, Kawase R, Yuasa-Kawase M, Yamashita Y, Takada M, Tsubakio-Yamamoto K, Tochino Y, Koseki M, Matsuura F, Nishida M, Kawamoto T, Ishigami M, Hori M, Shimomura I, Yamashita S (2009) Chylomicron remnants are increased in the postprandial state in CD36 deficiency. J Lipid Res 50:999–1011

McCormack DN, Clyburn VL, Pittman DW (2006) Detection of free fatty acids following a conditioned taste aversion in rats. Physiol Behav 87:582–594

Mela DJ (1988) Sensory assessment of fat content in fluid dairy products. Appetite 10:37–44

Miranda M, Sorkin A (2007) Regulation of receptors and transporters by ubiquitination: new insights into surprisingly similar mechanisms. Mol Interv 7:157–167

Mitchell RW, On NH, Del Bigio MR, Miller DW, Hatch GM (2011) Fatty acid transport protein expression in human brain and potential role in fatty acid transport across human brain microvessel endothelial cells. J Neurochem 117:735–746

Nakagawa Y, Nagasawa M, Yamada S, Hara A, Mogami H, Nikolaev VO, Lohse MJ, Shigemura N, Ninomiya Y, Kojima I (2009) Sweet taste receptor expressed in pancreatic beta-cells activates the calcium and cyclic AMP signaling systems and stimulates insulin secretion. PLoS One 4:e5106

Nassir F, Wilson B, Han X, Gross RW, Abumrad NA (2007) CD36 is important for fatty acid and cholesterol uptake by the proximal but not distal intestine. J Biol Chem 282:19493–19501
Nauli AM, Nassir F, Zheng S, Yang Q, Lo CM, Vonlehmden SB, Lee D, Jandacek RJ, Abumrad NA, Tso P (2006) CD36 is important for chylomicron formation and secretion and may mediate cholesterol uptake in the proximal intestine. Gastroenterology 131:1197–1207
Niot I, Poirier H, Tran TT, Besnard P (2009) Intestinal absorption of long-chain fatty acids: evidence and uncertainties. Prog Lipid Res 48:101–115
Obici S, Feng Z, Morgan K, Stein D, Karkanias G, Rossetti L (2002) Central administration of oleic acid inhibits glucose production and food intake. Diabetes 51:271–275
Oomura Y, Nakamura T, Sugimori M, Yamada Y (1975) Effect of free fatty acid on the rat lateral hypothalamic neurons. Physiol Behav 14:483–486
Oquendo P, Hundt E, Lawler J, Seed B (1989) CD36 directly mediates cytoadherence of plasmodium falciparum parasitized erythrocytes. Cell 58:95–101
Petit V, Arnould L, Martin P, Monnot MC, Pineau T, Besnard P, Niot I (2007) Chronic high-fat diet affects intestinal fat absorption and postprandial triglyceride levels in the mouse. J Lipid Res 48:278–287
Poirier H, Degrace P, Niot I, Bernard A, Besnard P (1996) Localization and regulation of the putative membrane fatty-acid transporter (FAT) in the small intestine. Comparison with fatty acid-binding proteins (FABP). Eur J Biochem 238:368–373
Rac ME, Safranow K, Poncyljusz W (2007) Molecular basis of human CD36 gene mutations. Mol Med 13:288–296
Ren Y, Silverstein RL, Allen J, Savill J (1995) CD36 gene transfer confers capacity for phagocytosis of cells undergoing apoptosis. J Exp Med 181:1857–1862
Rodriguez de Fonseca F, Navarro M, Gomez R, Escuredo L, Nava F, Fu J, Murillo-Rodriguez E, Giuffrida A, LoVerme J, Gaetani S, Kathuria S, Gall C, Piomelli D (2001) An anorexic lipid mediator regulated by feeding. Nature 414:209–212
Ross AC (1993) Overview of retinoid metabolism. J Nutr 123:346–350
Schoeller C, Keelan M, Mulvey G, Stremmel W, Thomson AB (1995) Oleic acid uptake into rat and rabbit jejunal brush border membrane. Biochim Biophys Acta 1236:51–64
Schwartz GJ, Fu J, Astarita G, Li X, Gaetani S, Campolongo P, Cuomo V, Piomelli D (2008) The lipid messenger OEA links dietary fat intake to satiety. Cell Metab 8:281–288
Serrano A, Pavon FJ, Tovar S, Casanueva F, Senaris R, Dieguez C, de Fonseca FR (2011) Oleoylethanolamide: effects on hypothalamic transmitters and gut peptides regulating food intake. Neuropharmacology 60:593–601
Simons PJ, Kummer JA, Luiken JJ, Boon L (2010) Apical CD36 immunolocalization in human and porcine taste buds from circumvallate and foliate papillae. Acta Histochem 113(8):839–43
Smith JC, Fisher EM, Maleszewski V, McClain B (2000) Orosensory factors in the ingestion of corn oil/sucrose mixtures by the rat. Physiol Behav 69:135–146
Sovic A, Panzenboeck U, Wintersperger A, Kratzer I, Hammer A, Levak-Frank S, Frank S, Rader DJ, Malle E, Sattler W (2005) Regulated expression of endothelial lipase by porcine brain capillary endothelial cells constituting the blood–brain barrier. J Neurochem 94:109–119
Stewart JE, Feinle-Bisset C, Golding M, Delahunty C, Clifton PM, Keast RS (2010) Oral sensitivity to fatty acids, food consumption and BMI in human subjects. Br J Nutr 104:145–152
Takeda M, Imaizumi M, Fushiki T (2000) Preference for vegetable oils in the two-bottle choice test in mice. Life Sci 67:197–204
Takeda M, Sawano S, Imaizumi M, Fushiki T (2001) Preference for corn oil in olfactory-blocked mice in the conditioned place preference test and the two-bottle choice test. Life Sci 69:847–854
Tran TT, Poirier H, Clement L, Nassir F, Pelsers MM, Petit V, Degrace P, Monnot MC, Glatz JF, Abumrad NA, Besnard P, Niot I (2011) Luminal lipid regulates CD36 levels and downstream signaling to stimulate chylomicron synthesis. J Biol Chem 286(28):25201–10
Tsuruta M, Kawada T, Fukuwatari T, Fushiki T (1999) The orosensory recognition of long-chain fatty acids in rats. Physiol Behav 66:285–288
Wang R, Cruciani-Guglielmacci C, Migrenne S, Magnan C, Cotero VE, Routh VH (2006) Effects of oleic acid on distinct populations of neurons in the hypothalamic arcuate nucleus are dependent on extracellular glucose levels. J Neurophysiol 95:1491–1498

Intestinal Sensing of Nutrients

Gwen Tolhurst, Frank Reimann, and Fiona M. Gribble

Contents

Abstract Ingestion of a meal triggers a range of physiological responses both within and outside the gut, and results in the remote modulation of appetite and glucose homeostasis. Luminal contents are sensed by specialised chemosensitive cells scattered throughout the intestinal epithelium. These enteroendocrine and tuft cells make direct contact with the gut lumen and release a range of chemical mediators, which can either act in a paracrine fashion interacting with neighbouring cells and nerve endings or as classical circulating hormones. At the molecular level, the chemosensory machinery involves multiple and complex signalling pathways including activation of G-protein-coupled receptors and solute carrier transporters. This chapter will discuss our current knowledge of the molecular mechanisms

G. Tolhurst • F. Reimann • F.M. Gribble (✉)
Cambridge Institute for Medical Research, Wellcome Trust/MRC Building, Addenbrooke's Hospital, Hills Road, Cambridge CB2 0XY, UK
e-mail: fmg23@cam.ac.uk

H.-G. Joost (ed.), *Appetite Control*, Handbook of Experimental Pharmacology 209,
DOI 10.1007/978-3-642-24716-3_14,

underlying intestinal chemosensation with a particular focus on the relatively well-characterised nutrient-triggered secretion from the enteroendocrine system.

Keywords CCK • Enteroendocrine cells • GLP-1 • GIP • Nutrient transporters • PYY

Abbreviations

2-APB	2-Aminoethoxydiphenyl borate
5-HT	Serotonin
CaMK	Ca^{2+}/calmodulin-dependent protein kinases
cAMP	Cyclic adenosine monophosphate
CCK	Cholecystokinin
DGAT1	Diacylglycerol acyltransferase 1
EC	Enterochromaffin cell
FFA	Free fatty acids
GIP	Glucose-dependent insulinotropic polypeptide
GLP	Glucagon-like peptide
GPCR	G-protein-coupled receptor
LCFA	Long-chain fatty acids
MTP	Microsomal triglyceride transfer protein
NOPE	N-oleoyl-phosphatidylethanolamine
OEA	Oleoylethanolamide
PKA	Protein kinase A
PKC	Protein kinase C
PLC	Phospholipase C
PYY	Peptide YY
SCFA	Short-chain fatty acids
SLC	Solute carrier transporters

1 Introduction

Ingestion of a meal triggers a range of physiological responses to optimise the digestion, absorption, distribution and metabolism of nutrients, and to regulate further food intake. Indeed, the gut is largely able to handle anything that arrives in it, whether it be digestible, indigestible or even toxic. As foods vary widely in their nutrient composition, there is a corresponding spectrum of nutrient-sensing pathways in the gut, which are linked either directly or indirectly to responses such as secretion of enzymes and electrolytes, coordination of gastric emptying and gut motility, and activation of neuronal and hormonal signalling pathways to the brain and peripheral tissues. The gut–brain axis is now recognised to play a particularly important role in the regulation of appetite, as exemplified by the range of intestinal peptides that have been demonstrated to affect food intake in animals and humans. Both inhibitory and stimulatory modulators of food ingestion have been isolated from different regions of the intestine, which can target the brainstem and hypothalamus either directly in the form of classical circulating hormones or indirectly via modulation of afferent vagal nerve activity.

1.1 Endocrine Cells and Gut Hormones

We have known for over a century that the gut is able to sense changes in the luminal content and respond by releasing chemical signals. Bayliss and Starling observed in 1902 that increasing the acidity in the small intestinal lumen elicited pancreatic secretion (Bayliss and Starling 1902). They noted that this was mediated not via the nervous system but by a humoral factor that they termed 'secretin', which was produced by the gut epithelium. Physiological actions of secreted gut hormones are now recognised to include inhibition of gastric emptying and gastric acid secretion, stimulation of gastric secretion, enhanced pancreatic exocrine and endocrine secretion, intestinal fluid secretion and inhibition of food intake. In fact, the gastrointestinal tract has been estimated to be the largest endocrine organ in the body, producing over 20 hormones that act locally, peripherally and centrally (Rehfeld 2004).

Intestinal (entero-) endocrine cells are scattered throughout the epithelial layer of the gastrointestinal tract, but represent only a small fraction (<1%) of the total epithelial cell number. Individual cell types that make up the enteroendocrine family have distinct expression patterns throughout the gut. For example, enterochromaffin (EC) cells and somatostatin-secreting D cells are interspersed along the length of the gut, whereas the K cells that produce gastric inhibitory peptide (also known as glucose-dependent insulinotropic polypeptide or GIP) are found predominantly in the duodenum (Sjölund et al. 1983). Many enteroendocrine cells have processes extending to the lumen, leading to their description as "open type" endocrine cells (as opposed to their "closed type" counterparts that are cut off from the lumen by epithelial tight junctions). These gut endocrine cells are prime candidates to fulfil the role of intestinal nutrient sensors, linking the luminal composition to a variety of secreted chemical signals which in turn stimulate local nerve endings and paracrine targets or circulate in the bloodstream as classical hormones.

The study of chemosensing in enteroendocrine cells has been hindered by the scarcity of these cells in the gut epithelium and the inability to identify them accurately without prior fixation and staining. Much of our understanding has therefore come from work using cell lines such as GLUTag (Drucker et al. 1994), STC-1 (Rindi et al. 1990) and BON (Parekh et al. 1994). However, the generation of transgenic mice expressing fluorescent proteins under the control of hormonal promoters such as proglucagon, GIP and cholecystokinin (CCK), resulting in the cell-specific labelling of L, K and I cells (glucagon-like peptide secreting cells are classically defined as L cells, while CCK-secreting cells are known as I cells), respectively (Parker et al. 2009; Reimann et al. 2008; Wang et al. 2010), has led to renewed interest in and understanding of the chemosensing pathways in these cells. When coupled with protocols to purify fluorescent cells by flow cytometry and to maintain enteroendocrine cells in primary intestinal cultures, these provide the technology to interrogate for the first time the expression profile and functional characteristics of primary gut endocrine cells. Primary L cells, like GLUTag cells, exhibit electrical activity, as demonstrated by electrophysiological recordings (Rogers et al. 2011), coupled to Ca^{2+} entry via L and P/Q-type voltage-gated ion channels (Reimann et al. 2005).

1.2 Sensory Nerve Supply of the Intestine

Sensory afferent nerves terminate in the gut wall and convey chemo- and mechanosensory signals to target tissues both within and outside of the intestine. Branches of the afferent vagus are involved, for example, in linking signals from ingested food to the control of appetite. The field of mechanosensory transduction will not be covered specifically in this chapter, but it is noteworthy that the physical act of food entering the intestinal tract triggers vagal mechanosensors in the serosal and muscle layers of the gut. Indeed, increased gastric luminal pressure is one of the earliest physiological consequences of food ingestion and is rapidly reflected by changes in electrical activity in the vagus nerve.

As afferent nerve endings do not appear to innervate the mucosal layer (Powley et al. 2011), they are not believed themselves to act as the direct sensors of luminal nutrients, and chemosensing by intestinal neurons is rather believed to be an indirect process. Vagal afferent activity can be triggered by a number of enteroendocrine hormones, including serotonin (5-hydroxytryptamine, 5-HT), glucagon-like peptide (GLP)-1, GLP-2, PeptideYY (PYY) and CCK (reviewed in Dockray (2003)), and there are some well-established links between luminal stimuli, enteroendocrine secretion and vagal activity (for further reading, see Grundy (2004)). Lipid and protein activation of afferent fibres, for example, can be triggered by CCK release (reviewed in Raybould et al. (2006)), and 5-HT has been proposed to mediate glucose responses (Zhu et al. 2001). With our developing understanding of the enteroendocrine system, it is likely that other peptides will also be found to play a role in mediating nutrient stimulation of intestinal sensory nerves.

1.3 Tuft Cells

Tuft cells, also known as brush or caveolated cells, were first identified in the gastrointestinal tract in 1956 (Jarvi and Keyrilainen 1956). They are described as being polarised cells with a narrow apical pole and a tuft of microvilli extending towards the lumen. Morphologically, they differ from other secretory cells in the gut as they lack electron-dense secretory granules (Höfer and Drenckhahn 1996) but contain caveolae in the apical cytoplasm (Nabeyama and Leblond 1974). They were originally believed to be a subset of enteroendocrine cells, but recently a tuft cell lineage pathway that is distinct from the endocrine cells of the gut has been described (Gerbe et al. 2011). Tuft cells express components of the taste signalling system (α-gustducin, TRPM5), as well as β-endorphin, met-enkephalin and uroguanylin (Gerbe et al. 2011; Kokrashvili et al. 2009b), and are hypothesised to act as chemoreceptors in the gut. Because they exhibit heavy apical staining, it has been suggested that they may release peptides into the lumen (Kokrashvili et al. 2009a). They might therefore underlie the observed release of opioids into the gut lumen in response to nutrients, which in turn modulate gastric motility, emptying and intestinal secretions (Holzer 2009). Met-encephalin, for example, is released into the lumen following ingestion of fat (Money et al. 1988), and intestinal β-endorphin secretion triggered by hypertonic stimuli (but not fat) was dependent on TRPM5 (Kokrashvili et al. 2009b). Although opioids and components of the taste signalling pathway have been detected in tuft cells by immunostaining, direct sensing of macronutrients by tuft cells remains to be proven. Interestingly, the guanylate cyclase-C receptor, which is activated by uroguanylin and guanylin, is expressed and functional in enteroendocrine L cells (Friedlander et al. 2010), providing a potential pathway for luminal cross talk between tuft cells and enteroendocrine cells.

2 Nutrient-Sensing Machinery

Sensors of nutrient ingestion need to monitor either the luminal contents themselves or locally elevated concentrations of substances absorbed across the epithelial layer. Their activation triggers intracellular signalling pathways, including membrane depolarisation, elevated calcium levels and second messenger cascades that ultimately result in endpoints such as the release of a hormone or changes in gene expression. Sensory systems described to date range from the utilisation of surface membrane solute transporters and G-protein-coupled receptors to a requirement for intracellular receptor binding or metabolism.

2.1 *Solute Carrier Transporters*

The intestinal brush border is rich in solute carrier transporters (SLC) which act as gateways for the absorption of nutrients, ions and drugs. In humans, the SLC family comprises 248 genes, subdivided into 47 families (Hediger et al. 2004). Modes of action of individual transporters vary: they can facilitate diffusional equilibration, exchange substances bidirectionally across plasma membranes or couple solute transfer to the electrochemical gradient of ions such as Na^+ and H^+. If the sensing mechanism for the solute is localised intracellularly, transport might become rate limiting. Thus, some cells that detect concentration changes downstream of an increase of metabolic flux, like, for example, the pancreatic β-cell (see below), depend on high-capacity transport (the β-cell employs the high-capacity facilitative GLUT2 transporter). Exchangers and ion-coupled transporters, on the other hand, have the capacity to move solutes against their concentration gradient, utilising the energy gradient of the coupled ion or solute. Electrogenic transporters, like the sodium-coupled glucose transporter, SGLT1, generate currents capable of triggering action potentials and Ca^{2+} entry (Gribble et al. 2003). There are also examples of transporters that have additional receptor-like activity and trigger intracellular signalling pathways, as exemplified by GLUT2 which has been found to mediate sugar-dependent gene transcription independent of its role as a transporter (Guillemain et al. 2000).

Individual cells typically express numerous members of the SLC family, acting individually and in concert to shape an overall response. Taken together with the wide variety of cell types along the length of the intestinal epithelium and the range of chemical mediators within the enteroendocrine system, this allows for highly sensitive "sensing" of and responses to changes in the luminal nutrient composition.

2.2 G-Protein-Coupled Receptors

In recent years, members of the G-protein-coupled receptor (GPCR) family have been identified as nutrient sensors, and many are expressed in the gut (Wellendorph et al. 2010). GPCRs represent the largest family of cell surface receptors, responding to a diverse array of ligands. Binding to the extracellular face of the receptor activates intracellular G proteins, thereby triggering cascade-like transduction pathways. G proteins primarily couple to two signalling pathways: the cyclic adenosine monophosphate (cAMP) and phosphatidylinositol pathway. However, the diversity and sensitivity of the receptors are further amplified by their ability to homo- or heterodimerise, affecting not only the affinity and efficacy of the ligand, but also the recruitment of downstream signalling pathways (Prezeau et al. 2010; Vilardaga et al. 2010). Receptor activity can be further modulated by intracellular protein–protein interactions (Ritter and Hall 2009), extracellular allosteric modulation (May et al. 2007) and changes in the membrane potential (Martinez-Pinna et al. 2004), introducing the potential for interactions between different members of the GPCR and solute transporter families.

Oral sensing of nutrients is largely achieved via GPCR stimulation. The sweet, bitter and umami taste modalities, classically described in lingual taste buds, act through two subgroups of GPCRs of the Tas1 and Tas2 taste receptor families. The sweet taste receptor, a heterodimeric GPCR comprised of T1R2 and T1R3, responds to a wide array of natural and artificial sweeteners (Margolskee 2002) and is largely responsible for the oral sensation of sweet taste. The related receptor combination, T1R1/T1R3, detects glutamate and underlies the taste modality known as umami. Members of the Tas2 receptor family, which in mammals contains approximately 30 members (Behrens et al. 2009), are responsible for the bitter taste of a wide range of chemicals. Taste receptors interact with a subgroup of G proteins including α-gustducin, leading to activation of the phosphatidylinositol pathway via phospholipase C (PLC) β2. A consequent rise in intracellular calcium activates the monovalent cation channel TRPM5, leading to sodium entry and membrane depolarisation (reviewed in (Margolskee 2002). Many tastants trigger similar intracellular signalling cascades, and whether they are recognised as sweet, umami or bitter appears to be encoded by the particular cell type activated and its central connectivity (Carleton et al. 2010).

3 Sensing of Different Nutrient Classes in the Intestine

3.1 *Carbohydrates*

Carbohydrates in the intestine have been linked to a number of physiological responses, including intracellular trafficking of glucose transporters, stimulation of enteroendocrine secretion and activation of enteric neurons. A variety of signalling pathways have been implicated in these different responses highlighted in Fig. 1.

3.1.1 Sweet Taste Receptors

In the past few years, it has been proposed that populations of intestinal cells can "taste" sugars or sweet compounds, using sensory pathways similar to those classically described in lingual taste cells. Antibodies against Tas1 receptor subunits, α-gustducin, TRPM5 and PLCβ2 have detected expression of these signalling components in isolated cells within the intestinal epithelium (Bezençon et al. 2007; Höfer et al. 1996; Jang et al. 2007; Kokrashvili et al. 2009b). However, it is still not clear which cell types contain taste receptors, and whether all components of the pathway are located in the same cells. Several reports suggest that taste machinery resides in cells with the hallmarks of enteroendocrine cells, perhaps those producing GLP-1 and PYY (Dyer et al. 2005; Jang et al. 2007; Rozengurt et al. 2006). Using TRPM5 as a marker for "taste cells", this population has been purified from the gastrointestinal tract of mice carrying a fluorescent reporter driven by the *trpm5* promoter (Kokrashvili et al. 2009b). Transcriptomic analysis of such *trpm5* positive cell populations suggests that they are not typical of enteroendocrine cells, and rather have expression profiles characteristic of tuft cells.

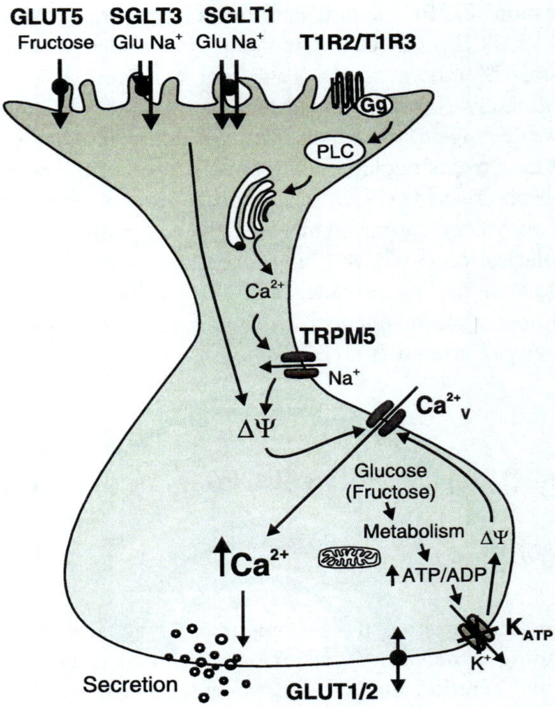

Fig. 1 Carbohydrate sensing in enteroendocrine cells Electrogenic transport of glucose (Glu) by SGLT1, or glucose binding to SGLT3, depolarises the membrane ($\Delta\Psi$) sufficiently to activate Ca^{2+} entry via voltage-gated Ca^{2+} channels ($Ca^{2+}{}_V$). The sweet taste receptor (T1R2/T1R3) elevates intracellular Ca^{2+} via α-gustducin (Gg), PLCβ2 activation and membrane depolarisation mediated by TRPM5. Elevated intracellular Ca^{2+} is a primary trigger of exocytosis. Glucose and fructose are transported across the membrane by GLUT1 and/or 2 and GLUT5, respectively. Metabolism results in elevated ATP and reduced MgADP, which inhibits K_{ATP} channels, thereby promoting membrane depolarisation and sensitisation to further stimuli

Sweet Taste Receptors in Enteroendocrine Cells

Despite the controversies about which cell types express taste receptor machinery, there are a number of reports that sweet taste receptors may underlie glucose sensing in gut endocrine cells, particularly those releasing the incretin hormones GLP-1 and GIP. Plasma GLP-1 and GIP levels following glucose gavage were reduced in α-gustducin knockout compared to wild-type mice (Jang et al. 2007). It was also reported that the artificial sweetener, sucralose, stimulated incretin release from GLUTag cells and that this was inhibited by gurmarin, a specific inhibitor of murine sweet taste receptors (Margolskee et al. 2007). However, sucralose and other sweeteners were not found to be stimuli of GLP-1 and GIP secretion in primary cultures of the murine intestinal epithelium, and expression profiling of purified L and K cells revealed no enrichment of mRNA for T1R2/T1R3 or α-gustducin (Parker et al. 2009; Reimann et al. 2008). Furthermore, artificial sweeteners have not been found to modify glucose absorption rates, plasma glucose or incretin levels or gastric emptying in humans (Little et al. 2009; Ma et al. 2009, 2010) or rodents (Fujita et al. 2009). Clearly, there is further work required to clarify the roles of sweet taste receptors in gut endocrine secretion.

Intracellular Trafficking of Glucose Transporters

The gastrointestinal mucosa cell expresses a number of monosaccharide transporters, employing both Na^+-coupled (e.g. SGLT1) and facilitative (e.g. GLUT1, GLUT2 and GLUT5) mechanisms. Expression of SGLT1 on the apical but not the basolateral surface of the epithelial layer (Hwang et al. 1991; Yoshida et al. 1995) promotes uphill glucose absorption out of the lumen, driven by the inward Na^+ gradient that is set up and maintained by basolateral Na^+/K^+-ATPases. SGLT1 mRNA and protein levels are regulated by dietary intake and diurnal rhythms (Balakrishnan et al. 2008) and may involve activity of the sweet taste receptor pathway. Mice fed a high-carbohydrate diet or supplemented with artificial sweeteners displayed elevated steady-state levels of SGLT1 mRNA and protein, and correspondingly higher rates of Na^+-dependent glucose transport in isolated brush border vesicles. This effect was not observed in either T1R3- or gustducin-knockout mice (Margolskee et al. 2007), and it has been suggested that the link may include a neural reflex, potentially activated by GLP-2 (Shirazi-Beechey et al. 2011). In support for the involvement of the afferent vagus, sugar/sweetener-dependent upregulation of SGLT1 was abolished in rats following chemical vagal deafferentation (Stearns et al. 2010).

Basolateral glucose efflux is facilitated by GLUT2 in the upper gastrointestinal tract and by GLUT1 more distally (Yoshikawa et al. 2011), thereby promoting the transcellular flux of glucose from the lumen into the bloodstream. It has been reported that GLUT2 is transiently recruited to the apical membrane surface after a meal, allowing the passive absorption of glucose at higher luminal sugar concentrations. Brush border SGLT1 activity saturates in vivo between 30 and 50 mM glucose (Debnam and Levin 1975), but apical GLUT2 insertion could be an energy saving measure, preventing the unnecessary run-down of the Na^+ gradient by SGLT1 when this is not energetically required. Apical GLUT2 insertion has been reported to occur rapidly following sugar exposure ($t_{½}$ 3.5 min, (Helliwell et al. 2003)) and may be controlled by Ca^{2+}, sweet taste receptors and hormones. The artificial sweetener sucralose, for example, doubled the rate of glucose absorption within minutes by increasing the expression of GLUT2 at the brush border (Mace et al. 2007). As the response to different sweet chemicals matched the known properties of sweet taste receptors, the authors proposed that T1R2/T1R3 acts as a glucose sensor at high concentrations (>30 mM) and controls the apical expression of GLUT2 and consequent increased capacity to absorb glucose. After a meal, when luminal glucose and dietary Ca^{2+} levels are high, L-type calcium channels ($Ca_V1.3$) contribute to Ca^{2+} entry (Morgan et al. 2007), and in this context, it was suggested that the Ca^{2+} response may in turn be controlled by the depolarising action of SGLT1, thus placing SGLT1 in the role of a candidate glucose sensor.

3.1.2 Glucose Sensing by Members of the Sodium-Coupled Glucose Transporter Family

The sodium-coupled glucose transporter family offers a distinct mode of sugar sensing involving the generation of small depolarising ion currents. Glucose transport by SGLT1 occurs with a fixed stoichiometry of 1 glucose molecule per 2 Na^+ ions (Chen et al. 1995), which directly generates a small inward Na^+ current. A similar picture, albeit with a different glucose to charge ratio, is observed for the related renal glucose transporter, SGLT2 (You et al. 1995). Human SGLT3, by contrast, reportedly generates large Na^+ currents without concomitant glucose transport (Diez-Sampedro et al. 2003), suggesting this may act as a glucose sensor that directly triggers membrane depolarisation. The mouse genome has 2 genes for SGLT3, known as mSGLT3a and mSGLT3b. Whilst mSGLT3a remains functionally uncharacterised, mSGLT3b has characteristics intermediate between SGLT1 and hSGLT3, displaying charge movement that is partially uncoupled from glucose transport (stoichiometry ~2.6 Na^+ to 1 glucose, (Díez-Sampedro and Barcelona 2011)).

SGLT-Dependent Secretion from Enteroendocrine Cells

It has been known for some time that luminal application of non-metabolisable as well as metabolisable sugars can stimulate GLP-1 and GIP release from the intact gut when added together with Na^+ ions. Hormonal responses to a range of sugar analogues were shown to match the selectivity of the intestinal uptake system and were blocked by the inhibitor of sodium-coupled uptake, phloridzin (Ritzel et al. 1997; Sykes et al. 1980). These data indicated firstly that metabolism is not a critical component of glucose-triggered incretin release, and secondly that the uptake pathway might itself hold the key to understanding the mechanism of sugar sensing.

The mechanistic link between SGLT activity and glucose sensing by enteroendocrine cells was demonstrated in the GLP-1-secreting cell line, GLUTag. As in the intact gut, GLUTag cells could be stimulated by non-metabolisable as well as metabolisable sugars, and using electrophysiological recordings, it was possible to demonstrate the appearance of sugar-dependent small depolarising currents of ~2 pA in magnitude. In agreement with the observed high input resistance of GLUTag cells, these currents were sufficient to trigger membrane depolarisation by 5–10 mV and action potential firing. The dose dependence of glucose-triggered GLP-1 release was submillimolar (~0.5 mM), similar to the K_m of SGLT1 (0.2 mM) (Diez-Sampedro et al. 2000) but well below the concentrations required to stimulate the sweet taste receptor pathway (30–100 mM). Similar results were observed in primary murine intestinal cultures (Reimann et al. 2008).

The role of SGLT1 in controlling GLP-1 release has been questioned because it has been difficult to detect the protein in enteroendocrine cells using traditional antibody-based techniques, and because the glucose-triggered pathway is also evident in the colon where SGLT1 expression is low. However, SGLT1 mRNA is found at relatively high levels in K cells and in L cells from both the small intestine and colon (Parker et al. 2009; Reimann et al. 2008), and protein expression has recently been demonstrated in the mouse colon (Yoshikawa et al. 2011). Although it may be argued that glucose is an unlikely physiological trigger of GLP-1 release in the colon, detectable glucose levels (0.2–2.1 mM) have been reported in the caecum from a number of species (Ferraris et al. 1990; Yoshikawa et al. 2011), and in humans, the glucose concentration at the ileo-caecal junction was found to rise as high as 10 mM after a meal (Stephen et al. 1983).

The role of SGLT3 in intestinal sugar sensing is less clear, although SGLT3 expression has been detected by PCR in GLUTag cells, rodent small intestinal epithelium and enteric neurons (Diez-Sampedro et al. 2003; Freeman et al. 2006a; Gribble et al. 2003). Like SGLT1, its substrates include glucose and αMG, but unlike SGLT1, it is not responsive to galactose. Significant species variability in the properties of SGLT3 became apparent when it was reported that human SGLT3, unlike pig SGLT3, had lost its capacity to transport sugars and operated rather as a sugar-sensitive Na^+ channel. In addition, it has recently been shown that hSGLT3, but not mSGLT3b, is activated by micromolar concentrations of imino sugars such as deoxynojirimycin (Aljure and Díez-Sampedro 2010).

SGLT3 has been implicated in glucose-triggered 5-HT release from rat EC cells, causing downstream activation of afferent vagal fibres via 5-HT_3 receptors (Freeman et al. 2006b; Zhu et al. 2001). Thus, using phosphorylated calcium/calmodulin-dependent kinase 2 as a marker of cells activated by elevated Ca^{2+}, Vincent and coworkers found that pCaMKII was detectable in rat EC cells, as well as cell bodies of the afferent vagus, following luminal application of glucose and imino sugars but not galactose (Vincent et al. 2011). However, full characterisation of the substrate preferences of mouse SGLT3a and the rat SGLT3 isoforms is clearly required before it will be possible to use substrate and inhibitor pharmacology as a reliable indicator of rodent SGLT3 activation.

3.1.3 Intracellular Sensing Mechanisms

Apart from sensing extracellular carbohydrates via surface receptors or the direct coupling of sugar uptake to membrane depolarisation, some cells exhibit changes in the metabolic rate in response to altered fuel supply. Metabolic regulation of ATP-sensitive K^+ (K_{ATP}) channel activity has been described in a number of glucose-sensitive tissues including pancreatic islets and brain. In pancreatic β cells metabolism of glucose generates ATP and simultaneously depletes MgADP, leading to closure of the K_{ATP} channels. This small reduction in the background K^+ current is sufficient to enable small inward currents to trigger membrane depolarisation, action potential generation and voltage-gated Ca^{2+} entry (Rorsman 1997). Primary murine L and K cells show enrichment, compared with their neighbouring enterocytes, of mRNAs encoding the K_{ATP} channel subunits, Kir6.2 and SUR1 (sulphonylurea receptor), and the enzyme glucokinase which regulates the glycolytic flux (Parker et al. 2009; Reimann et al. 2008). Functional K_{ATP} channels are evident in whole cell electrophysiological recordings from L cells, as well as in secretion assays that demonstrate modulation of GLP-1 release by openers and inhibitors of K_{ATP} currents (Reimann and Gribble 2002; Reimann et al. 2008; Rogers et al. 2011). In human intestine, Kir6.2 and glucokinase have been detected in L and K cells by immunostaining (Nielsen et al. 2007; Theodorakis et al. 2006). There is very little evidence, however, to suggest that K_{ATP} channel closure causes the peak of GLP-1 and GIP release following glucose ingestion in humans (El-Ouaghlidi et al. 2007; Murphy et al. 2009). Such a mechanism would predict that sulphonylureas, which stimulate insulin secretion by closing K_{ATP} channels, would similarly trigger GLP-1 secretion, and this is not the case. The role of K_{ATP} channels in enteroendocrine cells therefore remains an enigma, although they may provide an explanation for the observed link between gut hormone secretion and plasma glucose concentrations under certain conditions, as exemplified in the perfused pig intestine where GLP-1 release was enhanced at elevated vascular glucose concentrations in the presence of a slow luminal perfusion of nutrients (Hansen et al. 2004).

A subset of nerve fibres in the myenteric plexus of the gut express K_{ATP} channels and are reported to sense glucose directly (Liu et al. 1999). Immunostaining showed the presence of Kir6.2 and SUR1 subunits in cholinergic neurons of guinea pig ileum. Accordingly, the activity of these neurons was modulated by extracellular glucose and sensitive to the sulphonylurea, tolbutamide.

Dietary fructose has a number of metabolic effects, including stimulation of GLP-1 secretion and elevation of blood pressure by salt retention. Apical fructose transport in the intestine occurs via the facilitative sugar transporter GLUT5, activity of which is not electrogenic and has not been linked directly to activation of intracellular signalling pathways. GLUT5 appears necessary for the observed hypertensive effect of fructose, as demonstrated by the loss of fructose-triggered hypertension in GLUT5 knockout mice (Barone et al. 2009). Whilst the downstream mediators of fructose-triggered sodium uptake in the small intestine appear to include the ion transporters NHE3 and PAT1, the molecular identity of the fructose sensor linking these pathways remains obscure (Soleimani 2011). Similarly, it is not evident how fructose triggers GLP-1 release in vivo. L cells express high levels of GLUT5 mRNA (Reimann et al. 2008), suggesting they can take up fructose from the intestinal lumen. However, while absorbed fructose could enter the metabolic pathway, with consequent generation of ATP and closure of K_{ATP} channels, it seems unlikely that this accounts on its own for fructose-triggered GLP-1 release in vivo, as the simple closure of K_{ATP} channels by sulphonylureas is not a sufficient trigger for secretion (El-Ouaghlidi et al. 2007). Perhaps, in the case of fructose-triggered secretion, the enhanced metabolic rate may generate additional signals, including ATP itself, that enhance the exocytotic rate.

3.2 *Lipids*

Fat ingestion is a potent stimulus for the secretion of a number of enteroendocrine hormones, including CCK, GIP and GLP-1. Whilst much of the lipid in fat-rich food is in the form of triglycerides, other lipids, like cholesterol and phospholipids, are commonly found in fairly high concentrations in food but are additionally released into the small intestine as constituents of bile. Enterocytes absorb free fatty acids (FFA) and monoacylglycerides, resynthesise triglycerides and package them with apolipoproteins to form chylomicrons which are secreted into the lymph (Kindel et al. 2010). In principle, enteroendocrine cells could sense luminal triglycerides directly, but evidence points towards a prerequirement for lipolysis (Meyer et al. 1998). Long-chain fatty acids (LCFA), but also chylomicrons, inhibit gastric emptying (Glatzle et al. 2002; Hunt and Knox 1968) and induce satiety (Sakata et al. 1996), presumably by stimulating enteroendocrine secretion. Indeed, LCFA (>12 carbons) and chylomicrons are potent stimuli of CCK release (McLaughlin et al. 1998; Raybould et al. 1998), and it has long been known that CCK regulates gastric function. Indeed, vagal stimulation by LCFA has been attributed to an indirect process, whereby the FFA are sensed by I cells and the released CCK acts in a paracrine manner stimulating CCK-A receptors on the vagal afferents (Lal et al. 2001). Lipids and more specifically LCFA have also been shown to elevate plasma levels of GIP, GLP-1 and PYY (Aponte et al. 1988; Ellrichmann et al. 2008; Enç et al. 2009; Pilichiewicz et al. 2003), suggesting that lipid-sensing pathways may be common to enteroendocrine cells (Fig. 2).

3.2.1 Roles of Chylomicrons

For lipid-induced CCK release, chylomicron formation was found to be essential, as responses were impaired by Pluronic L-81 (Raybould et al. 1998) or specific inhibitors of microsomal triglyceride transfer protein, MTP, an enzyme responsible for assembly and secretion of triglyceride-rich apolipoprotein B lipoproteins (Hata et al. 2011). Knockout of diacylglycerol acyltransferase 1 (DGAT1), which is normally responsible for re-esterification of absorbed FFA and monoglycerides, similarly reduced secretion of GIP from the upper gastrointestinal tract (Okawa et al. 2009). These findings raise questions about the exact site of the nutrient sensor and the role of neighbouring enterocytes, as it is unknown whether endocrine cells are themselves capable of chylomicron synthesis.

In contrast to the inhibitory effects of MTP inhibitors on CCK secretion and of DGAT1 knockout on GIP release, both experimental paradigms resulted in elevated plasma levels of GLP-1 and PYY (Hata et al. 2011; Okawa et al. 2009). These data

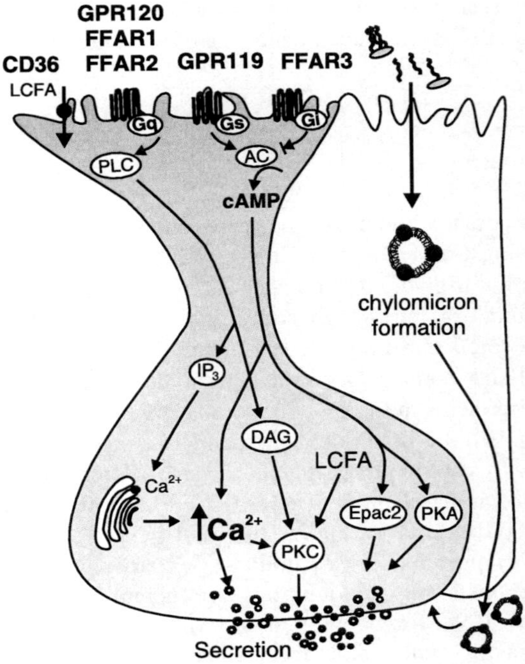

Fig. 2 Lipid sensing in the intestine A number of GPCRs have been identified that respond to lipids and are proposed to be involved in mediating lipid-induced gut hormone release from enteroendocrine cells (shaded cell). These receptors couple to either the phosphatidylinositol pathway (Gq) triggering the release of intracellular Ca^{2+} and stimulation of PKC, both of which are considered potent secretagogues, or modulate the cAMP pathway (Gs and Gi). cAMP and its downstream effectors, Epac2 and protein kinase A (PKA), have also been implicated in exocytosis mechanisms. The requirement for enterocytes (clear cell) and the exact cellular location of the lipid sensors are still unknown, as the formation of chylomicrons may contribute to lipid stimulation of some enteroendocrine cells. The fatty acid transporter CD36 has been proposed to act as a lipid sensor but may also transport LCFA into endocrine cells, which in turn may activate PKC

suggest that a chylomicron-dependent pathway is not responsible for fat-triggered GLP-1 and PYY release. One possible explanation for the findings is that inhibition of MTP or DGAT1 prevents lipid absorption in the upper intestinal tract and results in greater delivery of fat to more distal regions where there is a higher density of L cells.

3.2.2 Free Fatty Acid Receptors

In the last decade, a number of FFA-sensitive GPCRs have been identified, including FFAR1-3 and GPR120 (Briscoe et al. 2003; Brown et al. 2003; Hirasawa et al. 2005; Nilsson et al. 2003). A number of reports have further shown that mRNA messages for these receptors are enriched in gut endocrine cells (Liou et al. 2011a;

Parker et al. 2009; Reimann et al. 2008). FFAR1 (GPR40) and GPR120 respond to medium- to long-chain FFA, whereas FFAR 2–3 bind SCFA (see below). Both FFAR1 and GPR120 couple to the Gq–phosphatidylinositol pathway which elevates intracellular calcium and activates other key molecules required for exocytosis (Briscoe et al. 2003; Hirasawa et al. 2005). There is some evidence that these receptors may regulate gut hormone release and may explain some of the sensitivity of intestinal endocrine cell lines and primary cultures to FFAs (reviewed in (Reimann 2010). Thus, knockout of FFAR1 was associated with reduced GLP-1 and GIP responses to high-fat diet in vivo (Edfalk et al. 2008) and impaired linoleic-acid-stimulated CCK secretion and I cell Ca^{2+} responses in vitro (Liou et al. 2011a). In STC-1 cells, GPR120 was found to be important for GLP-1 and CCK secretion, as demonstrated by siRNA-mediated knockdown experiments (Hirasawa et al. 2005; Tanaka et al. 2008).

SCFA, the agonists identified for FFAR2 (GPR43) and FFAR3 (GPR41) (Brown et al. 2003; Le Poul et al. 2003), are found at high concentrations in the distal intestine. Dietary fibre composed of "indigestible carbohydrates" reaching the large intestine undergoes bacterial fermentation producing short-chain fatty acids (SCFA) which have been shown to increase GLP-1 and PYY secretion in humans (Freeland and Wolever 2010; Zhou et al. 2008). GPR43 is reported preferentially to bind C_2–C_3 chain length SCFA and to act via either the Gq–phosphatidylinositol pathway elevating intracellular calcium or the Gi/o pathway which decreases intracellular cAMP, whereas GPR41 reportedly couples only to Gi/o and preferentially binds C_3–C_5 chain length fatty acids (Nilsson et al. 2003). Given that immunostaining has located both these receptors to colonic L cells (Karaki et al. 2008; Tazoe et al. 2009), it is possible that SCFA may utilise these pathways to modulate the release of GLP-1. SCFA have been shown to stimulate colon motility via a serotonin receptor-dependent pathway, which has led to the hypothesis that serotonin-secreting EC cells may be involved (Fukumoto et al. 2003; Mitsui et al. 2005). However, it was shown by immunostaining that GPR43 colocalised with 5-HT containing mast cells, not EC cells (Karaki et al. 2006).

3.2.3 GPR119

GPR119, although not responsive to FFA per se, responds to lipid derivatives, particularly oleoylethanolamide (OEA) and lysophosphatidylcholine (Overton et al. 2006), although there remains some debate around whether these are the physiological GPR119 ligands. Like the FFA receptors, GPR119 is enriched in intestinal L and K cells (Parker et al. 2009; Reimann et al. 2008), consistent with the findings that a synthetic GPR119 ligand elevated plasma levels of GLP-1 and GIP in mice (Chu et al. 2008) and that OEA administration decreased food intake in rats (Rodriguez de Fonseca et al. 2001). It has been proposed that ingestion of nutrients stimulates the production of OEA in the small intestine, with local levels increasing greater than 100 pmol/g following a meal (Fu et al. 2007). Whilst it remains unclear whether locally elevated OEA levels are sufficient to stimulate GPR119 (EC_{50}

3 μM (Overton et al. 2006)), it is evident from the impairment of nutrient-stimulated GLP-1 release in GPR119 knockout mice (Lan et al. 2009) that this receptor contributes in some way to nutrient-stimulated incretin secretion.

3.2.4 CD36: An FFA Membrane Transporter

CD36, a transporter for free fatty acids, has been suggested as a sensor in gustatory taste papillae, where it is believed to trigger stored Ca^{2+} release and membrane depolarisation via a pathway involving Src-PTK-B, PLCβ2 and TRPM5 (Fukuwatari et al. 1997; Khan and Besnard 2009; Sclafani et al. 2007; Simons and Boon 2010). It has been hypothesised that CD36 acts not only as LCFA transporter in taste buds but also as a co-receptor for GPR120, whereby CD36 might act to trap and transfer LCFA to the lower affinity receptor GPR120 (Martin et al. 2011). CD36 may play a similar role in the intestine, where it is also found to be expressed in the brush border of the duodenum and jejunum.

Mice lacking CD36 exhibit reduced OEA production and OEA-induced satiety (Schwartz et al. 2008). In a slightly different model from that described in the lingual taste buds, it has been suggested that CD36 may transport fatty acids across the brush border, where they are required for synthesis of OEA and N-oleoyl-phosphatidylethanolamine (NOPE). Separate from the potential interaction of OEA with GPR119, there is evidence from knockout models that OEA and NOPE may influence satiety through a pathway involving activation of PPARα (Fu et al. 2003; Schwartz et al. 2008). It has also been suggested that elevated intracellular levels of LCFA may trigger GLP-1 release by direct activation of atypical protein kinase C (Iakoubov et al. 2007).

3.3 Protein and Amino Acids

Protein and its digested products elicit secretion of a number of gut hormones both in vivo (Elliott et al. 1993; Greenfield et al. 2009; Konturek et al. 1973) and in vitro (Conigrave and Brown 2006; Cordier-Bussat et al. 1998; Parker et al. 2009; Reimer et al. 2001; Reimer 2006; Tolhurst et al. 2011). Given the range of potential protein degradation products, ranging from individual amino acids and di/tripeptides to larger oligopeptides, the ability to sense this macronutrient is likely to be multifactorial. Inhibition of G_i-mediated pathways by pertussis toxin impaired peptone-induced CCK release from STC-1 cells and suggested the involvement of GPCR-evoked pathways (Choi et al. 2007). However, voltage-gated Ca^{2+} channels have also been implicated, suggesting that membrane potential changes play a role, perhaps triggered by electrogenic amino acid or peptide transporters (Nemoz-Gaillard et al. 1998). An overview of peptide- and amino-acid-evoked pathways is described in Fig. 3.

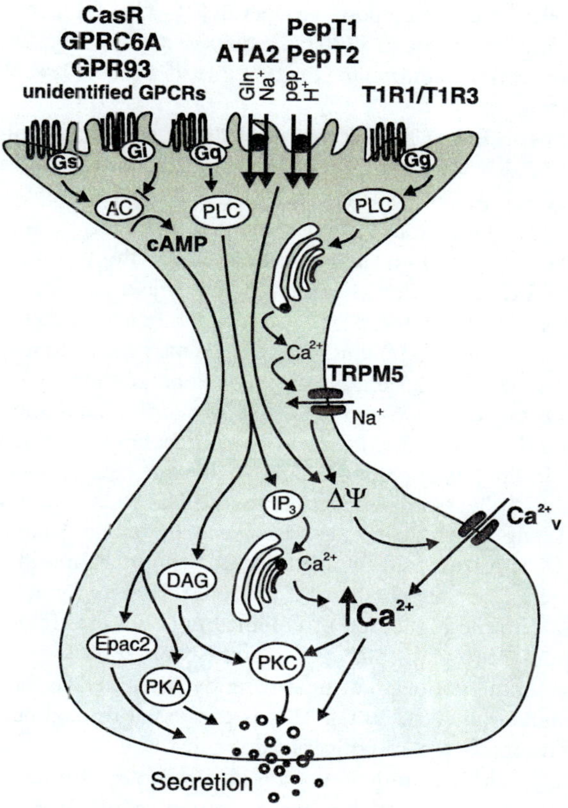

Fig. 3 Protein and amino acid sensing in enteroendocrine cells A number of GPCRs and electrogenic transporters have been identified as playing a role in the release of gut hormones in response to amino acids, peptides or peptones. The taste receptor T1R1/T1R3 signals through α-gustducin (Gg), releasing Ca^{2+} from intracellular stores, triggering membrane depolarisation (ΔΨ) via TRPM5 and thereby activating voltage-gated Ca^{2+} channels ($Ca_V{}^{2+}$). Other GPCR-coupled pathways which respond to amino acids act via the Gq pathway, elevating intracellular Ca^{2+} and PKC levels, or modulate cAMP levels with possible downstream effects on Epac2 and protein kinase A (PKA). Gq- and Gs-evoked pathways are common secretory mechanistic pathways in endocrine cell types. Coupled transport with ions of some amino acids (e.g. glutamine) or di-/tripeptides, through their respective carriers, ATA2 and PepT1 and/or PepT2, down an electrochemical gradient depolarises the membrane triggering Ca^{2+} entry, via voltage-gated Ca^{2+} ion channels ($Ca^{2+}{}_V$) and secretion of enteroendocrine hormones

3.3.1 Peptide and Amino-Acid-Sensitive G-Protein-Coupled Receptors

A number of GPCRs have been found to respond to small peptides or amino acids. GPR93 was proposed to link peptones to increased CCK expression and release from I cells or STC-1 cells (Nemoz-Gaillard et al. 1998), although endogenous levels of the receptor in STC-1 cells were insufficient to mediate robust responses.

Activation of GPR93 has been reported to recruit Gq-, Gs-, Gi- and $G_{12/13}$-mediated signalling pathways (Choi et al. 2007; Kotarsky et al. 2006; Lee et al. 2006). However, further work is required to clarify a role, if any, of GPR93 in physiological peptone sensing.

The heterodimeric T1R1/T1R3 receptor responds to a range of aliphatic, but not aromatic amino acids (Nelson et al. 2002), including the umami tastant, L-glutamate (Li et al. 2002). Although T1R1 and T1R3 subunits are reportedly expressed in discrete cell types throughout the gut, including endocrine cells (Bezençon et al. 2007), to date the only functional response attributed to this receptor in the intestine is regulation of GLUT2 and PepT1 trafficking in rat jejunum (Mace et al. 2009).

The calcium-sensing receptor, CaSR, was originally identified in the parathyroid gland (reviewed in Brown 2007), but was subsequently found to respond to a broad range of amino acids, especially aromatic compounds like phenylalanine (Conigrave and Hampson 2006), which act as allosteric modulators. CaSR couples to the phosphatidylinositol pathway, elevating intracellular Ca^{2+} (Rey et al. 2010). It is expressed in epithelial cells throughout the intestine (Chattopadhyay et al. 1998; Cheng et al. 2002; Gama et al. 1997) as well as CCK-expressing cells (Liou et al. 2011b; Wang et al. 2010), and has been linked to aromatic amino acid stimulation of CCK release and intracellular Ca^{2+} mobilisation in primary murine duodenal cultures. The intestinal CaSR is better known for its role in linking extracellular Ca^{2-} to the regulation of colonic epithelial proliferation, differentiation and development (reviewed in Geibel and Hebert 2009).

Very little is known about the most recently identified amino acid receptor, GPRC6A (Wellendorph et al. 2007). This is expressed throughout the gut, but to date there are no reports regarding its physiological role.

It seems likely that there remain unidentified receptors for protein degradation products. L-glutamine is a highly effective stimulus of GLP-1 secretion from primary murine colonic cultures and GLUTag cells, in part acting via elevation of intracellular cAMP (Tolhurst et al. 2011). This is unlikely to be mediated by the amino-acid-sensitive receptors mentioned above, since the profile of responses to other amino acids did not match the known data. In NCI-H716 cells, tetrapeptides stimulated GLP-1 release via a pathway dependent on 2-APB-sensitive Ca^{2+} mobilisation, potentially coupled to the opening of store operated Ca^{2+} channels (Le Nevé and Daniel 2011). To date, however, there are no receptors reported to respond to tetrapeptides, and for that matter, no peptide transporters known to transport them.

3.3.2 Peptide and Amino Acid Transporters

The proton-driven peptide transporters PepT1 and PepT2 (SLC15A1 and SLC15A2, respectively) and a number of Na^{+}-coupled amino acid transporters are electrogenic and thus potentially able to depolarise cell membranes and elicit voltage-dependent Ca^{2+} entry. Stimulation of hormone release from GLUTag, NCI-H716 and STC-1-derived cell lines by meat hydrolysates is partially inhibited by the voltage-gated Ca^{2+}-channel blocker verapamil (Cordier-Bussat et al. 1998; Nemoz-Gaillard et al. 1998), suggesting the possible involvement of peptide and amino acid transporters linked to membrane depolarisation. Glutamine stimulation of GLP-1 release from both GLUTag cells and primary colonic cultures also has an electrogenic component, hypothesised to be attributable to ATA2 (SNAT2, SLC38A2) and B^0AT1 (SLC6A19) activity (Reimann et al. 2004; Tolhurst et al. 2011).

4 Concluding Remarks

Nutrient sensors along the gastrointestinal tract trigger a range of physiological responses, enabling the gut, peripheral tissues and brain to act in coordination to optimise the absorption and utilisation of ingested foods. Specialised cell types, including enteroendocrine cells and tuft cells, are ideally placed to fulfil this role, and release chemical messengers that act locally or as classical circulating hormones. Although many receptors and transporters have been implicated, we still only have a rudimentary understanding of the chemosensory machinery employed by enteroendocrine cells and even less knowledge of specificity within the system. Nonetheless, as it is increasingly recognised that gut hormones and afferent vagal signalling play critical roles in the regulation of appetite and blood glucose, it is hoped that harnessing these pathways therapeutically may provide new avenues for the treatment of obesity and diabetes

Acknowledgements FMG, GT and FR are supported by grants from the Wellcome Trust ((#WT088357, #WT084210).

References

Aljure O, Díez-Sampedro A (2010) Functional characterization of mouse sodium/glucose transporter type 3b. Am J Physiol Cell Physiol 299:C58–C65

Aponte GW, Taylor IL, Soll AH (1988) Primary culture of PYY cells from canine colon. Am J Physiol 254:G829–G836

Balakrishnan A, Stearns AT, Rounds J, Irani J, Giuffrida M, Rhoads DB, Ashley SW, Tavakkolizadeh A (2008) Diurnal rhythmicity in glucose uptake is mediated by temporal periodicity in the expression of the sodium-glucose cotransporter (SGLT1). Surgery 143:813–818

Barone S, Fussell SL, Singh AK, Lucas F, Xu J, Kim C, Wu X, Yu Y, Amlal H, Seidler U, Zuo J, Soleimani M (2009) Slc2a5 (Glut5) is essential for the absorption of fructose in the intestine and generation of fructose-induced hypertension. J Biol Chem 284:5056–5066

Bayliss WM, Starling EH (1902) The mechanism of pancreatic secretion. J Physiol 28:325–353

Behrens M, Reichling C, Batram C, Brockhoff A, Meyerhof W (2009) Bitter taste receptors and their cells. Ann N Y Acad Sci 1170:111–115

Bezençon C, le Coutre J, Damak S (2007) Taste-signaling proteins are coexpressed in solitary intestinal epithelial cells. Chem Senses 32:41–49

Briscoe CP, Tadayyon M, Andrews JL, Benson WG, Chambers JK, Eilert MM, Ellis C, Elshourbagy NA, Goetz AS, Minnick DT, Murdock PR, Sauls HR Jr, Shabon U, Spinage LD, Strum JC, Szekeres PG, Tan KB, Way JM, Ignar DM, Wilson S, Muir AI (2003) The orphan G protein-coupled receptor GPR40 is activated by medium and long chain fatty acids. J Biol Chem 278:11303–11311

Brown EM (2007) The calcium-sensing receptor: physiology, pathophysiology and CaR-based therapeutics. Subcell Biochem 45:139–167

Brown AJ, Goldsworthy SM, Barnes AA, Eilert MM, Tcheang L, Daniels D, Muir AI, Wigglesworth MJ, Kinghorn I, Fraser NJ, Pike NB, Strum JC, Steplewski KM, Murdock PR, Holder JC, Marshall FH, Szekeres PG, Wilson S, Ignar DM, Foord SM, Wise A, Dowell SJ (2003) The Orphan G protein-coupled receptors GPR41 and GPR43 are activated by propionate and other short chain carboxylic acids. J Biol Chem 278:11312–11319

Carleton A, Accolla R, Simon SA (2010) Coding in the mammalian gustatory system. Trends Neurosci 33:326–334

Chattopadhyay N, Cheng I, Rogers K, Riccardi D, Hall A, Diaz R, Hebert SC, Soybel DI, Brown EM (1998) Identification and localization of extracellular Ca(2+)-sensing receptor in rat intestine. Am J Physiol 274:G122–G130

Chen XZ, Coady MJ, Jackson F, Berteloot A, Lapointe JY (1995) Thermodynamic determination of the Na+: glucose coupling ratio for the human SGLT1 cotransporter. Biophys J 69:2405–2414

Cheng SX, Okuda M, Hall AE, Geibel JP, Hebert SC (2002) Expression of calcium-sensing receptor in rat colonic epithelium: evidence for modulation of fluid secretion. Am J Physiol Gastrointest Liver Physiol 283:G240–G250

Choi S, Lee M, Shiu AL, Yo SJ, Halldén G, Aponte GW (2007) GPR93 activation by protein hydrolysate induces CCK transcription and secretion in STC-1 cells. Am J Physiol Gastrointest Liver Physiol 292:G1366–G1375

Chu ZL, Carroll C, Alfonso J, Gutierrez V, He H, Lucman A, Pedraza M, Mondala H, Gao H, Bagnol D, Chen R, Jones RM, Behan DP, Leonard J (2008) A role for intestinal endocrine cell-expressed g protein-coupled receptor 119 in glycemic control by enhancing glucagon-like Peptide-1 and glucose-dependent insulinotropic Peptide release. Endocrinology 149:2038–2047

Conigrave AD, Brown EM (2006) Taste receptors in the gastrointestinal tract. II. L-amino acid sensing by calcium-sensing receptors: implications for GI physiology. Am J Physiol Gastrointest Liver Physiol 291:G753–G761

Conigrave AD, Hampson DR (2006) Broad-spectrum L-amino acid sensing by class 3 G-protein-coupled receptors. Trends Endocrinol Metab 17:398–407

Cordier-Bussat M, Bernard C, Levenez F, Klages N, Laser-Ritz B, Philippe J, Chayvialle JA, Cuber JC (1998) Peptones stimulate both the secretion of the incretin hormone glucagon-like peptide 1 and the transcription of the proglucagon gene. Diabetes 47:1038–1045

Debnam ES, Levin RJ (1975) An experimental method of identifying and quantifying the active transfer electrogenic component from the diffusive component during sugar absorption measured in vivo. J Physiol 246:181–196

Díez-Sampedro A, Barcelona S (2011) Sugar binding residue affects apparent Na + affinity and transport stoichiometry in mouse sodium/glucose cotransporter type 3B. J Biol Chem 286:7975–7982

Diez-Sampedro A, Lostao MP, Wright EM, Hirayama BA (2000) Glycoside binding and translocation in Na(+)-dependent glucose cotransporters: comparison of SGLT1 and SGLT3. J Membr Biol 176:111–117

Diez-Sampedro A, Hirayama AB, Osswald C, Gorboulev V, Baumgarten K, Volk C, Wright E, Koepsell H (2003) A glucose sensor hiding in a family of transporters. Proc Natl Acad Sci 100:11753–11758

Dockray GJ (2003) Luminal sensing in the gut: an overview. J Physiol Pharmacol 54(Suppl 4):9–17

Drucker D, Jin T, Asa SL, Young TA, Brubaker PL (1994) Activation of proglucagon gene transcription by protein kinase-A in a novel mouse enteroendocrine cell line. Mol Endocrinol 8:1646–1655

Dyer J, Salmon KS, Zibrik L, Shirazi-Beechey SP (2005) Expression of sweet taste receptors of the T1R family in the intestinal tract and enteroendocrine cells. Biochem Soc Trans 33:302–305

Edfalk S, Steneberg P, Edlund H (2008) Gpr40 is expressed in enteroendocrine cells and mediates free fatty acid stimulation of incretin secretion. Diabetes 57:2280–2287

Elliott RM, Morgan LM, Tredger JA, Deacon S, Wright J, Marks V (1993) Glucagon-like peptide-1 (7-36)amide and glucose-dependent insulinotropic polypeptide secretion in response to nutrient ingestion in man: acute post-prandial and 24-h secretion patterns. J Endocrinol 138:159–166

Ellrichmann M, Kapelle M, Ritter PR, Holst JJ, Herzig KH, Schmidt WE, Schmitz F, Meier JJ (2008) Orlistat inhibition of intestinal lipase acutely increases appetite and attenuates post-prandial glucagon-like peptide-1-(7-36)-amide-1, cholecystokinin, and peptide YY concentrations. J Clin Endocrinol Metab 93:3995–3998

El-Ouaghlidi A, Rehring E, Holst JJ, Schweizer A, Foley J, Holmes D, Nauck MA (2007) The dipeptidyl peptidase 4 inhibitor vildagliptin does not accentuate glibenclamide-induced hypoglycemia but reduces glucose-induced glucagon-like peptide 1 and gastric inhibitory polypeptide secretion. J Clin Endocrinol Metab 92:4165–4171

Enç FY, Ones T, Akin HL, Dede F, Turoğlu HT, Ulfer G, Bekiroğlu N, Haklar G, Rehfeld JF, Holst JJ, Ulusoy NB, Imeryüz N (2009) Orlistat accelerates gastric emptying and attenuates GIP release in healthy subjects. Am J Physiol Gastrointest Liver Physiol 296:G482–G489

Ferraris RP, Yasharpour S, Lloyd KC, Mirzayan R, Diamond JM (1990) Luminal glucose concentrations in the gut under normal conditions. Am J Physiol 259:G822–G837

Freeland KR, Wolever TM (2010) Acute effects of intravenous and rectal acetate on glucagon-like peptide-1, peptide YY, ghrelin, adiponectin and tumour necrosis factor-alpha. Br J Nutr 103:460–466

Freeman SL, Bohan D, Darcel N, Raybould HE (2006a) Luminal glucose sensing in the rat intestine has characteristics of a sodium-glucose cotransporter. Am J Physiol Gastrointest Liver Physiol 291:G439–G445

Freeman SL, Glatzle J, Robin CS, Valdellon M, Sternini C, Sharp JW, Raybould HE (2006b) Ligand-induced 5-HT3 receptor internalization in enteric neurons in rat ileum. Gastroenterology 131:97–107

Friedlander RS, Moss CE, Mace J, Parker HE, Tolhurst G, Habib AM, Wachten S, Cooper DM, Gribble FM, Reimann F (2010) Role of phosphodiesterase and adenylate cyclase isozymes in murine colonic glucagon like peptide 1 secreting cells. Br J Pharmacol 163:261–271

Fu J, Gaetani S, Oveisi F, Lo Verme J, Serrano A, Rodríguez De Fonseca F, Rosengarth A, Luecke H, Di Giacomo B, Tarzia G, Piomelli D (2003) Oleylethanolamide regulates feeding and body weight through activation of the nuclear receptor PPAR-alpha. Nature 425:90–93

Fu J, Astarita G, Gaetani S, Kim J, Cravatt BF, Mackie K, Piomelli D (2007) Food intake regulates oleoylethanolamide formation and degradation in the proximal small intestine. J Biol Chem 282:1518–1528

Fujita Y, Wideman RD, Speck M, Asadi A, King DS, Webber TD, Haneda M, Kieffer TJ (2009) Incretin release from gut is acutely enhanced by sugar but not by sweeteners in vivo. Am J Physiol Endocrinol Metab 296:E473–E479

Fukumoto S, Tatewaki M, Yamada T, Fujimiya M, Mantyh C, Voss M, Eubanks S, Harris M, Pappas TN, Takahashi T (2003) Short-chain fatty acids stimulate colonic transit via intraluminal 5-HT release in rats. Am J Physiol Regul Integr Comp Physiol 284:R1269–R1276

Fukuwatari T, Kawada T, Tsuruta M, Hiraoka T, Iwanaga T, Sugimoto E, Fushiki T (1997) Expression of the putative membrane fatty acid transporter (FAT) in taste buds of the circumvallate papillae in rats. FEBS Lett 414:461–464

Gama L, Baxendale-Cox LM, Breitwieser GE (1997) Ca^{2+}-sensing receptors in intestinal epithelium. Am J Physiol 273:C1168–C1175

Geibel JP, Hebert SC (2009) The functions and roles of the extracellular Ca^{2+}-sensing receptor along the gastrointestinal tract. Annu Rev Physiol 71:205–217

Gerbe F, van Es JH, Makrini L, Brulin B, Mellitzer G, Robine S, Romagnolo B, Shroyer NF, Bourgaux JF, Pignodel C, Clevers H, Jay P (2011) Distinct ATOH1 and Neurog3 requirements define tuft cells as a new secretory cell type in the intestinal epithelium. J Cell Biol 192:767–780

Glatzle J, Kalogeris TJ, Zittel TT, Guerrini S, Tso P, Raybould HE (2002) Chylomicron components mediate intestinal lipid-induced inhibition of gastric motor function. Am J Physiol Gastrointest Liver Physiol 282:G86–G91

Greenfield JR, Farooqi IS, Keogh JM, Henning E, Habib AM, Blackwood A, Reimann F, Holst JJ, Gribble FM (2009) Oral glutamine increases circulating glucagon-like peptide 1, glucagon, and insulin concentrations in lean, obese, and type 2 diabetic subjects. Am J Clin Nutr 89:106–113

Gribble FM, Williams L, Simpson AK, Reimann F (2003) A novel glucose-sensing mechanism contributing to glucagon-like peptide-1 secretion from the GLUTag cell line. Diabetes 52:1147–1154

Grundy D (2004) What activates visceral afferents? Gut 53(Suppl 2):ii5–ii8

Guillemain G, Loizeau M, Pinçon-Raymond M, Girard J, Leturque A (2000) The large intracytoplasmic loop of the glucose transporter GLUT2 is involved in glucose signaling in hepatic cells. J Cell Sci 113:841–847

Hansen L, Hartmann B, Mineo H, Holst JJ (2004) Glucagon-like peptide-1 secretion is influenced by perfusate glucose concentration and by a feedback mechanism involving somatostatin in isolated perfused porcine ileum. Regul Pept 118:11–18

Hata T, Mera Y, Ishii Y, Tadaki H, Tomimoto D, Kuroki Y, Kawai T, Ohta T, Kakutani M (2011) JTT-130, a novel intestine-specific inhibitor of microsomal triglyceride transfer protein, suppresses food intake and gastric emptying with the elevation of plasma peptide YY and glucagon-like peptide-1 in a dietary fat-dependent manner. J Pharmacol Exp Ther 336:850–856

Hediger MA, Romero MF, Peng JB, Rolfs A, Takanaga H, Bruford EA (2004) The ABCs of solute carriers: physiological, pathological and therapeutic implications of human membrane transport proteins introduction. Pflugers Arch 447:465–468

Helliwell PA, Rumsby MG, Kellett GL (2003) Intestinal sugar absorption is regulated by phosphorylation and turnover of protein kinase C betaII mediated by phosphatidylinositol 3-kinase- and mammalian target of rapamycin-dependent pathways. J Biol Chem 278:28644–28650

Hirasawa A, Tsumaya K, Awaji T, Katsuma S, Adachi T, Yamada M, Sugimoto Y, Miyazaki S, Tsujimoto G (2005) Free fatty acids regulate gut incretin glucagon-like peptide-1 secretion through GPR120. Nat Med 11:90–94

Höfer D, Drenckhahn D (1996) Cytoskeletal markers allowing discrimination between brush cells and other epithelial cells of the gut including enteroendocrine cells. Histochem Cell Biol 105:405–412

Höfer D, Püschel B, Drenckhahn D (1996) Taste receptor-like cells in the rat gut identified by expression of alpha-gustducin. Proc Natl Acad Sci USA 93:6631–6634

Holzer P (2009) Opioid receptors in the gastrointestinal tract. Regul Pept 155:11–17

Hunt JN, Knox MT (1968) A relation between the chain length of fatty acids and the slowing of gastric emptying. J Physiol 194:327–336

Hwang ES, Hirayama BA, Wright EM (1991) Distribution of the SGLT1 Na+/glucose cotransporter and mRNA along the crypt-villus axis of rabbit small intestine. Biochem Biophys Res Commun 181:1208–1217

Iakoubov R, Izzo A, Yeung A, Whiteside CI, Brubaker PL (2007) Protein kinase Czeta is required for oleic acid-induced secretion of glucagon-like peptide-1 by intestinal endocrine L cells. Endocrinology 148:1089–1098

Jang HJ, Kokrashvili Z, Theodorakis MJ, Carlson OD, Kim BJ, Zhou J, Kim HH, Xu X, Chan SL, Juhaszova M, Bernier M, Mosinger B, Margolskee RF, Egan JM (2007) Gut-expressed gustducin and taste receptors regulate secretion of glucagon-like peptide-1. Proc Natl Acad Sci USA 104:15069–15074

Jarvi O, Keyrilainen O (1956) On the cellular structures of the epithelial invasions in the glandular stomach of mice caused by intramural application of 20-methylcholantren. Acta Pathol Microbiol Scand Suppl 39:72–73

Karaki S, Mitsui R, Hayashi H, Kato I, Sugiya H, Iwanaga T, Furness JB, Kuwahara A (2006) Short-chain fatty acid receptor, GPR43, is expressed by enteroendocrine cells and mucosal mast cells in rat intestine. Cell Tissue Res 324:353–360

Karaki S, Tazoe H, Hayashi H, Kashiwabara H, Tooyama K, Suzuki Y, Kuwahara A (2008) Expression of the short-chain fatty acid receptor, GPR43, in the human colon. J Mol Histol 39:135–142

Khan NA, Besnard P (2009) Oro-sensory perception of dietary lipids: new insights into the fat taste transduction. Biochim Biophys Acta 1791:149–155

Kindel T, Lee DM, Tso P (2010) The mechanism of the formation and secretion of chylomicrons. Atheroscler Suppl 11:11–16

Kokrashvili Z, Mosinger B, Margolskee RF (2009a) T1r3 and alpha-gustducin in gut regulate secretion of glucagon-like peptide-1. Ann N Y Acad Sci 1170:91–94

Kokrashvili Z, Rodriguez D, Yevshayeva V, Zhou H, Margolskee RF, Mosinger B (2009b) Release of endogenous opioids from duodenal enteroendocrine cells requires Trpm5. Gastroenterology 137:598–606

Konturek SJ, Radecki T, Thor P, Dembinski A (1973) Release of cholecystokinin by amino acids. Proc Soc Exp Biol Med 143:305–309

Kotarsky K, Boketoft A, Bristulf J, Nilsson NE, Norberg A, Hansson S, Owman C, Sillard R, Leeb-Lundberg LM, Olde B (2006) Lysophosphatidic acid binds to and activates GPR92, a G protein-coupled receptor highly expressed in gastrointestinal lymphocytes. J Pharmacol Exp Ther 318:619–628

Lal S, Kirkup AJ, Brunsden AM, Thompson DG, Grundy D (2001) Vagal afferent responses to fatty acids of different chain length in the rat. Am J Physiol Gastrointest Liver Physiol 281: G907–G915

Lan H, Vassileva G, Corona A, Liu L, Baker H, Golovko A, Abbondanzo SJ, Hu W, Yang S, Ning Y, Del Vecchio RA, Poulet F, Laverty M, Gustafson EL, Hedrick JA, Kowalski TJ (2009) GPR119 is required for physiological regulation of glucagon-like peptide-1 secretion but not for metabolic homeostasis. J Endocrinol 201:219–230

Le Nevé B, Daniel H (2011) Selected tetrapeptides lead to a GLP-1 release from the human enteroendocrine cell line NCI-H716. Regul Pept 167:14–20

Le Poul E, Loison C, Struyf S, Springael JY, Lannoy V, Decobecq ME, Brezillon S, Dupriez V, Vassart G, Van Damme J, Parmentier M, Detheux M (2003) Functional characterization of human receptors for short chain fatty acids and their role in polymorphonuclear cell activation. J Biol Chem 278:25481–25489

Lee CW, Rivera R, Gardell S, Dubin AE, Chun J (2006) GPR92 as a new G12/13- and Gq-coupled lysophosphatidic acid receptor that increases cAMP, LPA5. J Biol Chem 281:23589–23597

Li X, Staszewski L, Xu H, Durick K, Zoller M, Adler E (2002) Human receptors for sweet and umami taste. Proc Natl Acad Sci USA 99:4692–4696

Liou AP, Lu X, Sei Y, Zhao X, Pechhold S, Carrero RJ, Raybould HE, Wank S (2011a) The G-protein-coupled receptor GPR40 directly mediates long-chain fatty acid-induced secretion of cholecystokinin. Gastroenterology 140:903–912

Liou AP, Sei Y, Zhao X, Feng J, Lu X, Thomas C, Pechhold S, Raybould HE, Wank SA (2011b) The extracellular calcium sensing receptor is required for cholecystokinin secretion in response to L-phenylalanine in acutely isolated intestinal I cells. Am J Physiol Gastrointest Liver Physiol 300:538–546

Little TJ, Gupta N, Case RM, Thompson DG, McLaughlin JT (2009) Sweetness and bitterness taste of meals per se does not mediate gastric emptying in humans. Am J Physiol Regul Integr Comp Physiol 297:R632–R639

Liu M, Seino S, Kirchgessner AL (1999) Identification and characterization of glucoresponsive neurons in the enteric nervous system. J Neurosci 19:10305–10317
Ma J, Bellon M, Wishart JM, Young R, Blackshaw LA, Jones KL, Horowitz M, Rayner CK (2009) Effect of the artificial sweetener, sucralose, on gastric emptying and incretin hormone release in healthy subjects. Am J Physiol Gastrointest Liver Physiol 296:G735–G739
Ma J, Chang J, Checklin HL, Young RL, Jones KL, Horowitz M, Rayner CK (2010) Effect of the artificial sweetener, sucralose, on small intestinal glucose absorption in healthy human subjects. Br J Nutr 104:803–806
Mace OJ, Affleck J, Patel N, Kellett GL (2007) Sweet taste receptors in rat small intestine stimulate glucose absorption through apical GLUT2. J Physiol 582:379–392
Mace OJ, Lister N, Morgan E, Shepherd E, Affleck J, Helliwell P, Bronk JR, Kellett GL, Meredith D, Boyd R, Pieri M, Bailey PD, Pettcrew R, Foley D (2009) An energy supply network of nutrient absorption coordinated by calcium and T1R taste receptors in rat small intestine. J Physiol 587:195–210
Margolskee RF (2002) Molecular mechanisms of bitter and sweet taste transduction. J Biol Chem 277:1–4
Margolskee RF, Dyer J, Kokrashvili Z, Salmon KS, Ilegems E, Daly K, Maillet EL, Ninomiya Y, Mosinger B, Shirazi-Beechey SP (2007) T1R3 and gustducin in gut sense sugars to regulate expression of Na^+-glucose cotransporter 1. Proc Natl Acad Sci USA 104:15075–15080
Martin C, Chevrot M, Poirier H, Passilly-Degrace P, Niot I, Besnard P (2011) CD36 as a lipid sensor. Physiol Behav 105:36–42
Martinez-Pinna J, Tolhurst G, Gurung IS, Vandenberg JI, Mahaut-Smith MP (2004) Sensitivity limits for voltage control of P2Y receptor-evoked Ca^{2+} mobilization in the rat megakaryocyte. J Physiol 555:61–70
May LT, Leach K, Sexton PM, Christopoulos A (2007) Allosteric modulation of G protein-coupled receptors. Annu Rev Pharmacol Toxicol 47:1–51
McLaughlin JT, Lomax RB, Hall L, Dockray GJ, Thompson DG, Warhurst G (1998) Fatty acids stimulate cholecystokinin secretion via an acyl chain length-specific, Ca^{2+}-dependent mechanism in the enteroendocrine cell line STC-1. J Physiol 513:11–18
Meyer JH, Hlinka M, Khatibi A, Raybould HE, Tso P (1998) Role of small intestine in caloric compensations to oil premeals in rats. Am J Physiol 275:R1320–R1333
Mitsui R, Ono S, Karaki S, Kuwahara A (2005) Neural and non-neural mediation of propionate-induced contractile responses in the rat distal colon. Neurogastroenterol Motil 17:585–594
Money SR, Petroianu A, Gintzler AR, Jaffe BM (1988) Meal-stimulated release of methionine-enkephalin into the canine jejunal lumen. J Clin Invest 81:822–825
Morgan EL, Mace OJ, Affleck J, Kellett GL (2007) Apical GLUT2 and Cav1.3: regulation of rat intestinal glucose and calcium absorption. J Physiol 580:593–604
Murphy R, Tura A, Clark PM, Holst JJ, Mari A, Hattersley AT (2009) Glucokinase, the pancreatic glucose sensor, is not the gut glucose sensor. Diabetologia 52:154–159
Nabeyama A, Leblond CP (1974) "Caveolated cells" characterized by deep surface invaginations and abundant filaments in mouse gastro-intestinal epithelia. Am J Anat 140:147–165
Nelson G, Chandrashekar J, Hoon MA, Feng L, Zhao G, Ryba NJ, Zuker CS (2002) An amino-acid taste receptor. Nature 416:199–202
Nemoz-Gaillard E, Bernard C, Abello J, Cordier-Bussat M, Chayvialle JA, Cuber JC (1998) Regulation of cholecystokinin secretion by peptones and peptidomimetic antibiotics in STC-1 cells. Endocrinology 139:932–938
Nielsen LB, Ploug KB, Swift P, Orskov C, Jansen-Olesen I, Chiarelli F, Holst JJ, Hougaard P, Porksen S, Holl R, de Beaufort C, Gammeltoft S, Rorsman P, Mortensen HB, Hansen L (2007) Co-localisation of the Kir6.2/SUR1 channel complex with glucagon-like peptide-1 and glucose-dependent insulinotrophic polypeptide expression in human ileal cells and implications for glycaemic control in new onset type 1 diabetes. Eur J Endocrinol 156:663–671
Nilsson NE, Kotarsky K, Owman C, Olde B (2003) Identification of a free fatty acid receptor, FFA2R, expressed on leukocytes and activated by short-chain fatty acids. Biochem Biophys Res Commun 303:1047–1052

Okawa M, Fujii K, Ohbuchi K, Okumoto M, Aragane K, Sato H, Tamai Y, Seo T, Itoh Y, Yoshimoto R (2009) Role of MGAT2 and DGAT1 in the release of gut peptides after triglyceride ingestion. Biochem Biophys Res Commun 390:377–381

Overton HA, Babbs AJ, Doel SM, Fyfe MC, Gardner LS, Griffin G, Jackson HC, Procter MJ, Rasamison CM, Tang-Christensen M, Widdowson PS, Williams GM, Reynet C (2006) Deorphanization of a G protein-coupled receptor for oleoylethanolamide and its use in the discovery of small-molecule hypophagic agents. Cell Metab 3:167–175

Parekh D, Ishizuka J, Townsend CM, Haber B, Beauchamp RD, Karp G, Kim SW, Rajaraman S, Greeley G, Thompson JC (1994) Characterization of a human pancreatic carcinoid in vitro: morphology, amine and peptide storage, and secretion. Pancreas 9:83–90

Parker HE, Habib AM, Rogers GJ, Gribble FM, Reimann F (2009) Nutrient-dependent secretion of glucose-dependent insulinotropic polypeptide from primary murine K cells. Diabetologia 52:289–298

Pilichiewicz A, O'Donovan D, Feinle C, Lei Y, Wishart JM, Bryant L, Meyer JH, Horowitz M, Jones KL (2003) Effect of lipase inhibition on gastric emptying of, and the glycemic and incretin responses to, an oil/aqueous drink in type 2 diabetes mellitus. J Clin Endocrinol Metab 88:3829–3834

Powley TL, Spaulding RA, Haglof SA (2011) Vagal afferent innervation of the proximal gastrointestinal tract mucosa: chemoreceptor and mechanoreceptor architecture. J Comp Neurol 519:644–660

Prezeau L, Rives ML, Comps-Agrar L, Maurel D, Kniazeff J, Pin JP (2010) Functional crosstalk between GPCRs: with or without oligomerization. Curr Opin Pharmacol 10:6–13

Raybould HE, Meyer JH, Tabrizi Y, Liddle RA, Tso P (1998) Inhibition of gastric emptying in response to intestinal lipid is dependent on chylomicron formation. Am J Physiol 274: R1834–R1838

Raybould HE, Glatzle J, Freeman SL, Whited K, Darcel N, Liou A, Bohan D (2006) Detection of macronutrients in the intestinal wall. Auton Neurosci 125:28–33

Rehfeld JF (2004) A centenary of gastrointestinal endocrinology. Horm Metab Res 36:735–741

Reimann F (2010) Molecular mechanisms underlying nutrient detection by incretin-secreting cells. Int Dairy J 20:236–242

Reimann F, Gribble FM (2002) Glucose-sensing in glucagon-like peptide-1-secreting cells. Diabetes 51:2757–2763

Reimann F, Williams L, da Silva Xavier G, Rutter GA, Gribble FM (2004) Glutamine potently stimulates glucagon-like peptide-1 secretion from GLUTag cells. Diabetologia 47:1592–1601

Reimann F, Maziarz M, Flock G, Habib AM, Drucker DJ, Gribble FM (2005) Characterization and functional role of voltage gated cation conductances in the glucagon-like peptide-1 secreting GLUTag cell line. J Physiol 563:161–175

Reimann F, Habib AM, Tolhurst G, Parker HE, Rogers GJ, Gribble FM (2008) Glucose sensing in L cells: a primary cell study. Cell Metab 8:532–539

Reimer RA (2006) Meat hydrolysate and essential amino acid-induced glucagon-like peptide-1 secretion, in the human NCI-H716 enteroendocrine cell line, is regulated by extracellular signal-regulated kinase1/2 and p38 mitogen-activated protein kinases. J Endocrinol 191:159–170

Reimer RA, Darimont C, Gremlich S, Nicolas-Metral V, Ruegg UT, Mace K (2001) A human cellular model for studying the regulation of glucagon-like peptide-1 secretion. Endocrinology 142:4522–4528

Rey O, Young SH, Jacamo R, Moyer MP, Rozengurt E (2010) Extracellular calcium sensing receptor stimulation in human colonic epithelial cells induces intracellular calcium oscillations and proliferation inhibition. J Cell Physiol 225:73–83

Rindi G, Grant SG, Yiangou Y, Ghatei MA, Bloom SR, Bautch VL, Solcia E, Polak JM (1990) Development of neuroendocrine tumors in the gastrointestinal tract of transgenic mice. Heterogeneity of hormone expression. Am J Pathol 136:1349–1363

Ritter SL, Hall RA (2009) Fine-tuning of GPCR activity by receptor-interacting proteins. Nat Rev Mol Cell Biol 10:819–830

Ritzel U, Fromme A, Ottleben M, Leonhardt U, Ramadori G (1997) Release of glucagon-like peptide-1 (GLP-1) by carbohydrates in the perfused rat ileum. Acta Diabetol 34:18–21

Rodriguez de Fonseca F, Navarro M, Gomez R, Escuredo L, Nava F, Fu J, Murillo-Rodriguez E, Giuffrida A, LoVerme J, Gaetani S, Kathuria S, Gall C, Piomelli D (2001) An anorexic lipid mediator regulated by feeding. Nature 414:209–212

Rogers GJ, Tolhurst G, Ramzan A, Habib AM, Parker HE, Gribble FM, Reimann F (2011) Electrical activity-triggered glucagon-like peptide-1 secretion from primary murine L-cells. J Physiol 589:1081–1093

Rorsman P (1997) The pancreatic beta-cell as a fuel sensor: an electrophysiologist's viewpoint. Diabetologia 40:487–495

Rozengurt N, Wu SV, Chen MC, Huang C, Sternini C, Rozengurt E (2006) Colocalization of the alpha-subunit of gustducin with PYY and GLP-1 in L cells of human colon. Am J Physiol Gastrointest Liver Physiol 291:G792–G802

Sakata Y, Fujimoto K, Ogata S, Koyama T, Fukagawa K, Sakai T, Tso P (1996) Postabsorptive factors are important for satiation in rats after a lipid meal. Am J Physiol 271:G438–G442

Schwartz GJ, Fu J, Astarita G, Li X, Gaetani S, Campolongo P, Cuomo V, Piomelli D (2008) The lipid messenger OEA links dietary fat intake to satiety. Cell Metab 8:281–288

Sclafani A, Ackroff K, Abumrad NA (2007) CD36 gene deletion reduces fat preference and intake but not post-oral fat conditioning in mice. Am J Physiol Regul Integr Comp Physiol 293: R1823–R1832

Shirazi-Beechey SP, Moran AW, Bravo D, Al-Rammahi M (2011) Intestinal glucose sensing and regulation of glucose absorption: implications for swine nutrition. J Anim Sci 89:1854–1862

Simons PJ, Boon L (2010) Lingual CD36 and obesity: a matter of fat taste? Acta Histochem 113:765–767

Sjölund K, Sandén G, Håkanson R, Sundler F (1983) Endocrine cells in human intestine: an immunocytochemical study. Gastroenterology 85:1120–1130

Soleimani M (2011) Dietary fructose, salt absorption and hypertension in metabolic syndrome: towards a new paradigm. Acta Physiol (Oxf) 201:55–62

Stearns AT, Balakrishnan A, Rhoads DB, Tavakkolizadeh A (2010) Rapid upregulation of sodium-glucose transporter SGLT1 in response to intestinal sweet taste stimulation. Ann Surg 251:865–871

Stephen AM, Haddad AC, Phillips SF (1983) Passage of carbohydrate into the colon. Direct measurements in humans. Gastroenterology 85:589–595

Sykes S, Morgan LM, English J, Marks V (1980) Evidence for preferential stimulation of gastric inhibitory polypeptide secretion in the rat by actively transported carbohydrates and their analogues. J Endocrinol 85:201–207

Tanaka T, Katsuma S, Adachi T, Koshimizu TA, Hirasawa A, Tsujimoto G (2008) Free fatty acids induce cholecystokinin secretion through GPR120. Naunyn Schmiedebergs Arch Pharmacol 377:523–527

Tazoe H, Otomo Y, Karaki S, Kato I, Fukami Y, Terasaki M, Kuwahara A (2009) Expression of short-chain fatty acid receptor GPR41 in the human colon. Biomed Res 30:149–156

Theodorakis MJ, Carlson O, Michopoulos S, Doyle ME, Juhaszova M, Petraki K, Egan JM (2006) Human duodenal enteroendocrine cells: source of both incretin peptides, GLP-1 and GIP. Am J Physiol Endocrinol Metab 290:550–559

Tolhurst G, Zheng Y, Parker HE, Habib AM, Reimann F, Gribble FM (2011) Glutamine triggers and potentiates glucagon-like peptide-1 secretion by raising cytosolic Ca^{2+} and cAMP. Endocrinology 152:405–413

Vilardaga JP, Agnati LF, Fuxe K, Ciruela F (2010) G-protein-coupled receptor heteromer dynamics. J Cell Sci 123:4215–4220

Vincent KM, Sharp JW, Raybould HE (2011) Intestinal glucose-induced calcium-calmodulin kinase signaling in the gut-brain axis in awake rats. Neurogastroenterol Motil 23:282–293

Wang Y, Chandra R, Samsa LA, Gooch B, Fee BE, Cook JM, Vigna SR, Grant AO, Liddle RA (2010) Amino acids stimulate cholecystokinin release through the calcium-sensing receptor. Am J Physiol Gastrointest Liver Physiol 300:528–537

Wellendorph P, Burhenne N, Christiansen B, Walter B, Schmale H, Bräuner-Osborne H (2007) The rat GPRC6A: cloning and characterization. Gene 396:257–267

Wellendorph P, Johansen LD, Bräuner-Osborne H (2010) The emerging role of promiscuous 7TM receptors as chemosensors for food intake. Vitam Horm 84:151–184

Yoshida A, Takata K, Kasahara T, Aoyagi T, Saito S, Hirano H (1995) Immunohistochemical localization of Na(+)-dependent glucose transporter in the rat digestive tract. Histochem J 27:420–426

Yoshikawa T, Inoue R, Matsumoto M, Yajima T, Ushida K, Iwanaga T (2011) Comparative expression of hexose transporters (SGLT1, GLUT1, GLUT2 and GLUT5) throughout the mouse gastrointestinal tract. Histochem Cell Biol 135:183–194

You G, Lee WS, Barros EJ, Kanai Y, Huo TL, Khawaja S, Wells RG, Nigam SK, Hediger MA (1995) Molecular characteristics of Na(+)-coupled glucose transporters in adult and embryonic rat kidney. J Biol Chem 270:29365–29371

Zhou J, Martin RJ, Tulley RT, Raggio AM, McCutcheon KL, Shen L, Danna SC, Tripathy S, Hegsted M, Keenan MJ (2008) Dietary resistant starch upregulates total GLP-1 and PYY in a sustained day-long manner through fermentation in rodents. Am J Physiol Endocrinol Metab 295:E1160–E1166

Zhu JX, Zhu XY, Owyang C, Li Y (2001) Intestinal serotonin acts as a paracrine substance to mediate vagal signal transmission evoked by luminal factors in the rat. J Physiol 530:431–442

Part IV
Agents Modifying Food Intake

Reuptake Inhibitors of Dopamine, Noradrenaline, and Serotonin

Ulrich Kintscher

Contents

Abstract Pharmacological inhibition of monoamine reuptake transporters has been known for many years as an effective therapy to reduce food intake and body weight in obese subjects. However, most of the marketed drugs failed after a distinct period in clinical use and had to be withdrawn because of serious adverse effects resulting in a negative benefit–risk profile. The most common side effects for this drug class included increases in systemic or pulmonary blood pressure and/or heart rate, cardiac valvulopathies, higher cardiovascular event rates, psychiatric disorders, or high abuse potential. The recent withdrawal of sibutramine as result of its adverse actions on the cardiovascular system highlighted again the problems with this drug class in antiobesity therapy. Recent developments to combine reuptake inhibitors with other drug classes, for example, opioid antagonists seem to be a promising approach to improve the benefit–risk profile of these compounds.

This chapter will discuss the history of this drug class in appetite control, its mechanism of action, and the clinical effects of selected drugs from this class.

Keywords Appetite control • Monoamines With best regards • Reuptake inhibitors • Ulrich Kintscher

---

U. Kintscher (✉)
Charité – Universitätsmedizin, Institute of Pharmacology, Berlin, Germany
e-mail: ulrich.kintscher@charite.de

H.-G. Joost (ed.), *Appetite Control*, Handbook of Experimental Pharmacology 209,
DOI 10.1007/978-3-642-24716-3_15, © Springer-Verlag Berlin Heidelberg 2012

1 History of Drugs Targeting the Monoamine System for Appetite Control

The modulation of biogenic amines has been known for many years as a pharmacological strategy in the therapy of obesity.

Amphetamines such as amphetamine sulfate/benzedrine or methamphetamine were the first major class of sympathomimetic agents described for their antiobesity actions (Lesses and Myerson 1994). Amphetamines exhibit an indirect sympathomimetic effect (Sulzer et al. 2005). When applied to the central nervous system (CNS), they induce an enhanced monoamine release from nerve terminals (Sulzer et al. 2005). Amphetamines enter the presynaptic nerves via the monoamine uptake 1 transporter (Sulzer et al. 2005). Within the nerve terminal, amphetamines are taken up by presynaptic vesicles via the vesicular monoamine transporter in exchange for monoamines resulting in an increased cytosolic concentration of these amines. In part, cytosolic monoamines are then rapidly metabolized by the monoamine oxidase (MAO). The other part, however, is released in the synaptic cleft by the uptake 1 transporter in exchange for amphetamines (Sulzer et al. 2005). Finally, the released monoamines such as noradrenaline, dopamine, or serotonin (5-hydroxytryptamine, 5-HT) act on their corresponding postsynaptic receptors to modify body weight (mechanism described in detail below). (Samanin and Garattini 1993) Treatment with these substances resulted in significant body weight reduction when compared to placebo. (Bray and Greenway 1999) However, clinical application was limited by a broad range of side effects including adverse cardiovascular effects, CNS-stimulatory actions, and a high abuse potential.

In the 1960s, aminorex, an amphetamine-like drug, was approved as an appetite suppressant and anorectic medication in Germany, Switzerland, and Austria (Hadler 1967). The use of aminorex was associated with a marked increase in the incidence of pulmonary hypertension in these countries which resulted in the withdrawal of the drug in 1972 (Greiser 1973).

The use of amphetamine-like drugs in antiobesity treatment continued during the following years with the most popular drug named fen–phen introduced in the early 1990s. Fen–phen was a combination of fenfluramine and phentermine. Fenfluramine releases monoamines from the presynaptic nerve terminal, in particular serotonin and noradrenaline, by acting on vesicular transporters (Rothman et al. 2003). In addition, fenfluramine has been described to inhibit the serotonin reuptake transporter (Rothman and Baumann 2002). Phentermine is a congener of amphetamine and has a similar mechanism of action (Samanin and Garattini 1993). With regard to monoamine uptake inhibition and release in the synaptic cleft, phentermine preferentially targets noradrenaline followed by dopamine and serotonin (Rothman et al. 2001). In a double-blind placebo-controlled clinical trial with 81 obese subjects, the fen–phen combination therapy was compared to both drugs in monotherapy (Weintraub et al. 1984). The weight-reducing efficacy was similar in both groups (mean weight loss kg – placebo 4.4 vs. fen 7.5 vs. phen 10.0 vs. fen–phen 8.4). However, adverse events were less frequent in the group receiving the combination therapy (Weintraub et al. 1984). While this study showed promising effects of fen–phen on long-term weight loss, routine clinical usage of the drug revealed rare but serious side effects mediated by the treatment, in particular cardiac valvulopathy and pulmonary hypertension (Connolly et al. 1997; Fishman 1999). This problem led to withdrawal of the drug in 1997.

2 Monoamine Reuptake Inhibitors and Appetite: Mechanism of Action

Monoamine reuptake inhibitors block the neuronal uptake of noradrenaline, dopamine, or serotonin and increase the synaptic concentrations of these molecules. The effects of these drugs are well recognized from the pharmacological therapy of depressive disorders (Haenisch and Bonisch 2011; Schildkraut 1965). The potency on individual reuptake transporters differs among the different substances. Increased monoamines act on postsynaptic receptors and mediate multiple biological actions. The regulation of appetite and food intake is an important biological process modulated by monoamines in the CNS.

Monoamine action crucially depends on ligand–receptor interaction. For noradrenaline, dopamine, and serotonin, a marked number of receptors have been cloned and functionally characterized. Noradrenaline can mainly act on nine different receptors α1 (α_{1A}, α_{1B}, α_{1D}), α2 (α_{2A}, α_{2B}, α_{2C}), β1, β2, and β3. Its action on appetite and food intake depends on the type of activated receptor (Leibowitz 1970a). Activation of hypothalamic α1-adrenoceptors in the paraventricular nucleus decreases food intake. In contrast, activation of α2-adrenoceptors in the same brain region induces eating behavior and food intake (Goldman et al. 1985; Leibowitz et al. 1985; Tsujii and Bray 1992; Wellman et al. 1993). With respect to the importance of distinct α1- or α2-adrenoceptor isoforms in appetite regulation, only limited data are available. Intracerebroventricular (ICV) application of the α_{1A}-adrenoceptor antagonist 5-methylurapidil resulted in reduced food intake in rats (Clifford et al. 2007). In contrast, α_{1B}-adrenoceptor-deficient mice exhibit increased body weight under high-fat-diet feeding (Burcelin et al. 2004). Whether the observed weight difference was mediated by central or peripheral mechanism was not investigated. All three α2-adrenergic receptor isoforms are involved in the presynaptic feedback control of noradrenaline release (Hein 2006). However, currently, no consistent data are available characterizing the impact of the different α2-isoforms in centrally mediated appetite regulation. Taken together, a subtype-specific functional distinction of α1- or α2-adrenoceptors in the context of noradrenaline's action on appetite regulation remains to be established in the future.

Noradrenaline-mediated activation of β2-adrenergic receptors in the perifornical region has been shown to inhibit food intake (Leibowitz 1970b). It appears that in situations of increased catecholamines, release (e.g., with amphetamines or reuptake inhibitors) activation of adrenergic receptor subtypes involved in the suppression of food intake dominates over stimulatory adrenergic signals. These processes may provide a molecular mechanism of drugs modulating noradrenaline release in the CNS.

There are seven different types of serotonin (5-HT) receptors (5-HAT$_{1-7}$) including further subtypes of 5HT$_1$ and 5-HT$_2$ receptors. The receptors most critically involved in the regulation of food intake seem to be 5-HT$_{1B}$ and 5-HT$_{2C}$ receptors. These receptors will be discussed in more detail in Sect. 4.2. Evidence for the involvement of these receptors in pharmacological actions has come from experiments in knockout mice. For instance, 5-HT$_{1B}$-receptor-deficient mice do not exhibit the anorectic effect of fenfluramine (Lucas et al. 1998). Along this line, mice deficient for 5-HT$_{2C}$ receptors did show an attenuated anorectic response to dexfenfluramine, consistent with a central role of this receptor in the antiobesity action of serotoninergic drugs (Vickers et al. 1999).

Five dopamine receptors (D$_{1-5}$) have been described to be expressed in the CNS. They are involved in a variety of physiological and pathophysiological processes, for example, Parkinson's disease or schizophrenia. In the context of food intake, D$_1$- and D$_2$-receptor agonists have been demonstrated to reduce food intake either by affecting the frequency of feeding episodes or the local rate of eating (Terry et al. 1995).

Taken together, there is ample evidence that drugs acting on monoamine reuptake transporters modulate food intake by indirectly augmenting postsynaptic monoamine receptor signaling in hypothalamic neurocircuits involved in feeding responses.

3 Sibutramine

Sibutramine is a monoamine reuptake inhibitor with the highest potency on serotonin and noradrenaline transporters, and lower potency on dopamine reuptake transporters. Sibutramine's anorexigenic effects can be blocked by 5-HT$_{2A/2C}$-, α1-, and β1-adrenergic receptor antagonists indicating that the predominant actions on food intake are mediated via serotonin and noradrenaline reuptake inhibition (Jackson et al. 1997). Interestingly, the inhibitory action of sibutramine on reuptake transporters in vitro has been shown to be weak and in the micromolar range (Glick et al. 2000). Thus, for its in vivo action, two demethylated metabolites with IC$_{50}$ values in the nanomolar range have been characterized and likely mediate major parts of its inhibitory actions on monoamine reuptake transporters (Glick et al. 2000). The parent compound of sibutramine has a half-life of 1.1 h after oral administration; however, the metabolites show half-lives of 14 and 16 h providing a sustained therapeutic action (Bello and Liang 2011). In addition to its CNS effects on satiety and food intake, Hansen et al. described that sibutramine treatment resulted in increased energy expenditure in healthy men suggesting an accessory weight-reducing mechanism for the drug (Hansen et al. 1998).

Sibutramine was approved for weight reduction in patients who were unable to lose weight by diet and exercise alone. The drug had been tested in several clinical randomized controlled trials (Padwal and Majumdar 2007). In a study by Wirth and Krause (2001), sibutramine 15 mg was tested in over 1,000 obese subjects (BMI 30–40 kg/m^2) over a period of 48 weeks in comparison to placebo. Mean weight loss in the treatment group was 4% compared to a weight gain of 0.2% in the placebo group (Wirth and Krause 2001). In a meta-analysis of ten randomized controlled trials with over 2,623 patients treated with sibutramine 10–20 mg daily, a mean body weight loss of 3.7–5% was documented for sibutramine when compared to placebo (Rucker et al. 2007). These data confirm the clinical efficacy of sibutramine with regard to weight loss.

However, because of its mechanism as a monoamine reuptake inhibitor, sibutramine has the potential to raise blood pressure in certain patients which may counteract the beneficial actions on the cardiovascular system achieved by weight loss (Birkenfeld et al. 2002; Jordan et al. 2005). Thus, continuous concerns had been raised that sibutramine treatment may negatively impact blood pressure and cardiovascular outcome. In 2010, the Sibutramine Cardiovascular Outcome Trial (SCOUT) was published (James et al. 2010). SCOUT included 9,804 overweight or obese patients with preexisting cardiovascular disease, type 2 diabetes mellitus, or both. Patients were randomized to sibutramine or placebo and followed for a mean treatment duration of 3.4 years. The primary outcome was a composite endpoint including nonfatal myocardial infarction, nonfatal stroke, resuscitation after cardiac arrest, and cardiovascular death. Despite better weight loss in the sibutramine group, 561 of 4,906 (11.4%) patients treated with sibutramine exhibited the primary endpoint compared to 490 of 4,898 patients (10.0%) in the placebo group. This translates into a significant 16% risk increase of the primary outcome event in the sibutramine group compared with placebo ($p = 0.02$). Based on this result, sibutramine has been withdrawn from the market in 2010 and is no longer in clinical use.

4 Bupropion

The noradrenaline and dopamine reuptake inhibitor bupropion is approved for major depressive disorders and as an aid for smoking cessation. Bupropion has a modestly more potent action on dopamine reuptake transporters compared to noradrenaline transporters (K_i for human dopamine reuptake transporters 0.56 μM vs. >10 μM for noradrenaline reuptake transport) (Bymaster et al. 2002). During hepatic metabolization, bupropion metabolites are produced including *(S,S)* hydroxybupropion and *(R,R)*hydroxybupropion, both also acting as low-potency reuptake inhibitors. Data on the detailed mechanism of weight-reducing/anorexigenic actions of bupropion are limited. It has been postulated that bupropion exerts its action on food intake via the activation of POMC neurons in the arcuate nucleus (Greenway et al. 2009b).

In addition to its antidepressant effects, bupropion has long been associated with body weight reduction in obese patients with or without depression. In a meta-analysis published by Li Z et al. in 2005, three trials with bupropion were included with an observation period between 6 and 12 months (Li et al. 2005). The authors reported a mean difference in weight loss for bupropion when compared to placebo of 2.77 kg. A Cochrane review has been published about the efficacy of different interventions for preventing weight gain after smoking cessation (Parsons et al. 2009). In this analysis, bupropion (300 mg/day) was effective in limiting postcessation weight gain.

More recently, combination therapies containing bupropion as a drug partner were tested in clinical trials. Based on the action of bupropion on POMC neurons, and on the fact that activation of the opioid pathway is capable to block POMC signaling, it has been postulated that weight-reducing effectiveness of bupropion may be enhanced by antagonizing opioid action. This has led to the development of a combination of bupropion and the opioid antagonist naltrexone (product name: Contrave) (Greenway et al. 2009a; Katsiki et al. 2011). The Contrave Obesity Research (COR) program includes four randomized, double-blind, placebo-controlled phase III clinical trials. In the COR-I trial, 1,742 obese patients were randomized to the bupropion SR 360 mg/naltrexone SR 32 mg combination or placebo. Active treatment resulted in a weight loss of 8.1% after 56 weeks compared with −1.8% in the placebo group (Greenway et al. 2010). Combination therapy was associated with a transient increase in systolic blood pressure of 1.5 mmHg during the first 8 weeks of treatment. Additional data from this trial program are awaited to be published in the near future.

Bupropion is currently also tested in a combination with the antiepileptic drug zonisamide (product name: Empatic). In phase II clinical trials, combinations of bupropion 360 mg with different doses of zonisamide 120 mg and 360 mg resulted in significantly greater weight loss than placebo (Valentino et al. 2010). Detailed information will be provided in Sect. 4.6.

The efficacy of the bupropion combination therapy appears to be very promising, but additional safety data are required to finally assess the benefit–risk profile of such combinations.

5 Tesofensine

Tesofensine is a novel triple monoamine reuptake inhibitor which blocks noradrenaline, serotonin, and dopamine transporters resulting in increased concentrations of these amines in relevant CNS regions (see above). The agent mediates hypophagia and body weight loss in rodents (Axel et al. 2010; Hansen et al. 2010). Its hypophagic actions are completely blocked by $\alpha 1$-adrenoceptor antagonists and attenuated by dopamine D_1-receptor antagonists, indicating that both noradrenaline and dopamine are involved (Axel et al. 2010). Interestingly, ritanserin, a 5-$HT_{2A/C}$ receptors antagonist, failed to reverse the anorectic effect of tesofensine in rats, suggesting that serotonin signaling may not be involved in its weight-regulating effects (Axel et al. 2010).

Furthermore, a randomized, double-blind, placebo-controlled phase II trial was conducted with tesofensine (Astrup et al. 2008). The drug was administered to 203 obese patients in three different doses: 0.25 mg, 0.5 mg, and 1.0 mg over a 24-week period. In parallel, all patients were subjected to an energy-restricted diet. Mean weight losses in the different groups were −4.5% (tesofensine 0.25 mg), −9.2% (tesofensine 0.5 mg), −10.6% (tesofensine 1.0 mg), and −2.0% (placebo). Tesofensine-induced weight reduction was associated with a decrease in waist circumference, fat mass, and plasma triglycerides, as well as an increase in adiponectin plasma levels. Most common side effects were dry mouth, nausea, constipation, hard stools, diarrhea, and insomnia. The highest dose of tesofensine (1 mg) induced a significant rise in systolic and diastolic blood pressure (+6.8 mmHg and +5.8 mmHg, respectively vs. +1.3 mmHg and +1.5 mmHg for placebo). The authors reported no serious psychiatric reactions such as mood disorders or anxiety. These data appear promising, but the long-term safety of this agent still has to be established.

References

Astrup A, Madsbad S, Breum L, Jensen TJ, Kroustrup JP, Larsen TM (2008) Effect of tesofensine on bodyweight loss, body composition, and quality of life in obese patients: a randomised, double-blind, placebo-controlled trial. Lancet 372:1906–1913

Axel AM, Mikkelsen JD, Hansen HH (2010) Tesofensine, a novel triple monoamine reuptake inhibitor, induces appetite suppression by indirect stimulation of alpha1 adrenoceptor and dopamine D1 receptor pathways in the diet-induced obese rat. Neuropsychopharmacology 35:1464–1476

Bello NT, Liang NC (2011) The use of serotonergic drugs to treat obesity – is there any hope? Drug Des Devel Ther 5:95–109

Birkenfeld AL, Schroeder C, Boschmann M, Tank J, Franke G, Luft FC, Biaggioni I, Sharma AM, Jordan J (2002) Paradoxical effect of sibutramine on autonomic cardiovascular regulation. Circulation 106:2459–2465

Bray GA, Greenway FL (1999) Current and potential drugs for treatment of obesity. Endocr Rev 20:805–875

Burcelin R, Uldry M, Foretz M, Perrin C, Dacosta A, Nenniger-Tosato M, Seydoux J, Cotecchia S, Thorens B (2004) Impaired glucose homeostasis in mice lacking the alpha1b-adrenergic receptor subtype. J Biol Chem 279:1108–1115

Bymaster FP, Katner JS, Nelson DL, Hemrick-Luecke SK, Threlkeld PG, Heiligenstein JH, Morin SM, Gehlert DR, Perry KW (2002) Atomoxetine increases extracellular levels of norepinephrine and dopamine in prefrontal cortex of rat: a potential mechanism for efficacy in attention deficit/hyperactivity disorder. Neuropsychopharmacology 27:699–711

Clifford PS, Davis KW, Elliott AE, Wellman PJ (2007) Effects of ICV administration of the alpha1A-adrenoceptor antagonist 5-methylurapidil on concurrent measures of eating and locomotion after cocaine in the rat. Life Sci 81:1059–1065

Connolly HM, Crary JL, McGoon MD, Hensrud DD, Edwards BS, Edwards WD, Schaff HV (1997) Valvular heart disease associated with fenfluramine-phentermine. N Engl J Med 337:581–588

Fishman AP (1999) Aminorex to fen/phen: an epidemic foretold. Circulation 99:156–161
Glick SD, Haskew RE, Maisonneuve IM, Carlson JN, Jerussi TP (2000) Enantioselective behavioral effects of sibutramine metabolites. Eur J Pharmacol 397:93–102
Goldman CK, Marino L, Leibowitz SF (1985) Postsynaptic alpha 2-noradrenergic receptors mediate feeding induced by paraventricular nucleus injection of norepinephrine and clonidine. Eur J Pharmacol 115:11–19
Greenway FL, Dunayevich E, Tollefson G, Erickson J, Guttadauria M, Fujioka K, Cowley MA (2009a) Comparison of combined bupropion and naltrexone therapy for obesity with monotherapy and placebo. J Clin Endocrinol Metab 94:4898–4906
Greenway FL, Fujioka K, Plodkowski RA, Mudaliar S, Guttadauria M, Erickson J, Kim DD, Dunayevich E (2010) Effect of naltrexone plus bupropion on weight loss in overweight and obese adults (COR-I): a multicentre, randomised, double-blind, placebo-controlled, phase 3 trial. Lancet 376:595–605
Greenway FL, Whitehouse MJ, Guttadauria M, Anderson JW, Atkinson RL, Fujioka K, Gadde KM, Gupta AK, O'Neil P, Schumacher D, Smith D, Dunayevich E, Tollefson GD, Weber E, Cowley MA (2009b) Rational design of a combination medication for the treatment of obesity. Obesity (Silver Spring) 17:30–39
Greiser E (1973) Epidemiologic studies on the relation between use of appetite depressants and primary vascular pulmonary hypertension. Internist (Berl) 14:437–442
Hadler AJ (1967) Studies of aminorex, a new anorexigenic agent. J Clin Pharmacol J New Drugs 7:296–302
Haenisch B, Bonisch H (2011) Depression and antidepressants: insights from knockout of dopamine, serotonin or noradrenaline re-uptake transporters. Pharmacol Ther 129:352–368
Hansen DL, Toubro S, Stock MJ, Macdonald IA, Astrup A (1998) Thermogenic effects of sibutramine in humans. Am J Clin Nutr 68:1180–1186
Hansen HH, Hansen G, Tang-Christensen M, Larsen PJ, Axel AM, Raben A, Mikkelsen JD (2010) The novel triple monoamine reuptake inhibitor tesofensine induces sustained weight loss and improves glycemic control in the diet-induced obese rat: comparison to sibutramine and rimonabant. Eur J Pharmacol 636:88–95
Hein L (2006) Adrenoceptors and signal transduction in neurons. Cell Tissue Res 326:541–551
Jackson HC, Bearham MC, Hutchins LJ, Mazurkiewicz SE, Needham AM, Heal DJ (1997) Investigation of the mechanisms underlying the hypophagic effects of the 5-HT and noradrenaline reuptake inhibitor, sibutramine, in the rat. Br J Pharmacol 121:1613–1618
James WP, Caterson ID, Coutinho W, Finer N, Van Gaal LF, Maggioni AP, Torp-Pedersen C, Sharma AM, Shepherd GM, Rode RA, Renz CL (2010) Effect of sibutramine on cardiovascular outcomes in overweight and obese subjects. N Engl J Med 363:905–917
Jordan J, Scholze J, Matiba B, Wirth A, Hauner H, Sharma AM (2005) Influence of Sibutramine on blood pressure: evidence from placebo-controlled trials. Int J Obes (Lond) 29:509–516
Katsiki N, Hatzitolios AI, Mikhailidis DP (2011) Naltrexone sustained-release (SR) + bupropion SR combination therapy for the treatment of obesity: 'A new kid on the block'? Ann Med 43:249–258
Leibowitz SF (1970a) Hypothalamic beta-adrenergic "satiety" system antagonizes an alpha-adrenergic "hunger" system in the rat. Nature 226:963–964
Leibowitz SF (1970b) Reciprocal hunger-regulating circuits involving alpha- and beta-adrenergic receptors located, respectively, in the ventromedial and lateral hypothalamus. Proc Natl Acad Sci USA 67:1063–1070
Leibowitz SF, Brown O, Tretter JR, Kirschgessner A (1985) Norepinephrine, clonidine, and tricyclic antidepressants selectively stimulate carbohydrate ingestion through noradrenergic system of the paraventricular nucleus. Pharmacol Biochem Behav 23:541–550
Lesses MF, Myerson A (1994) Human autonomic pharmacology. XVI. Benzedrine sulfate as an aid in the treatment of obesity. 1938. Obes Res 2:286–292

Li Z, Maglione M, Tu W, Mojica W, Arterburn D, Shugarman LR, Hilton L, Suttorp M, Solomon V, Shekelle PG, Morton SC (2005) Meta-analysis: pharmacologic treatment of obesity. Ann Intern Med 142:532–546

Lucas JJ, Yamamoto A, Scearce-Levie K, Saudou F, Hen R (1998) Absence of fenfluramine-induced anorexia and reduced c-Fos induction in the hypothalamus and central amygdaloid complex of serotonin 1B receptor knock-out mice. J Neurosci 18:5537–5544

Padwal RS, Majumdar SR (2007) Drug treatments for obesity: orlistat, sibutramine, and rimonabant. Lancet 369:71–77

Parsons AC, Shraim M, Inglis J, Aveyard P, Hajek P (2009) Interventions for preventing weight gain after smoking cessation. Cochrane Database Syst Rev: CD006219

Rothman RB, Baumann MH (2002) Therapeutic and adverse actions of serotonin transporter substrates. Pharmacol Ther 95:73–88

Rothman RB, Baumann MH, Dersch CM, Romero DV, Rice KC, Carroll FI, Partilla JS (2001) Amphetamine-type central nervous system stimulants release norepinephrine more potently than they release dopamine and serotonin. Synapse 39:32–41

Rothman RB, Clark RD, Partilla JS, Baumann MH (2003) (+)-Fenfluramine and its major metabolite, (+)-norfenfluramine, are potent substrates for norepinephrine transporters. J Pharmacol Exp Ther 305:1191–1199

Rucker D, Padwal R, Li SK, Curioni C, Lau DC (2007) Long term pharmacotherapy for obesity and overweight: updated meta-analysis. BMJ 335:1194–1199

Samanin R, Garattini S (1993) Neurochemical mechanism of action of anorectic drugs. Pharmacol Toxicol 73:63–68

Schildkraut JJ (1965) The catecholamine hypothesis of affective disorders: a review of supporting evidence. Am J Psychiatry 122:509–522

Sulzer D, Sonders MS, Poulsen NW, Galli A (2005) Mechanisms of neurotransmitter release by amphetamines: a review. Prog Neurobiol 75:406–433

Terry P, Gilbert DB, Cooper SJ (1995) Dopamine receptor subtype agonists and feeding behavior. Obes Res 3(Suppl 4):515S–523S

Tsujii S, Bray GA (1992) Food intake of lean and obese Zucker rats following ventricular infusions of adrenergic agonists. Brain Res 587:226–232

Valentino MA, Lin JE, Waldman SA (2010) Central and peripheral molecular targets for antiobesity pharmacotherapy. Clin Pharmacol Ther 87:652–662

Vickers SP, Clifton PG, Dourish CT, Tecott LH (1999) Reduced satiating effect of d-fenfluramine in serotonin 5-HT(2C) receptor mutant mice. Psychopharmacology (Berl) 143:309–314

Weintraub M, Hasday JD, Mushlin AI, Lockwood DH (1984) A double-blind clinical trial in weight control. Use of fenfluramine and phentermine alone and in combination. Arch Intern Med 144:1143–1148

Wellman PJ, Davies BT, Morien A, McMahon L (1993) Modulation of feeding by hypothalamic paraventricular nucleus alpha 1- and alpha 2-adrenergic receptors. Life Sci 53:669–679

Wirth A, Krause J (2001) Long-term weight loss with sibutramine: a randomized controlled trial. JAMA 286:1331–1339

5-HT_{2C} Receptor Agonists and the Control of Appetite

Jason C.G. Halford and Joanne A. Harrold

Contents

Abstract The role of serotonin (5-HT) in appetite control is well recognised. 5-HT drugs reduce food intake in rodents in a manner consistent with an enhancement of satiety. In humans, they have been shown to reduce caloric intake, an effect associated with reduced hunger and increased satiety. These effects appear to be mediated, at least in part, by the 5-HT_{2C} receptor subtype. 5-HT-acting drugs such as fenfluramine, d-fenfluramine, and sibutramine have provided effective anti-obesity treatments in the past. However, more selective agents are needed that produce the same changes in eating behaviour and induce weight loss without unacceptable side effects. Lorcaserin, a selective 5-HT_{2C} receptor agonist, is a novel anti-obesity agent that reduces both energy intake and body weight. The effects of lorcaserin on eating behaviour remain to be characterised as does its behavioural specificity.

Keywords Appetite • Behaviour • Food intake • Obesity • Satiety

---

J.C.G. Halford (✉) • J.A. Harrold
Kissileff Laboratory for the Study of Human Ingestive Behaviour, School of Psychology, University of Liverpool, Liverpool L69 7ZA, UK
e-mail: j.c.g.halford@liverpool.ac.uk

H.-G. Joost (ed.), *Appetite Control*, Handbook of Experimental Pharmacology 209,
DOI 10.1007/978-3-642-24716-3_16, © Springer-Verlag Berlin Heidelberg 2012

1 Background

Changes in diet and lifestyle are fundamental to successful weight control, yet pharmacological approaches to challenge obesity are varied. Drugs can induce weight loss by interfering in the process of fat deposition or enhancing the utilisation of energy reserves. Digestion and absorption within the gastrointestinal tract can also be inhibited to reduce the energy available from ingested nutrients. Finally, inhibitory signals originating in the periphery or anorexigenic circuits within the brain can be targeted to suppress appetite. All these approaches can modify energy balance to produce weight loss. However, not all tackle the behaviours associated with overconsumption or the negative psychological consequences of caloric restriction. Targeting appetite addresses the aetiology of obesity, the inability to control eating behaviour in an environment of abundant, readily available, appealing food.

Appetite-suppressing drugs reduce energy intake by changing eating behaviour through selectively reducing the motivation to consume and the pleasure derived from ingestion and//or increasing the impact of ingestion on meal-derived inhibitory feedback. Targeting these components of appetite expression could (1) reduce the number of eating episode initiated, (2) reduce the duration of a meal, (3) decrease the intensity of consumption (eating rating), (4) increase the impact of ingestion on appetite, and/or (5) increase the time taken to the next meal. Appetite-suppressing drugs should be behaviourally specific. Reductions in energy intake should not be achieved through sedation or hyper-stimulation that disrupt eating behaviour producing unwanted behavioural side effects, or through malaise or feelings of nausea.

Prior to the introduction of serotonergic anti-obesity drugs, pharmacological solutions to weight loss lacked behavioural specificity. Amphetamines were used to control hunger despite their psychological effects and abuse potential. Other monoaminergic drugs with lower abuse potential such as phentermine, diethylpropion, and phenylpropanolamine were also used, but insomnia, anxiety, and irritability remained an issue with many of these drugs. In contrast, the 5-HT releaser and reuptake inhibitor fenfluramine appeared equally effective and lacked the behavioural side effects as well as the abuse potential of its catecholamine predecessors. All forms of fenfluramine were withdrawn due to valvular heart disease and pulmonary hypertension (Abeniam et al. 1996). Nonetheless, serotonergic drugs (fenfluramine, d-fenfluramine, and sibutramine) remain the only successful class of appetite suppressants to date. They were the first to modulate human appetite expression without producing any unacceptable behavioural effects. Consequently, 5-HT drugs remain in clinical development over 40 years after fenfluramine was first used to treat obesity.

2 Appetite Control

Early studies demonstrated that augmenting (administering the precursor 5-HTP) or disrupting neuronal 5-HT function (lesioning) produced profound effects on rodent feeding (Halford et al. 2007). However, it was Blundell (1977) that first proposed that the 5-HT system not only had an inhibitory role in feeding but also was a key satiety factor. Serotonin function is now associated with within-meal satiation and post-meal satiety, the processes that naturally bring a meal to an end and inhibit further consumption. This is crucial in the meal-by-meal regulation of energy intake, and critical to both the appetite fluctuations and patterns of eating behaviour experienced throughout the day (Halford and Blundell 2000).

Currently, 14 different 5-HT receptor subtypes have been identified (Hoyer and Martin 1997) of which post-synaptic 5-HT_{2C} receptors appear to mediate the effects of many serotonergic drugs on eating behaviour (Blundell and Halford 1998). 5-HT neurons project up towards the hypothalamic region, an area critical to energy regulation. These systems include the hypothalamic anorexigenic neuropeptide melanocortin (MC) system, a system that modulates the effects of 5-HT drugs on feeding behaviour (Heisler et al. 2002, 2003, 2006). Activation of both 5-HT_{2C} and 5-HT_{1B} receptors produce hypophagia by promoting the release of the endogenous agonist alpha-melanocyte-stimulating hormone (α-MSH) and inhibiting the release of the endogenous antagonist of the MC4R agouti-related peptide (AgRP).

3 5-HT Receptors

3.1 *Feeding Behaviour and Body Weight in Rodents*

The role of 5-HT in the control of appetite was initially examined with the 5-HT releasing and reuptake-inhibiting drug fenfluramine and its selective enantiomer d-fenfluramine. Both drugs produced changes in feeding behaviour normally brought about by ingestion rather than hyperactivity, sedation, or malaise (Blundell and Latham 1978, 1980; Blundell and McArthur 1981; Halford and Blundell 1993). The use of selective 5-HT receptor antagonists demonstrated that these effects of d-fenfluramine on eating behaviour were mediated at least in part by 5-HT_{2C} receptors (Clifton 1994; Vickers et al. 1999, 2001). The serotonergic and noradrenergic reuptake inhibitor (SNRI) sibutramine also produced similar changes in rodent feeding behaviour (Halford et al. 1998), an effect also mediated by 5-HT_{2c} receptors (Higgs et al. 2011). Selectively, agonising 5-HT_{2C} receptors also produced changes in feeding behaviour consistent with the operation of satiety (Kennett and Curzon 1988a, b; Halford et al. 1998). However, drugs which agonise other 5-HT receptor subtypes disrupt the BSS by inducing hyperactivity, a critical side effect (Kitchener and Dourish 1994; Halford et al. 1998; Hewitt et al. 2002). In rodent models of obesity, it appears that direct agonism of 5-HT_{2C} receptors, including lorcaserin

(Bjenning et al. 2004), reduced and even prevented diet-induced weight gain (Vickers et al. 2000, 2003; Hayashi et al. 2004). Behavioural genetics also provided support for the role of 5-HT_{2C} receptors in the hypophagic effects of endogenous 5-HT. Mice lacking functional 5-HT_{2C} receptors displayed marked hyperphagia, leading to the development of obesity (Tecott et al. 1995; Nonogaki et al. 2003).

3.2 *Eating Behaviour and Weight Control in Humans*

Rogers and Blundell (1979) demonstrated that fenfluramine significantly reduced food intake, eating rate, and desire to eat when given to normal-weight volunteers. D-fenfluramine, fluoxetine, and sibutramine all produced similar changes to human appetite in healthy, normal-weight volunteers (Goodall and Silverstone 1988; McGuirk and Silverstone 1990; Hansen et al. 1998; Chapelot et al. 2000). Importantly, all four drugs brought about similar changes in appetite and energy intake in obese individuals (Blundell and Hill 1990; Pijl et al. 1991; Lawton et al. 1995; Ward et al. 1999; Rolls et al. 1998; Halford et al. 2010a, b). The 5-$HT_{1B/2C}$ receptor preferential agonist chlorophenylpiperazine (mCPP) also reduced hunger and food intake in healthy, normal-weight volunteers (Walsh et al. 1994; Cowen et al. 1995) and hunger and body weight in the obese (Sargent et al. 1997).

5-HT drugs effectively induced weight loss in the obese. Licensed treatments such as fenfluramine (Pondimin, Ponderax, and Adifax), d-fenfluramine (Redux), and the SNRI sibutramine (Meridia, Reductil) produced placebo-subtracted weight loss (PSWL) ranging between 2.4 and 4.45 kg (Haddock et al. 2002; Arterburn et al. 2004; Padwal et al. 2003) over 6–12 months. These drugs, however, were not side effect free. Fenfluramine and d-fenfluramine were withdrawn due to primary pulmonary hypertension (Abeniam et al. 1996). Sibutramine was in turn withdrawn due to cardiovascular side effects (James et al. 2010). Drug development continued to focus on selectively targeting the 5-HT receptors implicated in satiety (Vickers and Dourish 2004). Over the last 15 years, considerable effort has been expended in developing a new generation of side-effect-free selective 5-HT_{2C} drugs.

4 Lorcaserin

Lorcaserin ([1R]-8-chloro-2,3,4,5-tetrahydro-1-methyl-1H-3-benzazepine) is a selective 5-HT_{2C} receptor agonist developed to target human appetite expression (Smith et al. 2009). Lorcaserin possesses a functional selectivity for the 5-HT_{2C} over the 5-HT_{2B} receptor (100-fold) subtypes. This selectivity should ensure it has no functional activity at other receptor types, specifically the cardiovascular side effects including valvulopathy, associated with activity at peripheral 5-HT_{2B} receptors. The drug has completed a number of clinical trials in the quest to receive regulatory approval.

Martin et al. (2011) examined the effects of 10 mg lorcaserin on energy balance in 39 overweight and obese participants in an 8-week trial. Compared to baseline, on day 7, treatment produced a reduction in lunch and dinner intake of 286 kcal, which was larger (but not significantly different) than the reduction observed in the placebo group (147 kcal). On day 8, a 600-kcal-deficit weight-reducing diet and exercise plans were introduced. Energy intake was measured again on day 56. At that time point, lunch and dinner energy intake was reduced by 470 kcals in the lorcaserin condition. This was significantly greater than the reduction in energy intake of 205 kcal observed in the placebo. Few real-time effects on appetite were found, but a significant reduction in one retrospective measure of appetite ("how much could you have eaten") was noted. No effects of treatment on blood pressure, respiratory quotient, or energy expenditure were observed at day 7 or day 56. Over the 8-week treatment period, lorcaserin produced a significantly larger reduction in body weight (3.8 kg or 3.92% of body weight) than the placebo group (2.2 kg or 2.19%), resulting in a placebo-subtracted weight loss (PSWL) of 1.6 kg. These data show that lorcaserin significantly reduced energy intake rather than altering energy expenditure. However, the effects on appetite and weight loss appeared small in magnitude compared to that produced by previous serotonergic drugs.

In a 12-week trial, the effects of lorcaserin 10 mg once daily, 15 mg, and 10 mg twice daily on body mass in the obese were compared against placebo (Smith et al. 2009). Weight loss reported in the placebo condition was 0.2 kg (0.2%), compared to 1.7 kg in the 10 mg (1.7%, PSWL = 1.5 kg), 2.2 kg in the 15 mg (2.2% PSWL = 2.0 kg), and 3.1 kg in the 10 mg twice daily (3.1%, PSWL = 2.9 kg) conditions. The reductions in body mass in all active treatment conditions were significantly greater than in the placebo. In a larger scale, phase 3 trial (BLOOM: *B*ehavioral modification and *l*orcaserin for *O*verweight and *O*besity *M*anagement), 3,182 overweight and obese patients received 10 mg of lorcaserin or placebo (Smith et al. 2010). Over the first year of treatment, average weight loss was 5.8 kg in the lorcaserin group compared with 2.2 kg in the placebo group (PSWL = 3.6 kg). Notably, more than twice the number of lorcaserin-treated than placebo-treated patients lost >5% of their initial body weight, and twice as many lost >10% (22.6% lorcaserin versus 7.7% placebo). Lorcaserin treatment for 1 year improved levels of fasting glucose, fasting insulin, total cholesterol, LDL cholesterol, and triglycerides, and reduced waist circumference. No major increases in mood-related adverse effects were reported and rates of cardiac valvulopathy, although slightly higher in year 1 in the lorcaserin than the placebo group, were the same between the groups over the 2-year duration of the trial. These data support the notion that selectively targeting the 5-HT_{2C} receptor is a safe treatment option over a period of 2 years. However, the dropout rate was high. Critically, 50% of patients treated with lorcaserin failed to lose at least 5% of their initial body weight, and many of these may have lost nothing or even gained weight. Nevertheless, while the extent of weight loss may seem modest, it compares favourably to that produced by the lipase inhibitor orlistat, the only currently available treatment option.

5 Summary

Serotonergic drugs have been proven to reduce energy intake and suppress appetite through their selective effects on satiety. Drugs that stimulate hypothalamic 5-HT_{2C} receptors in rodents produce both changes in the structure of feeding behaviour and reductions in food intake that are consistent with the satiety process and prevent weight gain in rodent models of obesity. In humans, 5-HT drugs enhance the post-meal satiety of fixed caloric loads and reduce pre-meal appetite as well as food intake at ad libitum meals in both the lean and obese. A number of these agents have proven to be useful anti-obesity agents in the past, but their unacceptable side-effect profiles have necessitated their withdrawal.

A new generation of selective 5-HT_{2C} agonists have been developed, and one of those, lorcaserin, has passed through late-stage clinical development. Lorcaserin can provide a safe and effective serotonergic treatment. However, the behavioural specificity of lorcaserin remains unknown (Halford et al. 2010a, b). The effects of this drug on psychological functioning or behavioural expression remain largely uncharacterised. Specifically, no study to date clearly demonstrates its purported satiety-enhancing effect. Moreover, its weight loss-inducing effects appear moderate. For some, lorcaserin treatment failed to induce any weight loss. A clearer understanding of the drug's behavioural effects may indicate who lorcaserin treatment may most benefit. For those individuals, the drug may prove highly effective. The clinical data clearly demonstrate that even modest reductions in body weight significantly reduce the risk for some of the secondary complications of obesity such as diabetes. Therefore, the limited clinical data available support the notion that selectively targeting the 5-HT_{2C} receptor may be an effective treatment option.

References

Abeniam L, Moride Y, Brenot F et al (1996) Appetite suppressant drugs and the risk of primary pulmonary hypertension. N Engl J Med 335:609–616

Arterburn DE, Crane PK, Veenstra DL (2004) The efficacy and safety of sibutramine for weight loss: a systematic review. Arch Intern Med 164:994–1003

Bjenning C, Williams J, Whelan K et al (2004) Chronic oral administration of APD356 significantly reduces body weight and fat mass in obesity-prone (DIO) male and female rats. Int J Obes 28(suppl 1):214

Blundell JE (1977) Is there a role for serotonin (5-hydroxytryptamine) in feeding? Int J Obes 1:15–42

Blundell JE, Halford JCG (1998) Serotonin and appetite regulation: implications for the treatment of obesity. CNS Drugs 9:473–495

Blundell JE, Hill AJ (1990) Sensitivity of the appetite control system in obese subjects to nutritional and serotoninergic challenges. Int J Obes 14:219–233

Blundell JE, Latham CJ (1978) Pharmacological manipulation of feeding behaviour: possible influences of serotonin and dopamine on food intake. In: Garattini S, Samanin R (eds) Central mechanisms of anorectic drugs. Raven, NY, pp 83–109

Blundell JE, Latham CJ (1980) Characteristic adjustments to the structure of feeding behaviour following pharmacological treatments: effects of amphetamine and fenfluramine and the antagonism by pimozide and metergoline. Pharmacol Biochem Behav 12:717–722
Blundell JE, McArthur RA (1981) Behavioural flux and feeding: continuous monitoring of food intake and food selection, and the video-recording of appetitive and satiety sequences for the analysis of drug action. In: Samanin R, Garattini S (eds) Anorectic agents: mechanisms of action and tolerance. Raven, NY, pp 19–43
Chapelot D, Mamonier C, Thomas F et al (2000) Modalities of the food intake-reducing effect of sibutramine in humans. Physiol Behav 68:299–308
Clifton PG (1994) The neuropharmacology of meal patterning. In: Cooper SJ (ed) Ethology and psychopharmacology. Wiley, Chichester, pp 313–328
Cowen PJ, Sargent PA, Williams C et al (1995) Hypophagic, endocrine and subjective responses to m-chlorophenylpiperazine in healthy men and women. Hum Psychopharmacol 10:385–391
Goodall E, Silverstone T (1988) Differential effect of d-fenfluramine and metergoline on food intake in human subjects. Appetite 11:215–288
Haddock CK, Poston WSC, Dill PL et al (2002) Pharmacotherapy for obesity: a quantitative analysis of four decades of published randomized clinical trials. Int J Obes 26:262–273
Halford JCG, Blundell JE (1993) 5-Hydroxytryptaminergic drugs compared on the behavioural sequence associated with satiety. Br J Pharmacol 100:95
Halford JCG, Boyland EJ, Cooper SJ et al (2010a) The effects of sibutramine on the microstructure of eating behaviour and energy expenditure in obese women. J Psychopharmacol 24:99–109
Halford JCG, Boyland EJ, Blundell JE et al (2010b) Pharmacological management of human appetite expression: evaluating the effects of novel anti-obesity agents on eating behaviour. Nat Endocrine Rev 6:255–269
Halford JCG, Harrold JA, Boyland EJ, Lawton CL, Blundell JE (2007) Serotonergic drugs: effects on appetite expression and use for the treatment of obesity. Drugs 67:27–55
Halford JCG, Blundell JE (2000) Separate systems for serotonin and leptin in appetite control. Ann Med 32:222–232
Halford JCG, Wanninayake SCD, Blundell JE (1998) Behavioural satiety sequence (BSS) for the diagnosis of drug action on food intake. Pharmacol Biochem Behav 61:159–168
Hansen DL, Toubro S, Stock MJ et al (1998) Thermogenic effects of sibutramine in humans. Am J Clin Nutr 68:1180–1186
Hayashi A, Sonoda R, Kimura Y et al (2004) Antiobesity effect of YM348, a novel 5-HT_{2C} receptor agonist, in Zucker rats. Brain Res 1011:221–227
Heisler LK, Cowley MA, Tecott LH et al (2002) Activation of Central Melanocortin pathways by Fenfluramine. Science 297:609–611
Heisler LK, Cowley MA, Kishi T et al (2003) Central serotonin and melanocortin pathways regulating energy homeostasis. N Y Acad Sci 994:169–174
Heisler LK, Jobst EE, Sutton GM et al (2006) Serotonin reciprocally regulates melanocortin neurons to modulate food intake. Neuron 51:239–249
Hewitt KN, Lee MD, Dourish CT et al (2002) Serotonin 2C receptor agonists and the behavioural satiety sequence in mice. Pharmacol Biochem Behav 71:691–700
Higgs S, Cooper AJ, Barnes NM (2011) Reversal of sibutramine-induced anorexia with a selective 5-HT_{2C} receptor antagonist. Psychopharmacology 214:941–947
Hoyer D, Martin G (1997) 5-HT receptor classification and nomenclature: towards a harmonization with the human genome. Neuropharmacology 36:419–428
James WP, Caterson ID, Coutinho W, Finer N, Van Gaal LF, Maggioni AP, Torp-Pedersen C, Sharma AM, Shepherd GM, Rode RA, Renz CL, SCOUT Investigators (2010) Effect of sibutramine on cardiovascular outcomes in overweight and obese subjects. N Engl J Med 363:905–917
Kennett GA, Curzon G (1988a) Evidence that the hypophagia induced by mCPP and TFMPP requires 5-HT_{1C} and 5-HT_{1B} receptors; hypophagia induced by RU-24969 only requires 5-HT_{1B} receptors. Psychopharmacology 96:93–100

Kennett GA, Curzon G (1988b) Evidence that mCPP may have behavioural effects mediated by central 5-HT_{1C} receptors. Br J Pharmacol 94:137–147
Kitchener SJ, Dourish CT (1994) An examination of the behavioural specificity of hypophagia induced by 5-HT1B, 5-HT1C and 5-HT2 receptor agonists using the post-prandial sequence in rats. Psychopharmacology 113:368–377
Lawton CL, Wales JK, Hill AJ et al (1995) Serotoninergic manipulation, meal-induced satiety and eating patterns. Obes Res 3:345–356
Martin CK, Redman LM, Zhang J et al (2011) Lorcaserin, a 5-HT_{2C} receptor agonist, reduced body weight by decreasing energy intake within influencing energy expenditure. J Clin Endocrinol Metab 96:837–45
McGuirk J, Silverstone T (1990) The effect of 5-HT re-uptake inhibitor fluoxetine on food intake and body weight in healthy male subjects. Int J Obes 14:361–372
Nonogaki K, Abdullah L, Goulding EH et al (2003) Hyperactivity and reduced energy cost of physical activity in serotonin 5-HT_{2C} receptor mutant mice. Diabetes 52:315–320
Padwal R, Li SK, Lau DCW (2003) Long-term pharmacotherapy for overweight and obesity: a systematic review and meta-analysis of randomized controlled trials. Int J Obes 27:1437–46
Pijl H, Koppeschaar HPF, Willekens FLA et al (1991) Effect of serotonin re-uptake inhibition by fluoxetine on body weight and spontaneous food choice in obesity. Int J Obes 15:237–242
Rogers PJ, Blundell JE (1979) Effect of anorexic drugs on food intake and the micro-structure of eating in human subjects. Psychopharmacology 66:159–165
Rolls BJ, Shide DJ, Thorward ML et al (1998) Sibutramine reduces food intake in non-dieting women with obesity. Obes Res 6:1–11
Sargent PA, Sharpley AL, Williams C et al (1997) 5-HT_{2C} receptor activation decreases appetite and body weight in obese subjects. Psychopharmacology 133:309–312
Smith SR, Prosser WA, Donahue DF et al (2009) Lorcaserin (APD356), a selective 5-HT_{2C} agonist, reducing body weight in obese men and women. Obesity 17:494–500
Smith SR, Wiessman NJ, Anderson CM et al (2010) Multicenter, placebo-controlled trial for Lorcaserin for weight management. N Engl J Med 363:245–56
Tecott LH, Sun LM, Akanna SF et al (1995) Eating disorder and epilepsy in mice lacking 5-HT_{2C} serotonin receptors. Nature 374:542–546
Vickers SP, Benwell KR, Porter RH et al (2000) Comparative effects of continuous infusion of *m*CPP, Ro 60–0175 and *d*-fenfluramine on food intake, water intake, body weight and locomotor activity in rats. Br J Pharmacol 130:1305–1314
Vickers SP, Clifton PG, Dourish CT et al (1999) Reduced satiating effect of d-fenfluramine in serotonin 5-HT_{2C} receptor mutant mice. Psychopharmacology 143:309–314
Vickers SP, Dourish CT (2004) Serotonin receptor ligands and the treatment of obesity. Curr Opin Investig Drugs 5:377–388
Vickers SP, Dourish CT, Kennett GA (2001) Evidence that hypophagia induced by *d*-fenfluramine and *d*-norfenfluramine in the rat is mediated by 5-HT_{2C} receptors. Neuropharmacology 41:200–209
Vickers SP, Easton N, Webster LJ et al (2003) Oral administration of the 5-HT_{2C} receptor agonist, mCPP, reduces body weight gain in rats over 28 days as a result of maintained hypophagia. Psychopharmacology 167:274–280
Walsh AE, Smith KA, Oldman AD (1994) M-Chlorophenylpiperazine decreases food intake in a test meal. Psychopharmacology 116:120–122
Ward AS, Comer SD, Haney M et al (1999) Fluoxetine-maintained obese humans: effect on food intake and body weight. Physiol Behav 66:815–821

Central and Peripheral Cannabinoid Receptors as Therapeutic Targets in the Control of Food Intake and Body Weight

Stefan Engeli

Contents

Abstract The endocannabinoid system consists of lipid-derived agonists that activate cannabinoid (CB) receptors. CB receptor agonists, namely, the phytocannabinoid Δ^9-THC and the endocannabinoid anandamide, increase hunger sensation and food intake. These discoveries led to the clinical use of Δ^9-THC derivatives for the treatment of cancer and HIV-related nausea and cachexia. Animal studies clarified the important role of CB1 receptors in the hypothalamus and in the limbic system in mediating orexigenic effects. In parallel, data on CB1-specific blockade either by drugs or by genetic ablation further demonstrated that CB1 inhibition protects against weight gain induced by high-fat feeding and reduces body weight

---

S. Engeli (✉)
Institute of Clinical Pharmacology, Hannover Medical School, Carl-Neuberg-Straße 1, 30625 Hannover, Germany
e-mail: engeli.stefan@mh-hannover.de

H.-G. Joost (ed.), *Appetite Control*, Handbook of Experimental Pharmacology 209, DOI 10.1007/978-3-642-24716-3_17, © Springer-Verlag Berlin Heidelberg 2012

in obese animals and humans. The mechanisms of weight reduction by CB1 blockade are complex: they comprise interactions with several orexigenic and anorexigenic neuropeptides and hormones, regulation of sympathetic activity, influences on mitochondrial function, and on lipogenesis. Although these mechanisms appear to be mainly mediated by the CNS, weight loss also occurs when drugs that do not reach CNS concentrations sufficient to inhibit CB1 signaling are used. The development of peripherally restricted CB1 inverse agonists and antagonists opened new routes in CB1 pharmacology because centrally acting CB1 inverse agonists, e.g., rimonabant and taranabant, exerted unacceptable side effects that precluded their further development and application as weight loss drugs. Tissue and circulating endocannabinoid concentrations are often increased in animal models of obesity and in obese humans, especially those with visceral fat accumulation. Thus, further research on CB1 inhibition is still promising to treat human obesity.

Keywords Endocannabinoids • Anandamide • Tetrahydrocannabinol • Cannabinoid receptor • Rimonabant • Taranabant • Obesity • Cachexia • Food intake

1 Introduction

The endocannabinoid system (ECS) consists of lipid-derived agonists activating cannabinoid (CB) receptors. Membrane phospholipid precursors, enzymes, cell membrane carriers, and transport proteins are involved in the signaling and bioavailability of endocannabinoids. The cannabinoid receptors (CB1 and CB2) are encoded by two different genes on human chromosomes: 6q14-q15 (*CNR1*) and 1p36.11 (*CNR2*). Both are heptahelical $G_{i/o}$ protein-coupled receptors. Typical intracellular events following cannabinoid receptor activation are decreased cAMP synthesis, increased K^+ efflux, and decreased Ca^{2+} influx. The cannabinoid receptors have a distinct expression pattern with rare overlap in a given cell type (Howlett et al. 2002; Pertwee et al. 2010).

In general, the ECS exerts damping actions in situations of stress or injury to facilitate cellular repair and regeneration (Pacher et al. 2006). These protective actions result from presynaptic CB1 activity in the brain modulating the release of neurotransmitters (e.g., GABA, glutamine, dopamine, norepinephrine, serotonin). Full CB1 activation results in a distinct behavioral pattern including ataxia, hypothermia, analgesia, and short-term memory impairment. For general insight into the physiology of the ECS, the reader is referred to reviews published in a previous issue of the *Handbook* (Abood 2005; Di Marzo et al. 2005; Howlett 2005; Pertwee 2005).

Part of the protective role of the ECS is facilitation of energy intake and storage, which in modern times may promote the development of obesity (Pagotto et al. 2006). Several animal models of obesity and obese humans exhibit increased availability of endocannabinoids in tissues and blood, and changes of expression of key genes coding for CB1 and the fatty acid amide hydrolase, FAAH (Blüher et al. 2006; Coté et al. 2007; Engeli et al. 2005; Matias et al. 2006). Whether ECS dysregulation is a consequence of the development of obesity, or a primary cause predisposing individuals to weight gain, is an unsolved matter of debate that has been discussed elsewhere (Engeli 2008a).

The phytocannabinoid Δ^9-tetrahydrocannabinol (Δ^9-THC) has been identified in 1964, but the endocannabinoid research history is short (see Table 1). The phenomenon that CB1 agonists such as Δ^9-THC increase hunger sensation and food intake led to the first clinical use of cannabis for the treatment of nausea and cachexia (Abel 1975; Kirkham and Williams 2001). In parallel, CB1 inverse agonists, originally developed as therapeutics against substance abuse, showed weight-reducing activities.

The first synthetic drug modulating the ECS and approved for weight loss was rimonabant (SR141716). Approval, however, was granted only in Europe, not in the USA. Two years later, rimonabant was withdrawn from the European market because of psychiatric side effects that prompted a negative recommendation by the EMA. This development led to a halt in research and development activities of several large pharmaceutical companies, which mostly abandoned antiobesity drugs and centrally acting CB1 inverse agonists from their agenda. This chapter will describe the molecular and physiological mechanisms by which CB1 regulates body weight and will also provide a prospect on recent developments in the field of CB1 antagonism that may generate a second wave of drugs against human obesity and associated metabolic disease.

2 Activation of CB1 Signaling Increases Hunger, Food Intake, and Body Weight

2.1 *Experimental Animal Studies*

A series of experiments in laboratory rodents revealed that Δ^9-THC, as well as the endocannabinoids anandamide and 2-AG, increased food intake when administered orally, subcutaneously, or centrally (Kirkham and Williams 2001). A comparison between peripherally administered Δ^9-THC and anandamide revealed striking similarities in the orexigenic effects. The most prominent findings were the rapid onset of feeding behavior after administration of each drug and the observation that food intake was induced in presatiated rats. Other features of feeding behavior were not different from control experiments without drugs and after fasting (Williams and Kirkham 2002). Another repeatedly reported finding was that induction of feeding behavior did not follow a strict direct dose-effect relationship. The most likely explanation is the sedative effect of higher Δ^9-THC and anandamide doses which counteracts the orexigenic drive. This rather simple explanation may help to

Table 1 History of ECS research with a focus on body weight regulation

Year	Discovery	References
1964	Δ^9-THC identified as psychoactive ingredient of *Cannabis sativa*	Gaoni and Mechoulam (1964)
1971	Δ^9-THC increased appetite	Hollister (1971)
1988–1990	CB1 identified as the "brain" CB receptor subtype	Devane et al. (1988) and Matsuda et al. (1990)
1992	Anandamide (AEA) identified in porcine brain as the first endogenous CB receptor agonist	Devane et al. (1992)
1993	CB2 cloned as the "peripheral" CB receptor subtype	Munro et al. (1993)
1994	CB1 inverse agonist SR141716 = rimonabant described	Rinaldi-Carmona et al. (1994)
1995	2-arachidonoylglycerol (2-AG) identified in canine gut as another endocannabinoid	Mechoulam et al. (1995)
1998	Rimonabant reduced appetite and body weight in rats	Colombo et al. (1998)
1999–2002	Description of three mouse CB1 knockout strains	Ledent et al. (1999), Marsicano et al. (2002), and Zimmer et al. (1999)
2001	Increased food intake after anandamide administration mediated by hypothalamic CB1 receptors	Jamshidi and Taylor (2001)
2003–2004	CB1 knockout mice are lean and protected against high-fat feeding	Cota et al. (2003) and Ravinet Trillou et al. (2004)
2005–2006	Trial data of the "Rimonabant in Obesity (RIO)" program published	Després et al. (2005), Pi-Sunyer et al. (2006), Scheen et al. (2006), and van Gaal et al. (2005)
2006	Taranabant described as another CB1 inverse agonist	Lin et al. (2006)
2006	Rimonabant approved in Europe for weight loss	Jones (2008)
2007	FDA panel voted against approval of rimonabant	Jones (2008)
2008	Rimonabant withdrawn in response to EMA advice	Jones (2008)
2008–2010	CB1 activation impaired and CB1 inhibition increased mitochondrial activity in adipocytes	Tedesco et al. (2008, 2010)
2009	URB447, a mixed CB1 antagonist/CB2 agonist, reduced food intake and body weight without entering the CNS	LoVerme et al. (2009)
2010	CB1 signaling in forebrain and sympathetic neurons identified as key component of body weight regulation	Quarta et al. (2010)
2010	AM6545, a peripheral inverse CB1 agonist, reduced body weight in obese mice	Tam et al. (2010)

(continued)

Table 1 (continued)

Year	Discovery	References
2010	TM38837, a peripheral inverse CB1 agonist, reduced body weight in obese rodents studies and passed phase I	Oral presentation T2: OS2.4 at ICO 2010
2011	Description of a peripheral neutral CB1 antagonist that reduced body weight in obese mice	Personal communication at ECO 2011

CB cannabinoid receptor, *Δ⁹-THC* Δ9-tetrahydrocannbinol, *ECO* European Congress on Obesity, *EMA* European Medicines Agency, *FDA* Food and Drug Administration, *ICO* International Congress on Obesity

understand why earlier studies failed to conclusively show effects of cannabinoids in animals and humans (Abel 1975).

The orexigenic effects of peripherally administered anandamide in presatiated rats were prevented by prior administration of rimonabant, demonstrating for the first time the involvement of CB1 in mediating cannabinoid-stimulated food intake (Williams and Kirkham 1999). Rimonabant, however, inhibits brain and peripheral CB1; thus further proof was needed to identify the site of action. Evidence came from a study that employed direct injection of anandamide and rimonabant into the ventromedial hypothalamus in presatiated rats. Again, anandamide injection induced hyperphagia, but 30-min pretreatment with the CB1-selective inverse agonist rimonabant completely abolished this effect. Rimonabant alone had no effect on acute food intake in these experiments (Jamshidi and Taylor 2001). In animals, rimonabant also inhibited specific characteristics of cannabinoid-induced feeding, such as preferred intake of sugar and palatable food, and abolished interactions with rewarding behavior such as alcohol drinking (Arnone et al. 1997; Simiand et al. 1998). Of special importance was the finding that 2-AG injection into the nucleus accumbens shell also increased food intake in a dose- and CB1-dependent manner (Kirkham et al. 2002). The nucleus accumbens is crucial in determining the motivation to eat. By direct injection of an inhibitor of the anandamide-degrading enzyme FAAH into the nucleus accumbens, or into the parabrachial region involved in integrating gustatory signals with hypothalamic nuclei, food intake was also stimulated, and preference for palatable food increased. These effects were blocked by the specific CB1 inverse agonist AM251 (Dipatrizio and Simansky 2008; Soria-Gomez et al. 2007). Together, these studies further strengthened the important role of direct and indirect stimulation of central CB1 receptors as the site of action of cannabinoid-mediated orexigenic effects.

2.2 *The Human Experience*

Sporadic reports on the effects of cannabis/marijuana on increasing hunger sensation and food intake date back to ancient times and were then retrieved and repeated during the nineteenth century. Soldiers were among the first individuals participating in an observational study to prove orexigenic effects of plant cannabinoids in 1933 (Siler et al. 1933). Later on, these early findings were verified by more thoroughly controlled and standardized studies (e.g., standardized for Δ^9-THC intake). Their results suggested that the orexigenic effects of marijuana in volunteers were more pronounced in the fed state than under fasting conditions. Stimulation of food intake varied with respect to time course depending on the Δ^9-THC dose: under acute conditions, low-dose Δ^9-THC always stimulated food intake, whereas high doses initially suppress hunger sensation, although later on, food intake was stimulated. Other common results were that quantity of food eaten increased in association with the psychoactive effects and more palatable food qualities (e.g., sweet, spicy, solid) were preferred (Gagnon and Elie 1975; Greenberg et al. 1976; Tart 1970).

Under controlled chronic conditions over 2 weeks, repeated marijuana consumption with standardized Δ^9-THC doses increased body weight and food intake, although the effect on food intake was shorter and less pronounced than the gain in body weight (Foltin et al. 1988). These findings suggest that CB receptor activation not only acts in the brain to modulate hunger sensation, but may also regulate body weight on other levels. Placebo-controlled studies demonstrated that the social surroundings (e.g., being alone, having to fulfill tasks, being in social contact with others) are of crucial importance for the marijuana effects on food intake. Thus, Δ^9-THC is not only able to interact with orexigenic pathways but also plays a special role for hedonistic eating behavior. This finding, if also true for endocannabinoids, might be of special importance for the development of obesity in humans because interactions between endocannabinoid signaling and the reward system may provide some explanation for the obesity epidemic in modern societies.

Ancient experience and accumulated data in healthy human subjects on the stimulating effects of Δ^9-THC on eating behavior have led to the exploration of cannabis-related drugs as adjunctive treatment in disease conditions such as nausea and vomiting, unintentional weight loss, and cachexia. Nabilone is a synthetic Δ^9-THC analogue with activity on CB1 and CB2 receptors, dronabinol is synthetic Δ^9-THC, and Sativex® is a 1:1 mixture of plant-derived Δ^9-THC and cannabidiol (Thakur et al. 2009). Nabilone and dronabinol have been evaluated for the treatment of chemotherapy-induced nausea and vomiting (Sallan et al. 1980) and are approved for this indication in some countries (Robson 2005). Cancer and HIV-related cachexia are other conditions for which the use of nabilone and dronabinol has been approved in some countries (e.g., the USA and Canada) in order to increase appetite and body weight of the patients (Beal et al. 1997; Plasse et al. 1991). Increased body weight has also been observed in patients with Alzheimer's disease treated with dronabinol as adjunct therapy (Volicer et al. 1997). Given the psychotropic actions of these drugs, abuse and side effects clearly present matters of concern, but the safety profiles appear to be acceptable given the severity of conditions treated by these drugs, and abuse appears to be an uncommon issue (Gorter et al. 1992; Robson 2005; Ware and St Arnaud-Trempe 2010).

A recent report demonstrated that in HIV-positive marijuana smokers, the approved dronabinol dose may be too small to maintain increased appetite and food intake in the long term. Whether this finding points to tolerance development against the used dose of dronabinol, or tolerance development against the orexigenic effect of dronabinol, remains to be determined because mood effects of dronabinol were maintained at the same dose over longer periods (Bedi et al. 2010). As will be described in Sect. 3, these findings support the now accepted notion that weight changes by modulation of CB1 signaling are not only a matter of altered food intake.

3 Inhibition of CB1 Signaling Decreases Body Weight

3.1 *Experimental Animal Studies*

Since the description of rimonabant as a selective brain-penetrating CB1 antagonist in 1994 (Rinaldi-Carmona et al. 1994), several animal studies described the reduction of food intake and body weight by the drug. Treatment of lean Wistar rats with 10 mg/kg i.p. rimonabant significantly reduced food intake and body weight. The anorectic effect of rimonabant, however, was already diminished after 5 days, whereas the reduced body weight was maintained over 14 days of treatment (Rinaldi-Carmona et al. 1994). In contrast, lost body weight was rapidly regained if drug treatment was stopped (Carai et al. 2006). These findings were further substantiated in diet-induced obese (DIO) mice. Rimonabant 10 mg/kg p.o. significantly reduced food intake only during the first week of a 5-week treatment period. Decreased body weight was again maintained during the complete treatment period (Ravinet Trillou et al. 2003). The differential effect on food intake and body weight reduction was also demonstrated by comparing short-term rimonabant treatment with a pair-fed control group and by studying animals during a 24-h fasting period treated with rimonabant or untreated. Both experimental designs demonstrated that weight loss was more pronounced with the drug (Ravinet Trillou et al. 2003). Rimonabant was also effective in genetic models of obesity, with the *fa*/*fa* obese Zucker rat being the most widely studied model (Bensaid et al. 2003; Gary-Bobo et al. 2007; Jbilo et al. 2005). Rimonabant treatment of *fa*/*fa* rats was effective in reducing adipose tissue mass and was associated with increased adiponectin plasma concentrations, decreased inflammatory markers, decreased liver steatosis, and increased insulin sensitivity. In leptin-deficient *ob*/*ob* mice, rimonabant similarly reduced food intake and body weight. The finding that there was no difference between rimonabant treated and pair-fed animals was most likely due to the short duration of the study of only 7 days (Liu et al. 2005). However, this matter is not completely resolved because a study in DIO rats with 2 weeks' duration also observed no differences in the reduction of body weight between rimonabant-treated animals and the pair-feeding control group (Thornton-Jones et al. 2006). Taranabant effectively reduced body weight in DIO rats over a 2-week treatment period, and somewhat different to rimonabant, the reduction in food intake appeared to be prolonged. Weight and fat mass reduction was achieved with partial CB1 receptor occupancy, and the drug's efficiency to reduce body weight was clearly correlated to CB1 receptor occupancy in the brain (Fong et al. 2007).

Another line of evidence was developed by studying mice strains with genetic ablation of CB1 receptors. Importantly, rimonabant and taranabant had no effects on feeding behavior and body weight in $CB1^{-/-}$ mice (Di Marzo et al. 2001; Fong et al. 2007; Ravinet Trillou et al. 2003), and adipose tissue gene expression signatures were remarkably similar between rimonabant-treated wild-type mice and untreated $CB1^{-/-}$ mice (Jbilo et al. 2005). $CB1^{-/-}$ mice were leaner than wild-type littermates, with a specific reduction of visceral adipose tissue mass. The lean phenotype in young animals was clearly associated with a reduction of food intake, whereas in older animals, leanness and food intake were again dissociated (Cota et al. 2003). When challenged with a high-fat diet, $CB1^{-/-}$ mice did not develop obesity and insulin resistance and maintained a low feeding efficiency (Ravinet Trillou et al. 2004). A common finding with rimonabant treatment or $CB1^{-/-}$ mice studies was the reduction of plasma leptin. This finding is not surprising per se because a reduction in adipose tissue mass clearly leads to reduced circulating leptin. However, injection of leptin in wild-type mice on standard chow reduced body weight by 4.7%, whereas leptin reduced body weight by 7.5% in $CB1^{-/-}$ mice on the same diet. This important finding demonstrates that lack of CB1 signaling increases leptin sensitivity under these experimental conditions (Ravinet Trillou et al. 2004).

In summary, experimental animal data obtained by either pharmacological or genetic inhibition of CB1 signaling without doubt demonstrated a pronounced effect on body weight and adipose tissue mass. This weight-reducing effect is only partially accompanied by a reduction of food intake, suggesting that CB1 receptors have additional effects on energy metabolism.

3.2 Clinical Experience with CB1 Inverse Agonists

Two brain-penetrating inverse CB1 agonists, rimonabant and taranabant, have been thoroughly studied for their clinical efficacy during the last years. Weight loss was the primary endpoint in all clinical trials.

The efficacy data of rimonabant submitted to regulatory authorities were based on 6,600 overweight and obese subjects in four randomized trials that lasted for one (RIO-Lipids, RIO-Diabetes) or 2 years (RIO-Europe, RIO-North America) and tested 5 mg/day and 20 mg/day rimonabant against placebo. According to the study protocols, all subjects should have reduced caloric intake by 600 kcal/day throughout the placebo run-in and treatment periods. Primary endpoint in all trials was weight reduction. All trials followed a similar study design so that data could be combined to a large degree (Després et al. 2005; Pi-Sunyer et al. 2006; Scheen et al. 2006; van Gaal et al. 2005). The placebo-subtracted effects of 20 mg/day rimonabant were a reduction of approximately 5 kg body weight and reduction of 4 cm of waist circumference. In RIO-Diabetes, reductions of weight and waist circumference were less pronounced, but still significantly larger compared to placebo. This is a common finding in weight loss trials that recruited diabetic patients on metformin or sulfonylureas.

Weight loss with rimonabant was accompanied by favorable changes in triglycerides (−14% in addition to placebo), HDL (+8%), and fasting insulin (−4 μU/ml in nondiabetic patients). Also, a significant reduction in insulin secretion during an oral glucose load occurred. High-sensitive C-reactive protein decreased by 30%, and adiponectin increased by 40%. In RIO-Diabetes, the absolute change of hemoglobin A_{1c} (HbA_{1c}) compared with placebo was −0.7%, independent of concomitant oral antidiabetic drug treatment. Metabolic changes and weight loss lasted as long as rimonabant was taken. Once patients were rerandomized to placebo after 1-year treatment in RIO-North America, the treatment effects were lost, and body weight rose to the level of the ever-placebo treated group. This finding demonstrates the chronic nature of obesity and obesity-associated metabolic disease. Using the placebo data as a calibrator and analysis of covariance (ANCOVA), 50% of the favorable changes in above-mentioned metabolic variables were calculated to be due to weight loss, whereas 50% were due to rimonabant-specific effects independent of weight loss. These statistics have never been verified by controlled experiments, but point to metabolic effects of endocannabinoids and CB1 receptors (Engeli 2008b).

Before European marketing authorization of rimonabant in 2006, other phase III trials were started to broaden the knowledge on metabolic and cardiovascular effects of the drug. The randomized, double-blind, and placebo-controlled Comprehensive Rimonabant Evaluation Study of Cardiovascular Endpoints and Outcomes (CRESCENDO) included women and men ≥55 years with abdominal obesity and a history of cardiovascular disease, or at least two major cardiovascular risk factors (Topol et al. 2010). The primary endpoint was occurrence of myocardial infarction, stroke, or cardiac death. CRESCENDO enrolled 9,381 in the rimonabant (20 mg/day) group and 9,314 in the placebo group. After mean follow-up of 13.8 months, regulatory authorities requested premature discontinuation of the study due to the EMA decision to suspend marketing authorization. At this point, about half of the events required had occurred. Overall, rimonabant did not reduce occurrence of the primary endpoint. Although the event rate appeared to diverge after 1 year, the number of patients was not sufficient to assess potential beneficial actions of the drug. Gastrointestinal and neuropsychiatric side effects were of particular concern and occurred more common with rimonabant, including four completed suicides in the rimonabant and one in the placebo group. No data on body weight, waist circumference, blood pressure, glucose metabolism, or any other cardiovascular risk factors were reported (Jordan et al. 2011; Topol et al. 2010).

Reduced food intake in human subjects treated with rimonabant has only been reported in the form of abstracts, but a crossover study with taranabant was published. A high single dose of taranabant (12 mg) reduced cumulative 24-h food intake by 22% in comparison to placebo, whereas sibutramine reduced food intake by 12% compared to placebo. Reduced food intake with the high taranabant dose was present during all meals over the day, and no specific changes in macronutrient choice were observed (Addy et al. 2008).

Efficacy data of taranabant in approximately 5,850 overweight and obese patients enrolled in randomized, placebo-controlled, double-blind studies were published several months after the decision of the company to suspend further development of the drug for weight loss. In a low-dose study over 52 weeks, taranabant 2 mg/day decreased body weight by 5 kg more than placebo treatment. The proportion of patients losing 5% or 10% of initial body weight was significantly higher with taranabant 2 mg/day than with placebo and lower doses. Fat mass reduction and a reduction of waist circumference was observed accordingly (Proietto et al. 2010). In a high-dose study over 104 weeks, both higher-dose arms (4 mg/day and 6 mg/day) were prematurely stopped during the first or second year due to safety concerns, and patients of these arms were switched to 2 mg or placebo until the end of year 2. Weight reduction at week 104 was 5 kg (2 mg/day) or 6.2 kg (4 mg/day) more than with placebo (Aronne et al. 2010). Weight loss in patients with type 2 diabetes over 52 weeks with taranabant 2 mg/d was significantly greater than with placebo, but smaller than in overweight and obese patients without type 2 diabetes (Kipnes et al. 2010). In a very interesting approach different from other weight loss trials, taranabant was also tested against placebo in a randomized double-blind study after initial successful weight loss with low-calorie diet for 6 weeks (−9.6 kg). During the following year, patients on placebo gained 1.7 kg again, whereas patients on 2 mg/day taranabant lost an additional 1.2 kg (Wadden et al. 2010). These data point to the strong effects of CB1 inverse agonists on body weight regulation. On the other hand, these data also demonstrate the efficacy of well-conducted and controlled lifestyle interventions.

The rimonabant and taranabant trials clearly proved that pharmacological inhibition of CB1 signaling is efficacious to reduce and maintain lower body weight in overweight and obese patients. However, the use of brain-penetrating CB1 inverse agonists was associated with significant unwanted effects. First, gastrointestinal symptoms occurred in many patients (e.g., nausea, vomiting, diarrhea). These symptoms may result from blockade of not only central but also gastrointestinal CB1 receptors. Typically, however, most patients recovered quickly, and the symptoms disappeared, maybe because tolerance developed. Thus, weight loss by these drugs could not be explained by gastrointestinal side effects. More serious were CNS-related side effects, typically, anxiety, depressed mood, clinical relevant depression, and suicidal tendencies. As seen in the taranabant trials, these CNS side effects were dose dependent. The overall impression was that rigorous patient selection might have reduced the number of adverse CNS events and might have improved patient safety. Nevertheless, lessons learned from other new drugs clearly suggest that once a drug is marketed, rigorous patient selection rapidly vanishes in routine daily practice. Furthermore, the CNS side effects of centrally acting CB1 inverse agonists have a well-known physiological basis (Moreira et al. 2009). Both points have ultimately determined the failed introduction of this drug class into clinical practice.

4 Mechanisms of CB1-Mediated Body Weight Regulation

4.1 Food Intake and Endocannabinoid Bioavailability

Anandamide and 2-AG were measured in rat hypothalamus, in the limbic forebrain, and in the cerebellum in response to fasting and feeding. Fasting significantly increased AEA and 2-AG concentrations in the limbic forebrain and 2-AG concentrations in the hypothalamus. In response to feeding, hypothalamic 2-AG rapidly declined. No changes of endocannabinoid tissue concentrations were observed in the cerebellum, a brain region not involved in the regulation of food intake (Kirkham et al. 2002). The rapid response may reflect a decrease of CB1 signaling that, together with other signals, determines meal duration. The question then is which signals regulate postprandial endocannabinoid concentrations. A role of leptin has been discussed because leptin administration decreased elevated hypothalamic endocannabinoid concentrations in obese animal models (Di Marzo et al. 2001). However, leptin is not an acute satiety signal but rather a long-term regulator of caloric intake in relationship to fat and body mass. Also, the investigated animal models mostly had genetic defects in leptin signaling (*ob*/*ob* mice, *db*/*db* mice, *fa*/*fa* rats). Increased tissue endocannabinoid concentrations have also been described in peripheral tissues of these particular models (Maccarrone et al. 2005). Thus, the described reduction of endocannabinoids by leptin merely reflects the correction of the hormonal deficiency in these models, rather than providing a mechanistic explanation for endocannabinoid regulation by food intake. The same authors, however, also demonstrated that leptin administration is able to reduce hypothalamic anandamide and 2-AG in a normal-weight Sprague–Dawley rats. Dependency of this effect on feeding condition or daytime was not reported (Di Marzo et al. 2001).

A rapid decrease of blood endocannabinoids in response to a test meal has also been described in a small human study (Matias et al. 2006). We have reproduced this finding in a larger study with lean and obese subjects (Fig. 1, data not published). The reduction of anandamide, but not 2-AG, was observed as early as 30 min after starting food intake and was still observed 2 h after meal intake.

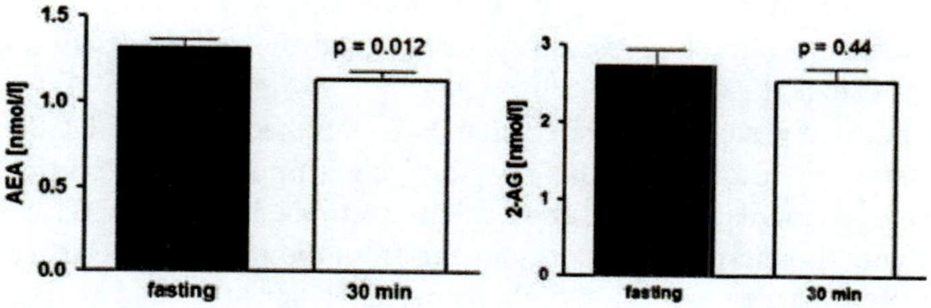

Fig. 1 Postprandial reduction of AEA, but not 2-AG, 30 min after meal initiation. Data are compiled from $n = 56$ test meals in lean and obese human volunteers and are shown as mean ± SEM, group comparison by unpaired *t* test

The peak reduction of anandamide was associated with the peaks of blood glucose and insulin. Consequently, a role of insulin as a negative regulator of anandamide was recently reported (Di Marzo et al. 2009a). In the light of increased endocannabinoid availability in obese subjects and their changes with long-term weight reduction, a detailed further exploration of the role of insulin and insulin resistance on endocannabinoid availability clearly is warranted (Blüher et al. 2006; Coté et al. 2007; Di Marzo et al. 2009b; Engeli et al. 2005; Matias et al. 2006). In the hypothalamus of lean Wistar rats, however, insulin failed to acutely alter endocannabinoid concentrations (Matias et al. 2008). Thus, tissue- and species-specific mechanisms may play a role for postprandial endocannabinoid regulation. Other possible candidates for postprandial endocannabinoid changes are polyunsaturated fatty acids (Watanabe et al. 2003). In summary, disturbed postprandial regulation of CB1 endogenous agonists may represent one possible mechanism leading to increased CB1 signaling and the development of obesity.

4.2 *Influence on Central Neuropeptides Regulating Food Intake*

CB1 is the G-protein-coupled receptor with the largest abundance in the mammalian brain (Pertwee et al. 2010). Imaging studies identified several brain regions with high CB1 density in humans (Burns et al. 2007), and ultrastructural analyses demonstrated presynaptic CB1 expression on hypothalamic neurons in the mouse brain (Wittmann et al. 2007). Also, coexpression of CB1 and important neuropeptides (cocaine- and amphetamine-related transcript, CART; corticotrophin-releasing hormone, CRH; and, to a much lesser extent, melanocortin-concentrating hormone, MCH) in hypothalamic neurons was reported in the mouse brain (Cota et al. 2003). Thus, endocannabinoids may act as modulators of orexigenic and anorexigenic neurotransmitters and neuropeptides by presynaptic regulation of their release, as described in other brain regions and other neural functions (Wilson and Nicoll 2001). Specifically, the presynaptic reduction of norepinephrine and serotonin release by CB1 activation may have a profound effect on hunger and satiety (Piomelli 2003). As an example, sibutramine, one of the weight loss drugs of the last decade, is both a norepinephrine and serotonin reuptake inhibitor. Sibutramine's efficacy clearly suggested a profound role of these neurotransmitters in the suppression of hunger and the rapid initiation of satiety after a meal (Stock 1997).

Autoradiography studies revealed that DIO rats had decreased CB1 receptor density in several extrahypothalamic regions, e.g., hippocampus, cortical layers I and VII, entopeduncular nucleus, and nucleus accumbens. Lower CB1 density was correlated with the cumulative energy intake from the palatable, fat-enriched diet. Consistent with earlier reports, hypothalamic CB1 density was low compared to other brain regions, but not influenced by DIO (Harrold et al. 2002). Thus, orexigenic effects of CB1 activation in the hypothalamus might be preserved with the development of obesity. The authors suggested that CB1 downregulation through high-fat feeding is the result of increased endocannabinoid availability in extrahypothalamic regions that are at least partly involved in the preference for palatable foods. But endocannabinoid concentrations were not measured in this study; thus the mechanisms of CB1 downregulation with DIO remain uncertain.

In CB1$^{-/-}$ mice, hypothalamic expression of the anorexigenic peptide CRH was increased, whereas expression of the orexigenic peptide CART was decreased (Cota et al. 2003). Other studies in mice also demonstrated that CB1 activation increased CART expression in hypothalamic nuclei and the nucleus accumbens, again suggesting that CART is an important downstream mediator of orexigenic endocannabinoid effects (Osei-Hyiaman et al. 2005a). For a long time, disturbances of the hypothalamic–pituitary–adrenal (HPA) axis have been discussed as a possible cause of human obesity (Wallerius et al. 2003). Interactions between CB1 signaling and the HPA axis that go beyond changes of hypothalamic CRH expression were found in CB$^{-/-}$ mice, but if and how these interactions may contribute to CB1-mediated weight regulation was not studied (Fig. 2).

Direct electrophysiological recordings in neurons from the lateral hypothalamus demonstrated an interaction between endocannabinoid and leptin signaling. These neurons are important because they express the orexigenic MCH and orexin neuropeptides. Stimulation of these neurons leads to CB1-mediated suppression of inhibition of local hypothalamic circuits. These effects were inhibited by leptin. Suppression of inhibition via CB1 was strongly enhanced in leptin-deficient mice (Jo et al. 2005).

4.3 Energy Metabolism

Studies in metabolic chambers did not reveal differences in body temperature, locomotor activity, or energy expenditure between CB$^{-/-}$ and wild-type control mice (Cota et al. 2003). This finding was unexpected because at the age of the mice during the experiment, weight differences between CB1$^{-/-}$ and wild-type mice could not be explained by differences in food intake. In other experiments, a single dose of rimonabant resulted in an acute increase of oxygen consumption that was not associated with increased locomotor activity in normal-weight Sprague–Dawley rats. No changes in respiratory quotient were observed. The effect was not replicated in CB1$^{-/-}$ mice, but the more important finding was the rapid development of tolerance, because the second dose of rimonabant already failed to increase energy expenditure (Kunz et al. 2008). These findings are in contrast to experiments with *ob/ob* mice. Here, 7 days of treatment with rimonabant rather robustly increased basal oxygen consumption of the animals (Liu et al. 2005), although pair feeding was associated with similar weight reduction (see Sect. 3.1). Unfortunately, energy expenditure was not measured in the pair-fed animals in this study. Rimonabant treatment enhanced skeletal muscle glucose uptake in *ob/ob* mice, which may enhance insulin sensitivity. In other studies, AMP-kinase did not mediate the insulin-desensitizing effect of cannabinoids in rat skeletal muscle (Kola et al. 2005), but blockade of CB1 receptors in human skeletal muscle myotubes increased AMP-kinase mRNA expression (Cavuoto et al. 2007).

One indirect calorimetry study was conducted in humans to determine effects of a single dose of taranabant on energy expenditure. This placebo-controlled, double-blind, crossover study in overweight and moderately obese men revealed that a 12-mg single dose taranabant increased resting energy expenditure significantly versus placebo, whereas a 30-mg single dose sibutramine did not. Furthermore, respiratory quotient decreased with taranabant, suggesting a preference for fat oxidation under these experimental conditions (Addy et al. 2008). Although the differences were small, the data point to an influence of CB1 inverse agonists on energy metabolism. Whether the effect was transmitted by central or peripheral effects of the drug could not have been solved by the study design.

Investigators in a recently published study knocked out CB1 receptors in forebrain and sympathetic neurons. The conditional knockout resulted in a lean phenotype, although global $CB1^{-/-}$ were still leaner, and protection against high-fat feeding induced obesity which was even stronger than in $CB1^{-/-}$ global knockout animals. Of great importance was the observation that rimonabant effects on food intake, body weight, and respiratory quotient were not anymore evident in the conditional knockout model, although these animals still expressed a reasonable number of CB1 receptors in several brain regions and peripheral organs (Quarta et al. 2010). Metabolic changes typically seen in DIO mice did not occur in the conditional knockout, suggesting that several metabolic disturbances of obesity are subject to central control. Overall, this model suggests that presynaptic CB1 receptors inhibit sympathetic activity. The conditional knockout of CB1 receptors in sympathetic neurons then led to sympathetic activation, which protected animals against high-fat feeding. The effect was partly mediated by increased brown adipose tissue thermogenesis. These data clearly point to the great importance of central regulation of energy metabolism, both in general and in mediating CB1 effects on body weight.

Nevertheless, direct effects of CB1 signaling have also been described in adipocyte mitochondria. These studies reported that CB1 activation decreased mitochondrial respiration, whereas CB1 blockade increased mitochondriogenesis and oxygen consumption in murine adipocytes (Tedesco et al. 2008, Tedesco et al. 2010). We obtained similar data in human adipocytes and observed a decrease of the oxygen consumption rate with the CB1 agonist HU210 (Fig. 3, unpublished data). Gene expression data in a murine brown adipocyte model suggested that CB1 activation led to decreased expression of the key gene uncoupling protein 1 (UCP1) (Perwitz et al. 2006). This finding points to an energy-saving effect of CB1 receptors that is not mediated by the CNS.

4.4 Gastrointestinal Mechanisms

In the gastrointestinal tract, CB1 receptors are primarily expressed in the enteric nervous system. Activation of CB1 receptors decreases gastric secretion, decreases

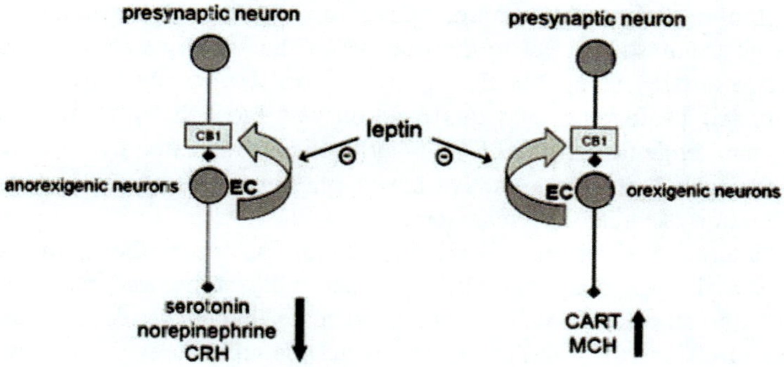

Fig. 2 Presynaptic CB1 receptors regulate the release of orexigenic and anorexigenic neuropeptides. Endocannabinoids (EC) are produced by postsynaptic orexigenic and anorexigenic neurons. Whereas the effects of changes in CB1 signaling on the depicted neuropeptides were described in several studies using $CB1^{-/-}$ mice models or CB1 antagonists, the proposed mechanism by retrograde inhibition via CB1 receptors has as yet only been proved for MCH in specific prefornical lateral hypothalamus neurons (Jo et al. 2005). Thus, no further details are given concerning specific hypothalamic nuclei and specific neuron populations. Leptin interacts with retrograde endocannabinoid signaling by inhibition of EC synthesis (Di Marzo et al. 2001)

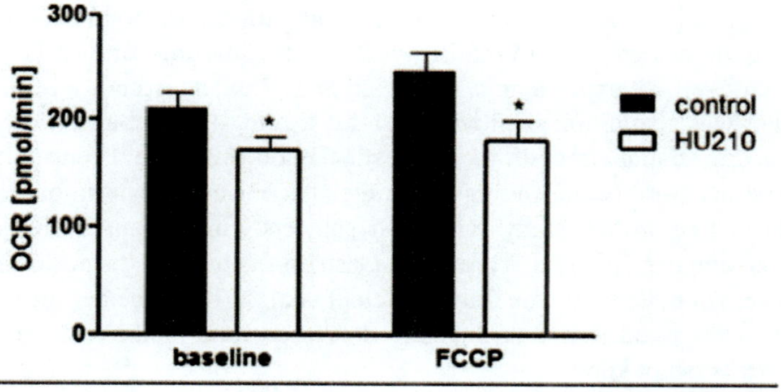

Fig. 3 Energy metabolism of differentiated human SGBS adipocytes under control conditions or after 72-h treatment with the CB1 agonist HU210. HU210 significantly decreased oxygen consumption rate (OCR) under resting conditions (baseline) and after uncoupling with FCCP. Data are mean ± SEM, $n = 4$ independent experiments with 60 wells for each column. Group comparison by 2-way ANOVA. Significant differences are shown for the control vs. HU210 comparison ($*p < 0.05$)

acetylcholine release, and delays gastric emptying (Storr and Sharkey 2007). In rat small intestine, anandamide concentrations increased with fasting and decreased with feeding (Gomez et al. 2002). In the same study, induction of feeding by peripherally administered CB1 agonists was inhibited through vagal ablation with capsaicin. The influence of endocannabinoids on feeding behavior can to some extent also be explained by peripheral interactions with gastrointestinal hormones. CB1 receptors and receptors for gastrointestinal hormones such as cholecystokinin

(CCK), ghrelin, and orexins are colocalized on vagal nerve terminals projecting from the gastrointestinal tract to the nucleus of the solitary tract (Burdyga et al. 2004, 2006, 2010). Vagal CB1 receptors are upregulated by fasting and downregulated by feeding, and this reaction was blocked by CCK antagonists. Thus, the anorexigenic action of CCK may be in part mediated by downregulation of orexigenic CB1 receptors. Activation of ghrelin receptors also prevented the downregulation of vagal CB1 receptors.

Other interactions between the ECS and ghrelin, one of the main orexigenic gastrointestinal hormones, were also described. CB1 activation enhanced ghrelin release from the stomach. In addition, ghrelin increased hypothalamic endocannabinoid concentrations. Consequently, rimonabant reduced ghrelin's orexigenic activity when injected into the hypothalamus and reduced circulating ghrelin in fasted and fed rats. In contrast, no interactions were observed with glucagon-like peptide, a hormone that may lead to weight reduction (Cani et al. 2004; Tucci et al. 2004). The orexigenic effects of ghrelin and CB1 signals appear to be mediated by stimulation of AMP-activated protein kinase (AMPK) in the hypothalamus (Kola et al. 2005). Whether the influence of the ECS on vagal input to the nucleus of the solitary tract contributes to CB1-mediated regulation of body weight remains to be determined.

In DIO mice, anandamide concentrations of the stomach decreased with the development of obesity, and CB1 gene expression also decreased, but gastric emptying decreased as well (Di Marzo et al. 2008). In contrast, anandamide concentrations decreased, and motility increased in the intestine of DIO mice. In this second set of experiments, the sensitivity of intestinal motility against rimonabant was also diminished with high-fat feeding (Izzo et al. 2009). Thus, the ECS has a differential effect on gastrointestinal endocannabinoid concentrations and gastrointestinal motility, depending on the site of action and the nutritional status. In a small placebo-controlled study with lean subjects, rimonabant decreased gastric accommodation in response to a test meal. Gastric sensitivity to distension and gastric motility were not altered by the drug (Ameloot et al. 2010). Whether the small effect on gastric accommodation contributes to decreased food intake and weight loss in obese patients is not known.

Another line of evidence was investigated in a rat model of the metabolic syndrome and associated cardiovascular disease, the JCR:LA-*cp* rat. The obese, disease-prone phenotype is due to the *cp* mutation that results in a stop codon in the extracellular domain of the leptin receptor (Russell et al. 2010). Treatment with rimonabant resulted in a transient reduction in food intake, and in diminished weight gain with aging, a finding consistent with several studies that have been summarized before in this paper. The most striking finding in this study was the reduction of fasting triglycerides (as seen in clinical studies as well) and the marked reduction of postprandial lymphatic apolipoprotein B48. This finding suggests a direct influence of pharmacological CB1 blockade on intestinal fatty acid resorption or enterocyte triglyceride/chylomicron synthesis. But again, whether this process contributes to a negative energy balance remains to be determined.

5 Peripheral CB1 Inhibition: New Players, More Mechanisms?

Most of the presented data in this chapter clearly point to the predominant role of the CNS in mediating CB1 effects on body weight. A similar predominance of the CNS has also been suggested by some authors with a more focused view on the regulation of glucose and lipid metabolism by modulation of CB1 signaling (Fong and Heymsfield 2009; O'Hare et al. 2011; Quarta et al. 2010). Nevertheless, other published reports speak for additional peripheral mechanisms. In a study with DIO rats, this was demonstrated by comparing ICV and sc administration of rimonabant (Nogueiras et al. 2008). Weight loss was more pronounced with systemic than CNS administration. The observed changes in adipose tissue function and gene expression were closely related to the hypophagic effect of CNS administration of rimonabant, as controlled for by pair feeding. In clear contrast to another study (O'Hare et al. 2011), glucose metabolism and insulin sensitivity was only modified by peripheral administration of rimonabant, independent of the effects on food intake (Nogueiras et al. 2008). Other authors have shown that the development of hepatic steatosis, inflammatory responses, and insulin resistance with high-fat feeding were prevented by selective knockout of hepatic CB1 receptors (Osei-Hyiaman et al. 2005b, 2008). The hepatic CB1 knockout did not mediate protection against the development of obesity with high-fat feeding, but these data nevertheless fostered the development of new CB1 drugs with restricted action in the periphery.

URB447 is a combined CB1 neutral antagonist/CB2 agonist with a 50× larger IC_{50} for CB1 receptors compared to rimonabant. After systemic administration, brain concentrations were below 10 pmol/g which was the lower limit of detection of the drug (LoVerme et al. 2009). Typical CNS-mediated effects of CB1 agonists such as catalepsy or hypothermia were not prevented by URB447. Interestingly, the reduction of food intake of equimolar doses of peripherally administered rimonabant and URB447 was similar, and weight reduction occurred with URB447 that was slower in onset than with rimonabant, but reached the same level at the end of the study. Weight gain in *ob/ob* mice was also prevented by URB447 at the same magnitude as with rimonabant (LoVerme et al. 2009).

AM6545 is a CB1 neutral antagonist with high transport capacity for members of the multidrug resistance (MDR) family. MDR transporter activity decreased brain concentrations of AM6545 to rather low values. Consequently, CB1-mediated catalepsy and hypothermia were prevented by peripherally administered rimonabant, but not by AM6545 (Tam et al. 2010). Body weight reduction in DIO mice was stronger with rimonabant, but AM6545 significantly reduced body weight as well, and pair feeding demonstrated that the body weight change with AM6545 was independent of a change in food intake. Respiratory quotient was shifted towards fat oxidation by the drug, and this effect was not observed in $CB1^{-/-}$ mice. Enhanced insulin sensitivity and glucose tolerance as well as amelioration of liver steatohepatitis with AM6545 treatment were also independent of decreased food intake (Tam et al. 2010).

If we accept that peripheral CB1 receptors in liver, skeletal muscle, adipose tissue, and pancreas are involved in the reported metabolic activities of peripherally restricted CB1 antagonists (Engeli 2008b), the question remains, how weight loss is mediated by these drugs. The gastrointestinal effects of the ECS (see Sect. 4.4) of course may be involved. Another possible explanation would be that peripheral effects are mediated by presynaptic CB1 receptors on sympathetic neurons (see Sect. 4.3). Sympathetic disinhibition by blockade of these receptors might occur (Quarta et al. 2010). If this is indeed the case, the clinical safety of such a drug would be of major concern.

6 Summary

CB receptor agonists increase hunger sensation and food intake. This discovery led to the clinical use of Δ^9-THC derivatives for the treatment of cancer and HIV-related nausea and cachexia. Animal studies clarified the important role of CB1 receptors in the hypothalamus and in the limbic system in mediating orexigenic effects. In parallel, CB1-specific blockade either by drugs or by genetic ablation proved to protect against weight gain induced by high-fat feeding and to reduce body weight in obese animals and humans. The weight-reducing mechanisms of CB1 blockade are more complex than simply decreasing food intake. These mechanisms involve interactions with several orexigenic and anorexigenic neuropeptides and hormones, regulation of sympathetic activity, and influence on mitochondrial function, on lipogenesis, and on gastrointestinal functions including lipid absorption. Although weight reduction is to a large part mediated by the CNS, weight loss also occurs if drugs that do not reach CNS concentrations sufficient to inhibit CB1 signaling are used. The development of peripherally restricted CB1 inverse agonists and antagonists opened new routes in CB1 pharmacology because centrally acting CB1 inverse agonists, e.g., rimonabant and taranabant, were associated with unacceptable adverse reactions precluding further development and clinical application. Tissue and circulating endocannabinoids are often increased in animal models of obesity and in obese humans, especially those with visceral fat accumulation. Thus, further research on CB1 inhibition is still promising to treat obesity. But given the recent experience with CNS penetrating CB1 antagonistic drugs, a well-designed preclinical and clinical safety program has to be performed.

References

Abel EL (1975) Cannabis: effects on hunger and thirst. Behav Biol 15:255–281

Abood ME (2005) Molecular biology of cannabinoid receptors. Handb Exp Pharmacol 168:81–115

Addy C, Wright H, van Laere K, Gantz I, Erondu N, Musser BJ, Lu K, Yuan J, Sanabria-Bohorquez SM, Stoch A, Stevens C, Fong TM, De Lepeleire I, Cilissen C, Cote J, Rosko K, Gendrano IN III, Nguyen AM, Gumbiner B, Rothenberg P, de Hoon J, Bormans G, Depre M, Eng WS, Ravussin E, Klein S, Blundell J, Herman GA, Burns HD, Hargreaves RJ, Wagner J, Gottesdiener K, Amatruda JM, Heymsfield SB (2008) The acyclic CB1R inverse agonist taranabant mediates weight loss by increasing energy expenditure and decreasing caloric intake. Cell Metab 7:68–78

Ameloot K, Janssen P, Scarpellini E, Vos R, Boesmans W, Depoortere I, Vanden BP, Tack J (2010) Endocannabinoid control of gastric sensorimotor function in man. Aliment Pharmacol Ther 31:1123–1131

Arnone M, Maruani J, Chaperon F, Thiebot MH, Poncelet M, Soubrie P, Le Fur G (1997) Selective inhibition of sucrose and ethanol intake by SR 141716, an antagonist of central cannabinoid (CB1) receptors. Psychopharmacology (Berl) 132:104–106

Aronne LJ, Tonstad S, Moreno M, Gantz I, Erondu N, Suryawanshi S, Molony C, Sieberts S, Nayee J, Meehan AG, Shapiro D, Heymsfield SB, Kaufman KD, Amatruda JM (2010) A clinical trial assessing the safety and efficacy of taranabant, a CB1R inverse agonist, in obese and overweight patients: a high-dose study. Int J Obes (Lond) 34:919–935

Beal JE, Olson R, Lefkowitz L, Laubenstein L, Bellman P, Yangco B, Morales JO, Murphy R, Powderly W, Plasse TF, Mosdell KW, Shepard KV (1997) Long-term efficacy and safety of dronabinol for acquired immunodeficiency syndrome-associated anorexia. J Pain Symptom Manage 14:7–14

Bedi G, Foltin RW, Gunderson EW, Rabkin J, Hart CL, Comer SD, Vosburg SK, Haney M (2010) Efficacy and tolerability of high-dose dronabinol maintenance in HIV-positive marijuana smokers: a controlled laboratory study. Psychopharmacology (Berl) 212:675–686

Bensaid M, Gary-Bobo M, Esclangon A, Maffrand JP, Le Fur G, Oury-Donat F, Soubrie P (2003) The cannabinoid CB1 receptor antagonist SR141716 increases Acrp30 mRNA expression in adipose tissue of obese fa/fa rats and in cultured adipocyte cells. Mol Pharmacol 63:908–914

Blüher M, Engeli S, Klöting N, Berndt J, Fasshauer M, Batkai S, Pacher P, Schön MR, Jordan J, Stumvoll M (2006) Dysregulation of the peripheral and adipose tissue endocannabinoid system in human abdominal obesity. Diabetes 55:3053–3060

Burdyga G, Lal S, Varro A, Dimaline R, Thompson DG, Dockray GJ (2004) Expression of cannabinoid CB1 receptors by vagal afferent neurons is inhibited by cholecystokinin. J Neurosci 24:2708–2715

Burdyga G, Varro A, Dimaline R, Thompson DG, Dockray GJ (2006) Ghrelin receptors in rat and human nodose ganglia: putative role in regulating CB-1 and MCH receptor abundance. Am J Physiol Gastrointest Liver Physiol 290:G1289–G1297

Burdyga G, Varro A, Dimaline R, Thompson DG, Dockray GJ (2010) Expression of cannabinoid CB1 receptors by vagal afferent neurons: kinetics and role in influencing neurochemical phenotype. Am J Physiol Gastrointest Liver Physiol 299:G63–G69

Burns HD, van Laere K, Sanabria-Bohorquez S, Hamill TG, Bormans G, Eng WS, Gibson R, Ryan C, Connolly B, Patel S, Krause S, Vanko A, van Hecken A, Dupont P, De Lepeleire I, Rothenberg P, Stoch SA, Cote J, Hagmann WK, Jewell JP, Lin LS, Liu P, Goulet MT, Gottesdiener K, Wagner JA, de Hoon J, Mortelmans L, Fong TM, Hargreaves RJ (2007) [18F]MK-9470, a positron emission tomography (PET) tracer for in vivo human PET brain imaging of the cannabinoid-1 receptor. Proc Natl Acad Sci U S A 104:9800–9805

Cani PD, Montoya ML, Neyrinck AM, Delzenne NM, Lambert DM (2004) Potential modulation of plasma ghrelin and glucagon-like peptide-1 by anorexigenic cannabinoid compounds, SR141716A (rimonabant) and oleoylethanolamide. Br J Nutr 92:757–761

Carai MA, Colombo G, Maccioni P, Gessa GL (2006) Efficacy of rimonabant and other cannabinoid CB1 receptor antagonists in reducing food intake and body weight: preclinical and clinical data. CNS Drug Rev 12:91–99

Cavuoto P, McAinch AJ, Hatzinikolas G, Cameron-Smith D, Wittert GA (2007) Effects of cannabinoid receptors on skeletal muscle oxidative pathways. Mol Cell Endocrinol 267:63–69

Colombo G, Agabio R, Diaz G, Lobina C, Reali R, Gessa GL (1998) Appetite suppression and weight loss after the cannabinoid antagonist SR 141716. Life Sci 63:L113–L117
Cota D, Marsicano G, Tschöp M, Grubler Y, Flachskamm C, Schubert M, Auer D, Yassouridis A, Thöne-Reineke C, Ortmann S, Tomassoni F, Cervino C, Nisoli E, Linthorst AC, Pasquali R, Lutz B, Stalla GK, Pagotto U (2003) The endogenous cannabinoid system affects energy balance via central orexigenic drive and peripheral lipogenesis. J Clin Invest 112:423–431
Coté M, Matias I, Lemieux I, Petrosino S, Almeras N, Després JP, Di Marzo V (2007) Circulating endocannabinoid levels, abdominal adiposity and related cardiometabolic risk factors in obese men. Int J Obes (Lond) 31:692–699
Després JP, Golay A, Sjöström L (2005) Effects of rimonabant on metabolic risk factors in overweight patients with dyslipidemia. N Engl J Med 353:2121–2134
Devane WA, Dysarz FA III, Johnson MR, Melvin LS, Howlett AC (1988) Determination and characterization of a cannabinoid receptor in rat brain. Mol Pharmacol 34:605–613
Devane WA, Hanus L, Breuer A, Pertwee RG, Stevenson LA, Griffin G, Gibson D, Mandelbaum A, Etinger A, Mechoulam R (1992) Isolation and structure of a brain constituent that binds to the cannabinoid receptor. Science 258:1946–1949
Di Marzo V, Goparaju SK, Wang L, Liu J, Batkai S, Jarai Z, Fezza F, Miura GI, Palmiter RD, Sugiura T, Kunos G (2001) Leptin-regulated endocannabinoids are involved in maintaining food intake. Nature 410:822–825
Di Marzo V, De Petrocellis L, Bisogno T (2005) The biosynthesis, fate and pharmacological properties of endocannabinoids. Handb Exp Pharmacol 168:147–185
Di Marzo V, Capasso R, Matias I, Aviello G, Petrosino S, Borrelli F, Romano B, Orlando P, Capasso F, Izzo AA (2008) The role of endocannabinoids in the regulation of gastric emptying: alterations in mice fed a high-fat diet. Br J Pharmacol 153:1272–1280
Di Marzo V, Verrijken A, Hakkarainen A, Petrosino S, Mertens I, Lundbom N, Piscitelli F, Westerbacka J, Soro-Paavonen A, Matias I, Van GL, Taskinen MR (2009a) Role of insulin as a negative regulator of plasma endocannabinoid levels in obese and nonobese subjects. Eur J Endocrinol 161:715–722
Di Marzo V, Coté M, Matias I, Lemieux I, Arsenault BJ, Cartier A, Piscitelli F, Petrosino S, Almeras N, Després JP (2009b) Changes in plasma endocannabinoid levels in viscerally obese men following a 1 year lifestyle modification programme and waist circumference reduction: associations with changes in metabolic risk factors. Diabetologia 52:213–217
Dipatrizio NV, Simansky KJ (2008) Inhibiting parabrachial fatty acid amide hydrolase activity selectively increases the intake of palatable food via cannabinoid CB1 receptors. Am J Physiol Regul Integr Comp Physiol 295:R1409–R1414
Engeli S (2008a) Dysregulation of the endocannabinoid system in obesity. J Neuroendocrinol 20 (Suppl 1):110–115
Engeli S (2008b) Peripheral metabolic effects of endocannabinoids and cannabinoid receptor blockade. Obes Facts 1:8–15
Engeli S, Böhnke J, Feldpausch M, Gorzelniak K, Janke J, Batkai S, Pacher P, Harvey-White J, Luft FC, Sharma AM, Jordan J (2005) Activation of the peripheral endocannabinoid system in human obesity. Diabetes 54:2838–2843
Foltin RW, Fischman MW, Byrne MF (1988) Effects of smoked marijuana on food intake and body weight of humans living in a residential laboratory. Appetite 11:1–14
Fong TM, Heymsfield SB (2009) Cannabinoid-1 receptor inverse agonists: current understanding of mechanism of action and unanswered questions. Int J Obes (Lond) 33:947–955
Fong TM, Guan XM, Marsh DJ, Shen CP, Stribling DS, Rosko KM, Lao J, Yu H, Feng Y, Xiao JC, Van der Ploeg LH, Goulet MT, Hagmann WK, Lin LS, Lanza TJ Jr, Jewell JP, Liu P, Shah SK, Qi H, Tong X, Wang J, Xu SS, Francis B, Strack AM, MacIntyre DE, Shearman LP (2007) Antiobesity efficacy of a novel cannabinoid-1 receptor inverse agonist, N-[(1S,2S)-3-(4-chlorophenyl)-2-(3-cyanophenyl)-1-methylpropyl]-2-methyl-2-[[5-(t rifluoromethyl)pyridin-2-yl]oxy]propanamide (MK-0364), in rodents. J Pharmacol Exp Ther 321:1013–1022

Gagnon MA, Elie R (1975) Effects of marijuana and D-amphetamine on the appetite, food consumption and various cardio-respiratory variables in man. Union Med Can 104:914–921

Gaoni Y, Mechoulam R (1964) Isolation, structure and partial synthesis of an active constituent of hashish. J Am Chem Soc 86:1646

Gary-Bobo M, Elachouri G, Gallas JF, Janiak P, Marini P, Ravinet Trillou C, Chabbert M, Cruccioli N, Pfersdorff C, Roque C, Arnone M, Croci T, Soubrie P, Oury-Donat F, Maffrand JP, Scatton B, Lacheretz F, Le Fur G, Herbert JM, Bensaid M (2007) Rimonabant reduces obesity-associated hepatic steatosis and features of metabolic syndrome in obese Zucker fa/fa rats. Hepatology 46:122–129

Gomez R, Navarro M, Ferrer B, Trigo JM, Bilbao A, Del Arco I, Cippitelli A, Nava F, Piomelli D, Rodriguez dF (2002) A peripheral mechanism for CB1 cannabinoid receptor-dependent modulation of feeding. J Neurosci 22:9612–9617

Gorter R, Seefried M, Volberding P (1992) Dronabinol effects on weight in patients with HIV infection. AIDS 6:127

Greenberg I, Kuehnle J, Mendelson JH, Bernstein JG (1976) Effects of marihuana use on body weight and caloric intake in humans. Psychopharmacology (Berl) 49:79–84

Harrold JA, Elliott JC, King PJ, Widdowson PS, Williams G (2002) Down-regulation of cannabinoid-1 (CB-1) receptors in specific extrahypothalamic regions of rats with dietary obesity: a role for endogenous cannabinoids in driving appetite for palatable food? Brain Res 952:232–238

Hollister LE (1971) Hunger and appetite after single doses of marihuana, alcohol, and dextroamphetamine. Clin Pharmacol Ther 12:44–49

Howlett AC (2005) Cannabinoid receptor signaling. Handb Exp Pharmacol 168:53–79

Howlett AC, Barth F, Bonner TI, Cabral G, Casellas P, Devane WA, Felder CC, Herkenham M, Mackie K, Martin BR, Mechoulam R, Pertwee RG (2002) International Union of Pharmacology. XXVII. Classification of cannabinoid receptors. Pharmacol Rev 54:161–202

Izzo AA, Piscitelli F, Capasso R, Aviello G, Romano B, Borrelli F, Petrosino S, Di Marzo V (2009) Peripheral endocannabinoid dysregulation in obesity: relation to intestinal motility and energy processing induced by food deprivation and re-feeding. Br J Pharmacol 158:451–461

Jamshidi N, Taylor DA (2001) Anandamide administration into the ventromedial hypothalamus stimulates appetite in rats. Br J Pharmacol 134:1151–1154

Jbilo O, Ravinet Trillou C, Arnone M, Buisson I, Bribes E, Peleraux A, Penarier G, Soubrie P, Le Fur G, Galiegue S, Casellas P (2005) The CB1 receptor antagonist rimonabant reverses the diet-induced obesity phenotype through the regulation of lipolysis and energy balance. FASEB J 19:1567–1569

Jo YH, Chen YJ, Chua SC Jr, Talmage DA, Role LW (2005) Integration of endocannabinoid and leptin signaling in an appetite-related neural circuit. Neuron 48:1055–1066

Jones D (2008) End of the line for cannabinoid receptor 1 as an anti-obesity target? Nat Rev Drug Discov 7:961–962

Jordan J, Schlaich M, Redon J, Narkiewicz K, Luft FC, Grassi G, Dixon J, Lambert G, Engeli S (2011) European Society of Hypertension Working Group on Obesity: obesity drugs and cardiovascular outcomes. J Hypertens 29:189–193

Kipnes MS, Hollander P, Fujioka K, Gantz I, Seck T, Erondu N, Shentu Y, Lu K, Suryawanshi S, Chou M, Johnson-Levonas AO, Heymsfield SB, Shapiro D, Kaufman KD, Amatruda JM (2010) A one-year study to assess the safety and efficacy of the CB1R inverse agonist taranabant in overweight and obese patients with type 2 diabetes. Diabetes Obes Metab 12:517–531

Kirkham TC, Williams CM (2001) Endogenous cannabinoids and appetite. Nutr Res Rev 14:65–86

Kirkham TC, Williams CM, Fezza F, Di Marzo V (2002) Endocannabinoid levels in rat limbic forebrain and hypothalamus in relation to fasting, feeding and satiation: stimulation of eating by 2-arachidonoyl glycerol. Br J Pharmacol 136:550–557

Kola B, Hubina E, Tucci SA, Kirkham TC, Garcia EA, Mitchell SE, Williams LM, Hawley SA, Hardie DG, Grossman AB, Korbonits M (2005) Cannabinoids and ghrelin have both central and peripheral metabolic and cardiac effects via AMP-activated protein kinase. J Biol Chem 280:25196–25201

Kunz I, Meier MK, Bourson A, Fisseha M, Schilling W (2008) Effects of rimonabant, a cannabinoid CB1 receptor ligand, on energy expenditure in lean rats. Int J Obes (Lond) 32:863–870

Ledent C, Valverde O, Cossu G, Petitet F, Aubert JF, Beslot F, Bohme GA, Imperato A, Pedrazzini T, Roques BP, Vassart G, Fratta W, Parmentier M (1999) Unresponsiveness to cannabinoids and reduced addictive effects of opiates in CB1 receptor knockout mice. Science 283:401–404

Lin LS, Lanza TJ Jr, Jewell JP, Liu P, Shah SK, Qi H, Tong X, Wang J, Xu SS, Fong TM, Shen CP, Lao J, Xiao JC, Shearman LP, Stribling DS, Rosko K, Strack A, Marsh DJ, Feng Y, Kumar S, Samuel K, Yin W, Van der Ploeg LH, Goulet MT, Hagmann WK (2006) Discovery of N-[(1S,2S)-3-(4-Chlorophenyl)-2-(3-cyanophenyl)-1-methylpropyl]-2-methyl-2-{[5-(trifluoromethyl)pyridin-2-yl]oxy}propanamide (MK-0364), a novel, acyclic cannabinoid-1 receptor inverse agonist for the treatment of obesity. J Med Chem 49:7584–7587

Liu YL, Connoley IP, Wilson CA, Stock MJ (2005) Effects of the cannabinoid CB1 receptor antagonist SR141716 on oxygen consumption and soleus muscle glucose uptake in Lep(ob)/Lep(ob) mice. Int J Obes (Lond) 29:183–187

LoVerme J, Duranti A, Tontini A, Spadoni G, Mor M, Rivara S, Stella N, Xu C, Tarzia G, Piomelli D (2009) Synthesis and characterization of a peripherally restricted CB1 cannabinoid antagonist, URB447, that reduces feeding and body-weight gain in mice. Bioorg Med Chem Lett 19:639–643

Maccarrone M, Fride E, Bisogno T, Bari M, Cascio MG, Battista N, Finazzi AA, Suris R, Mechoulam R, Di Marzo V (2005) Up-regulation of the endocannabinoid system in the uterus of leptin knockout (ob/ob) mice and implications for fertility. Mol Hum Reprod 11:21–28

Marsicano G, Wotjak CT, Azad SC, Bisogno T, Rammes G, Cascio MG, Hermann H, Tang J, Hofmann C, Zieglgansberger W, Di Marzo V, Lutz B (2002) The endogenous cannabinoid system controls extinction of aversive memories. Nature 418:530–534

Matias I, Gonthier MP, Orlando P, Martiadis V, De Petrocellis L, Cervino C, Petrosino S, Hoareau L, Festy F, Pasquali R, Roche R, Maj M, Pagotto U, Monteleone P, Di Marzo V (2006) Regulation, function, and dysregulation of endocannabinoids in models of adipose and beta-pancreatic cells and in obesity and hyperglycemia. J Clin Endocrinol Metab 91:3171–3180

Matias I, Vergoni AV, Petrosino S, Ottani A, Pocai A, Bertolini A, Di Marzo V (2008) Regulation of hypothalamic endocannabinoid levels by neuropeptides and hormones involved in food intake and metabolism: insulin and melanocortins. Neuropharmacology 54:206–212

Matsuda LA, Lolait SJ, Brownstein MJ, Young AC, Bonner TI (1990) Structure of a cannabinoid receptor and functional expression of the cloned cDNA. Nature 346:561–564

Mechoulam R, Ben-Shabat S, Hanus L, Ligumsky M, Kaminski NE, Schatz AR, Gopher A, Almog S, Martin BR, Compton DR (1995) Identification of an endogenous 2-monoglyceride, present in canine gut, that binds to cannabinoid receptors. Biochem Pharmacol 50:83–90

Moreira FA, Grieb M, Lutz B (2009) Central side-effects of therapies based on CB1 cannabinoid receptor agonists and antagonists: focus on anxiety and depression. Best Pract Res Clin Endocrinol Metab 23:133–144

Munro S, Thomas KL, Abu-Shaar M (1993) Molecular characterization of a peripheral receptor for cannabinoids. Nature 365:61–65

Nogueiras R, Veyrat-Durebex C, Suchanek PM, Klein M, Tschöp J, Caldwell C, Woods SC, Wittmann G, Watanabe M, Liposits Z, Fekete C, Reizes O, Rohner-Jeanrenaud F, Tschöp MH (2008) Peripheral, but not central, CB1 antagonism provides food intake-independent metabolic benefits in diet-induced obese rats. Diabetes 57:2977–2991

O'Hare JD, Zielinski E, Cheng B, Scherer T, Buettner C (2011) Central endocannabinoid signaling regulates hepatic glucose production and systemic lipolysis. Diabetes 60:1055–1062

Osei-Hyiaman D, Depetrillo M, Harvey-White J, Bannon AW, Cravatt BF, Kuhar MJ, Mackie K, Palkovits M, Kunos G (2005a) Cocaine- and amphetamine-related transcript is involved in the orexigenic effect of endogenous anandamide. Neuroendocrinology 81:273–282

Osei-Hyiaman D, Depetrillo M, Pacher P, Liu J, Radaeva S, Batkai S, Harvey-White J, Mackie K, Offertaler L, Wang L, Kunos G (2005b) Endocannabinoid activation at hepatic CB(1) receptors stimulates fatty acid synthesis and contributes to diet-induced obesity. J Clin Invest 115:1298–1305

Osei-Hyiaman D, Liu J, Zhou L, Godlewski G, Harvey-White J, Jeong WI, Batkai S, Marsicano G, Lutz B, Buettner C, Kunos G (2008) Hepatic CB1 receptor is required for development of diet-induced steatosis, dyslipidemia, and insulin and leptin resistance in mice. J Clin Invest 118:3160–3169

Pacher P, Batkai S, Kunos G (2006) The endocannabinoid system as an emerging target of pharmacotherapy. Pharmacol Rev 58:389–462

Pagotto U, Marsicano G, Cota D, Lutz B, Pasquali R (2006) The emerging role of the endocannabinoid system in endocrine regulation and energy balance. Endocr Rev 27:73–100

Pertwee RG (2005) Pharmacological actions of cannabinoids. Handb Exp Pharmacol 168:1–51

Pertwee RG, Howlett AC, Abood ME, Alexander SP, Di Marzo V, Elphick MR, Greasley PJ, Hansen HS, Kunos G, Mackie K, Mechoulam R, Ross RA (2010) International Union of Basic and Clinical Pharmacology. LXXIX. Cannabinoid receptors and their ligands: beyond CB1 and CB2. Pharmacol Rev 62:588–631

Perwitz N, Fasshauer M, Klein J (2006) Cannabinoid receptor signaling directly inhibits thermogenesis and alters expression of adiponectin and visfatin. Horm Metab Res 38:356–358

Piomelli D (2003) The molecular logic of endocannabinoid signalling. Nat Rev Neurosci 4:873–884

Pi-Sunyer FX, Aronne LJ, Heshmati HM, Devin J, Rosenstock J (2006) Effect of rimonabant, a cannabinoid-1 receptor blocker, on weight and cardiometabolic risk factors in overweight or obese patients: RIO-North America: a randomized controlled trial. JAMA 295:761–775

Plasse TF, Gorter RW, Krasnow SH, Lane M, Shepard KV, Wadleigh RG (1991) Recent clinical experience with dronabinol. Pharmacol Biochem Behav 40:695–700

Proietto J, Rissanen A, Harp JB, Erondu N, Yu Q, Suryawanshi S, Jones ME, Johnson-Levonas AO, Heymsfield SB, Kaufman KD, Amatruda JM (2010) A clinical trial assessing the safety and efficacy of the CB1R inverse agonist taranabant in obese and overweight patients: low-dose study. Int J Obes (Lond) 34:1243–1254

Quarta C, Bellocchio L, Mancini G, Mazza R, Cervino C, Braulke LJ, Fekete C, Latorre R, Nanni C, Bucci M, Clemens LE, Heldmaier G, Watanabe M, Leste-Lassere T, Maitre M, Tedesco L, Fanelli F, Reuss S, Klaus S, Srivastava RK, Monory K, Valerio A, Grandis A, De Giorgio R, Pasquali R, Nisoli E, Cota D, Lutz B, Marsicano G, Pagotto U (2010) CB(1) signaling in forebrain and sympathetic neurons is a key determinant of endocannabinoid actions on energy balance. Cell Metab 11:273–285

Ravinet Trillou C, Arnone M, Delgorge C, Gonalons N, Keane P, Maffrand JP, Soubrie P (2003) Anti-obesity effect of SR141716, a CB1 receptor antagonist, in diet-induced obese mice. Am J Physiol Regul Integr Comp Physiol 284:R345–R353

Ravinet Trillou C, Delgorge C, Menet C, Arnone M, Soubrie P (2004) CB1 cannabinoid receptor knockout in mice leads to leanness, resistance to diet-induced obesity and enhanced leptin sensitivity. Int J Obes Relat Metab Disord 28:640–648

Rinaldi-Carmona M, Barth F, Heaulme M, Shire D, Calandra B, Congy C, Martinez S, Maruani J, Neliat G, Caput D (1994) SR141716A, a potent and selective antagonist of the brain cannabinoid receptor. FEBS Lett 350:240–244

Robson P (2005) Human studies of cannabinoids and medicinal cannabis. Handb Exp Pharmacol 168:719–756

Russell JC, Kelly SE, Diane A, Wang Y, Mangat R, Novak S, Vine DF, Proctor SD (2010) Rimonabant-mediated changes in intestinal lipid metabolism and improved renal vascular dysfunction in the JCR:LA-cp rat model of prediabetic metabolic syndrome. Am J Physiol Gastrointest Liver Physiol 299:G507–G516

Sallan SE, Cronin C, Zelen M, Zinberg NE (1980) Antiemetics in patients receiving chemotherapy for cancer: a randomized comparison of delta-9-tetrahydrocannabinol and prochlorperazine. N Engl J Med 302:135–138

Scheen AJ, Finer N, Hollander P, Jensen MD, van Gaal LF (2006) Efficacy and tolerability of rimonabant in overweight or obese patients with type 2 diabetes: a randomised controlled study. Lancet 368:1660–1672
Siler JF, Sheep WL, Bates LB, Clark GF, Cook GW, Smith WH (1933) Marihuana smoking in Panama. Mil Surg 73:269–280
Simiand J, Keane M, Keane PE, Soubrie P (1998) SR 141716, a CB1 cannabinoid receptor antagonist, selectively reduces sweet food intake in marmoset. Behav Pharmacol 9:179–181
Soria-Gomez E, Matias I, Rueda-Orozco PE, Cisneros M, Petrosino S, Navarro L, Di Marzo V, Prospero-Garcia O (2007) Pharmacological enhancement of the endocannabinoid system in the nucleus accumbens shell stimulates food intake and increases c-Fos expression in the hypothalamus. Br J Pharmacol 151:1109–1116
Stock MJ (1997) Sibutramine: a review of the pharmacology of a novel anti- obesity agent. Int J Obes Relat Metab Disord 21(Suppl 1):S25–S29
Storr MA, Sharkey KA (2007) The endocannabinoid system and gut-brain signalling. Curr Opin Pharmacol 7:575–582
Tam J, Vemuri VK, Liu J, Batkai S, Mukhopadhyay B, Godlewski G, Osei-Hyiaman D, Ohnuma S, Ambudkar SV, Pickel J, Makriyannis A, Kunos G (2010) Peripheral CB1 cannabinoid receptor blockade improves cardiometabolic risk in mouse models of obesity. J Clin Invest 120:2953–2966
Tart CT (1970) Marijuana intoxication common experiences. Nature 226:701–704
Tedesco L, Valerio A, Cervino C, Cardile A, Pagano C, Vettor R, Pasquali R, Carruba MO, Marsicano G, Lutz B, Pagotto U, Nisoli E (2008) Cannabinoid type 1 receptor blockade promotes mitochondrial biogenesis through endothelial nitric oxide synthase expression in white adipocytes. Diabetes 57:2028–2036
Tedesco L, Valerio A, Dossena M, Cardile A, Ragni M, Pagano C, Pagotto U, Carruba MO, Vettor R, Nisoli E (2010) Cannabinoid receptor stimulation impairs mitochondrial biogenesis in mouse white adipose tissue, muscle, and liver: the role of eNOS, p38 MAPK, and AMPK pathways. Diabetes 59:2826–2836
Thakur GA, Tichkule R, Bajaj S, Makriyannis A (2009) Latest advances in cannabinoid receptor agonists. Expert Opin Ther Pat 19:1647–1673
Thornton-Jones ZD, Kennett GA, Benwell KR, Revell DF, Misra A, Sellwood DM, Vickers SP, Clifton PG (2006) The cannabinoid CB1 receptor inverse agonist, rimonabant, modifies body weight and adiponectin function in diet-induced obese rats as a consequence of reduced food intake. Pharmacol Biochem Behav 84:353–359
Topol EJ, Bousser MG, Fox KA, Creager MA, Després JP, Easton JD, Hamm CW, Montalescot G, Steg PG, Pearson TA, Cohen E, Gaudin C, Job B, Murphy JH, Bhatt DL (2010) Rimonabant for prevention of cardiovascular events (CRESCENDO): a randomised, multicentre, placebo-controlled trial. Lancet 376:517–523
Tucci SA, Rogers EK, Korbonits M, Kirkham TC (2004) The cannabinoid CB1 receptor antagonist SR141716 blocks the orexigenic effects of intrahypothalamic ghrelin. Br J Pharmacol 143:520–523
van Gaal LF, Rissanen AM, Scheen AJ, Ziegler O, Röossner S (2005) Effects of the cannabinoid-1 receptor blocker rimonabant on weight reduction and cardiovascular risk factors in overweight patients: 1-year experience from the RIO-Europe study. Lancet 365:1389–1397
Volicer L, Stelly M, Morris J, McLaughlin J, Volicer BJ (1997) Effects of dronabinol on anorexia and disturbed behavior in patients with Alzheimer's disease. Int J Geriatr Psychiatry 12:913–919
Wadden TA, Fujioka K, Toubro S, Gantz I, Erondu NE, Chen M, Suryawanshi S, Carofano W, Johnson-Levonas AO, Shapiro DR, Kaufman KD, Heymsfield SB, Amatruda JM (2010) A randomized trial of lifestyle modification and taranabant for maintaining weight loss achieved with a low-calorie diet. Obesity (Silver Spring) 18:2301–2310
Wallerius S, Rosmond R, Ljung T, Holm G, Björntorp P (2003) Rise in morning saliva cortisol is associated with abdominal obesity in men: a preliminary report. J Endocrinol Invest 26:616–619

Ware MA, St Arnaud-Trempe E (2010) The abuse potential of the synthetic cannabinoid nabilone. Addiction 105:494–503
Watanabe S, Doshi M, Hamazaki T (2003) n-3 Polyunsaturated fatty acid (PUFA) deficiency elevates and n-3 PUFA enrichment reduces brain 2-arachidonoylglycerol level in mice. Prostaglandins Leukot Essent Fatty Acids 69:51–59
Williams CM, Kirkham TC (1999) Anandamide induces overeating: mediation by central cannabinoid (CB1) receptors. Psychopharmacology (Berl) 143:315–317
Williams CM, Kirkham TC (2002) Observational analysis of feeding induced by Delta9-THC and anandamide. Physiol Behav 76:241–250
Wilson RI, Nicoll RA (2001) Endogenous cannabinoids mediate retrograde signalling at hippocampal synapses. Nature 410:588–592
Wittmann G, Deli L, Kallo I, Hrabovszky E, Watanabe M, Liposits Z, Fekete C (2007) Distribution of type 1 cannabinoid receptor (CB1)-immunoreactive axons in the mouse hypothalamus. J Comp Neurol 503:270–279
Zimmer A, Zimmer AM, Hohmann AG, Herkenham M, Bonner TI (1999) Increased mortality, hypoactivity, and hypoalgesia in cannabinoid CB1 receptor knockout mice. Proc Natl Acad Sci USA 96:5780–5785

Antiobesity Effects of Melanin-Concentrating Hormone Receptor 1 (MCH-R1) Antagonists

Hyae Gyeong Cheon

Contents

Abstract Despite remarkable progress in the elucidation of energy balance and regulation, the development of new antiobesity drugs is still at the stage of infancy. This review describes the MCH and MCH receptor system with regard to its involvement in energy homeostasis and summarizes the pharmacological profiles of selected small molecule MCH-R1 antagonists that are relevant for their development as antiobesity drugs. Although their clinical value still has to be demonstrated, and challenges with regard to unwanted side effects remain to be resolved, MCH-R1 antagonists may provide an effective pharmacotherapy for the treatment of obesity in the near future.

Keywords Melanin-concentrating hormone • Obesity • Energy intake • Energy expenditure • Pharmacotherapy

H.G. Cheon (✉)
Department of Pharmacology and Pharmaceutical Sciences, Gachon University of Medicine and Science, Yeonsu-3 Dong, Yeonsu-Gu, Incheon 406-799, Republic of Korea
e-mail: hgcheon@gachon.ac.kr

H.-G. Joost (ed.), *Appetite Control*, Handbook of Experimental Pharmacology 209,
DOI 10.1007/978-3-642-24716-3_18, © Springer-Verlag Berlin Heidelberg 2012

1 Introduction

Chronic inappropriate balance between energy intake and energy expenditure often results in the development of obesity, which is commonly associated with the secondary complications of the metabolic syndrome including non-insulin-dependent diabetes, cardiovascular disease, and certain cancers (Kopelman 2000). Furthermore, obese patients often suffer from some psychosocial discrimination causing depression and anxiety. Due to the western lifestyle (low activity and high caloric food), the epidemic of obesity is rapidly progressing, and up to 30% of adult US population is classified as obese (Ogden et al. 2006). The increasing incidence of obese children and adolescents is of even greater concern (Hedley et al. 2004) because poor eating habits are often established during childhood. Thus, there is an urgent need for an effective pharmacotherapy at present.

Obesity is usually defined as a body mass index (BMI, calculated as body weight in kilograms divided by height in meters squared) bigger than 30 kg/m^2, although the BMI alone does not sufficiently predict its detrimental effects. In addition to lifestyle modifications such as diet and exercise, pharmacotherapy is necessary to obtain sufficient weight loss in order to reduce the incidence of secondary complications of the metabolic syndrome. As a guideline for antiobesity drugs suggested by the US Food and Drug Administration (FDA), at least 5% weight loss over a year versus placebo should occur upon drug treatment (Hanif and Kumar 2002). In parallel, other comorbid conditions present in obese patients should be improved without long-term safety issues (Staten 2007). Until recently, only two drugs were approved in the United States for the long-term treatment of obesity: orlistat (Xenical, Roche) and sibutramine (Meridia, Abbott) (McNeely and Goa 1998; Hvizdos and Markham 1999). Orlistat, an inhibitor of gastric and pancreatic lipases, plus a low-calorie diet produced a loss of 3–4% of body weight as compared with diet alone in a 2-year period. The weight reduction was sustained as long as the treatment was continued (Foxcrogt and Milne 2000), but side effects such as fecal urgency/incontinence, flatulence, and steatorrhea limited its use. Sibutramine is a centrally acting appetite suppressant that inhibits reuptake of noradrenaline, serotonin, and dopamine in central neuronal synapses. Chronic treatment with sibutramine resulted in an average reduction of 4 kg over 44–54 weeks treatment, a value comparable to that seen with orlistat (Arterburn et al. 2004). However, because of several cardiovascular side effects (tachycardia and hypertension) induced by its peripheral action, sibutramine is contraindicated in patients with elevated cardiovascular risk such as hypertension (Finer 2002). In Europe and the United States, the drug was withdrawn because it increased mortality in the SCOUT trial designed to investigate its long-term efficacy and safety. Thus, development of more efficacious and safer antiobesity drugs than current pharmacotherapy is needed. Among the many targets investigated for the development of antiobesity drugs, this review will solely focus on the action of melanin-concentrating hormone (MCH) and on the development of small molecule antagonists of melanin-concentrating hormone receptor 1 (MCH-R1) as potential new antiobesity drugs.

2 Importance of MCH in Energy Homeostasis

Over the past decade, a number of neuropeptides have been identified as regulators of food intake and energy metabolism, most of them originating from the hypothalamus. Examples include alpha-melanocyte stimulating hormone (α-MSH) and cocaine- and amphetamine-regulated transcript (CART) as anorexigenic peptides, and neuropeptide Y (NPY), agouti-gene-related peptide (AgRP), orexin, and melanin-concentrating hormone (MCH) as orexigenic ones. Among these, MCH drew great attention as a major player in feeding and energy homeostasis based on numerous studies.

Following discovery of MCH from teleost fish as a skin-paling factor (Kawauchi et al. 1983), mammalian MCHs have also been identified and showed 100% sequence identity between mouse, rat, rabbit, and human (Vaughan et al. 1989; Presse et al. 1990). Whereas fish MCH consisted of 17 amino acids, mammalian MCH is a 19-amino-acid cyclic peptide synthesized from precursor protein, prepro-MCH, by posttranslational cleavage along with the production of two additional peptides named neuropeptide E-I (NEI) and neuropeptide G-E (NGE) (Breton et al. 1993). It was found that residues Arg6, Met8, Arg11, and Tyr13 as well as a disulfide bond between cysteine residues are critical for the biological activity of MCH (MacDonald et al. 2000). MCH is almost exclusively expressed in hypothalamic neurons located in the zona incerta and in the lateral hypothalamic area (LHA), from which these neurons project extensively throughout the brain (Bittencourt et al. 1992; Skofitsch et al. 1985). MCH is also expressed in the peripheral nervous system and in other tissues and cells such as the intestine, the reproductive system, and immune cells (Hervieu and Nahon 1995; Viale et al. 1997).

In mammals, MCH has been implicated in a variety of physiological functions including sensory processing (Miller et al. 1993), stress response by regulating hypothalamic–pituitary–adrenal axis (Jezova et al. 1992; Presse et al. 1992), and learning (McBride et al. 1994). Recent evidence also points to a key role of MCH in the regulation of feeding behavior and associated pathologies such as obesity. The lateral hypothalamic area where MCH neurons are located has been known as an important center regulating feeding behavior, as stimulation produces overeating, whereas lesions result in hypophagia and body weight reduction (Anand and Brobeck 1951). In addition, either acute or chronic administration of MCH increased body weight along with higher food intake. For example, acute intracerebroventricular injection of MCH, or direct injection of MCH into the paraventricular nucleus (PVN), arcuate nucleus (ARC), or dorsomedial nucleus (DMH) (Abbott et al. 2003; Rossi et al. 1999), produces a transient increase in food intake in a dose-dependent manner. In parallel with increased food intake, MCH administration increased the expression of the orexigenic peptides NPY and AgRP, but decreased expression of anorexic peptides α-MSH and CART from hypothalamic explants. These results suggest that increase in food intake by MCH is at least in part mediated by the regulation of these peptides. Chronic effects of MCH also demonstrated that central infusion of MCH increases food intake, body weight, and adiposity (Della-Zuana et al. 2002; Ito et al. 2003). Importantly, beyond its role in feeding, MCH appears to affect energy expenditure and thermogenesis since animals exposed to MCH reduced their core body temperature in parallel with a decrease of the expression of genes involved in thermogenesis (Pereira-da-Silva et al. 2003).

Further evidence that MCH acts as a major factor in the control of energy balance comes from the animal studies with genetic manipulation of the MCH gene. Targeted deletion of MCH in mice resulted in hypophagia, increased metabolic activity, and weight loss (Shimada et al. 1998); conversely, overexpression of MCH in transgenic mice led to an increased body weight with hyperphagia, hyperleptinemia, hyperinsulinemia, and hyperglycemia when mice were fed a high-fat diet (Ludwig et al. 2001). Furthermore, it was found that MCH mRNA and pro-MCH-derived peptide levels were upregulated in various obese rodents including *ob/ob* mice (Qu et al. 1996), *db/db* mice (Huang et al. 1999), *fat/fat* mice (Rovere et al. 1996), $A^{y/a}$ (agouti) mice (Hanada et al. 2000), and *fa/fa* rat (Stricker-Krongrad et al. 2001), suggesting that the level of MCH expression may be linked to feeding behavior and obesity. Similarly, a threefold increased expression of MCH mRNA and peptide was observed in obese humans as compared to lean subjects, implicating the involvement of MCH in human feeding disorders (Zhang et al. 1999).

Various conditions can regulate the expression of MCH. For example, starvation increases MCH mRNA levels in rodents (Herve and Fellmann 1997; Tritos et al. 1998), whereas leptin negatively regulates MCH expression. Leptin replacement in *ob/ob* mice and fasted rats reduced increased MCH mRNA levels to normal, and leptin injection abolished hyperphagia elicited by the central administration of MCH (Sahu 1998). This effect may be mediated by a direct action on MCH neurons in the LHA or by an indirect action on projections from areas such as the ARC (Elias et al. 1998, 1999). In contrast, MCH expression was shown to be positively regulated by leptin in dietary obese rats (Elliott et al. 2004). Interestingly, MCH mRNA was significantly elevated upon feeding of a palatable diet. These data suggest that MCH may have a specific role in stimulating appetite for palatable food, which can override mechanisms counteracting the development of obesity.

Recently, peripheral effects of MCH were described in endocrine pancreas and adipose tissue. MCH appears to stimulate leptin release in adipocytes (Bradley et al. 2000) and to modulate insulin secretion (Tadayyon et al. 2000). These findings indicate that MCH also acts within the periphery via regulation of other hormones.

3 Characterization of MCH Receptors

Two receptors for MCH in humans have recently been characterized. The first MCH receptor (MCH-R1) was identified as the orphan G-protein-coupled receptor SLC-1 (Bachner et al. 1999; Chambers et al. 1999; Saito et al. 1999) with high affinity for human MCH in the low nanomolar range and which is not activated by any other known peptide (Chambers et al. 1999). MCH-R1 consists of 353 amino acids with ~32% amino acid identity with members of the somatostatin receptor family. The gene is localized on chromosome 22q13.3. MCH-R1 has a high sequence identity (higher than 95%) and a similar tissue distribution between species (Tan et al. 2002), with particularly high abundance in the hypothalamus, thalamus, olfactory cortex, amygdala, and hippocampus. This pattern of expression corresponds with the areas of MCH immunoreactive implicated in feeding regulation (Hervieu et al. 2000). Some peripheral tissues including the adipose tissue are also reported to express MCH-R1 (Bradley et al. 2000; Saito et al. 1999). It appears that MCH-R1 signaling pathways are diverse via multiple G proteins (Bachner et al. 1999; Chambers et al. 1999; Hawes et al. 2000; Lembo et al. 1999; Shimomura et al. 1999). For example, MCH increased calcium concentration via both pertussis toxin-sensitive and pertussis toxin-insensitive GTP-binding proteins including $G_{\alpha i}$, $G_{\alpha 0}$, and $G_{\alpha q}$ (Hawes et al. 2000). In addition, MCH inhibited adenylyl cyclase more potently than stimulating calcium mobilization (Lembo et al. 1999).

Similar to the effects on MCH expression, fasting also increased the expression of MCH-R1. Likewise, *ob/ob* mice and dietary obese rats were shown to have higher level of MCH-R1 (Elliott et al. 2004). To the contrary, MCH-R1 expression is unchanged in MCH knockout mice (Kokkotou et al. 2001), which appears to be distinguishable from other G protein-coupled receptors regulated by their own ligands. Importantly, targeted disruption of the MCH-R1 gene results in resistance to diet-induced obesity and hyperphagia (Marsh et al. 2002), leanness, hyperactivity, and increased metabolic rate (Chen et al. 2002), a phenotype similar to that of MCH knockout mice. These results suggest that an increased metabolic activity as well as a decreased food intake may mediate reduced weight gain in MCH-R1 KO mice (Marsh et al. 2002; Chen et al. 2002) and that the effects of MCH on energy balance are mediated by MCH-R1. The recent reports of the anorectic and antiobesity effects of the MCH-R1 antagonists further confirmed the importance of MCH in the regulation of body weight.

A second MCH receptor referred to as MCH-R2 was found independently by several laboratories (An et al. 2001; Mori et al. 2001; Rodriguez et al. 2001; Sailer et al. 2001; Wang et al. 2001). Overall sequence of MCH-R2 displays an unusually low sequence identity with MCH-R1 (~38%) but a significant homology to the core region of MCH-R1. MCH-R2 is a 340-amino-acid membrane protein which binds MCH in the nanomolar range and has a similar but distinct expression profile in human brain and peripheral tissues. Levels of MCH-R2 mRNA in adipose tissue are significantly higher than those of MCH-R1 (Hill et al. 2001). The coupling mechanism of MCH-R2 is different from that of MCH-R1: MCH-R2 preferentially activates G_q, whereas MCH-R1 activates both $G_{i/o}$ and G_q (Hawes et al. 2000). Uniquely, MCH-R2 is not expressed in rodent species including mice, rats, rabbits, hamsters, or guinea pigs (Tan et al. 2002), which renders it difficult to determine the functional importance of MCH-R2. On the other hand, nonhuman species such as dogs, ferrets, and rhesus monkeys are known to have a functional MCH-R2 (Tan et al. 2002).

4 Dysregulation of MCH and MCH-R1 in Obesity

Although many genetic analyses have been carried out to elucidate the role of genetic variants of MCH and MCH-R1, no conclusive evidence for an association with obesity has been obtained so far. For example, a single amino acid substitution (R248Q) in the MCH-R1 gene co-segregated with obesity across more than one generation (Gibson et al. 2004), but this mutation was also present in lean subjects. Two other nonconservative missense variations (R317Q and T305M) cosegregated with obesity, but there was no evidence for a functional relevance (Gibson et al. 2004; Wermter et al. 2005). Another mutation in the promoter region of the MCH-R1 gene is associated with a reduced risk of obesity (Bell et al. 2005). At present, it is not conclusive that certain sequence variations in MCH-R1 are associated with obesity, although a combination with other minor variations in different genes may contribute to the susceptibility to obesity. Further studies in large populations and/or in genetically modified animal models would clarify the relationship of genetic variants of MCH and/or MCH-R1 to obesity.

5 MCH-R1 Antagonists in Discovery and Development

After the discovery of MCH-R1, numerous reports on the identification and optimization of small molecule and peptide antagonists of MCH-R1 for the potential treatment of obesity have been published (Carpenter and Hertzog 2002; Collins and Kym 2003; Handlon and Zhou 2006; Luthin 2007; McBriar 2007; Rokosz 2007; Rivera et al. 2008). However, despite extensive research efforts to develop small molecule MCH-R1 antagonists, so far, only four compounds entered human clinical trials as shown in Table 1, and only ALB-127158 is currently under active development.

In this section, the review will focus on the small molecule antagonists of MCH-R1, categorized with regard to companies and/or frequently encountered substructures, and will describe their pharmacological characteristics. Based on the patent literature and to other publications reported so far, common structural characteristics of the MCH-R1 antagonists included a central scaffold to which an aryl or heteroaryl group and a basic amino group are attached, which appears to be essential for antagonistic activity on MCH-R1. On the other hand, these structural characteristics are also common in hERG blockers, which can induce QT prolongation frequently associated with potentially lethal arrhythmias in clinical use. Therefore, interaction with hERG channel has been often a problem observed with MCH-R1 antagonists.

Neurogen researchers have reported several small molecule antagonists of MCH-R1, and among those, a piperazine compound **1** (NGD-4715) is most advanced to phase 1 study. Based on the phase 1 study, NGD-4715 seemed to be safe and well tolerated at all doses studied in the trial. In addition, Neurogen

Table 1 MCH-R1 antagonists in clinical studies

Name	Structure	Organization	Status
AMG-076	Undisclosed	Amgen	Phase I (2004)
GW-856464	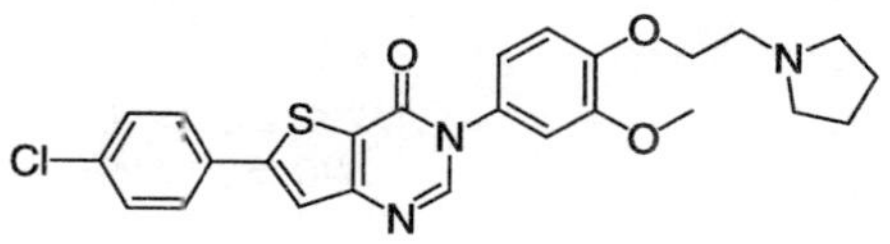	GSK	Phase I (2004)
NGD-4715	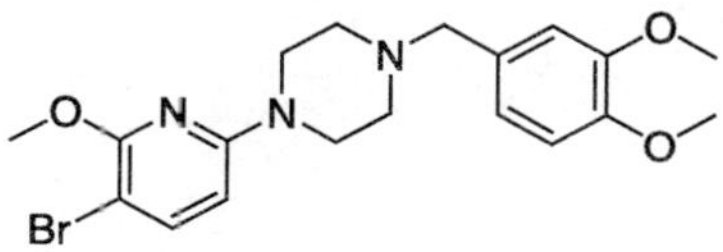	Neurogen	Phase I (2006)
ALB-127158	Undisclosed	AMRI	Phase I (2010)

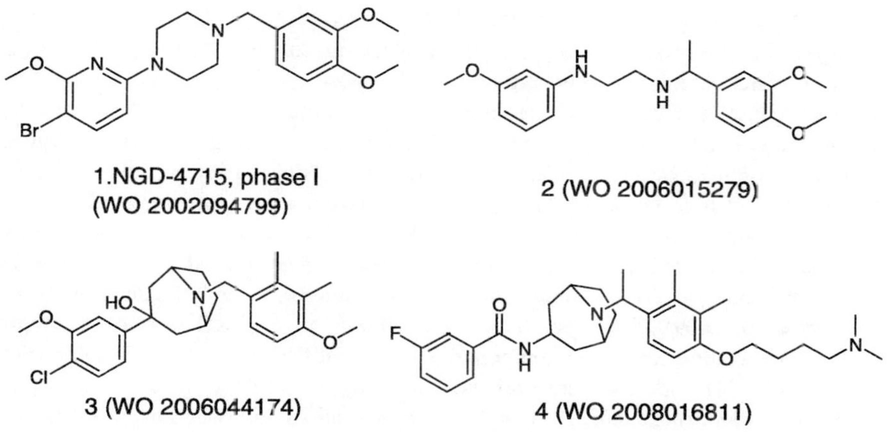

Fig. 1 Neurogen's representative compounds

disclosed diaminoethane (compound **2**) and 8-azabicyclo[3.2.1]octane (compounds **3** and **4**) derivatives as variants of the piperazine-based compounds (Fig. 1).

GlaxoSmithKline disclosed various scaffolds as MCH-R1 antagonists as summarized in Fig. 2. After an activity of the biphenyl carboxamide skeleton was identified in a high-throughput screening, compound **5** (SB-568849) was synthesized, and its pharmacological characteristics were investigated. SB-568849 exhibited good affinity (pK_i = 7.7) in the FLIPR assay with >30-fold selectivity over a wide range of monoamine receptors. In an effort to discover more rigid analogues of the compound SB-568849, the thienopyrimidinone compound **6** (GW-803430, GW-856464) was developed as a candidate for clinical trials. Compound **6** was found to be a potent

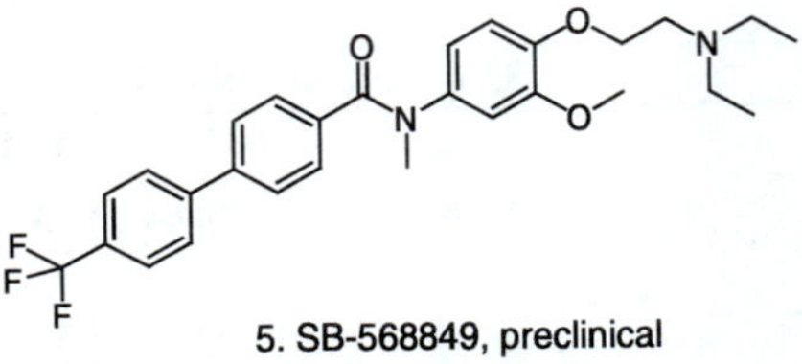

5. SB-568849, preclinical
(WO 2002010146)

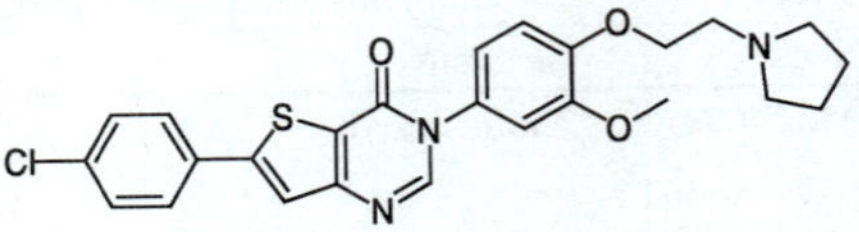

6. GW-803430, GW-856464, phase I
(WO 2003033476)

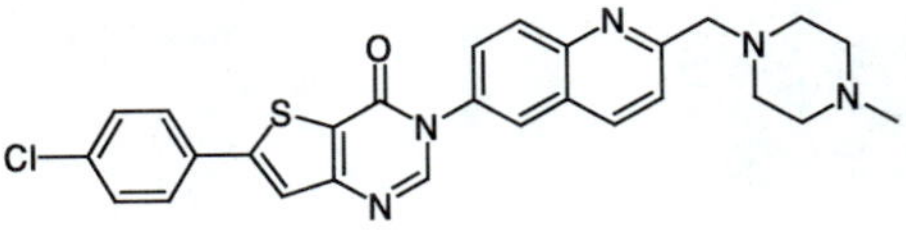

7. preclinical
(WO 2004092181)

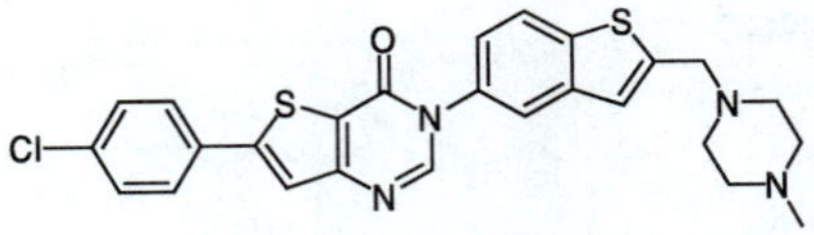

8. preclinical
(WO 2004092181)

Fig. 2 GSK's representative compounds

MCH-R1 antagonist ($pIC_{50} = 9.3$) and was highly selective (>100×) as tested with a battery of G protein-coupled receptors, ion channels, and enzymes (Hertzog et al. 2006). Its maleate salt exhibited good pharmacokinetic properties ($F = 31\%$, $t_{1/2} = 11$ h) in mice, with high brain penetration (brain/plasma concentration = 6:1). Oral administration of the compound at 0.3, 3, and 15 mg/kg once daily during a 12-day treatment caused a sustained dose-dependent weight loss of −6.2%, −12.1%, and −13.1%, respectively, in high-fat-diet-induced obesity of AKR/J mice (Hertzog et al. 2006). Preclinical studies with compound **6** revealed that it also produced an antidepressant response in the mouse forced-swim test and in the tail suspension test. Thus, compound **6** (GW-803430) was proven to be active as both antiobesity agent and antidepressant (Gehlert et al. 2009). However, the agent exhibited some hERG activity. Through further optimization study, the compound **7** was identified to be a potent antagonist ($IC_{50} = 3.1$ nM) with excellent oral bioavailability ($F = 98\%$, $t_{1/2} = 7.7$ h, $Cl_{total} = 39.7$ mL/min/kg) (Tavares et al. 2006a). Compound **7** had no significant off-target activity including hERG channel as determined by patch clamp assay ($pIC_{50} = 4.66$) and was selective over human MCH-R2 (Tavares et al. 2006a). Consistent with its pharmacokinetic profile, oral administration of the compound **7** at 1, 3, and 10 mg/kg once daily during a 26-day treatment caused a sustained dose-dependent weight loss of −4.8%, −9.4%, and −16.9%, respectively, in high-fat-diet-induced obesity of AKR/J mice, with good brain partitioning properties (brain/serum = 31) (Tavares et al. 2006a). The benzothiophene compound **8,** an analogue of compound **7,** was also found to be a potent antagonist ($IC_{50} = 5.7$ nM) and had a good systemic exposure ($F = 66\%$, $t_{1/2} = 9.1$ h) with moderate clearance ($Cl_{total} = 39$ mL/min/kg) and good solubility (Tavares et al. 2006b). Oral administration of compound **8** at 1, 3, and 10 mg/kg once daily during a 21-day treatment caused a dose-dependent weight loss of −1.1%, −3.2%, and −11.8%, respectively, in high-fat-diet-induced obesity of AKR/J mice, comparable to the effect of a CB1 receptor inverse agonist rimonabant (−8.4%)

(Tavares et al. 2006b). Furthermore, compound **8** was shown to have negligible hERG activity, but no further progress on this compound was reported.

The biaryl carboxamide moiety in compound **5** (SB-568849) in Fig. 2 is often encountered in many other representative compounds. For example, the tetrahydronaphthalene compound **9** (T-226296, $IC_{50} = 5.5$ and 8.6 nM at human and rat MCH-R1), the first small molecule MCH-R1 antagonist reported by Takeda (Takekawa et al. 2002), exhibited selectivity over a host of other receptors including MCH-R2 (>100 nM affinity) and was shown to be efficacious in suppressing food intake induced by intracerebroventricular injection of MCH in rats when orally administered at the dose of 30 mg/kg (Takekawa et al. 2002; Kowalski and McBriar 2004). Procter & Gamble pharmaceuticals reported compound **10** (DABA-821) as a potent MCH-R1 antagonist ($K_i = -39.3$ nM and $IC_{50} = 14.0$ nM). The compound showed ~25-fold selectivity over $5HT_{2C}$ receptor and caused significant body weight and fat mass reduction in rodents (Hu et al. 2008). Further optimization of the compound by decreasing hERG activity while retaining potent MCH-R1 antagonist activity ($K_i = 16$ nM) was achieved (compound **11**) but with concurrent loss of in vivo activity due to poor brain penetration (brain/plasma = 0.3). Neurocrine reported that the 3-aminopyrrolidine compound **12** was a potent antagonist ($K_i = 2.3$ nM) with good oral bioavailability in rats ($F = 32\%$, $t_{1/2} = 2.7$ h). In vivo efficacy of the compound was examined, and the orally, either acutely or chronically, administered compound decreased food intake as well as body weight in rats (Huang et al. 2005). In an effort to improve hERG activity of the Neurocrine compounds, compound **13** was identified. The compound exhibited hERG selectivity (MCH-R1 $K_i = 7.3$ nM versus hERG $IC_{50} = 2.6$ μM), but other undesirable characteristics such as inhibition of cytochrome P450 CYP2D6 ($IC_{50} = 3.3$ μM) and poor brain penetration (brain/plasma ≪1) hampered further studies on this compound (Fig. 3).

After the compounds **7** and **8** had shown improved hERG selectivity, in vivo activity on food intake, and desirable pharmacokinetic properties, analogous compounds have often been disclosed in related patents by other companies including Neurocrine (Fig. 4). The thienopyridazinone compound **14** (NBI-845) comprising a 3-aminopyrrolidine residue was identified by Neurocrine to be a potent MCH-R1 antagonist ($K_i = 3.3$ nM, and $IC_{50} = 7.7$ nM) and to be metabolically stable as assessed with human liver microsomes (Dyck et al. 2006). The agent showed a good

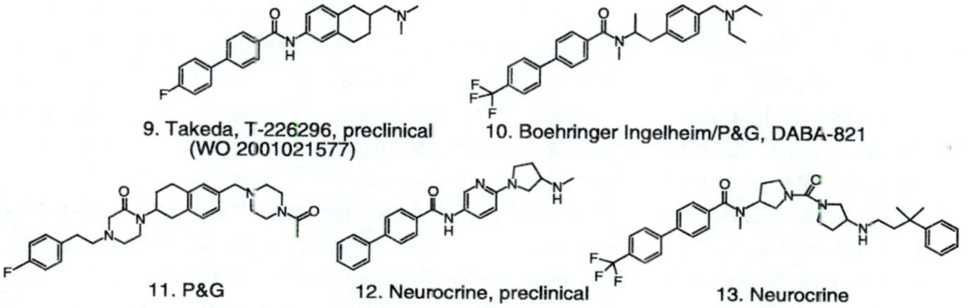

Fig. 3 Biaryl carboxamide derivatives

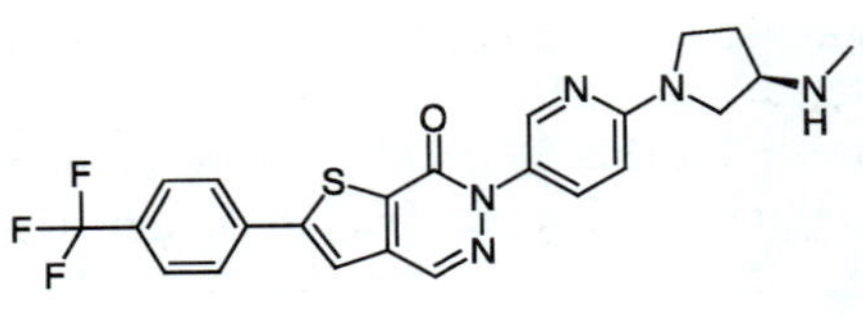

14. Neurocrine, NBI-845
(WO 2005103039)

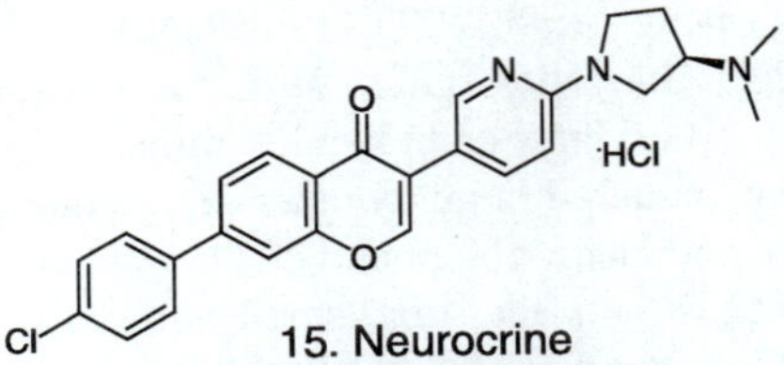

15. Neurocrine
(WO 2005103039)

Fig. 4 Bicyclic analogues

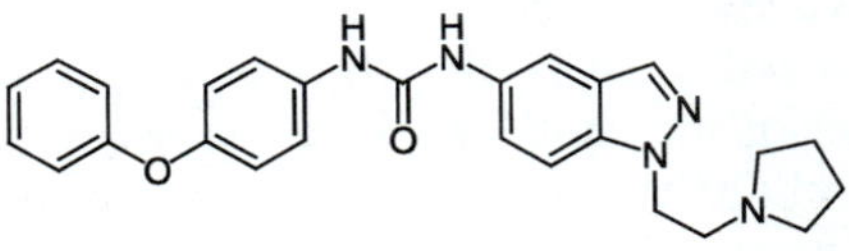

16. Abbott, preclinical
(US 2005137243)

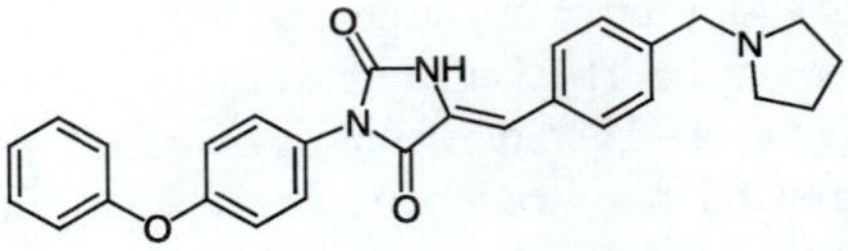

17. Cerep, preclinical
(WO 2007012670)

Fig. 5 Biaryl ether and the related compounds

pharmacokinetic profile ($F = 24\%$, $t_{1/2} = 11$ h, brain/plasma $= 1.4 \sim 3.1$), inhibited food intake in vivo, and induced weight loss in DIO male Sprague–Dawley rats (Dyck et al. 2006). Compound **14** was further evaluated in a selectivity screen (only modest cross-reactivity with M1R $K_i = 260$ nM; M3R $K_i = 160$ nM), CYP2D6/3A4 inhibition (<10% at 5 μM), hERG affinity ($IC_{50} = 1.8$ μM), and preliminary toxicological studies in rats (Dyck et al. 2006). Similarly, the chromone-based compound **15** was reported to be a potent MCH-R1 antagonist ($K_i = 1.6$ nM, and $IC_{50} = 20$ nM) and was further characterized in a pharmacokinetic study with Sprague–Dawley rats ($F = 66\%$, $t_{1/2} = 5.0$ h, brain/plasma $= 9.5 \sim 25$) (Dyck et al. 2006).

Biaryl ether and the related compounds were shown to have antagonistic effects on MCH-R1, and their structures are represented in Fig. 5. The indazole compound **16** comprising an ureido linkage was reported to be a potent MCH-R1 antagonist ($K_i = 12.0$ nM, and $IC_{50} = 104$ nM) and was also evaluated in pharmacokinetic and oral efficacy studies in mice fed a high-fat diet (Souers et al. 2005). Cerep reported that a representative compound **17** with a hydantoin skeleton was a MCH-R1 ligand ($K_i = 220$ nM). Intraperitoneal administration of the compound **17** (30 mg/kg) reduced cumulative food intake compared to the vehicle group and also showed antidepressant activity comparable to the effect of a clinically used antidepressant, imipramine (Alavoine et al. 2007). However, the compound inhibited the hERG channel (73% current amplitude inhibition at 1 μM in a standard in vitro electrophysiology assay) (Alavoine et al. 2007).

Amgen researchers reported several small molecule antagonists of MCH-R1 as represented in Fig. 6. In addition to a series of polycyclic compounds based on the indole moiety as exemplified in compound **18**, they disclosed the tetrazole compound **20** and biaryl ether compounds **21** and **22**. Compound **21** was reported to

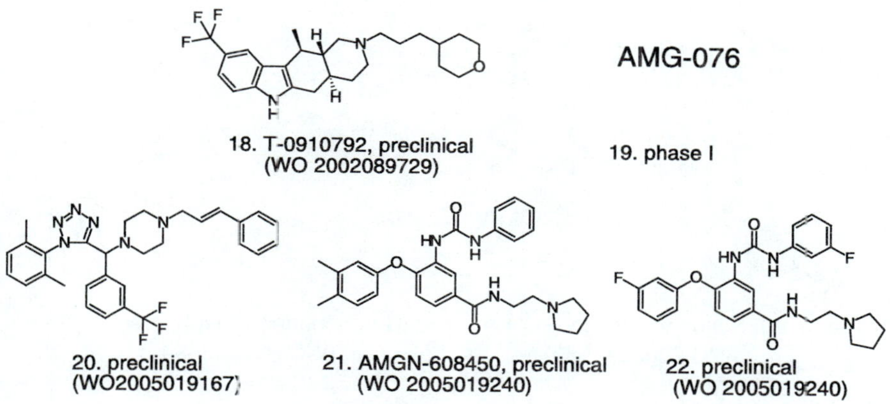

Fig. 6 Amgen's representative compounds

have potent antagonistic activity ($IC_{50} = 1.0$ nM, $t_{1/2}$ 1.5 h, C_{max} in brain 31.5 ng/g). The compound AMG-076 was reported to enter into phase 1 study in 2004 (Mendez-Andino and Wos 2007), with undisclosed structure. Little information on the biological profile of these compounds is available. Recently, novel pyrrolidine MCH-R1 antagonists with reduced hERG inhibition were reported (Fox et al. 2011).

Researchers at Schering-Plough/Pharmacopeia claimed many variants of biaryl urea derivatives, with the highly mutagenic biarylamine unit. To avoid the mutagenic risk of diphenyl aniline moiety, bicyclo[3,1,0]hexyl urea compound **23** (SCH-A, $K_i = 2.7$ nM, $K_b = 1.9$ nM, $F = 27\%$) was synthesized and tested. This compound exhibited low nanomolar potency for MCH-R1, with good selectivity against MCH-R2 and other receptors, enzymes, and transporters. With 27% of oral bioavailability, in vivo administration of the compound **23** (30 mg/kg, po) to obese mice resulted in a significant reduction of food intake as well as body weight, primarily due to the loss of fat mass (Guo et al. 2005; McBriar et al. 2006). In contrast, no effects of SCH-A on energy expenditure were observed. The inhibitory action of SCH-A on hERG channel ($IC_{50} = 52$ nM in patch clamp assay) led to attempts to improve hERG selectivity. The bicyclo[4,1,0]heptyl aminobenzimidazole derivative **24** was shown to be a potent MCH-R1 ligand ($K_i = 2.2$ nM), to have good ex vivo binding (>81%) with good pharmacokinetic properties (AUC = 965 ng · h/mL at 10 mg/kg po dose in rat), and to be efficacious in fasted obese mice (9–19% inhibition of cumulative food intake at 30 mg/kg po dose) (Sasikumar et al. 2006). However, the activity on hERG channel was not reported (Fig. 7).

Abbott has disclosed several MCH-R1 antagonists with 4-aminopiperidine scaffold as summarized in Fig. 8. From structure–activity relationship (SAR) investigation on a screening hit, benzamides **25** (binding $IC_{50} = 3$ nM, functional $IC_{50} = 90$ nM) and **26** (binding $IC_{50} = 2$ nM, functional $IC_{50} = 16$ nM), an indazole **27** (binding $IC_{50} = 20$ nM, functional $IC_{50} = 260$ nM), and a coumarin **28** (binding $IC_{50} = 2$ nM, functional $IC_{50} = 28$ nM) were identified as potent

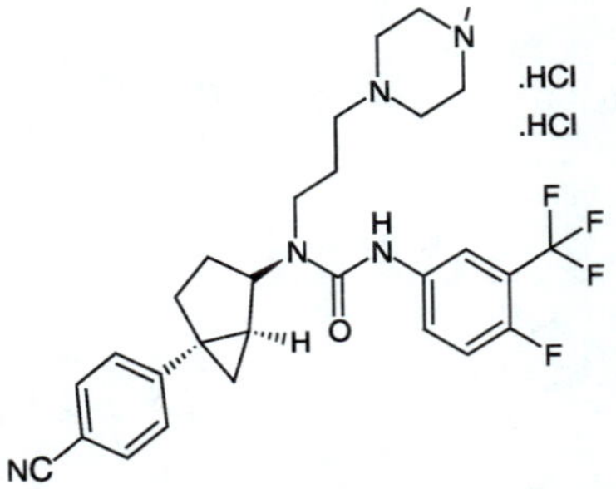

23. Schering-Plough, SCH-A
(WO 2003047568)

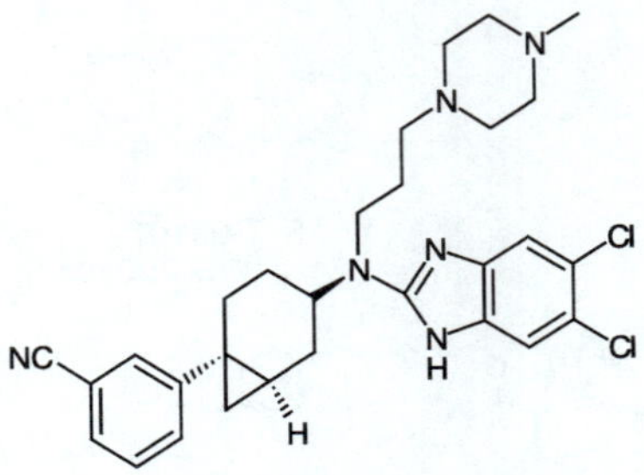

24. Schering-Plough, preclinical
(WO 2005034947)

Fig. 7 Representative compounds of Schering-Plough

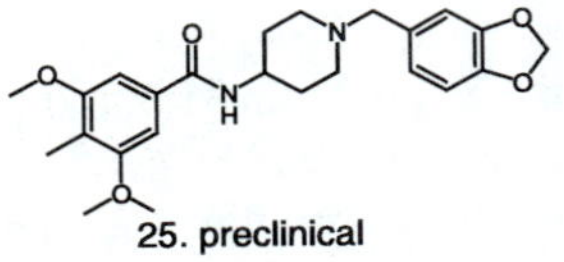

25. preclinical

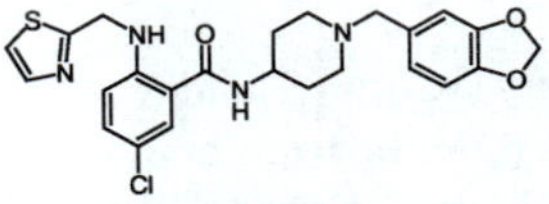

26. preclinical

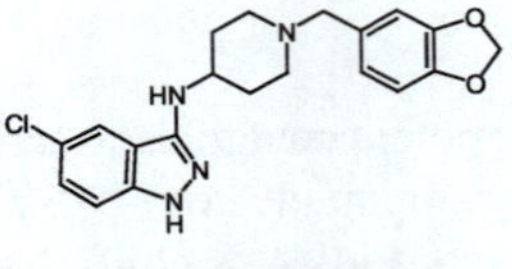

27. A-777903, preclinical

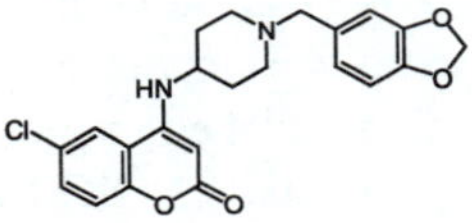

28. preclinical
(WO 2003106452)

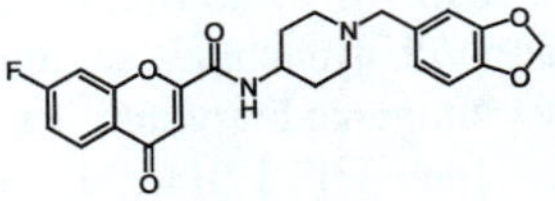

29. A-761, preclinical

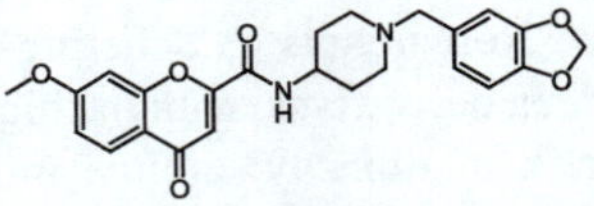

30. preclinical

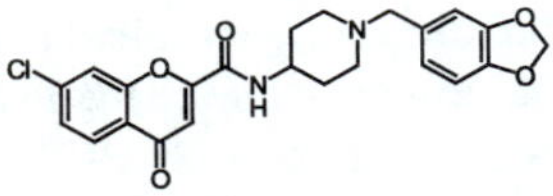

31

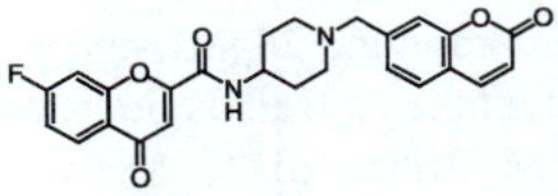

32

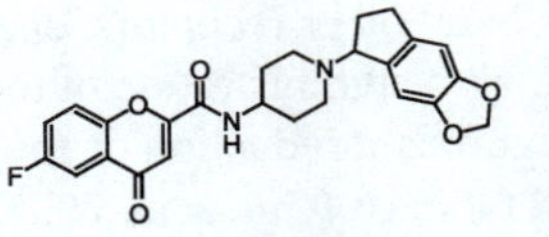

33. preclinical

Fig. 8 Abbott's representative compounds

antagonists. Compound **28** showed a high brain/plasma concentration ratio and produced a decrease in body weight without significant loss of lean body mass in obese mice (30% body weight reduction at 30 mg/kg, po) (Vasudevan et al. 2005; Kym et al. 2005). However, the compounds showed adverse hemodynamic effects as revealed in an anesthetized dog model of cardiovascular safety (Souers et al. 2007). In an attempt to reduce the effects on hERG, a series of chromone derivatives such as **29–31** was synthesized and was shown to possess in vivo efficacy in diet-induced obese mice with a high brain/plasma distribution ratio. These compounds still showed unacceptable affinity for hERG channel and caused QT prolongation in dogs at low doses (Lynch et al. 2005). On the other hand, compounds **32** and **33** produced a 7% weight loss in obese mice after 2 weeks of treatment (3 mg/kg, po) while the compounds exhibited good hERG selectivity (Iyengar et al. 2007; Souers et al. 2007).

Arena Pharmaceuticals and Taisho Pharmaceuticals collaborated in a discovery program of MCH-R1 antagonists and described the aminoquinazoline derivative, *N*-(cis-4-{[4-(dimethylamino)quinazolin-2-yl]amino}cyclohexyl)-3,4-difluorobenzamide hydrochloride (ATC-0175, compound **34**) to exhibit potent antagonistic activities on MCH-R1 ($IC_{50} = 7$ nM). ATC-0175 showed selectivity against MCH-R2 but had a moderate affinity for 5-HT_{2B} and 5-HT_{1A} receptors. In the diet-induced obesity mouse model, ATC-0175 reduced body weight significantly without loss in lean body mass (Semple et al. 2004). Interestingly, ATC-0175 exhibited anxiolytic effects in numerous animal models of anxiety (Chaki et al. 2005), with adequate pharmacokinetic profile and well-tolerated toxicity, suggesting potential utility for the treatment of depression and/or anxiety disorders. Recently, Taisho has disclosed additional compounds with this scaffold, although there have been no reports of cardiovascular effects of these compounds (Fig. 9).

Lundbeck/Synaptic reported an arylpiperidine derivative, SNAP-7941 (compound **35**), which is one of the earliest examples of efficacious MCH-R1 antagonists. SNAP-7941 showed high affinity to MCH-R1 (K_i of 15 nM), >1,000× selectivity against other receptors, and was able to inhibit MCH-induced food intake when injected intraperitoneally but not when administered orally. Chronic intraperitoneal treatment with the compound produced sustained weight loss in obese rat (26% reduction versus vehicle treatment), accompanied by a modest reduction of food intake, suggesting that energy expenditure may also contribute to the reduction of body weight in this model. In addition to its effects on food intake, SNAP-7941 improved anxiety and depression in animal models (Borowsky et al. 2002). The follow-up compound SNAP-94847 (compound **36**) exhibited improved bioavailability, resulting in the inhibition of body weight gain in normal rats after oral application. However, relatively moderate selectivity over other targets was described with no published data on hERG selectivity (Fig. 9).

Based on the results reported so far, a large number of small molecule MCH-R1 antagonists were discovered. Some of those showed good in vivo oral efficacy (weight loss accompanied by reduced food intake and/or increased energy expenditure) in animal models, with good selectivity against a panels of receptors, enzymes and transporters. So far, limited success to obtain hERG selectivity was achieved (Judd et al. 2008), and major safety issues on cardiovascular toxicity due

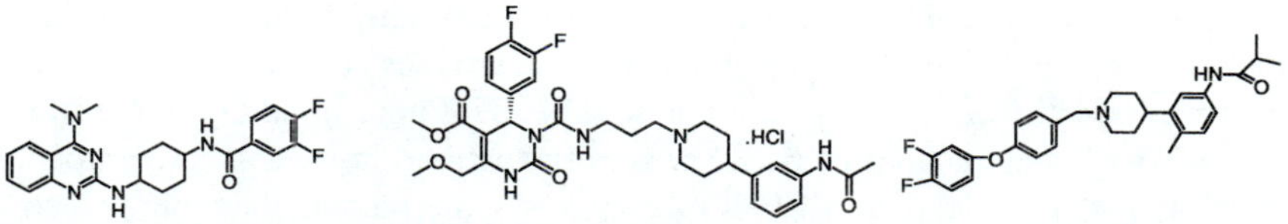

34. Taisho/Arena, ATC-0175, precliical   35. Lundbeck, SNAP-7941, preclinical (WO 2003069261)   36. SNAP-94847

Fig. 9 Taisho and Lundbeck's representative compounds

to hERG binding still need to be resolved in order to proceed to clinical studies. Since structural characteristics of many MCH-R1 antagonists are shared by classical hERG binding agents, novel scaffolds need to be identified and tested.

6 Perspectives

Besides the effects on food intake and regulation of energy balance, a number of diverse functions of MCH have been reported, which is expected by the broad distribution of MCH fibers in the CNS. Examples of the effects of MCH include (1) reduction of pentylenetetrazole- and kainic acid-induced seizure (Knigge and Wagner 1997), (2) enhancement of luteinizing hormone-releasing hormone (Pissios et al. 2006), (3) reduction of thyroid hormone (Kennedy et al. 2003), and (4) increase in NMDA receptor-mediated long-term potentiation (Varas et al. 2003), all of which may cause adverse effects when using MCH-R1 antagonists. Among many effects elicited by MCH, the effects of MCH on corticosterone levels appear to be very interesting with respect to the development of MCH-R1 antagonists as antiobesity agents. MCH modulates the hypothalamic–pituitary –adrenal axis, resulting in increased plasma corticosterone levels, thus producing anxiety-related responses (Smith et al. 2006). In addition, the findings that MCH reduces serotonin levels in discrete hypothalamic nuclei (Gonzalez et al. 1998), and that 5-HT receptors are expressed in MCH containing neurons, suggest that a reciprocal interaction between MCH and serotonergic pathways may exist (Collin et al. 2002). Indeed, an anxiolytic and antidepressant effect of MCH-R1 antagonists was reported in animal models (Lucki 1997; Borowsky et al. 2002). Since obesity patients are often accompanied with depression, the possibility that MCH-R1 antagonists have anxiolytic and antidepressant properties will definitely offer an additional benefit for the pharmacotherapy of obesity. Furthermore, since rimonabant, a cannabinoid CB1 receptor antagonist, was discontinued due to the side effects of depression (Despres et al. 2005), the potential of anxiolytic and antidepressant effects of MCH-R1 antagonists will provide an exciting clinical utility as antiobesity drugs, although clinical outcome needs to be proven.

The issue of cardiovascular safety became important because several classes of the MCH-R1 antagonists interacted with the hERG channel (Lynch et al. 2005), and early screening of the compounds for hERG liability was necessary in the discovery programs. To avoid the interaction with hERG channel, unique structural motifs that possess the desired pharmacological characteristics such as potency and good pharmacokinetic and safety profiles should be identified. In addition, the fact that the MCH-R2 gene is nonfunctional or is not expressed in rodents may impede the development of selective MCH-R1 antagonists for human therapies, since any actions of MCH-R1 antagonists on MCH-R2 may have adverse effects not detected in studies on rodents. Alternative animal models such as primate, dog, and ferret species with functional MCH-R2 receptors may be amenable to development.

7 Conclusion

Obesity represents a key disease area with considerable therapeutic needs and market potential. However, currently, only few drugs are available for its pharmacotherapy. Therefore, there are considerable opportunities for pharmaceutical companies in the development of effective antiobesity drugs to treat the worldwide obesity epidemic. Since MCH is a central anabolic regulator of appetite and energy balance, MCH-R1 antagonists are among the most thoroughly investigated agents for their efficacy to reduce food intake and increase energy expenditure in animal models. This review has presented an overview on the small molecule MCH-R1 antagonists, based on the published agents and data. While substantial challenges such as the validation in clinical trials and the issues of drug safety remain, it is anticipated that the development of small molecule MCH-R1 antagonists may provide key therapeutic options for the treatment of human obesity and mood disorders.

References

Abbott CR, Kennedy AR, Wren AM, Rossi M, Murphy KG, Seal LJ, Todd JF, Ghatei MA, Small CJ, Bloom SR (2003) Identification of hypothalamic nuclei involved in the orexigenic effect of melanin-concentrating hormone. Endocrinology 144:3943–3949

An S, Cutler G, Zhao JJ, Huang SG, Tian H, Li W, Liang L, Rich M, Bakleh A, Du J, Chen JL, Dai K (2001) Identification and characterization of a melanin-concentrating hormone receptor. Proc Natl Acad Sci U S A 98:7576–7581

Anand BK, Brobeck JR (1951) Localization of a feeding center in the hypothalamus of the rat. Proc Soc Exp Biol Med 77:323–324

Arterburn DE, Crane PK, Veenstra DL (2004) The efficacy and safety of sibutramine for weight loss: a systemic review. Arch Intern Med 164:994–1003

Bachner D, Kreienkamp H, Weise C, Buck F, Richter D (1999) Identification of melanin concentrating hormone (MCH) as the natural ligand for the orphan somatostatin-like receptor 1 (SLC-1). FEBS Lett 457:522–524

Balavoine F, Malabre P, Alleaume T, Rey A, Cherfils V, Jeanneton O, Seigneurin VS, Revah F (2007) Design and synthesis of novel hydantoin-containing melanin concentrating hormone receptor antagonists. Bioorg Med Chem Lett 17:3754–3759

Bell CG, Meyre D, Samson C, Boyle C, Lecoeur C, Tauber M, Jouret B, Jaquet D, Levy-Marchal C, Charles MA, Weill J, Gilson F, Mein CA, Froquel P, Walley AJ (2005) Association of melanin-concentrating hormone receptor 1 5′polymorphism with early onset extreme obesity. Diabetes 54:3049–3055

Bittencourt JC, Presse F, Arias C, Peto C, Vaughan J, Nahon JL, Vale W, Sawchenko PE (1992) The melanin-concentrating hormone system of the rat brain: an immuno- and hybridization histochemical characterization. J Comp Neurol 319:218–245

Borowsky B, Durkin MM, Ogozalek K, Marzabadi MR, DeLeon J, Heurich R, Lichtblau H, Shaposhnik Z, Daniewska I, Blackburn TP, Branchek TA, Gerald C, Vaysse PJ, Forray C (2002) Antidepressant, anxiolytic and anorectic effects of a melanin concentrating hormone-1 receptor antagonist. Nat Med 8:825–830

Bradley RL, Kokkotou EG, Matatos-Flier E, Cheatam B (2000) Melanin concentrating hormone (MCH) regulates leptin synthesis and secretion in rat adipocytes. Diabetes 49:1073–1077

Breton C, Schorpp M, Nahon JL (1993) Isolation and characterization of the human melanin concentrating hormone gene and a variant gene. Brain Res Mol Brain Res 18:297–310

Carpenter AJ, Hertzog DL (2002) Melanin-concentrating hormone receptor antagonists as potential antiobesity agents. Expert Opin Ther Patents 12:1639–1646

Chambers J, Ames RS, Bergsma D, Muir A, Fitzgerald LR, Hervieu G, Dytko GM, Foley JJ, Martin J, Liu WS, Park J, Ellis C, Ganguly S, Konchar S, Cluderay J, Leslie R, Wilson S, Sarau HM (1999) Melanin-concentrating hormone is the cognate ligand for the orphan G protein-coupled receptor SLC-1. Nature 400:261–265

Chaki S, Yamaguchi J, Yamada H, Thomsen W, Tran TA, Semple G, Sekiguchi Y (2005) ATC0175; an orally active melanin-concentrating hormone receptor 1 antagonist for the potential treatment of depression and anxiety. CNS Drug Rev 11:341–352

Chen Y, Hu C, Hsu CK, Zhang Q, Bi C, Asnicar M, Hsiung HM, Fox N, Slieker LJ, Yang DD, Heiman ML, Shi Y (2002) Targeted disruption of the melanin-concentrating hormone receptor-1 results in hyperphagia and resistance to diet-induced obesity. Endocrinology 143:2469–2477

Collin M, Backberg M, Onnestam K, Meister B (2002) 5-HT1A receptor immunoreactivity in hypothalamic neurons involved in body weight control. Neuroreport 13:945–951

Collins CA, Kym PR (2003) Prospects for obesity treatment: MCH receptor antagonists. Curr Opin Investig Drugs 4:386–394

Della-Zuana O, Presse F, Ortola C, Duhault J, Nahon JL, Levens N (2002) Acute and chronic administration of melanin-concentrating hormone enhances food intake and body weight in Wistar and Sprague–Dawley rats. Int J Obes Relat Metab Disord 26:1289–1295

Despres JP, Golay A, Sjostrom L (2005) Effects of rimonabant on metabolic risk factors in overweight patients with dyslipidemia. N Engl J Med 353:2121–2134

Dyck B, Markison S, Zhao L, Tamiya J, Grey J, Rowbottom MW, Zhang M, Vickers T, Sorensen K, Norton C, Wen J, Heise CE, Saunders J, Conlon P, Madan A, Schwarz D, Goodfellow VS (2006) A thienopyridazinone-based melanin concentrating hormone receptor 1 antagonist with potent in vivo anorectic properties. J Med Chem 49:3753–3756

Elias CF, Aschkenasi C, Lee C, Kelly J, Ahima RS, Bjorbaek C, Flier JS, Saper CB, Elmquist JK (1999) Leptin differentially regulates NPY and POMC neurons projecting to the lateral hypothalamic area. Neuron 23:775–786

Elias CF, Saper CB, Maratos-Flier E, Tritos NA, Lee C, Kelly J, Tatro JB, Hoffman GE, Ollmann MM, Barsh GS, Sakurai T, Yanagisawa M, Elmquist JK (1998) Chemically defined projections linking the mediobasal hypothalamus and the lateral hypothalamic area. J Comp Neurol 402:442–459

Elliott JC, Harrold JA, Brodin P, Enquist K, Backman A, Bystrom M, Lindgren K, King P, Williams G (2004) Increases in melanin-concentrating hormone and MCH receptor levels in the hypothalamus of dietary-obese rat. Brain Res Mol Brain Res 128:150–159

Finer N (2002) Sibutramine: its mode of action and efficacy. Int J Obes Relat Metab Disord 26 (Suppl 4):S29–S33

Fox BM, Natero R, Richard K, Connors R, Roveto PM, Beckmann H, Haller K, Golde J, Xiao SH, Kayser F (2011) Novel pyrrolidine melanin-concentrating hormone receptor 1 antagonists with reduced hERG inhibition. Bioorg Med Chem Lett 21:2460–2467

Foxcroft DR, Milne R (2000) Orlistat for the treatment of obesity: rapid review and cost effectiveness model. Obes Rev 1:121–126

Gehlert DR, Rasmussen K, Shaw J, Li X, Ardayfio P, Craft L, Coskun T, Zhang HY, Chen Y, Witkin JM (2009) Preclinical evaluation of melanin-concentrating hormone receptor 1 antagonism for the treatment of obesity and depression. J Pharmacol Exp Ther 329:429–438

Gibson WT, Pissios P, Trombly DJ, Luan J, Keogh J, Wareham NJ, Maratos-Flier E, O'Rahilly S, Farooqi IS (2004) Melanin-concentrating hormone receptor mutations and human obesity: functional analysis. Obes Res 12:743–749

Gonzalez MI, Baker BI, Hole DR, Wilson CA (1998) Behavioral effects of neuropeptide E-I(NEI) in the female rat: interaction with alpha-MSH, MCH and dopamine. Peptides 19:1007–1016

Guo T, Shao Y, Qian G, Rokosz LL, Stauffer TM, Hunter RC, Babu SD, Gu H, Hobbs DW (2005) Discovery and SAR of biaryl piperidine MCH1 receptor antagonists through solid-phase encoded combinatorial synthesis. Bioorg Med Chem Lett 15:3696–3700

Hanada R, Nakazato M, Mutsukura S, Murakami N, Yoshimatu H, Sakata T (2000) Differential regulation of MCH and orexin genes in the agouti-related protein/melanocortin-4 (AgRP/MC4) receptor system. Biochem Biophys Res Commun 268:88–91

Handlon AL, Zhou H (2006) Melanin-concentrating hormone-1 receptor antagonists for the treatment of obesity. J Med Chem 49:4017–4022

Hanif MW, Kumar S (2002) Pharmacological management of obesity. Expert Opin Pharmacother 3:1711–1718

Hawes BE, Kil E, Green B, O'Neill K, Fried S, Graziano MP (2000) The melanin concentrating hormone receptor couples to multiple G proteins to activate diverse intracellular signaling pathways. Endocrinology 141:4524–4532

Hedley AA, Ogden CL, Johnson CL, Carroll MD, Curtin LR, Flegal KM (2004) Prevalence of overweight and obesity among US children, adolescents, and adults, 1999–2002. J Am Med Assoc 291:2847–2850

Hervieu G, Nahon JL (1995) Pro-melanin-concentrating hormone messenger ribonucleic acid and peptides expression in peripheral tissues of the rat. Neuroendocrinology 61:348–364

Herve C, Fellmann D (1997) Changes in rat melanin-concentrating hormone and dynorphin messenger ribonucleic acids induced by food deprivation. Neuropeptides 31:237–242

Hervieu GJ, Cluderay JE, Harrison D, Meakin J, Maycox P, Nasir S, Leslie RA (2000) The distribution of the mRNA and protein products of the melanin-concentrating hormone (MCH) receptor gene, slc-1, in the central nervous system of the rat. Eur J Neurosci 12:1194–1216

Hertzog DL, Al-Barazanji KA, Bigham EC, Bishop MJ, Britt CS, Carlton DL, Cooper JP, Daniels AJ, Garrido DM, Goetz AS, Grizzle MK, Guo YC, Handlon AL, Ignar DM, Morgan RO, Peat AJ, Tavares FX, Zhou H (2006) The discovery and optimization of pyrimidinone-containing MCH R1 antagonists. Bioorg Med Chem Lett 16:4723–4727

Hill J, Duckworth M, Murdock P, Rennie G, Sabido-David C, Ames RS, Szekeres P, Wilson S, Bergsma DJ, Gloger IS, Levy DS, Chambers JK, Muir AI (2001) Molecular cloning and functional characterization of MCH2, a novel human MCH receptor. J Biol Chem 276:20125–20129

Hu XE, Wos JA, Dowty ME, Suchanek PM, Ji W, Chambers JB, Benoit SC, Clegg DJ, Reizes O (2008) Small-molecule melanin-concentrating hormone-1 receptor antagonists require brain penetration for inhibition of food intake and reduction in body weight. J Pharmacol Exp Ther 324:206–213

Huang Q, Viale A, Picard F, Nahon JL, Richard D (1999) Effects of leptin on Melanin-concentrating hormone (MCH) expression in the brain of lean and obese Lep(ob)/Lep(ob) mice. Neuroendocrinology 69:145–153

Huang CQ, Baker T, Schwarz D, Fan J, Heise CE, Zhang M, Goodfellow VS, Markison S, Gogas KR, Chen T, Wang XC, Zhu YF (2005) 1-(4-Amino-phenyl) pyrrolidin-3-yl-amine and 6-(3-amino-pyrrolidin-1-yl)-pyridin-3-yl-amine derivatives as melanin-concentrating hormone receptor-1 antagonists. Bioorg Med Chem Lett 15:3701–3706

Hvizdos KM, Markham A (1999) Orlistat: a review of its use in the management of obesity. Drugs 58:743–760

Ito M, Gomori A, Ishihara A, Oda Z, Mashiko S, Matsushita H, Yumoto M, Ito M, Sano H, Tokita S, Moriya M, Iwaasa H, Kanatani A (2003) Characterization of MCH-mediated obesity in mice. Am J Physiol Endocrinol Metabol 284:E940–E945

Iyengar RR, Lynch JK, Mulhern MM, Judd AS, Freeman JC, Gao J, Souers AJ, Zhao G, Wodka D, Falls HD, Brodjian S, Dayton BD, Reilly RM, Swanson S, Su Z, Martin RL, Leitza ST, Houseman KA, Diaz G, Collins CA, Sham HL, Kym PR (2007) An evaluation of 3,4-methylenedioxy phenyl replacements in the aminopiperidine chromone class of MCHr1 antagonists. Bioorg Med Chem Lett 17:874–878

Jezova D, Bartanusz V, Westergren I, Johansson BB, Rivier J, Vale W, Rivier C (1992) Rat melanin-concentrating hormone stimulates adrenocorticotropin secretion: evidence for a site of action in brain regions protected by the blood-brain barrier. Endocrinology 130:1024–1029

Judd AS, Souers AJ, Kym PR (2008) Lead optimization of melanin concentrating hormone receptor 1 antagonists with low hERG channel activity. Curr Top Med Chem 8:1152–1157

Kawauchi H, Kawazoe I, Tsubokawa M, Kishida M, Baker BI (1983) Characterization of melanin concentrating hormone in chum salmon pituitaries. Nature 305:321–323

Kennedy AR, Todd JF, Dhillo WS, Seal LJ, Ghatei MA, O'Toole CP, Jones M, Witty D, Winborne K, Riley G, Hervieu G, Wilson S, Bloom SR (2003) Effects of direct injection of melanin-concentrating hormone into the paraventricular nucleus: further evidence for a stimulatory role in the adrenal axis via SLC-1. J Neuroendocrinol 15:268–272

Knigge KM, Wagner JE (1997) Melanin-concentrating hormone (MCH) involvement in pentylenetetrazole (PTZ)-induced seizure in rat and guinea pig. Peptides 18:1095–1097

Kokkotou EG, Tritos NA, Mastaitis JW, Slieker L, Maratos-Flier E (2001) Melanin concentrating hormone receptor is a target of leptin action in the mouse brain. Endocrinology 142:680–686

Kopelman PG (2000) Obesity as a medical problem. Nature 404:635–643

Kowalski TJ, Mcbriar MD (2004) Therapeutic potential of melanin-concentrating hormone-1 receptor antagonists for the treatment of obesity. Expert Opin Investig Drugs 13:1113–1122

Kym PR, Iyengar R, Souers AJ, Lynch JK, Judd AS, Gao J, Freeman J, Mulhern M, Zhao G, Vasudevan A, Wodka D, Blackburn C, Brown J, Che JL, Cullis C, Lai SJ, LaMarche MJ, Marsilje T, Roses J, Sells T, Geddes B, Govek E, Patane M, Fry D, Dayton BD, Brodjian S, Falls D, Brune M, Bush E, Shapiro R, Knourek-Segel V, Fey T, McDowell C, Reinhart GA, Preusser LC, Marsh K, Hernandez L, Sham HL, Collins CA (2005) Discovery and characterization of aminopiperidinecoumarin melanin concentrating hormone receptor 1 antagonists. J Med Chem 48:5888–5891

Lembo PM, Grazzini E, Cao J, Hubatsch DA, Pelletier M, Hoffert C, St-Onge S, Pou C, Labrecque J, Groblewski T, O'Donnell D, Payza K, Ahmad S, Walker P (1999) The receptor for the orexigenic peptide melanin-concentrating hormone is a G protein-coupled receptor. Nat Cell Biol 1:267–271

Lucki I (1997) The forced swimming test as a model for core and component behavioral effects of antidepressant drugs. Behav Pharmacol 8:523–532

Ludwig DS, Tritos NA, Mastaitis JW, Kulkarni R, Kokkotou E, Elmquist J, Lowell B, Flier JS, Maratos-Flier E (2001) Melanin-concentrating hormone overexpression in transgenic mice leads to obesity and insulin resistance. J Clin Invest 107:379–386

Luthin DR (2007) Anti-obesity effects of small molecule melanin-concentrating hormone receptor1 (MCHR1) antagonists. Life Sci 81:423–440

Lynch JK, Collins CA, Freeman JC, Gao J, Iyengar RR, Judd AS, Kym PR, Mulhern MM, Sham HL, Souers AJ, Zhao G, Wodka D (2005) Preparation of piperidinyl chromenecarboxamides as antagonists of melanin concentrating hormone effects on the melanin concentrating hormone receptor. US 2005/209274

Macdonald D, Mugolo N, Zhang R, Durkin JP, Yao X, Strader CD, Graziano MP (2000) Molecular characterization of the melanin-concentrating hormone/receptor complex: identification of critical residues involved in binding and activation. Mol Pharmacol 58:217–225

Marsh DJ, Weingarth DT, Novi DE, Chen HY, Trumbauer ME, Chen ASM, Guan XM, Jiang MM, Feng Y, Camacho RE, Shen Z, Frazier EG, Yu H, Metzger JM, Kuca SJ, Shearman LP, Gopal-Truter S, MacNeil DJ, Strack AM, MacIntyre DE, Van der Ploeg LH, Qian S (2002) Melanin-concentrating hormone 1 receptor-deficient mice are lean, hyperactive, and hyperphagic and have altered metabolism. Proc Natl Acad Sci U S A 99:3240–3245

McBride RB, Beckwith BE, Swenson RR, Sawyer TK, Hadley ME, Matsunaga TO, Hruby VJ (1994) The actions of melanin-concentrating hormone (MCH) on passive avoidance in rats: a preliminary study. Peptides 15:757–759

McBriar MD (2007) Melanin concentrating hormone receptor antagonists as antiobesity agents: from M2 to MCHR-1. Curr Top Med Chem 7:1440–1454

McBriar MD, Guzik H, Shapiro S, Paruchova J, Xu R, Palani A, Clader JW, Cox K, Greenlee WJ, Hawes BE, Kowalski TJ, O'Neill K, Spar BD, Weig B, Weston DJ, Farley C, Cook J (2006) Discovery of orally efficacious melanin-concentrating hormone receptor-1 antagonists as antiobesity agents. Synthesis, SAR, and biological evaluation of bicyclo[3.1.0]hexyl ureas. J Med Chem 49:2294–2310

McNeely W, Goa KL (1998) Sibutramine: a review of its contribution to the management of obesity. Drugs 56:1093–1124

Mendez-Andino JL, Wos JA (2007) MCH-R1 antagonists: what is keeping most research programs away from the clinic? Drug Discov Today 12:972–979

Miller CL, Hruby VJ, Matsunaga TO, Bickford PC (1993) Alpha-MSH and MCH are functional antagonists in a Central nervous system (CNS) auditory gating paradigm. Ann NY Acad Sci 680:571–574

Mori M, Harada M, Terao Y, Sugo T, Watanabe T, Shimomura Y, Abe M, Shintani Y, Onda H, Nishimura O, Fujino M (2001) Cloning of a novel G protein-coupled receptor, SLT, a subtype of the melanin-concentrating hormone receptor. Biochem Biophys Res Commun 283:1013–1018

Ogden CL, Carroll MD, Curtin LR, McDowell MA, Tabak CJ, Flegal KM (2006) Prevalence of overweight and obesity in the United States, 1999–2004. J Am Med Assoc 295:1549–1555

Pereira-da-Silva M, Torsoni MA, Nourani HV, Augusto VD, Souza CT, Gasparetti AL, Carvalheira JB, Ventrucci G, Marcondes MC, Cruz-Neto AP, Saad MJ, Boschero AC, Carneiro EM, Velloso LA (2003) Hypothalamic melanin-concentrating hormone is Induced by cold exposure and participates in the control of energy expenditure in rats. Endocrinology 144:4831–4840

Pissios P, Bradley RL, Maratos-Flier E (2006) Expanding the scales: the multiple roles of MCH in regulating energy balance and other biological functions. Endocr Rev 27:606–620

Presse F, Nahon JL, Fischer WH, Vale W (1990) Structure of the human melanin concentrating hormone mRNA. Mol Endocrinol 4:632–637

Presse F, Hervieu G, Imaki T, Sawchenko PE, Vale W, Nahon JL (1992) Rat melanin-concentrating hormone messenger ribonucleic acid expression: marked changes during development and after stress and glucocorticoid stimuli. Endocrinology 131:1241–1250

Qu D, Ludwig DS, Gammeltoft S, Piper M, Pelleymounter MA, Cullen MJ, Mathes WF, Przypek R, Kanarek R, Maratos-Flier E (1996) A role for melanin-concentrating hormone in the central regulation of feeding behavior. Nature 380:243–247

Rivera G, Bocanegra-Garcia V, Galiano S, Cirauqui N, Ceras J, Pérez S, Aldana I, Monge A (2008) Melanin-concentrating hormone receptor 1 antagonists: a new perspective for the pharmacologic treatment of obesity. Curr Med Chem 15:1025–1043

Rodriguez M, Beauverger P, Naime I, Rique H, Ouvry C, Souchaud S, Dromaint S, Nagel N, Suply T, Audinot V, Boutin JA, Galizzi JP (2001) Cloning and molecular characterization of the novel human melanin-concentrating hormone receptor MCH2. Mol Pharmacol 60:632–639

Rokosz LL (2007) Discovery and development of melanin-concentrating hormone receptor 1 antagonists for the treatment of obesity. Expert Opin Drug Discov 2:1301–1327

Rossi M, Beak SA, Choi SJ, Small CJ, Morgan DG, Ghatei MA, Smith DM, Bloom SR (1999) Investigation of the feeding effects of melanin concentrating hormone on food intake-action independent of galanin and the melanocortin receptors. Brain Res 846:164–170

Rovere C, Viale A, Nahon JL, Kitabgi P (1996) Impaired processing of brain pro-neurotensin and pro-melanin-concentrating hormone in obese fat/fat mice. Endocrinology 137:2954–2958

Sahu A (1998) Leptin decreases food intake induced by melanin-concentrating hormone (MCH), galanin (GAL) and neuropeptide Y (NPY) in the rat. Endocrinology 139:4739–4742

Sailer AW, Sano H, Zeng Z, McDonald TP, Pan J, Pong SS, Feighner SD, Tan CP, Fukami T, Iwaasa H, Hreniuk DL, Morin NR, Sadowski SJ, Ito M, Bansal A, Ky B, Figueroa DJ, Jiang Q, Austin CP, MacNeil DJ, Ishihara A, Ihara M, Kanatani A, Vander Ploeg LH, Howard AD, Liu Q (2001) Identification and characterization of a second melanin-concentrating hormone receptor, MCH-2R. Proc Natl Acad Sci U S A 98:7564–7569

Saito Y, Nothacker HP, Wang Z, Lin SH, Leslie F, Civelli O (1999) Molecular characterization of the melanin-concentrating hormone receptor. Nature 400:265–269

Sasikumar TK, Qiang L, Burnett DA, Greenlee WJ, Hawes BE, Kowalski TJ, O'Neill K, Spar BD, Weig B (2006) Novel aminobenzimidazoles as selective MCH-R1 antagonists for the treatment of metabolic diseases. Bioorg Med Chem Lett 16:5427–5431

Semple G (2004) Identification and biological activity of a new series of antagonists of hMCH-R1. Presented at the 228th National American Chemical Society Meeting, Philadelphia, PA, August 22, Medi 007

Shimada M, Tritos NA, Lowell BB, Flier JS, Maratos-Flier E (1998) Mice lacking melanin concentrating hormone are hypophagic and lean. Nature 396:670–674

Shimomura Y, Mori M, Sugo T, Ishibashi Y, Abe M, Kurokawa T, Onda H, Nishimura O, Sumino Y, Fujino M (1999) Isolation and identification of melanin-concentrating hormone as the endogenous ligand of the SLC-1 receptor. Biochem Biophys Res Commun 261:622–626

Skofitsch G, Jacobowitz DM, Zamir N (1985) Immunohistochemical localization of a melanin concentrating hormone-like peptide in the rat brain. Brain Res Bull 15:635–649

Smith DG, Davis RJ, Rorick-Kehn L, Morin M, Witkin JM, McKinzie DL, Nomikos GG, Gehlert DR (2006) Melanin-concentrating hormone-1 receptor modulates neuroendocrine, behavioral, and corticolimbic neurochemical stress responses in mice. Neuropsychopharmacol 31:1135–1145

Souers AJ, Gao J, Wodka D, Judd AS, Mulhern MM, Napier JJ, Brune ME, Bush EN, Brodjian SJ, Dayton BD, Shapiro R, Hernandez LE, Marsh KC, Sham HL, Collins CA, Kym PR (2005) Synthesis and evaluation of urea-based indazoles as melanin concentrating hormone receptor 1 antagonists for the treatment of obesity. Bioorg Med Chem Lett 15:2752–2757

Souers AJ, Iyengar RR, Judd AS, Beno DW, Gao J, Zhao G, Brune ME, Napier JJ, Mulhern MM, Lynch JK, Freeman JC, Wodka D, Chen CJ, Falls HD, Brodjian S, Dayton BD, Diaz GJ, Bush EN, Shapiro R, Droz BA, Knourek-Segel V, Hernandez LE, Marsh KC, Reilly RM, Sham HL, Collins CA, Kym PR (2007) Constrained 7-fluorocarboxychromone-4-aminopiperidine based melanin concentrating hormone receptor 1 antagonists: the effects of chirality on substituted indan 1-ylamines. Bioorg Med Chem Lett 17:884–889

Staten MA (2007) Challenges in the discovery and development of new agents for the treatment of obesity. Clin Pharmacol Ther 81:753–755

Stricker-Krongrad A, Dimitrov T, Beck B (2001) Central and peripheral dysregulation of melanin concentrating hormone in obese Zucker rats. Brain Res Mol Brain Res 92:43–48

Tadayyon M, Welters HJ, Haynes AC, Cluderay JE, Hervieu G (2000) Expression of melanin concentrating hormone receptors in insulin-producing cells: MCH stimulates insulin release in RINm5F and CRI-G1 cell lines. Biochem Biophys Res Commun 275:709–712

Takekawa S, Asami A, Ishihara Y, Terauchi J, Kato K, Shimomura Y, Mori M, Murakoshi H, Kato K, Suzuki N, Nishimura O, Fujino M (2002) T-226296: a novel, orally active and selective melanin-concentrating hormone receptor antagonist. Eur J Pharmacol 438:129–135

Tan CP, Sano H, Iwaasa H, Pan J, Sailer AW, Hreniuk DL, Feighner SD, Palyha OC, Pong SS, Figueroa DJ, Austin CP, Jiang MM, Yu H, Ito J, Ito M, Guan XM, MacNeil DJ, Kanatani A, Van der Ploeg LHT, Howard AD (2002) Melanin-concentrating hormone receptor subtypes 1 and 2: species-specific gene expression. Genomics 79:785–792

Tavares FX, Al-Barazanji KA, Bigham EC, Bishop MJ, Britt CS, Carlton DL, Feldman PL, Goetz AS, Grizzle MK, Guo YC, Handlon AL, Hertzog DL, Ignar DM, Lang DG, Ott RJ, Peat AJ, Zhou HQ (2006a) Potent, selective, and orally efficacious antagonists of melanin-concentrating hormone receptor 1. J Med Chem 49:7095–7107

Tavares FX, Al-Barazanji KA, Bishop MJ, Britt CS, Carlton DL, Cooper JP, Feldman PL, Garrido DM, Goetz AS, Grizzle MK, Hertzog L, Ignar DM, Lang DG, McIntyre MS, Ott RJ, Peat AJ, Zhou HQ (2006b) 6-(4-Chlorophenyl)-3-substituted-thieno[3,2-d]pyrimidin-4(3H)-one-based melanin concentrating hormone receptor 1 antagonist. J Med Chem 49:7108–7118

Tritos NA, Elmquist JK, Mastaitis JW, Flier JS, Maratos-Flier E (1998) Characterization of expression of hypothalamic appetite-regulating peptides in obese hyperleptinemic brown adipose tissue-deficient (uncoupling protein-promoter-driven diphtheria toxin A) mice. Endocrinology 139:4634–4641

Varas MM, Perez MF, Ramirez OA, de Barioglio SR (2003) Increased susceptibility to LTP generation and changes in NMDA-NR1 and NR-2B subunits mRNA expression in rat hippocampus after MCH administration. Peptides 24:1403–1411

Vasudevan A, Souers AJ, Freeman JC, Verzal MK, Gao J, Mulhern MM, Wodka D, Lynch JK, Engstrom KM, Wagaw SH, Brodjian S, Dayton B, Falls DH, Bush E, Brune M, Shapiro RD, Marsh KC, Hernandez LE, Collins CA, Kym PR (2005) Aminopiperidine indazoles as orally efficacious melanin concentrating hormone receptor-1 antagonists. Bioorg Med Chem Lett 15:5293–5297

Vaughan JM, Fischer WH, Hoeger C, Rivier J, Vale W (1989) Characterization of melanin concentrating hormone from rat hypothalamus. Endocrinology 125:1660–1665

Viale A, Zhixing Y, Breton C, Pedeutour F, Coquerel A, Jordan D, Nahon JL (1997) The melanin-concentrating hormone gene in human: flanking region analysis, fine chromosome mapping, and tissue-specific expression. Brain Res Mol Brain Res 46:243–255

Wang S, Behan J, O'Neill K, Weig B, Fried S, Laz T, Bayne M, Gustafson E, Hawes BE (2001) Identification and pharmacological characterization of a novel human melanin concentrating hormone receptor, MCH-R2. J Biol Chem 276:34664–34670

Wermter AK, Reichwald K, Buch T, Geller F, Platzer C, Huse K, Hess C, Remschmidt H, Gudermann T, Preibisch G, Siegfried W, Goldschmidt HP, Li WO, Price RA, Biebermann H, Krude H, Vollmert C, Wichmann HE, Illig T, Sorensen TI, Astrup A, Larsen LH, Pedersen O, Eberle D, Clement K, Blundell J, Wabitsch M, Schafer H, Platzer M, Hinney A, Hebebrand J (2005) Mutation analysis of the MCHR1 gene in human obesity. Eur J Endocrinol 152:851–862

Wu WL, Burnett DA, Caplen MA, Domalski MS, Bennett C, Greenlee WJ, Hawes BE, O'Neill K, Weig B, Weston D, Spar B, Kowalski T (2006) Design and synthesis of orally efficacious benzimidazoles as melanin-concentrating hormone receptor 1 antagonists. Bioorg Med Chem Lett 16:3674–3678

Zhang, P, Liand, JD, Samdusky, GE (1999) Hypothalamic MCH mRNA and protein are increased in human obesity. Satellite Symposium; 9th International Congress of Obesity France, Proceedings of the Symposium. Int J Obes, p 51

Appetite-Modifying Effects of Bombesin Receptor Subtype-3 Agonists

Ishita Deb Majumdar and H. Christian Weber

Contents

Abstract Studies on bombesin-like peptides (BLP) and their respective mammalian receptors (Bn-r) have demonstrated a significant biological impact on a broad array of physiological and pathophysiological conditions. Pharmacological experiments in vitro and in vivo as well as utilization of genetic rodent models of the gastrin-releasing peptide receptor (GRP-R/BB2-receptor), neuromedin B receptor (NMB-R/BB1-receptor), and the bombesin receptor subtype-3 (BRS-3/BB3-receptor) further delineated their role in health and disease. All three mammalian bombesin receptors have been shown to possess some role in the regulation of energy balance and appetite and satiety. Compelling experimental evidence has accumulated indicating that the orphan BRS-3 is an important regulator of body weight, energy expenditure, and glucose homeostasis. BRS-3 possesses no high affinity to the endogenous bombesin-like peptides (BLP) bombesin, GRP, and NMB, and its endogenous ligand remains unknown. Recently, the synthesis of

I. Deb Majumdar
Section of Gastroenterology, Boston University School of Medicine, Boston, MA, USA

H.C. Weber (✉)
Section of Gastroenterology, Boston University School of Medicine, Boston, MA, USA

Department of Medicine and Department of Pathology and Laboratory Medicine, Boston University School of Medicine, 650 Albany Street, EBRC Room 515, Boston, MA 02118, USA
e-mail: christian.weber@bmc.org

H.-G. Joost (ed.), *Appetite Control*, Handbook of Experimental Pharmacology 209,
DOI 10.1007/978-3-642-24716-3_19,

novel, selective high-affinity BRS-3 agonists and antagonists has been accomplished and has demonstrated that BRS-3 regulates energy balance independent of other established pathways. Accordingly, the availability of new BRS-3 selective agonists and antagonists will facilitate further elucidation of its role in energy homeostasis and provides a potential approach for the pharmacological treatment of obesity.

Keywords Bombesin receptor subtype-3 • Bombesin-like peptides • Energy homeostasis • Gastrin-releasing peptide • Mammalian bombesin receptors • Neuromedin B • Obesity • Synthetic BRS-3 agonist

1 Introduction

The family of G protein-coupled mammalian bombesin receptors (Bn-r) consists of three receptors: the 384-amino-acid gastrin-releasing peptide receptor (GRP-R or BB2-receptor), the 390-amino-acid neuromedin B receptor (NMB-R or BB1-receptor), and the 399-amino-acid orphan bombesin receptor subtype 3 (BRS-3 or BB3-receptor). This family of cognate G protein-coupled receptors mediates biological effects through high-affinity binding of bombesin-like peptides (BLP) mainly by paracrine and neurocrine mechanisms of actions. The amphibian regulatory peptide bombesin and its mammalian homologue, gastrin-releasing peptide (GRP), bind with high affinity to the GRP-R, whereas neuromedin B (NMB) selectively binds to the NMB-preferring NMB-R. At present, no endogenous ligand has been identified for the orphan BRS-3. Both GRP and NMB are processed from larger precursor molecules and are found widely distributed in the central nervous system (CNS) and in various peripheral neuronal cells, e.g., in the gastrointestinal submucosal and myenteric plexus. GRP stimulates the exocrine secretion from the pancreas, the release of gastrointestinal hormones, and smooth muscle contraction, and modulates the function of immune cells (Gonzalez et al. 2008a; Ischia et al. 2009; Jensen et al. 2008; Majumdar and Weber 2011). Furthermore, GRP affects functions of the CNS such as the circadian rhythm, anxiety, fear responses (Shumyatsky et al. 2002), and thermoregulation (Jensen et al. 2008). Because of its expression in the spinal cord, GRP has also been associated with the chronic itch sensation and male reproductive functions (Sun and Chen 2007; Sun et al. 2009; Sakamoto et al. 2008; Sakamoto and Kawata et al. 2009; Sakamoto et al. 2009). GRP and NMB both exert effects on growth, migration, and invasion of cells derived from various epithelial tumors (Gonzalez et al. 2008a; Ischia et al. 2009; Jensen et al. 2008; Majumdar and Weber 2011; Patel et al. 2006).

Studies on the biology of the orphan BRS-3 are hampered due to the lack of the natural ligand. Because of its 51 and 47% amino acid homology with the GRP-R and NMB-R, respectively, it has been classified as a mammalian bombesin receptor (Jensen et al. 2008; Fathi et al. 1993). Although its expression in the central nervous system (CNS) and peripheral tissues (Fathi et al. 1993, 1996; Ohki-Hamazaki et al.

1997a; Porcher et al. 2005; Sano et al. 2004; Ohki-Hamazaki et al. 1997b; Liu et al. 2002; Jennings et al. 2003) has been reported previously, its roles in normal human physiology and disease remain largely unknown. Human BRS-3 cDNA was isolated and cloned from a small cell lung carcinoma (SCLC) cell line, and its expression was shown in a panel of human lung and breast carcinoma cell lines as well as other epithelial tumors (Fathi et al. 1993, 1996; Gorbulev et al. 1994; Reubi et al. 2002; Schulz and Rocken 2006), carcinoid tumors (Kuiper et al. 2010), and in the pregnant human uterus (Gorbulev et al. 1994). In the guinea pig, its expression was detected in the pregnant uterus and the brain (Gorbulev et al. 1992) but was confined to the testes and secondary spermatocyte in the rat. (Fathi et al. 1993; Liu et al. 2002). A selective synthetic peptide agonist of the human BRS-3 was used to show a role in lung tumor invasiveness (Hou et al. 2006), MAP kinase activation (Weber et al. 2001), and lung injury (Tan et al. 2006, 2007; Wang et al. 2007).

Currently, studies on BRS-3 are receiving increased attention, chiefly because its genetic disruption in rodent models revealed a phenotype of obesity, insulin resistance, and hypertension, reminiscent of the metabolic syndrome in men (Ohki-Hamazaki et al. 1997b). Subsequent studies suggested a role in regulation of hyperphagia (Ladenheim et al. 2008; Maekawa et al. 2004), insulin release (Matsumoto et al. 2003; Nakamichi et al. 2004), food consumption (Yamada et al. 2000), taste preference (Yamada et al. 1999), and gastrointestinal motility (Porcher et al. 2005). Taken together, these studies suggested a biologically significant role of BRS-3 in the regulation of energy balance and glucose homeostasis. Moreover, the availability of selective agonists with high affinity for the human BRS-3 provided the pharmacological tools to examine binding properties, second messenger molecules, and some downstream intracellular signaling pathways (Weber et al. 2001; Mantey et al. 2004, 2001, 1997; Ryan et al. 1998a, b; Sancho et al. 2010; Pradhan et al. 1998). Additional putative BRS-3 agonists were reported but with limited pharmacological characterization (Boyle et al. 2005; Carlton et al. 2008; Weber et al. 2003, 2002; Zhang et al. 2009; Lammerich et al. 2003). Recently, a series of small molecules were synthesized with agonist and antagonist properties on BRS-3 in various mammalian species, and with no appreciable binding to NMB-R and GRP-R. They produced the desired metabolic effects such as weight loss and decrease in food intake, indicating their potential use as antiobesity agents in humans (Coll 2010; Guan et al. 2010, 2011; Guo et al. 2010; Hadden et al. 2010; Liu et al. 2010; Metzger et al. 2010; He et al. 2010; Lo et al. 2011).

2 BRS-3 Genes

The G protein-coupled receptor BRS-3 was first identified in the guinea pig's uterus by Gorbulev et al. (Gorbulev et al. 1992), later in human lung cancer cells (Fathi et al. 1993). The cDNA clone encoded a protein showing the highest amino acid similarity with GRP-R (52%) and NMB-R (47%) and was designated as bombesin receptor subtype-3 (BRS-3) (Jensen et al. 2008). The human BRS-3 gene was mapped to chromosome Xq25, and the mouse gene to X-chromosome XA7.1–7.2. The mouse BRS-3 gene encodes a protein of 399 amino acids, which is identical in size compared to its human and guinea pig homologues (Ohki-Hamazaki et al. 1997b; Gorbulev et al. 1994; Weber et al. 1998). The overall homology of the mouse BRS-3 protein sequence compared to human and guinea pig BRS-3 is 85 and 83%, respectively, with the carboxyl terminus being the most divergent protein sequence. The human BRS-3 gene contains two introns and three exons similar to the sheep (Whitley et al. 1999), rhesus (Sano et al. 2004), mouse (Ohki-Hamazaki et al. 1997a; Weber et al. 1998), and rat BRS-3 gene (Liu et al. 2002). There are two TATA-like motifs present in the 5′ flanking region of the mouse BRS-3 gene, located at 492–486 bp and more than 1 kb upstream of the translation start. Comparison of the nucleotide sequence of the 5′ flanking region in the human and mouse BRS-3 genes indicated 58% homology within 1,000 bp and 72% homology within the immediate 500 bp upstream of the translation initiation codon ATG (Weber et al. 1998). One study screened 104 Japanese obese men for defects in BRS-3 gene, but no mutations or polymorphisms were found, suggesting that BRS-3 gene mutations are unlikely to be the major cause of obesity in humans (Hotta et al. 2000).

3 Biology of BRS-3

BRS-3-dependent pathways have been identified as critical for proper regulation of energy homeostasis, but its natural ligand has not been identified as of now. Thus far, studies have been hampered by the lack of potent selective agonists and antagonists for this receptor in rodent animal models. Synthetic peptides or small molecules lacked either potency or selectivity on the BRS-3 (Jensen et al. 2008), whereas a bombesin-derived peptide showed agonist activity on the human BRS-3 only (Mantey et al. 2004, 2001, 1997; Ryan et al. 1998a, b; Sancho et al. 2010; Pradhan et al. 1998).

Extensive studies were performed with mice lacking functional BRS-3 after its initial report in 1997 (Ohki-Hamazaki et al. 1997b). It was observed that disruption of functional BRS-3 produced mild obesity associated with hypertension and impairment of glucose metabolism. Moreover, the BRS-3 KO mice exhibited a reduced metabolic rate, increased feeding efficiency, and hyperphagia, indicating that BRS-3 is required for the regulation of energy balance and adiposity (Ohki-Hamazaki et al. 1997b). Expression of BRS-3 in the parabrachial nucleus, the medial and central nuclei of the amygdala and the hypothalamic region, responsible of taste perception, also suggested a role in food choice and preference (Yamada et al. 1999). BRS-3 KO mouse exhibited an upregulation of the hypothalamic m-RNA of the melanin-concentrating hormone (MCH) receptor, thereby leading to an enhanced hyperphagic response to exogenous MCH administration. These findings further indicated that hyperphagia is a major factor leading to increased body weight and hyperinsulinemia in BRS-3 KO mice (Maekawa et al. 2004). Previous results demonstrated a small but significant decrease in oxygen consumption in BRS-3 KO mice prior to the onset of obesity (Ladenheim et al. 2008). The sibutramine sensitivity assay revealed that BRS-3-deficient mice consume 11% less oxygen than wild-type mice in the resting state, and supports the notion that a defect in energy expenditure might contribute to the progress of obesity in these BRS-3 KO mice (Matsumoto and Iijima 2003). BRS-3 are present in pancreatic islets (Fleischmann et al. 2000), and BRS-3 knockout animals were reported to have altered plasma insulin levels and altered glucose transport (Matsumoto et al. 2003; Nakamichi et al. 2004).

A series of papers have now reported the synthesis of novel small molecule selective antagonists and agonists on the BRS-3. Antagonists increased food intake and body weight, whereas agonists increased metabolic rate and reduced food intake and body weight. (Coll 2010; Guan et al. 2010, 2011; Guo et al. 2010; Hadden et al. 2010; Liu et al. 2010; Metzger et al. 2010; He et al. 2010; Lo et al. 2011; Furutani et al. 2010). BRS-3 binding sites were identified in the hypothalamus, caudal brain stem, and several midbrain nuclei. The observed changes in energy metabolism were absent in animals genetically deficient in BRS-3 but present in knockout mice of other relevant molecules including NPY, melanocortin receptor 4, Agouti-related protein, and leptin receptor (Guan et al. 2010).

In a different follow-up study, the authors suggested that BRS-3 also modulates body temperature, presumably secondary to its effect on energy metabolism including effects on the sympathetic tone. It was reported that BRS-3 knockout mice have a reduced body temperature when compared to the wild type. Treatment with the BRS-3 agonist Bag-1 increases body temperature that is reduced by fasting or during resting (light) phase. Moreover, Bag-1 treatment causes a slight increase of body temperature in the dark (active) phase, but this increase did not exceed the daily maximum. Thus, it is predicted that BRS-3 agonists reverse the reduction of energy expenditure in the fasted state, and provide a new approach to the treatment of obesity (Metzger et al. 2010). Other studies suggested that BRS-3 agonist increased the intracellular calcium concentration of orexin neurons and induced depolarization in the presence of GABA receptor blockers, suggesting that BRS-3 agonist might indirectly inhibit orexin neurons through GABAergic input and directly activate orexin neurons (Furutani et al. 2010).

Other studies (Tan et al. 2006, 2007) demonstrated that BRS-3 are expressed in the airway in response to ozone injury and that wound repair and proliferation of bronchial epithelial cells is accelerated by BRS-3 activation, suggesting that it may mediate wound repair. It was found that ozone induced binding of activator protein-2α and peroxisome proliferator-activated receptor-α to the promoter of the BRS-3 gene, suggesting that these transcription factors are involved in regulation of BRS-3 expression (Tan et al. 2007).

Finally, the role of BRS-3 in cancer biology is apparently of importance, since BRS-3 activation stimulates tyrosine kinases and phospholipase C and D (Ryan et al. 1998a, b). Receptor activation stimulates tyrosine phosphorylation of the cytosolic focal adhesion kinase, $p125^{FAK}$ which regulates cell growth and motility

(Ryan et al. 1998a). In addition, activation of BRS-3 stimulates MAP kinase activation, resulting in rapid tyrosine phosphorylation of both its 42- and 44-kDa forms (Weber et al. 2001). In human NCI-1299 lung cancer cells transfected with BRS-3, selective activation of BRS-3 with peptide agonist [D-Phe^{6}, β-Ala^{11}, Phe^{13}, Nle^{14}]bombesin(6–14) resulted in stimulation of Elk-1 in a MEK-1-dependent manner as well as in a 47-fold increase of c-fos mRNA. These results demonstrated that BRS-3 activation stimulates pathways of cell proliferation such as activation of MAP kinase and Elk-1, and increase in nuclear proto-oncogene expression (Weber et al. 2001). More recently, it was demonstrated in the same cell model that BRS-3 agonists may stimulate lung cancer growth as a result of EGFR transactivation, and that the transactivation is regulated by BRS-3 in a Src-, reactive oxygen-, and matrix metalloprotease-dependent manner (Moody et al. 2011).

4 Structure–Activity Relationship and Pharmacology of Agents Targeting the BRS-3

Because of its role in the regulation of energy metabolism, tumor invasiveness (Jensen et al. 2008), adhesion and/or metastasis (Hou et al. 2006), lung development, bronchial epithelial cell proliferation and lung injury (Tan et al. 2006; Shan et al. 2004), taste perception (Yamada et al. 1999), social response/anxiety (Yamada et al. 2000, 2002), and a putative role in gastrointestinal motility (Porcher et al. 2005), pharmacological studies on the orphan BRS-3 are critical but were hampered by the absence of selective agonists and antagonists.

Discovery of a high-affinity ligand for the hBRS-3 demonstrated that it has a unique pharmacology compared with other mammalian bombesin receptors, and that none of the existing endogenous BLPs are the natural ligand for this receptor (Mantey et al. 1997). It was reported that none of the 15 naturally occurring bombesin-related peptides had an affinity of >1 μM for the human BRS-3 (hBRS-3). Furthermore, none of the 26 synthetic analogs that acted as GRP-R or NMB-R agonists or antagonists had a high affinity for the BRS-3 (Jensen et al. 2008). Detailed analysis also showed that hBRS-3 has a greater affinity for NMB, litorin, and ranatensin than GRP or bombesin (Mantey et al. 1997). However, the peptide agonist [D-Phe^{6}, β-Ala^{11}, Phe^{13}, Nle^{14}]bombesin(6–14) demonstrated relatively high affinity for the hBRS-3 (Table 1). [D-Phe^{6}, β-Ala^{11}, Phe^{13}, Nle^{14}]bombesin (6–14) stimulated a two- to threefold increase in $[Ca^{2+}]_i$ and a threefold increase in tyrosine phosphorylation of p125^{FAK} with an EC_{50} of 0.2–0.7 nM, but failed to either increase cyclic AMP or inhibit forskolin-stimulated increase. Similarly, no high-affinity Bn receptor antagonists had a high affinity for the hBRS-3 receptor although two low-affinity antagonists for GRP and NMB receptors, [D-Arg1, D-Trp7,9, Leu11]substance P and [D-Pro4, D-Trp7,9,10]substance P-(4–11), inhibited hBRS-3 receptor activation. Furthermore, the NMB-R-specific

Table 1 Summary of binding data of various agonist peptides on human bombesin receptors

Binding properties of K_i [nM]

	hBRS-3	hNMB-R	hGRP-R	Cell system and references
[D-Phe⁶, β-Ala¹¹, Phe¹³, Nle¹⁴]Bn (6–14)	8.9 ± 0.7	ND	ND	hBRS-3/Balb3T3 (Mantey et al. 1997)
	4.2 ± 1.0			hBRS-3/H1299 cells (Pradhan et al. 1998)
Pep-1 [D-Try⁶, β-Ala¹¹, Phe¹³, Nle¹⁴]Bn (6–14)	0.82 ± 0.1	5.9 ± 0.4	0.5 ± 0.04	Balb/c 3 T3 cells (Mantey et al. 2004)
	0.85 ± 0.05	1.48 ± 0.3	0.07 ± 0.01	hBRS-3/Balb, hNMB-R/Balb, hGRP-R/Balb (Sancho et al. 2010)
¹²⁵I-[D-Try⁶, β-Ala¹¹, Phe¹³, Nle¹⁴]Bn(6–14)	0.4–4.2	ND	ND	hBRS-3/Balb3T3 cells (Pradhan et al. 1998)
Pep-2 [D-Try⁶, R-Apa¹¹, Phe¹³, Nle¹⁴]Bn (6–14)	2.8 ± 0.1	2,400 ± 96	151 ± 6	Balb/c 3 T3 cells(Mantey et al. 2004)
	3.39 ± 0.15	158 ± 8	110 ± 5	hBRS-3/Balb, hNMB-R/Balb, hGRP-R/Balb (Sancho et al. 2010)
Pep-3 [D-Try⁶, (R) -Apa¹¹- 4Cl, Phe¹³, Nle¹⁴]Bn (6–14)	8.2 ± 1.1	7,200 ± 840	1,860 ± 69	Balb/c 3 T3 cells(Mantey et al. 2004)
	2.09 ± 0.06	589 ± 32	258 ± 6	hBRS-3/Balb, hNMB-R/Balb, hGRP-R/Balb (Sancho et al. 2010)
Pep-4 Ac-Phe, Trp, Ala, His(τBzl), Nip, Gly, Arg-NH₂	63–570	12–150-fold less selective than hBRS-3	4–150-fold less selective than hBRS-3	Balb/c 3 T3 cells (Mantey et al. 2004)
	63.1 ± 3.1	>10,000	>10,000	hBRS-3/Balb, hNMB-R/Balb, hGRP-R/Balb (Sancho et al. 2010)
Pep-16a Phenylacetyl-Ala, D-Trp-Phenthyl amide	3,000 to > 10,000	Nonselective >10,000	Nonselective >10,000	Balb/c 3 T3 cells (Mantey et al. 2004)
	6,026 ± 148			hBRS-3/Balb, hNMB-R/Balb, hGRP-R/Balb (Sancho et al. 2010)

ND not determined, *hGRP-R* human gastrin-releasing peptide receptor, *hNMB-R* human neuromedin B receptor, *hBRS-3* human bombesin receptor subtype-3

antagonist D-Nal,Cys,Tyr,D-Trp, Lys,Val,Cys,Nal-NH_2 inhibited hBRS-3 activation in a competitive fashion ($K_i = 0.5$ μM) (Ryan et al. 1998a).

In a subsequent study (Ryan et al. 1998b), two human lung cancer cells lines, NCI-N417 and NCI-N720, were found to possess sufficient wild-type BRS-3 to allow assessment of the pharmacology of the native BRS-3 using the ^{125}I-[D-Tyr^6, β-Ala^{11}, Phe^{13}, Nle^{14}]bombesin$_{6–14}$. Pharmacology for all agonists of the native BRS-3 was found to be similar to that reported previously with the BRS-3-transfected cell lines (Mantey et al. 1997) with the only agonist, [D-Phe^6, β-Ala^{11}, Phe^{13}, Nle^{14}]bombesin$_{6–14}$, demonstrating high affinity (K_i 7.4 nM) (Jensen et al. 2008). It was noticed that in both the cell lines, [D-Phe^6, β-Ala^{11}, Phe^{13}, Nle^{14}] bombesin$_{6–14}$ stimulated phospholipase D activity to generate diacylglycerol and phospholipase C for a concentration-dependent release of [^{3}H]inositol phosphate and intracellular calcium. Detailed study showed that hBRS-3 activation was not coupled to changes in adenylate cyclase activity, [^{3}H]-thymidine incorporation or cell proliferation (Ryan et al. 1998a, b).

Another synthetic selective BRS-3 potent agonist (compound 16a) was reported to promote ß-arrestin translocation to the cell membrane of rat, mouse, and human, though no visible internalization of the receptor was noticed following 2 h of agonist administration. It was reported that when applied to hBRS-3-transfected HEK293 cells, the compound caused dose-dependent increase in intracellular-free calcium concentration, with an EC_{50} of 14.15 ± 0.13 nM (Zhang et al. 2009) (Tables 1 and 2).

Table 2 Summary of second messenger data of various peptide agonists on the human BRS-3

Name of peptides	*[^{3}H]IP3 assay (EC_{50} in nM)	Ca^{2+} assay (EC_{50} in nM)	Cell system and references
[D-Phe^6, β-Ala^{11}, Phe^{13}, Nle^{14}]Bn(6–14)	20–35 21 ± 2.1 35 ± 5.0	7.21 ± 0.18	*NCI-H1229 and hBRS-3/ Balb 3T3(Ryan et al. 1998a) hBRS-3/CHO cells (Weber et al. 2002) *hBRS-3/Balb 3T3 (Ryan et al. 1998a) *hBRS-3/NCI-H1229 cells(Ryan et al. 1998a)
Pep-1 [D-Try^6, β-Ala^{11}, Phe^{13}, Nle^{14}]Bn(6–14)	1.1 ± 0.1	7.10 ± 0.17	*hBRS-3/Balb (Sancho et al. 2010) hBRS-3/HEK293T-cells (Zhang et al. 2009)
Pep-2 [D-Try^6, R-Apa^{11}, Phe^{13}, Nle^{14}]Bn(6–14)	5.9 ± 0.2	ND	*hBRS-3/Balb (Sancho et al. 2010)
Pep-3 [D-Try^6, (R) -Apa^{11}- 4Cl, Phe^{13}, Nle^{14}]Bn(6-14	3.6 ± 0.1	ND	*hBRS-3/Balb (Sancho et al. 2010)
Pep-4 Ac-Phe, Trp, Ala ,His (†Bzl), Nip, Gly, Arg-NH_2	21 ± 1	6.9	*hBRS-3/Balb (Sancho et al. 2010) hBRS-3/Balb (Boyle et al. 2005)
Pep-16a Phenylacetyl-Ala, D-Trp-phenthyl amide	>10,000	2.1–14 14.15 ± 0.13	*hBRS-3/Balb (Sancho et al. 2010) CHO K1 or HEK293 cells (Sancho et al. 2010) hBRS-3/HEK293T (Zhang et al. 2009)
VV-H-7 (VVYPWTQRF)		45 ± 15 μM 19 ± 6 μM	CHO-$G_{\alpha16}$/hBRS-3cells NCI-N417 cells (Lammerich et al. 2003)
LVV-H-7 (LVVYPWTQRF)		183 ± 60 μM 38 ± 18 μM	CHO-$G_{\alpha16}$/hBRS-3cells NCI-N417 cells (Lammerich et al. 2003)

ND not determined, *hBRS-3* human bombesin receptor subtype 3

With synthetic selective BRS-3 ligands and mutational analyses of the main ligand-binding pocket, epitopes determining the agonist properties were mapped. It was concluded that BRS-3 activation is dependent upon an epitope in the main ligand-binding pocket at the interface between the extracellular segments of transmembrane TM-III, TM-VI, and TM-VII (Gbahou et al. 2010).

In order to understand the molecular determinants of selectivity/high affinity of BRS-3, a mutagenesis approach was employed. With the synthetic peptide agonists #2, 3, and 4, it was demonstrated that [Val^{101}, His^{107}, Gly^{112}, Arg^{127}] in the EC2/adjacent upper TMs of BRS-3 are critical for the high BRS-3 selectivity of peptide #4. The role of His^{107} in EC2 appears important for BRS-3 activation, suggesting that amino-aromatic ligand/receptor interactions with peptide #4 are critical for both binding and activation. It also demonstrates that while the BRS-3 selective agonists were developed from the same template peptide, [D-Phe^{6}, ß-Ala^{11}, Phe^{13}, Nle^{14}]Bn-(6–14), their molecular determinants of selectivity varied considerably (Gonzalez et al. 2008b).

In another set of experiments, the role of an exchange of arginine for histidine at position 294 was investigated. Arg^{294} appears important in receptor–ligand affinity of ovine BRS-3. Thus, it was concluded that ovine BRS-3 may have binding characteristics different from those of the human, mouse, and guinea pig BRS-3 (Whitley et al. 1999).

In a separate set of experiments, GRP-R antagonists [(D-Phe^{6}, Leu$^{®}$-p-chloro-Phe^{14}]bombesin(6–14) and [D-Phe, Leu-NHEt^{13}, des-Met^{14}]bombesin(6–14) did not inhibit Ca^{2+} elevation induced by VV-H-7(a naturally occurring agonist from human placenta). In contrast, the somatostatin analog, a NMB-R-specific antagonist [D-Nal,Cys,Tyr,D-Trp, Lys,Val,Cys,Nal-NH_2] (Orbuch et al. 1993), inhibited the VV-H-7-induced response significantly (Pradhan et al. 1998; Lammerich et al. 2003). Detailed studies suggested that [D-Phe^{6}, β-Ala^{11}, Phe^{13}, Nle^{14}]bombesin (6–14) (a universal ligand for all the four types of bombesin receptors) can inhibit the binding of ^{125}I-[D-Try^{6}, β-Ala^{11}, Phe^{13}, Nle^{14}]bombesin(6–14) in a dose-dependent manner. Partial inhibition is seen in 0.1-nM range, half maximal inhibition at 5.0 and 9.0 nM in hBRS-3-transfected H1299 and BALB 3T3 cells, respectively, and complete inhibition at 1 μM. These data provided evidence that ^{125}I-[D-Try^{6}, β-Ala^{11}, Phe^{13}, Nle^{14}]bombesin(6–14) possesses properties as a universal ligand for all the bombesin subtype receptors, with affinities ranging between 0.4 and 4.2 nM for each of the four subtypes of bombesin receptors (Pradhan et al. 1998).

Due to its nonselectivity and high affinity for all the human Bn receptors subtypes, the usefulness of [D-Try^{6}, β-Ala^{11}, Phe^{13}, Nle^{14}]bombesin(6–14) (peptide #1) and its analog D-Phe^{6} to study the role of hBRS-3 in various processes appeared limited (Reubi et al. 2002; Mantey et al. 1997; Ryan et al. 1998a; Pradhan et al. 1998; Katsuno et al. 1999).

In a separate attempt to develop a hBRS-3 ligand with greater selectivity, Mantey et al. (2001) replaced the β-Ala^{11} of the nonselective high-affinity hBRS-3 ligand, [D-Try^{6}, β-Ala^{11}, Phe^{13}, Nle^{14}]bombesin(6–14), with conformationally restricted amino acids. Two peptides with an (R)- or (S)-amino-3-phenylpropionic acid (Apa) substitution for β-Ala^{11} were reported with some selectivity for the hBRS-3. Of these two, (R)-Apa^{11} analog (peptide #2) (Sancho et al. 2010) had 50-fold and (S)-Apa^{11} had 13-fold selectivity for the hBRS-3 over the hGRP-R and hNMB-R, and for stimulating phospholipase C activity. Further improvement of selectivity was achieved by insertion of the chlorine residue in the para position of the phenyl ring in the (R)-Apa^{11} analog, thus rendering [D-Try^{6}, (R)Apa^{11}-4Cl, Phe^{13}, Nle^{14}] bombesin(6–14) analog (peptide #3) (Sancho et al. 2010) very selective and more potent for hBRS-3. Binding studies showed that [D-Try^{6}, (R)Apa^{11}-4Cl, Phe^{13}, Nle^{14}]bombesin(6–14) analog had 227- and 880-fold selectivity for hBRS-3 over hGRP-R and hNMB-R, and 90- and 20-fold greater potency for activating the hBRS-3, respectively (Mantey et al. 2004) (Tables 1 and 2).

In the quest of discovery of a more selective BRS-3 agonist, extensive structure–activity relationship (SAR) studies on the unselective BRS-3 agonist [H-D-Phe^{6}, β-Ala^{11}, Phe^{13}, Nle^{14}]bombesin(6–14)-nonapeptide amide (agonist **1**) highlighted the structural features important for BRS-3 activity. A radically modified heptapeptide agonist (peptide #4) (Sancho et al. 2010), Ac-Phe-Trp-Ala-His{τBzl}-Nip-Gly-Arg-NH_2 (heptapeptide **34**), carrying only the Trp-Ala moiety of the parent [H-D-Phe^{6}, β-Ala^{11}, Phe ^{13}, Nle^{14}]bombesin(6–14)-peptide amide (agonist **1**) and a very different carboxyl terminal region, gained importance due to its selectivity for the BRS-3 over the NMB or GRP receptor. Heptapeptide 34 was potent and selective for BRS-3 with an EC_{50} of 6.9 nM and was significantly different from the other bombesin-related BRS-3 ligands already reported (Boyle et al. 2005) (Tables 1 and 2).

Similarly, the synthetic peptidomimetic compound, N1-(2-Phenylethyl)-(2R)-2-{[(1S)-1-(benzylcarboxamido)ethyl] carboxamido}-3-(1H-3-indolyl)propanamide **16a,** was also found to be a selective and highly potent agonist for human, rat, and mouse BRS-3. Studies on human, rat, and mouse cells showed that there is a dose-dependent increase of intracellular calcium concentration with an EC_{50} of 14.15 ± 0.13 nM, 109.90 ± 0.21 nM, and 33.30 ± 0.14 nM, respectively. Moreover, no activity was seen with NMB-R, GRP-R, or on various other GPCRs, rendering it a very selective and potent agonist (Zhang et al. 2009).

Previous reports confirmed that the synthetic peptides #2, 3, 4, and 16a are very selective and highly potent agonists of hBRS-3 as compared with hGRP-R or hNMB-R. However, a recent thorough study of the pharmacology of these synthetic peptides including binding studies, alterations in cellular signaling (PLC, PKD), and changes in cellular function showed that two peptides (#2 and #3) had nM affinities/potencies for hBRS-3, peptide #4 had low affinity/potency, and that of peptide #16a was very low (>3,000 nM). In terms of selectivity for hBRS-3, it was found that peptide #3 was more selective (100-fold) followed by peptides #2 and #4 with lower selectivity (threefold lower and 157-fold lower, respectively, than peptide #1). The selectivity of peptide 16a could not be determined because of its lower affinity/potencies for all the Bn receptors. The above described findings demonstrated that peptides #2, #3, and possibly #4, but not peptide #16a, have sufficient affinities and potencies for hBRS-3 receptors to be potentially useful, and confirmed the high affinity of the hBRS-3 selective ligand peptide #3 has a, similar to that of the universal ligand, peptide #1 (Sancho et al. 2010).

Despite the development of the high-affinity ligand [D-Try^{6}, β-Ala^{11}, Phe^{13}, Nle^{14}]bombesin(6–14) and its analogs, additional novel compounds were needed to test the usefulness of BRS-3 as a drug target. Several studies of the structure—activity relationship of [D-Try^{6}, β-Ala^{11}, Phe^{13}, Nle^{14}]bombesin(6–14) assessed its ability to mobilize intracellular calcium in the BRS-3-transfected CHOGα-16 cells by a fluorometric imaging plate reader (FLIPR) assay. By a strategy similar to the "peptoid" approach, a selective short peptide agonist, H-D-Phe-Gln-D-Trp-1-(2-phenylethyl)amide was developed for the human BRS-3. Various SAR studies on the N terminus, the D-Phe-Gln moiety of H-D-Phe-Gln-D-Trp-1-(2-phenylethyl) amide, showed that by incorporation of N-arylated glycine and alanine, azaglycine, piperazine, or piperidine residues as well as by the synthesis of semicarbazides and semicarbazones, several highly potent and selective agonists for hBRS-3 with a reduced number of peptide bonds and a higher metabolic stability were obtained. Thus, from the analysis of the SAR data, it was concluded that lipophilicity located at a certain distance from the tryptophan and combined with a basic residue between them was favorable for the potency of these compounds (Weber et al. 2003, 2002).

Recently, the gastric H+/K + −ATPase pump inhibitor omeprazole was identified as a weak partial BRS-3 agonist (BB3 EC_{50} = 14,000 nM, % Max = 27%) (Table 3) which suppressed body weight gain in rats. Its sulfide analog was found to be slightly more potent (BB3 EC_{50} = 3,900 nM) with better efficacy (%Max = 64%). Replacement of the pyrimidine moiety with different groups and the modification of the linker region increased affinity, metabolic stability, and eliminated the effect of omeprazole on the proton pump. Further modification to the benzimidazole revealed the importance of the NH residue as hydrogen donor and showed the importance of chloride and trifluoromethane analogs as submicromolar BRS-3 agonists. These analogs gained more importance as none of them exhibited any agonist efficacy on NMB-R and GRP-R, measured by intracellular calcium mobilization in a FLIPR-based assay (Carlton et al. 2008).

Development of other non-peptide BRS-3 agonists was based on a biarylethylimidazole pharmacophore. These agents were effective in lowering food intake and body weight in diet-induced obese mice (Hadden et al. 2010; Liu et al. 2010). Furthermore, substituted biphenyl imidazoles exhibited agonist potency on rodent and human BRS-3 (He et al. 2010). Recently, a high-throughput screening effort by Merck identified a non-peptidic racemic low-molecular-weight BRS-3 agonist, RY-337, which showed maximum potency on rodent BRS-3 compared to human BRS-3. Extensive SAR studies on the biphenyl imidazoles led to the discovery of various significantly improved agonists for both rodent and human BRS-3. Among these, compound 22e had significantly enhanced potency for both rodent and human BRS-3 (He et al. 2010) (Tables 4–6).

Merck scientists have also identified pyridinesulfonylureas and pyridinesulfonamides with a 4-(alkylamino) pyridine-3-sulfonamide core as potent and selective agonists for human and mouse BRS-3. In the sulfonylurea class, the potent

Table 3 Summary of second messenger data of various non-peptide agonists on the human BRS-3 (part I)

Compound pyridinesulfonylureas and pyridinesulfonamides 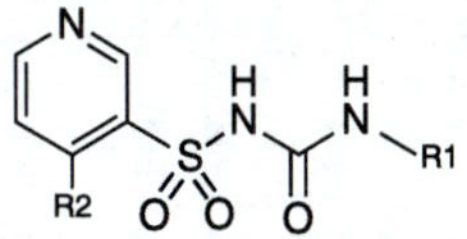	R_1	R_2	Calcium mobilization functional assay for hBRS-3 [EC_{50} (nM)]	Cell system and references
Compound 1 (lead compound)	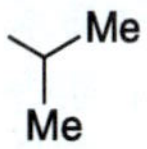	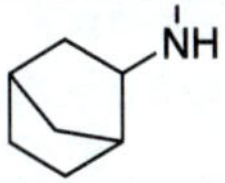	220	hBRS-3/HEK293AEQ cells (Lo et al. 2011)
Compound exo-2a	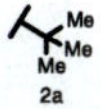	(±) 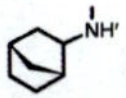	81 ± 12	hBRS-3/ HEK293AEQ cells (Lo et al. 2011)
Compound (1R, 2R, 4S)-2a		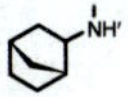	55 ± 4	hBRS-3/HEK293AEQ cells (Lo et al. 2011)
Compound 2b	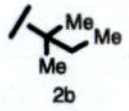	(±) 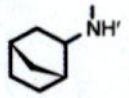	51 ± 11	hBRS-3/HEK293AEQ cells (Lo et al. 2011)
Compound 5e (pyridinesulfonamides) (+ or −)-5e 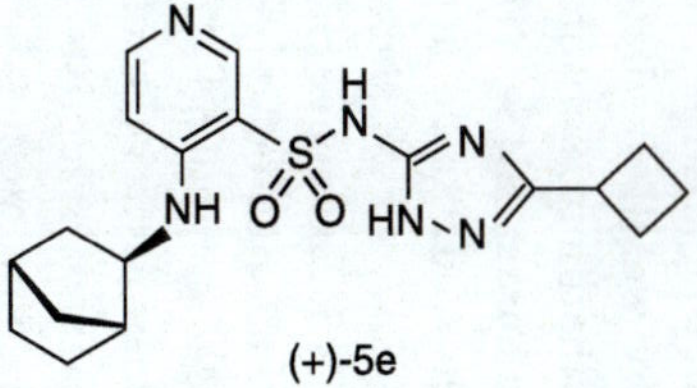 (+)-5e			380 ± 10	hBRS-3/HEK293AEQ cells (Lo et al. 2011)

Compound 1a Omeprazole (lead compound)	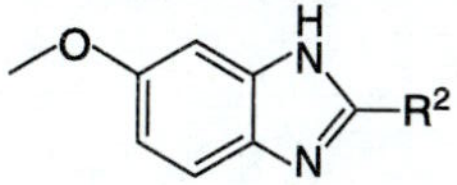	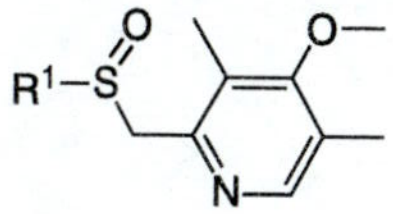	14,000	BB_3/HEK 293 cells (Carlton et al. 2008)
Compound 1b Sulfide analog	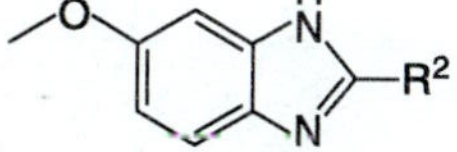	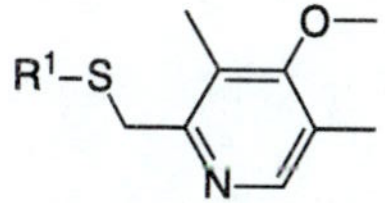	3,900	BB_3/HEK 293 cells (Carlton et al. 2008)
Compound 1h 1,3 benzodioxole (thioether analog)	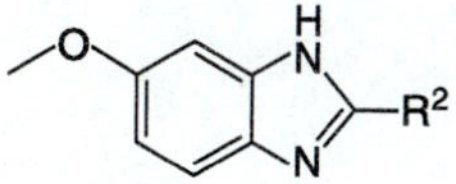	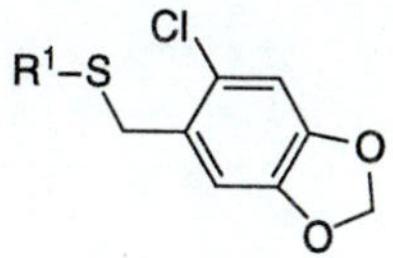	380	BB_3/HEK 293 cells (Carlton et al. 2008)
Compound 1k Ether-linked analog	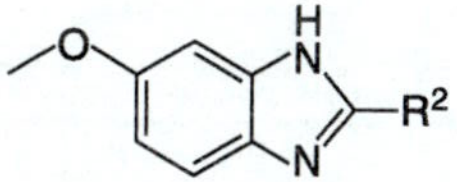	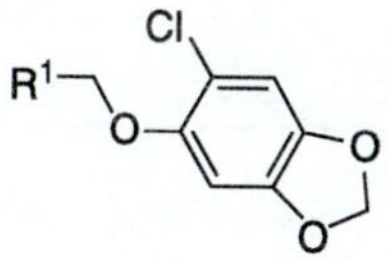	510	BB_3/HEK 293 cells (Carlton et al. 2008)
Compound 1l Carbon-linked analogs	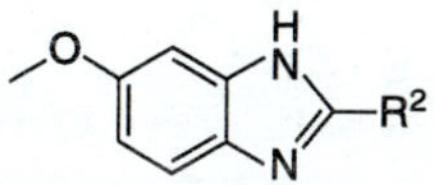	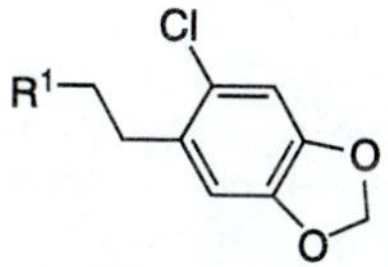	210	BB_3/HEK 293 cells (Carlton et al. 2008)

(continued)

Compound 1s Chloride-linked analog	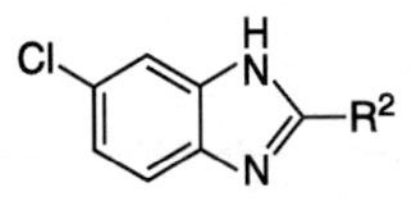	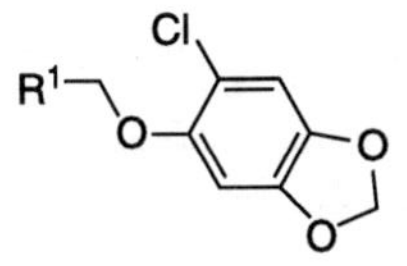	570	BB_3/HEK 293 cells (Carlton et al. 2008)
Compound 1t Trifluoromethane analogs	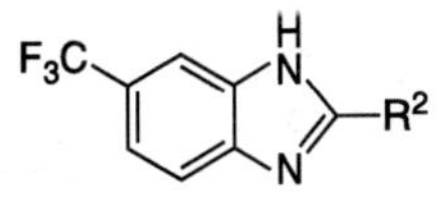	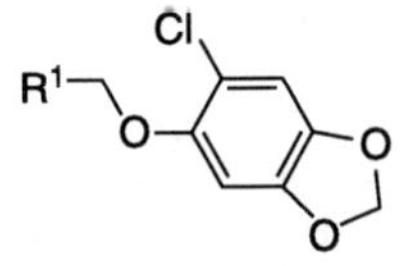	240	BB_3/HEK 293 cells (Carlton et al. 2008)

Table 4 Summary of binding data of various non-peptide agonists on the human BRS-3

Compound [biarylethylimidazole pharmacophore] 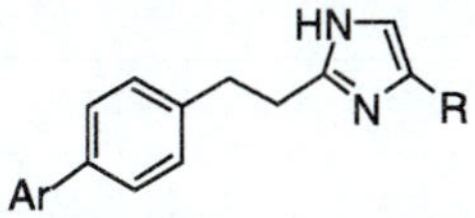	Ar	R	K_i (nM) hBRS-3 binding	Cell system and reference
Compound 9 Bag-1	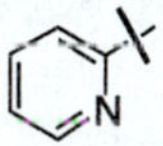		17.4	hBRS-3/CHO or HEK293cells (Guan et al. 2010)
Compound 10a	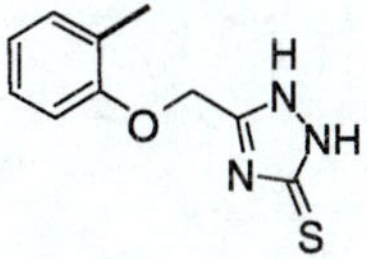		6.1	(Hadden et al. 2010)
Compound 11	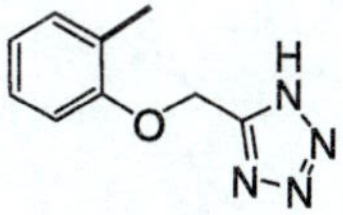		5.3	(Hadden et al. 2010)
Compound 14a	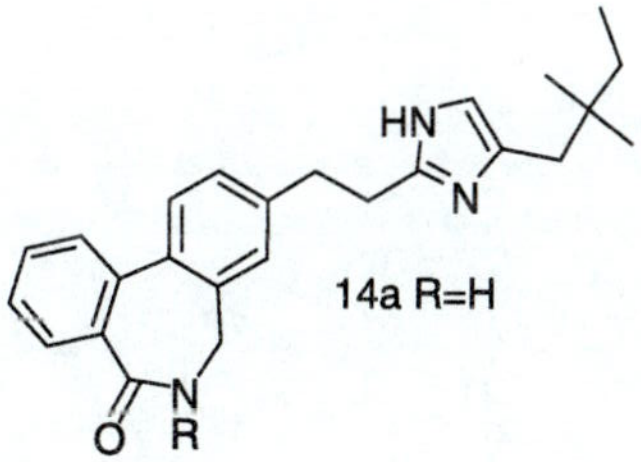 14a R=H		10	(Hadden et al. 2010)

(continued)

Compound 14b	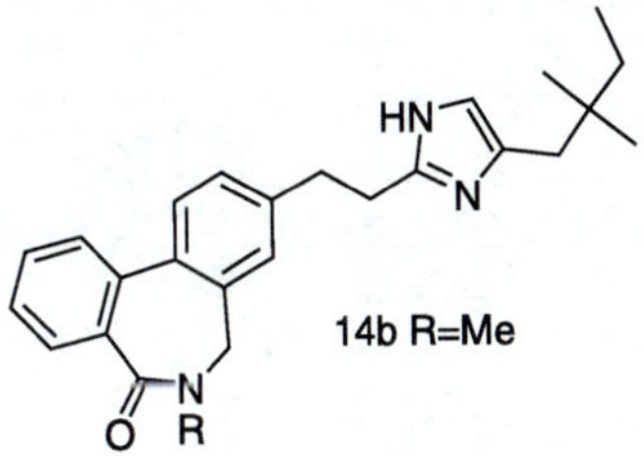		20	(Hadden et al. 2010)
Compound 17	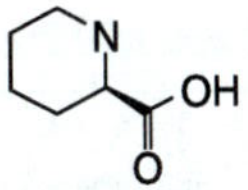		5.3	(Hadden et al. 2010)
Compound 19a	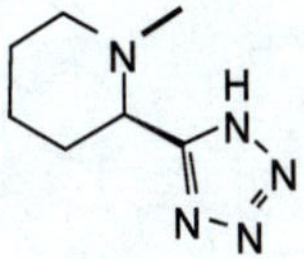		6.2	(Hadden et al. 2010)
Compound 19b	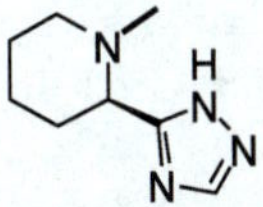		26	(Hadden et al. 2010)
Compound 24	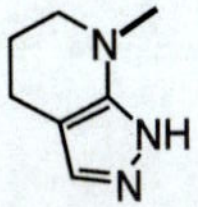		18	(Hadden et al. 2010)

Compound [biphenyl imidazole pharmacophore]	Ar	R	K_i (nM) hBRS-3 binding	Cell system and reference
Compound (RY-337) 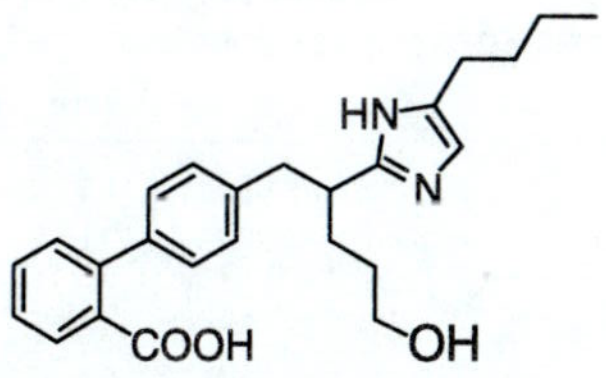 RY-337			3,676	NFAT-CHO cells (He et al. 2010)
Compound 22e 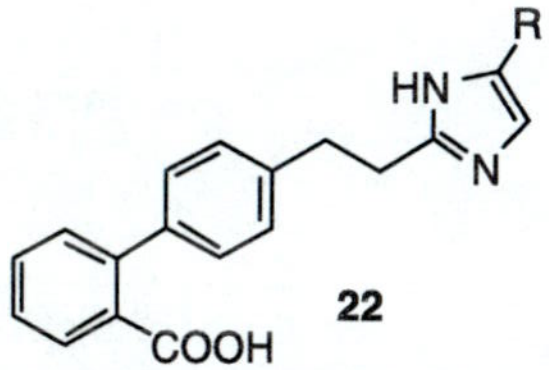 22		22e: 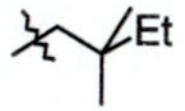	ND	NFAT-CHO cells (He et al. 2010)
Compound pyridinesulfonylureas and pyridinesulfonamides 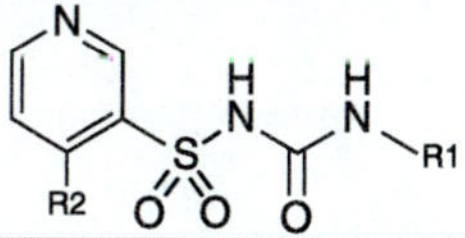	R_1	R_2	Ki (nM) hBRS-3 binding	Cell system and reference

(continued)

Compound 1 (lead compound) (pyridinesulfonylureas)	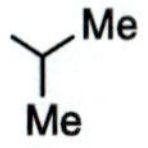	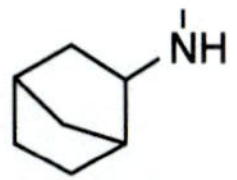	600	hBRS-3/HEK293AEQ cells (Lo et al. 2011)
Compound exo-2a	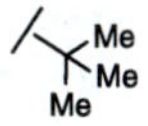 2a	(±) 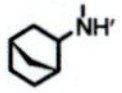	100 ± 20	hBRS-3/HEK293AEQ cells (Lo et al. 2011)
Compound (1R, 2R, 4S)-2a	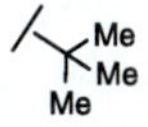 2a	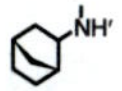	68 ± 7	hBRS-3/HEK293AEQ cells (Lo et al. 2011)
Compound 2b	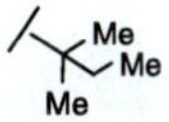 2b	(±) 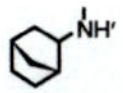	75 ± 5	hBRS-3/HEK293AEQ cells (Lo et al. 2011)
Compound 5e (pyridinesulfonamides) (+ or −)-5e 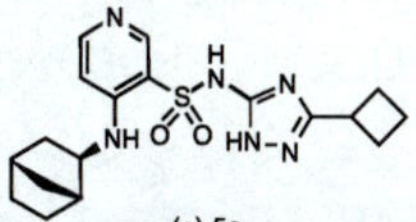 (+)-5e			68 ± 7	hBRS-3/HEK293AEQ cells (Lo et al. 2011)
MK-5046 (2S)-1,1,1-trifluoro-2-[4-(1H-pyrazol-1-yl)phenyl]-3-(4-{[1-(trifluoromethyl)cyclopropyl]methyl}-1H-imidazol-2-yl)propan-2-ol			25 nM	CHO or HEK293/hBRS-3 (Guan et al. 2011)

ND not determined

Table 5 Summary of second messenger data of various non-peptide agonists on the human BRS-3 (part II)

Compound [biarylethylimidazole pharmacophore] 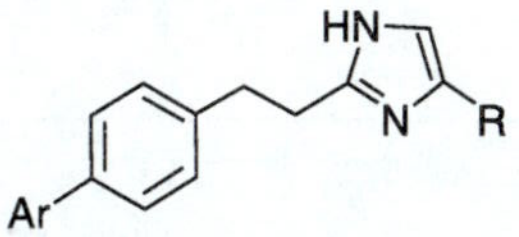	Ar	R	Calcium mobilization functional assay for hBRS-3 [EC_{50} (nM)]	Reference
Compound 9 Bag-1	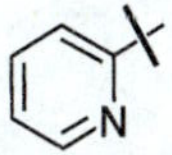		47.0 ± 4.4	(Guan et al. 2010)
Compound 10a	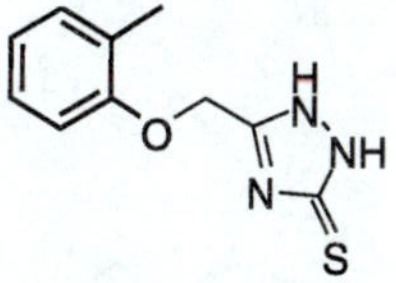		41	(Hadden et al. 2010)
Compound 11	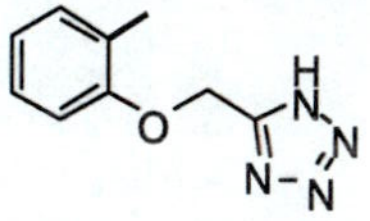		30	(Hadden et al. 2010)
Compound 14a	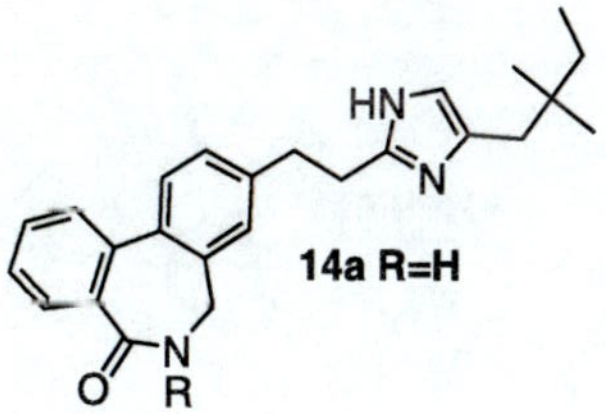 14a R=H		29	(Hadden et al. 2010)

(continued)

Compound 14b	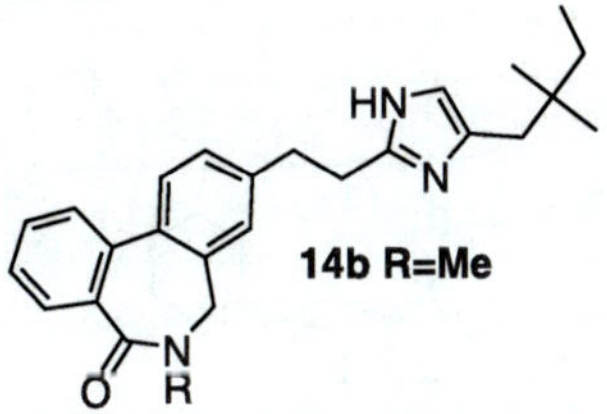		56	(Hadden et al. 2010)
Compound 17	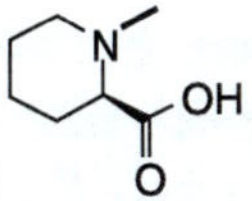		24	(Hadden et al. 2010)
Compound 19a	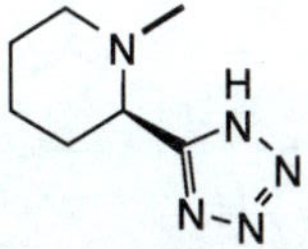		15	(Hadden et al. 2010)
Compound 19b	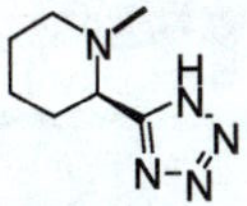		42	(Hadden et al. 2010)
Compound 24	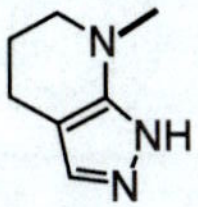		42	(Hadden et al. 2010)

Compound [biphenyl imidazole pharmacophore]			
Compound (RY-337) 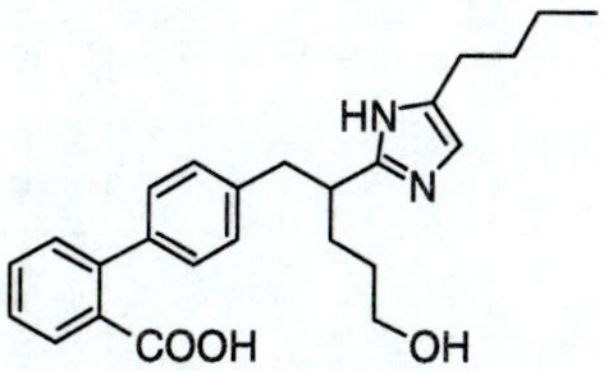 RY-337	–	1897	NFAT-CHO cells (He et al. 2010)
Compound 22e 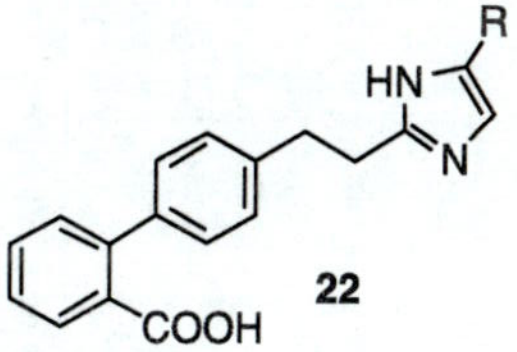 22	22e: 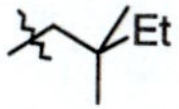	25	NFAT-CHO cells (He et al. 2010)

Table 6 Summary of second messenger data of various non-peptide agonists on the human BRS-3 (part III)

Compounds	R (R_1R_2N, R_3, R_4)	Calcium mobilization functional assay for hBRS-3 [EC_{50} (nM)]	References
Compound 2	R^1R^2N 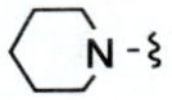	4,525	Guo et al. (2010)
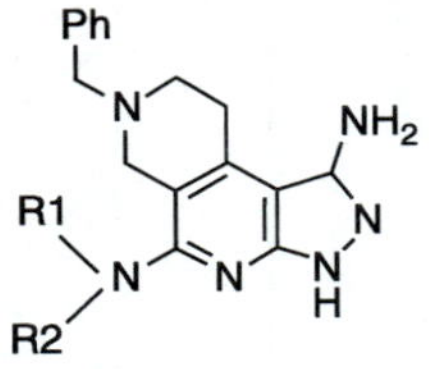 2			
Compound 10a	R3: 2-Pyridyl	6,779	Guo et al. (2010)
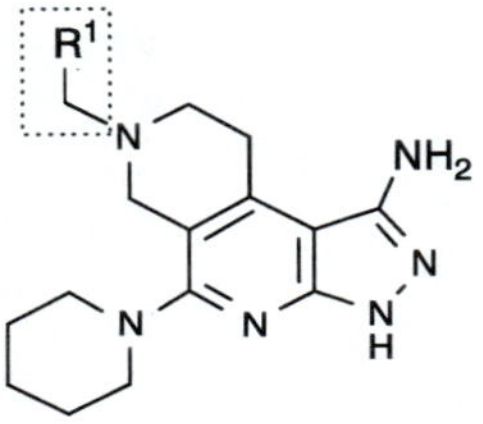 10			
Compound 10b	R3: 2-Cl-Ph	5,567	Guo et al. (2010)
Compound 11a	R4: CH_3–	3,164	Guo et al. (2010)
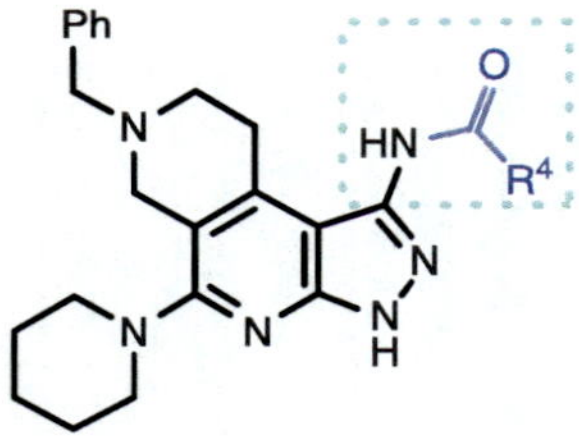 11			
Compound 11b	R4: CH_3CH_2–	3,474	Guo et al. (2010)
Compound 11c	R4: $(CH_3)_2CH$–	4,763	Guo et al. (2010)
Compound 11d	R4: $(CH_3)_3C$–	1,542	Guo et al. (2010)
Compound 11e	R4: $(CH_3)_3CCH_2$–	1,234	Guo et al. (2010)
Compound 11f	R4: Cyclopentyl	704	Guo et al. (2010)
Compound 11g	R4: $CH_3CH_2C(CH_3)_2$–	543	Guo et al. (2010)

Compounds 11d–g seemed to be most potent for hBRS-3. Otherwise all the other compounds like 7-benzyl-5-(piperidin-1-yl)-6,7,8,9-tetrahydro-3H-pyrazole[3,4-c]-(Ischia et al. 2009; Sun et al. 2009)naphthyridin-1-ylamine and its analogs are potent agonist for rBRS-3

human BRS-3 agonist compound 2a was shown to have good bioavailability and favorable pharmacokinetic parameters in rats. Replacement of the hydrogen donor/ acceptors in the sulfonylureas led to a series of sulfonamides that are effective as mouse BRS-3 agonist, and one of its enantiomers, 5e, displayed moderate human BRS-3 activity (Lo et al. 2011).

Moreover, the synthetic small molecule BRS-3 receptor agonist compound 2, 7-benzyl-5-(piperidin-1-yl)-6,7,8,9-tetrahydro-3H-pyrazole[3,4-c]-(Ischia et al. 2009; Sun et al. 2009)naphthyridin-1-ylamin, showed low bioavailability and a short half-life but achieved favorable drug levels in brain of rats. Modification of residues at positions 1, 3, 5, and 7 of compound 2 showed a tight SAR with only limited improvement in potency against BRS-3 receptor (Guo et al. 2010).

Besides the synthetic peptides and small molecule compounds, isolation of natural ligands received attention. Two naturally occurring low-affinity ligands, VV-hemorphin-7(VV-H-7) and LVV-hemorphin-7 (LVV-H-7) extracted from human placenta, were characterized as full agonists for the human BRS-3. Detailed analysis showed that these ligands induced a dose-dependent response in CHO cells overexpressing hBRS-3 and in NCI-N417 cells constitutively expressing the native hBRS-3. Functional studies of these ligands showed that VV-H-7 activates phospholipase C, resulting in release of calcium from an inositol triphosphate (IP_3)-sensitive store, followed by a capacitive entry of extracellular calcium. It was determined that the peptides VV-H-7 and LVV-H-7 increased intracellular calcium in a concentration-dependent manner in the CHO-$G_{\alpha 16}$–hBRS-3 cells with EC_{50} values of 45 ± 15 μM for VV-H-7 and 183 ± 60 μM for LVV-H-7. Moreover, both peptides increased intracellular Ca^{2+} dose dependently also in NCI-N417 cells with EC_{50} values of 19 ± 6 μM for VV-H-7 and 38 ± 18 μM for LVV-H-7. In addition, VV-H-7-induced hBRS-3 activation led to phosphorylation of p42/p44-MAP kinase, without stimulating cell proliferation or adhesion. Combination experiments of VV-H-7/LVV-H-7 with synthetic ligands also proved that the hemorphins are full agonists for hBRS-3 (Lammerich et al. 2003) (Table 2).

Although a large number of molecules activating human BRS-3 have been identified, including synthetic peptides (Mantey et al. 1997; Boyle et al. 2005; Weber et al. 2003; Zhang et al. 2009), natural peptides (Lammerich et al. 2003), and small molecules (Carlton et al. 2008), lack of selectivity and/or potency (Mantey et al. 2006), significant species preferences (Liu et al. 2002), or suboptimal pharmacokinetic properties of these compounds led to further searches for improved agents.

Recently, new potent selective agonist (Bag-1 and Bag-2) and antagonist (Bantag-1) ligands were developed to address the aforementioned issues. Bantag-1 was identified as a potent BRS-3 antagonist with a binding IC_{50} for human and rodent BRS-3 of 2–8 nM, and a ~1,000-fold lower affinity to hNMB-R and hGRP-R. Similarly, the selective BRS-3 agonist Bag-1/compound 9 (2-{2-[4-(pyridin-2-yl)phenyl] ethyl}-5-(2,2-dimethylbutyl)-1H-imidazole), generated by different aryl substitutions of a biarylethylimidazole pharmacophore, had a binding IC_{50} of 2.4–18 nM with corresponding K_i values of 17.4 and 3.7 nM for human and mouse, respectively, and a low affinity to NMB-R (human IC_{50} = 7,000 nM) and

GRP-R (human $IC_{50} = 6{,}400$ nM) (Table 4 and 5). Like other agonist BRS-3 ligands, Bag-1 reduced food intake, body weight, and body fat, and showed a remarkable increase in metabolic rate. A prolonged high level of receptor occupancy was also noticed. A detailed in vivo study with Bag-1(BRS-3 agonist) confirmed the effects of BRS-3 activation on body weight regulation and showed these pathways to be independent of other critical regulators such as NPY, leptin receptor, MCR-4, and CB1R, demonstrating the potential of BRS-3 in the therapy of obesity, either alone or in combination (Guan et al. 2010; Liu et al. 2010). Moreover, synthesis and SAR of heterocyclic carboxylic acids based on 2-biarylethylimidazole showed that replacement of the acidic moiety with a range of isosteres led to good in vitro potency. However, these agents failed to achieve effective drug levels in brain tissue in vivo (Hadden et al. 2010).

In a follow-up experiment, Guan et al. (2011) synthesized the optimized BRS-3 agonist, (2S)-1,1,1-trifluoro-2-[4-(1H-pyrazol-1-yl)phenyl]-3-(4-{[1-(trifluoromethyl) cyclopropyl]methyl}-1H-imidazol-2-yl)propan-2-ol (MK-5046) which reduced food intake and increased fasting metabolism in wild type but not in BRS-3 knockout mice. The compound reduced body weight, but exhibited undesired effects on body temperature, heart rate, and blood pressure which were short-lived upon repeated administration. Its improved properties comprised high and prolonged brain receptor occupancy with lower oral doses. MK-5046 was reported to have a high BRS-3 binding affinity (mouse $K_i = 1.6$ nM; human $K_i = 25$ nM) and exhibited no appreciable binding to NMB-R and GRP-R, and many other ion channels or enzymes (Table 4). Moreover, efficacy of MK-5046 and [D-Phe^6, beta-Ala^{11}, Phe^3, Nle^{14}]Bn-(6–14) in cell-based Ca^{2+} mobilization functional assay was found to be identical. Based on an overall assessment, MK-5046 appears currently the most suitable candidate compound for studies in human trials (Guan et al. 2011).

5 Conclusions

The orphan BRS-3, a G protein-coupled mammalian bombesin receptor, is an important regulator of body weight, energy expenditure, and glucose homeostasis whose mechanisms of action are independent of other relevant molecules including NPY, melanocortin receptor 4, Agouti-related protein, and leptin receptor. BRS-3 has a unique pharmacology compared with other mammalian bombesin receptors, and none of the existing endogenous BLPs are the natural ligand for this receptor. The synthesis of peptides and novel small molecule selective antagonists and agonists on the BRS-3 has now been accomplished. Antagonists increased food intake and body weight, whereas agonists increased metabolic rate and reduced food intake and body weight. Therefore, BRS-3 might be considered as a potential target for the pharmacological treatment of obesity and increased food intake.

References

Boyle RG, Humphries J, Mitchell T et al (2005) The design of a new potent and selective ligand for the orphan bombesin receptor subtype 3 (BRS3). J Pept Sci 11:136–41

Carlton DL, Collin-Smith LJ, Daniels AJ et al (2008) Discovery of small molecule agonists for the bombesin receptor subtype 3 (BRS-3) based on an omeprazole lead. Bioorg Med Chem Lett 18:5451–5

Coll AP (2010) Treating obesity? It's in the bag! Cell Metab 11:95–6

Fathi Z, Corjay MH, Shapira H et al (1993) BRS-3: a novel bombesin receptor subtype selectively expressed in testis and lung carcinoma cells. J Biol Chem 268:5979–84

Fathi Z, Way JW, Corjay MH et al (1996) Bombesin receptor structure and expression in human lung carcinoma cell lines. J Cell Biochem Suppl 24:237–46

Fleischmann A, Laderach U, Friess H et al (2000) Bombesin receptors in distinct tissue compartments of human pancreatic diseases. Lab Invest 80:1807–17

Furutani N, Hondo M, Tsujino N, Sakurai T (2010) Activation of bombesin receptor subtype-3 influences activity of orexin neurons by both direct and indirect pathways. J Mol Neurosci 42:106–11

Gbahou F, Holst B, Schwartz TW (2010) Molecular basis for agonism in the BB3 receptor: an epitope located on the interface of transmembrane-III, -VI, and -VII. J Pharmacol Exp Ther 333:51–9

Gonzalez N, Moody TW, Igarashi H et al (2008a) Bombesin-related peptides and their receptors: recent advances in their role in physiology and disease states. Curr Opin Endocrinol Diabetes Obes 15:58–64

Gonzalez N, Hocart SJ, Portal-Nunez S et al (2008b) Molecular basis for agonist selectivity and activation of the orphan bombesin receptor subtype 3 receptor. J Pharmacol Exp Ther 324:463–74

Gorbulev V, Akhundova A, Buchner H, Fahrenholz F (1992) Molecular cloning of a new bombesin receptor subtype expressed in uterus during pregnancy. Eur J Biochem 208:405–10

Gorbulev V, Akhundova A, Grzeschik KH, Fahrenholz F (1994) Organization and chromosomal localization of the gene for the human bombesin receptor subtype expressed in pregnant uterus. FEBS Lett 340:260–4

Guan XM, Chen H, Dobbelaar PH et al (2010) Regulation of energy homeostasis by bombesin receptor subtype-3: selective receptor agonists for the treatment of obesity. Cell Metab 11:101–12

Guan XM, Metzger JM, Yang L et al (2011) Antiobesity effect of MK-5046, a novel bombesin receptor subtype-3 agonist. J Pharmacol Exp Ther 336:356–64

Guo C, Guzzo PR, Hadden M et al (2010) Synthesis of 7-benzyl-5-(piperidin-1-yl)-6,7,8,9-tetrahydro-3H-pyrazolo[3,4-c][2,7]naphthyridin-1-ylamine and its analogs as bombesin receptor subtype-3 agonists. Bioorg Med Chem Lett 20:2785–9

Hadden M, Goodman A, Guo C et al (2010) Synthesis and SAR of heterocyclic carboxylic acid isosteres based on 2-biarylethylimidazole as bombesin receptor subtype-3 (BRS-3) agonists for the treatment of obesity. Bioorg Med Chem Lett 20:2912–5

He S, Dobbelaar PH, Liu J et al (2010) Discovery of substituted biphenyl imidazoles as potent, bioavailable bombesin receptor subtype-3 agonists. Bioorg Med Chem Lett 20:1913–7

Hotta K, Matsukawa Y, Nishida M et al (2000) Mutation in bombesin receptor subtype-3 gene is not a major cause of obesity in the Japanese. Horm Metab Res 32:33–4

Hou X, Wei L, Harada A, Tatamoto K (2006) Activation of bombesin receptor subtype-3 stimulates adhesion of lung cancer cells. Lung Cancer 54:143–8

Ischia J, Patel O, Shulkes A, Baldwin GS (2009) Gastrin-releasing peptide: different forms, different functions. Biofactors 35:69–75

Jennings CA, Harrison DC, Maycox PR et al (2003) The distribution of the orphan bombesin receptor subtype-3 in the rat CNS. Neuroscience 120:309–24

Jensen RT, Battey JF, Spindel ER, Benya RV (2008) International Union of Pharmacology. LXVIII. Mammalian bombesin receptors: nomenclature, distribution, pharmacology, signaling, and functions in normal and disease states. Pharmacol Rev 60:1–42

Katsuno T, Pradhan TK, Ryan RR et al (1999) Pharmacology and cell biology of the bombesin receptor subtype 4 (BB4-R). Biochemistry 38:7307–20

Kuiper P, Verspaget HW, Biemond I, de Jonge-Muller ES, van Eeden S, van Velthuysen ML, Taal BG, Lamers CB (2010) Expression and ligand binding of bombesin receptors in pulmonary and intestinal carcinoids The role of bombesin in carcinoids. J Endocrinol Invest. 2010 Nov 8. [Epub ahead of print] PubMed PMID: 21060250. PubMed accessed on November 13, 2011

Ladenheim EE, Hamilton NL, Behles RR et al (2008) Factors contributing to obesity in bombesin receptor subtype-3-deficient mice. Endocrinology 149:971–8

Lammerich HP, Busmann A, Kutzleb C et al (2003) Identification and functional characterization of hemorphins VV-H-7 and LVV-H-7 as low-affinity agonists for the orphan bombesin receptor subtype 3. Br J Pharmacol 138:1431–40

Liu J, Lao ZJ, Zhang J et al (2002) Molecular basis of the pharmacological difference between rat and human bombesin receptor subtype-3 (BRS-3). Biochemistry 41:8954–60

Liu J, He S, Jian T et al (2010) Synthesis and SAR of derivatives based on 2-biarylethylimidazole as bombesin receptor subtype-3 (BRS-3) agonists for the treatment of obesity. Bioorg Med Chem Lett 20:2074–7

Lo MM, Chobanian HR, Palyha O et al (2011) Pyridinesulfonylureas and pyridinesulfonamides as selective bombesin receptor subtype-3 (BRS-3) agonists. Bioorg Med Chem Lett 21:2040–3

Maekawa F, Quah HM, Tanaka K, Ohki-Hamazaki H (2004) Leptin resistance and enhancement of feeding facilitation by melanin-concentrating hormone in mice lacking bombesin receptor subtype-3. Diabetes 53:570–6

Majumdar ID, Weber HC (2011) Biology of mammalian bombesin-like peptides and their receptors. Curr Opin Endocrinol Diabetes Obes 18:68–74

Mantey SA, Weber HC, Sainz E et al (1997) Discovery of a high affinity radioligand for the human orphan receptor, bombesin receptor subtype 3, which demonstrates that it has a unique pharmacology compared with other mammalian bombesin receptors. J Biol Chem 272:26062–71

Mantey SA, Coy DH, Pradhan TK et al (2001) Rational design of a peptide agonist that interacts selectively with the orphan receptor, bombesin receptor subtype 3. J Biol Chem 276:9219–29

Mantey SA, Coy DH, Entsuah LK, Jensen RT (2004) Development of bombesin analogs with conformationally restricted amino acid substitutions with enhanced selectivity for the orphan receptor human bombesin receptor subtype 3. J Pharmacol Exp Ther 310:1161–70

Mantey SA, Gonzalez N, Schumann M et al (2006) Identification of bombesin receptor subtype-specific ligands: effect of *N*-methyl scanning, truncation, substitution, and evaluation of putative reported selective ligands. J Pharmacol Exp Ther 319:980–9

Matsumoto K, Iijima H (2003) Sibutramine sensitivity assay revealed a unique phenotype of bombesin BB3 receptor-deficient mice. Eur J Pharmacol 473:41–6

Matsumoto K, Yamada K, Wada E et al (2003) Bombesin receptor subtype-3 modulates plasma insulin concentration. Peptides 24:83–90

Metzger JM, Gagen K, Raustad KA et al (2010) Body temperature as a mouse pharmacodynamic response to bombesin receptor subtype-3 agonists and other potential obesity treatments. Am J Physiol Endocrinol Metab 299:E816–24

Moody TW, Sancho V, di Florio A et al (2011) Bombesin receptor subtype-3 agonists stimulate the growth of lung cancer cells and increase EGF receptor tyrosine phosphorylation. Peptides 32:1677–1684

Nakamichi Y, Wada E, Aoki K et al (2004) Functions of pancreatic beta cells and adipocytes in bombesin receptor subtype-3-deficient mice. Biochem Biophys Res Commun 318:698–703

Ohki-Hamazaki H, Wada E, Matsui K, Wada K (1997a) Cloning and expression of the neuromedin B receptor and the third subtype of bombesin receptor genes in the mouse. Brain Res 762:165–72

Ohki-Hamazaki H, Watase K, Yamamoto K et al (1997b) Mice lacking bombesin receptor subtype-3 develop metabolic defects and obesity. Nature 390:165–9
Orbuch M, Taylor JE, Coy DH et al (1993) Discovery of a novel class of neuromedin B receptor antagonists, substituted somatostatin analogues. Mol Pharmacol 44:841–50
Patel O, Shulkes A, Baldwin GS (2006) Gastrin-releasing peptide and cancer. Biochim Biophys Acta 1766:23–41
Porcher C, Juhem A, Peinnequin A, Bonaz B (2005) Bombesin receptor subtype-3 is expressed by the enteric nervous system and by interstitial cells of Cajal in the rat gastrointestinal tract. Cell Tissue Res 320:21–31
Pradhan TK, Katsuno T, Taylor JE et al (1998) Identification of a unique ligand which has high affinity for all four bombesin receptor subtypes. Eur J Pharmacol 343:275–87
Reubi JC, Wenger S, Schmuckli-Maurer J et al (2002) Bombesin receptor subtypes in human cancers: detection with the universal radioligand (125)I-[D-TYR(6), beta-ALA(11), PHE(13), NLE(14)] bombesin(6–14). Clin Cancer Res 8:1139–46
Ryan RR, Weber HC, Hou W et al (1998a) Ability of various bombesin receptor agonists and antagonists to alter intracellular signaling of the human orphan receptor BRS-3. J Biol Chem 273:13613–24
Ryan RR, Weber HC, Mantey SA et al (1998b) Pharmacology and intracellular signaling mechanisms of the native human orphan receptor BRS-3 in lung cancer cells. J Pharmacol Exp Ther 287:366–80
Sakamoto H, Matsuda K, Zuloaga DG, et al (2008) Sexually dimorphic gastrin releasing peptide system in the spinal cord controls male reproductive functions. Nat Neurosci 11:634–636
Sakamoto H, Kawata M (2009) Gastrin-releasing peptide system in the spinal cord controls male sexual behaviour. J Neuroendocrinol 21:432–435
Sakamoto H, Takanami K, Zuloaga DG, et al (2009) Androgen regulates the sexually dimorphic gastrin-releasing peptide system in the lumbar spinal cord that mediates male sexual function. Endocrinology 150:3672–3679
Sancho V, Moody TW, Mantey SA et al (2010) Pharmacology of putative selective hBRS-3 receptor agonists for human bombesin receptors (BnR): affinities, potencies and selectivity in multiple native and BnR transfected cells. Peptides 31:1569–78
Sano H, Feighner SD, Hreniuk DL et al (2004) Characterization of the bombesin-like peptide receptor family in primates. Genomics 84:139–46
Schulz S, Rocken C (2006) Immunohistochemical detection of bombesin receptor subtypes GRP-R and BRS-3 in human tumors using novel antipeptide antibodies. Virchows Arch 449:421–7
Shan L, Emanuel RL, Dewald D et al (2004) Bombesin-like peptide receptor gene expression, regulation, and function in fetal murine lung. Am J Physiol Lung Cell Mol Physiol 286: L165–73
Shumyatsky GP, Tsvetkov E, Malleret G et al (2002) Identification of a signaling network in lateral nucleus of amygdala important for inhibiting memory specifically related to learned fear. Cell 111:905–18
Sun YG, Chen ZF (2007) A gastrin-releasing peptide receptor mediates the itch sensation in the spinal cord. Nature 448:700–3
Sun YG, Zhao ZQ, Meng XL et al (2009) Cellular basis of itch sensation. Science 325:1531–4
Tan YR, Qi MM, Qin XQ et al (2006) Wound repair and proliferation of bronchial epithelial cells enhanced by bombesin receptor subtype 3 activation. Peptides 27:1852–8
Tan YR, Qin XQ, Xiang Y et al (2007) PPARalpha and AP-2alpha regulate bombesin receptor subtype 3 expression in ozone-stressed bronchial epithelial cells. Biochem J 405:131–7
Wang Y, Zhang M, Tan Y et al (2007) BRS-3 activation transforms the effect of human bronchial epithelial cells from PGE2 mediated inhibition to TGF-beta1 dependent promotion on proliferation and collagen synthesis of lung fibroblasts. Cell Biol Int 31:1495–500
Weber HC, Hampton LL, Jensen RT, Battey JF (1998) Structure and chromosomal localization of the mouse bombesin receptor subtype 3 gene. Gene 211:125–31

Weber HC, Walters J, Leyton J et al (2001) A bombesin receptor subtype-3 peptide increases nuclear oncogene expression in a MEK-1 dependent manner in human lung cancer cells. Eur J Pharmacol 412:13–20

Weber D, Berger C, Heinrich T et al (2002) Systematic optimization of a lead-structure identities for a selective short peptide agonist for the human orphan receptor BRS-3. J Pept Sci 8:461–75

Weber D, Berger C, Eickelmann P et al (2003) Design of selective peptidomimetic agonists for the human orphan receptor BRS-3. J Med Chem 46:1918–30

Whitley JC, Moore C, Giraud AS, Shulkes A (1999) Molecular cloning, genomic organization and selective expression of bombesin receptor subtype 3 in the sheep hypothalamus and pituitary. J Mol Endocrinol 23:107–16

Yamada K, Wada E, Imaki J et al (1999) Hyperresponsiveness to palatable and aversive taste stimuli in genetically obese (bombesin receptor subtype-3-deficient) mice. Physiol Behav 66:863–7

Yamada K, Ohki-Hamazaki H, Wada K (2000) Differential effects of social isolation upon body weight, food consumption, and responsiveness to novel and social environment in bombesin receptor subtype-3 (BRS-3) deficient mice. Physiol Behav 68:555–61

Yamada K, Santo-Yamada Y, Wada E, Wada K (2002) Role of bombesin (BN)-like peptides/ receptors in emotional behavior by comparison of three strains of BN-like peptide receptor knockout mice. Mol Psychiatry 7:113–7

Zhang L, Nothacker HP, Wang Z et al (2009) Pharmacological characterization of a selective agonist for bombesin receptor subtype-3. Biochem Biophys Res Commun 387:283–8

Weight-Reducing Side Effects of the Antiepileptic Agents Topiramate and Zonisamide

J. Antel and J. Hebebrand

Contents

Abstract Drug-induced weight alteration can be a serious side effect that applies to several therapeutic agents and must be referred to in the respective approved labeling texts. The side effect may become health threatening in case of significant weight change in either direction. Several antiepileptic drugs (AEDs) are associated with weight gain such as gabapentin, pregabalin, valproic acid, and vigabatrin and to some extent carbamazepine. Others are weight neutral such as lamotrigine, levetiracetam, and phenytoin or associated with slight weight loss as, e.g., felbamate. The focus of this chapter is on the two AEDs causing strong weight loss: topiramate and zonisamide. For both drugs, several molecular mechanisms of actions are published. We provide a review of these potential mechanisms, some of

J. Antel (✉)
Invited Lecturer at the Pharmaceutical Institute, University of Bonn, Lauenauerstrasse 63, 31848 Bad Münder, Germany
e-mail: jochen.antel@t-online.de

J. Hebebrand
Department of Child and Adolescent Psychiatry, LVR-Klinikum Essen, Kliniken/Institut der Universität Duisburg-Essen, Virchowstrasse 174, 45147 Essen, Germany

H.-G. Joost (ed.), *Appetite Control*, Handbook of Experimental Pharmacology 209,
DOI 10.1007/978-3-642-24716-3_20,

which are based on in vivo studies in animal models for obesity, and of clinical studies exploring these two drugs as single entities or in combinations with other agents.

Keywords AED • AMPA • Animal models • Anticonvulsant • Antiepileptic • Binge eating • Bupropion • Carbonic anhydrase • Clinical studies • Clozapine • Drug combinations • Drug induced • Empatic® • GABA • Ion channels • Mechanism of action • Molecular mechanism • Olanzapine • Phentermine • Potassium channel • QNEXA® • Side effect • Sodium channel • Topamax® • Topiramate • Weight gain • Weight loss • Weight modification • Weight reducing • Zonegran® • Zonisamide

1 Introduction

Obesity (BMI $\geq$ 30 kg/m^2) has become a major health problem in many industrial and developing countries. Prevalence rates in Europe in adult males and females ranged from 4.0 to 28.3% and from 6.2 to 36.5%, respectively, according to a recent review (Berghöfer et al. 2008). In the USA, roughly one third of the adult population is obese (Flegal et al. 2010). Obesity is a risk factor for several disorders including type 2 diabetes mellitus, hypertension, abnormal lipid profile, stroke, heart disease, diverse cancers, and arthritis. Higher grades of obesity are associated with excess mortality.

Genetic factors are presumed to account for 50% of the variance of the BMI in the general population; nonshared environment largely accounts for the other half of the variance. Thirty-two loci explaining 1–2% of the BMI variance have so far been detected in meta-analyses of several genome-wide association studies (Hebebrand et al. 2010). Weight gain as a side effect of drugs and in particular specific antipsychotics and anticonvulsants has been postulated to contribute to the obesity epidemic (McAllister et al. 2009). Single case studies of twin and sib-pairs have revealed that genetic factors contribute substantially to weight gain upon use of clozapine, olanzapine, and valproate (Theisen et al. 2001; Klein et al. 2005; Gebhardt et al. 2010).

Treatment of obesity remains challenging since attrition is typically high in weight-reduction programs. Whereas short-term weight loss in the range of several kilograms can be achieved by motivated patients, rates for long-term success are low. Currently, only orlistat is approved by the FDA and the EMA for long-term use. Weight loss after 1 year averages around 3 kg in excess of placebo (Davidson et al. 1999). While this weight loss is rather low and does not meet patient's expectations (Linne et al. 2002), attempts with more potent drugs like sibutramine (James et al. 2000) or CB_1 antagonists (Antel et al. 2006; Di Marzo et al. 2004) like rimonabant (Rinaldi-Carmona et al. 2004; Van Gaal et al. 2008) finally failed on the market because of safety concerns (Jones 2008; James et al. 2010; Scheen 2010) which led to a withdrawal of the marketing approval by regulatory authorities or a voluntary removal from the market by the respective pharmaceutical companies.

In this chapter, we take an in-depth look at the weight-reducing effects of two anticonvulsants, topiramate and zonisamide. We summarize the current hypotheses on their mechanisms of action and review the data obtained in clinical studies. In the case of zonisamide and the combination zonisamide plus bupropion (Empatic®), these include several randomized controlled phase II trials, and in the case of topiramate (monotherapy as well as in combination with phentermine (QNEXA®)), include even several phase III trials.

1.1 Weight-Modifying Effect of Antiepileptic Agents

The discovery of nearly all antiepileptics was serendipitous, and activity was first shown in one or more animal screening models (Rogawski 2006). Test results in a variety of models subsequently led to hypotheses about potential mechanisms. Hypotheses about target activities were often, years after the initial identification of activity, further refined through additional in vitro, in vivo, and ex vivo studies. In some cases, conventional targets were identified and thought to contribute, whereas in other cases, novel mechanisms were claimed to play a role. So far, there seems to be no single agent acting only via one target, and as time goes by and new biomolecular techniques and tools emerge, more and more subtle details about contributing molecular mechanisms show up. This not only applies to the mechanisms accounting for the antiepileptic properties of AEDs but even more to their weight-modifying potential.

The exact molecular mechanisms by which zonisamide and topiramate mediate weight changes and improve glucose and lipid abnormalities are largely unknown (Biton 2007; Kennett and Clifton 2010). Zonisamide and topiramate are thought to exert anticonvulsant actions via effects on sodium and calcium (T-type) channels (Oommen and Mathews 1999; Leppik 2004). However, this is, e.g., also the hypothesized mode of action of phenytoin, which is weight neutral (Ben-Menachem 2007), or carbamazepine which is even associated with moderate weight gain (Gaspari and Guerreiro 2010; Post et al. 2007).

Other mechanisms thought to contribute to the antiepileptic effects of topiramate (Angehagen et al. 2003a, b; Chengappa et al. 2001; Shank et al. 2000; Sills et al. 2000; Maryanoff 2009) include enhancement of GABA-mediated Cl^- fluxes into neurons and inhibition of kainate-mediated conductance at glutamate receptors of the AMPA/kainate type. Zonisamide interacts also with GABAergic and AMPAergic pathways. It binds allosterically to GABA receptors like the benzodiazepines (Mimaki et al. 1990a, b) and inhibits the uptake of the inhibitory neurotransmitter GABA while enhancing the uptake of the excitatory neurotransmitter glutamate (Ueda et al. 2003). However, several other AEDs like vigabatrin and gabapentin, which also act via GABAergic pathways (Kelly 1998; Rogawski and Löscher 2004; Rogawski 2006), did not cause weight loss but rather weight gain in clinical trials or postmarketing observations.

Furthermore, doses of zonisamide and topiramate that are anticonvulsant in rats increase extraneuronal levels of dopamine, noradrenaline, and 5-HT in the hippocampus (Yamamura et al. 2009). If similar effects occur in the hypothalamus, they might account for the hypophagic actions of topiramate (Richard et al. 2000).

Last but not least, there is proven evidence that topiramate as well as zonisamide inhibit several carbonic anhydrase (CA) isoenzymes (Dodgson et al. 2000; Masereel et al. 2002; Casini et al. 2003; Supuran 2008; Supuran et al. 2008), which might be a contributing factor for their weight-loss effects, since CA inhibition is clearly a common target of these two drugs which is not shared by the other weight-neutral or weight-gain-inducing AEDs.

2 Topiramate

Topiramate (Topamax™) (TPM) is a sulfamate-substituted fructose derivative (Maryanoff et al. 1987) that has been marketed in the USA for use as an AED since 1996 (Shank et al. 1994; Perucca 1997; Rosenfeld 1997; Elterman et al. 1999) and was recently approved (2004) for the prophylaxis of migraine headaches (Campistol et al. 2005) in adults at recommended doses up to 100 mg/day.

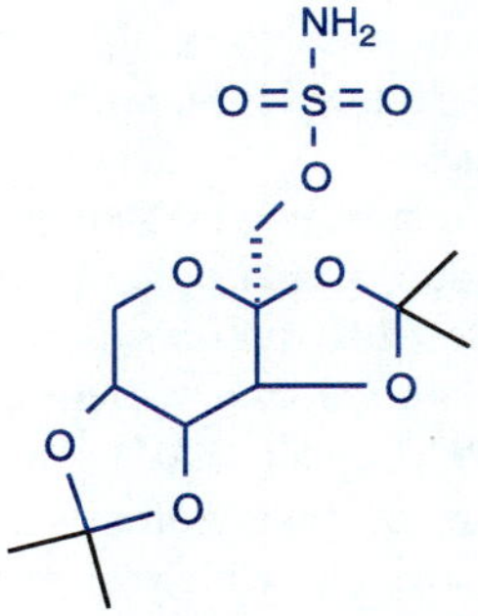

Topiramate was discovered by Maryanoff (2009) while searching for inhibitors of 1,6-fructose-bisphosphatase (FBPase), the rate-limiting enzyme of gluconeogenesis, with the goal to find drugs to reduce postprandial hyperglycemia caused by gluconeogenesis. Glucose lowering by FBPase inhibition is associated with an accumulation of hepatic d-fructose 1,6-bisphosphate and a reduction in hepatic d-fructose 6-phosphate.

Topiramate shows, aside of its labeled use as anticonvulsant and migraine headache prophylaxis, a broad range of other clinically relevant activities including analgesic (Raskin et al. 2004), neuroprotective (Schubert et al. 2005), and mood-stabilizing (McIntyre et al. 2005) properties. However, so far, there is insufficient evidence for recommendations to use topiramate in any phase of bipolar disorders (Vasudev et al. 2006; Kushner et al. 2006). Its antiobesity properties (Tonstad et al.

2005; Astrup and Toubro 2004; Wilding et al. 2004; Bray et al. 2003) will be discussed in more detail below.

2.1 Preclinical Observations

2.1.1 Potential Modes of Action

The mechanism(s) of action of this drug is as yet not fully understood (Chengappa et al. 2001; Shank et al. 2000; Sills et al. 2000). Several hypotheses have been discussed so far (Rosenfeld 1997), such as an enhancement of GABAergic transmission (Reis et al. 2002; Herrero et al. 2002; Kuzniecky et al. 2002), an antagonism of kainate/AMPA receptors (Skradski and White 2000; Gibbs et al. 2000; Zullino et al. 2003), and inhibition of the generation of action potentials in neurons via antagonism of the activation of Na^+ channels (Taverna et al. 1999; McLean et al. 2000).

Like topiramate, the antiepileptic drugs phenytoin, carbamazepine, and lamotrigine inhibit voltage-gated brain sodium channels (Catterall 1999) that are responsible for generation of the action potentials in central neurons. At clinically relevant concentrations, phenytoin as well as topiramate binds preferentially to the inactivated state of the channels and suppresses the sustained repetitive firing of neurons by inhibiting sodium flux through these channels. These effects are thought to be at least in part responsible for their therapeutic efficacy (Francis and Burnham 1992). However, in contrast to topiramate, phenytoin does not induce weight loss; carbamazepine can even entail minor weight gain (Ben-Menachem 2007). Brain sodium channels are target of many newer second-generation (fosphenytoin, oxcarbazepine, lamotrigine, felbamate, topiramate, zonisamide) and third-generation (eslicarbazepine, brivaracetam, carisbamate, fluorofelbamate, elpetrigine, lacosamide, rufinamide, safinamide, vinpocetine) antiepileptic drugs. Some of the newer drugs show either state-dependent antiepileptic action or sodium channel subtype selectivity, although most agents do not differentiate between these channel subtypes (Vohora et al. 2010).

Another mechanism that topiramate shares with other AEDs including zonisamide is the interaction with several subtypes of Ca^{2+} channels. Topiramate is thought to exert its anticonvulsant actions via effects on calcium (T-type) channels (Oommen and Mathews 1999; Leppik 2004), which is also the hypothesized mode of action for the weight-neutral phenytoin, (Ben-Menachem 2007) or for carbamazepine (Gaspari and Guerreiro 2010; Post et al. 2007). In addition, a modulation of neuronal L-type Ca^{2+}channel activity is thought to play an important role for topiramate's antiepileptic activity. Electrophysiological data suggest that topiramate may also interact with L-type Ca^{2+}channels, which was further supported by studying a concomitant administration of low L-type Ca^{2+} channel modulators (nifedipine, verapamil, Bay k 8644) on topiramate's antiepileptic properties (Russo et al. 2004). Furthermore, Kuzmiski and colleagues

(2005) examined whole-cell patch-clamp recordings from CA1 pyramidal neurons in rat hippocampal slices in order to investigate the actions of topiramate on the generation of cholinergic-dependent plateau potentials after cholinergic receptor stimulation. They reported that topiramate reduces ictal-like activity in CA1 hippocampal neurons through a novel inhibitory action of R-type calcium channels at therapeutically relevant concentrations of topiramate (50 μM). They also examined the effects of topiramate on recombinant Ca(V)2.3 calcium channel subunits. Complementary studies were performed by Weiergräber, concluding that topiramate, but also lamotrigine, targets the voltage-gated calcium channel (VGCC) Ca(v)2.3 (E/R-type) (Weiergräber et al. 2006).

A possible functional link between the interaction of topiramate with several Ca^{2+} channels and inhibition of several CA isoenzymes was discussed by McNaughton (McNaughton et al. 2004). They explored whether a range of carbonic anhydrase inhibitors (CAIs) interacts with high-voltage-activated voltage-sensitive calcium channels (VSCCs) encoded by the human alpha(1E) subunit. Interestingly, the anticonvulsant CAI ethoxyzolamide blocked alpha(1E)-mediated VSCC currents, with an IC(50) close to 1 μM, acetazolamide and benzolamide produced approximately 30–40% inhibition of alpha(1E)beta(3)-mediated Ca^{2+} currents at 10 μM, and topiramate inhibited these currents by 68 $\pm$ 7%. It remains however open whether a direct effect of CA inhibition contributes mainly to the antiepileptic potency of the aforementioned CAIs or whether this off-target activity of CAIs at VSCCs may be the main contributor. The other open question is how much every single of the multiple modes of action of topiramate and other AEDs contribute to an antiepileptic profile (Johannessen Landmark 2008) on the one hand and a weight-modifying profile on the other hand. Any conclusion is further complicated by observations about dose-dependent effects of AEDs on hippocampal extracellular levels of glutamate (Glu), GABA, norepinephrine (NE), dopamine (DA), and serotonin (5-HT), as determined by microdialysis with high-speed and high-sensitive extreme liquid chromatography (Yamamura et al. 2009). Therapeutically relevant doses of acetazolamide, topiramate, zonisamide, and carbamazepine increased hippocampal extracellular levels of GABA, NE, DA, and 5-HT. While phenytoin had no effect, supratherapeutic doses of acetazolamide, topiramate, zonisamide, phenytoin, and carbamazepine decreased extracellular levels of GABA, NE, DA, and 5-HT, without affecting Glu levels. Toxic doses of carbamazepine and phenytoin paradoxically produced seizures (determined by telemetric-electrocorticography). Topiramate and zonisamide did not produce seizures, even at toxic doses. In contrast, the coadministration of therapeutically relevant doses of acetazolamide, topiramate, and zonisamide even prevented the paradoxical intoxication by elevating the seizure-threshold doses of carbamazepine and phenytoin. The authors hypothesized that these results suggested that inhibition of carbonic anhydrases inhibits blockade of a high percentage of the population of voltage-dependent sodium channels by carbamazepine and phenytoin, which might be responsible for inducing paradoxical intoxication. However, also these observations do not allow to unambiguously deduce a molecular mechanism of action, since neither the relevant carbonic anhydrase isoenzyme pattern nor the sodium channel subtype can be finally described on a molecular level.

It was known from the very beginning that topiramate exerts an inhibitory effect on carbonic anhydrase (CA) isoenzymes, particularly CA-II and CA-IV (Maryanoff et al. 1987; Dodgson et al. 2000). This was later confirmed by X-ray crystallography (Casini et al. 2003, Antel et al. 2004) and by in vitro data obtained with other CA isoenzymes. The reported data on the CA inhibition are rather controversial. Initially, topiramate was classified as a very weak (millimolar) CA inhibitor (Maryanoff et al. 1987); later, the same group reported different results (Dodgson et al. 2000), showing that topiramate is a much stronger CA inhibitor (now in the micromolar range) in a different assay setup and with differing enzyme sources. These findings are supported by some clinically observed side effects (Kuo et al. 2002; Ribacoba Montero and Salas Puig 2002; Fakhoury et al. 2002), which are typical for strong sulfonamide CA inhibitors used as systemic antiglaucoma agents (such as acetazolamide, methazolamide) and include paresthesias, nephrolithiasis, and weight loss (Supuran 2008).

However, while a strong inhibition of CA isozymes by topiramate has unequivocally been proven, its anticonvulsant activity appears to differ mechanistically from that of acetazolamide, as demonstrated by some recently published in vivo investigations (Stringer 2000; Aribi and Stringer 2002). Previous observations had also shown that sulfamate derivatives of both topiramate (Maryanoff et al. 1987) and RWJ-37497 (Maryanoff et al. 1998) with a secondary sulfamate moiety show potent anticonvulsant activity but no inhibition of CA isozymes.

Therefore, the contribution of the inhibitory effect of topiramate on carbonic anhydrases (CA) to its anticonvulsive action is still discussed controversially (Shank et al. 2000; Stringer 2000; Perucca 1997; Dodgson et al. 2000; Masereel et al. 2002; Casini et al. 2003; Supuran 2008; Supuran et al. 2008). Herrero and colleagues (Herrero et al. 2002) hypothesized that topiramate-induced change of GABAergic depolarizations might be based on a decreased intracellular bicarbonate concentration, which could contribute to the anticonvulsant activity of topiramate via an inhibition of neuronal CA inhibition of carbonic anhydrase. Other CA inhibitors such as sulthiame and acetazolamide (AZM) have previously been shown to lower neuronal intracellular pH (pHi), which effectively reduced epileptiform activity in epilepsy model systems in vitro (Leniger et al. 2002). Leniger showed that topiramate lowers neuronal intracellular pH (pHi) most likely due to a combined effect on intracellular carbonic anhydrase isoenzymes and Na (+)-independent Cl^-/HCO_3^- exchange (Leniger et al. 2004). The apparent decrease of steady-state pHi may contribute to the anticonvulsive property of topiramate. It is hypothesized by some researchers that the anticonvulsant effects of topiramate or related sulfonamides may be due to CO_2 retention, secondary to inhibition of the red blood cell and brain isoenzymes (Masereel et al. 2002). By that means, inhibition of the brain CAs is thought to cause an increase of the cerebral blood flow and an increase of CO_2 partial pressure. Consequently, CAIs are used to treat conditions with increased intracranial pressure (Dhellemmes et al. 2008). Topiramate as well as zonisamide (ZNS) (Costa et al. 2010), which is also an inhibitor of carbonic anhydrase (see below), may influence neuronal activity via changes of pH (Biton 2007). However, this mechanism is controversially discussed with regard to its contribution to the antiepileptic activity of these drugs (Thone et al. 2008).

With regard to other mechanisms of action, initial hypotheses centered on the (Shank et al. 2000) binding of topiramate to phosphorylation sites of ion channels, resulting in inhibition of protein phosphorylation and an allosteric modulation of channel conductance.

As already mentioned, topiramate shows inhibitory effect on the AMPA and kainate subtypes of glutamate receptors. The nature of these effects at the molecular level has not been fully established, but it is hypothesized that topiramate binds to phosphorylation sites on AMPA and kainate receptors (Angehagen et al. 2004). The authors claim that topiramate binds only in the dephosphorylated state and thereby exerts an allosteric modulatory effect on channel conductance. Its interaction with glutamate-regulated pathways is further supported by the observed neuroprotective effects of topiramate in primary neuronal-astroglial cultures or astroglial-enriched cultures from newborn rats exposed to excitotoxic concentrations of glutamate (Glu) or kainate (Angehagen et al. 2003a, b). In these studies, valproate and phenytoin were used as reference AEDs. The same group further investigated the effect on AMPA-induced intracellular calcium ($[Ca^{2+}](i)$) responses in cultured rat cortical astrocytes and reported that topiramate (1–100 μm) inhibited AMPA-induced accumulation of Ca^{2+} in astrocytes, which was inversely related to the level of protein kinase A (PKA)-mediated phosphorylation of GluR1 subunits of channels activated by AMPA.

In order to identify genes that might be involved in the effects of topiramate on body weight and insulin sensitivity, Liang and colleagues (2006) determined the expression profiles in tissues from female Zucker diabetic fatty rats that were treated with topiramate for 7 days. Plasma glucose and triglycerides were significantly and dose dependently reduced, and messenger RNA profiles in liver, hypothalamus, white adipose tissue, and skeletal muscle were altered in the topiramate group. As expected, most pronounced effects were found for genes encoding metabolic enzymes or regulatory proteins involved in energy metabolism. They observed decreased messenger RNA amounts for sterol regulatory element binding protein-1c, stearoyl-coenzyme A desaturase-1, choline kinase (an enzyme involved in the synthesis of phosphatidylcholine), and fatty acid coenzyme A ligase long chain 4 (FACL4). Topiramate diminished the expression of acetyl-CoA carboxylase 2 (ACC2) and stearoyl-CoA desaturase-1 (SCD1) which as a consequence at least partially mimics the phenotypes of knockout mice of ACC2 and SCD1 by entailing lowered hepatic, plasma triglyceride concentrations, hepatic very-low-density lipoprotein secretion, and altered glycerolipid fatty acid composition. In addition, topiramate also upregulated three genes which are involved in cholesterol synthesis (isopentenyl-diphosphate d isomerase, squalene epoxidase, and 3-hydroxy-3-methylglutaryl-CoA synthase 1) independently of a reduced food intake (verified via a pair-feeding group). With respect to short-term effects (16 h after single administration), the expression of hepatic genes related with fatty acid synthesis, e.g., stearoyl-CoA desaturase and acetyl-CoA carboxylase, was significantly reduced. Topiramate also changed the expression levels of genes encoding enzymes of the fatty acid β-oxidation: the expression of 3-2-trans-enoyl-CoA isomerase and mitochondrial acyl-CoA thioesterase was increased, and of fatty acid CoA ligase (long chain 2 and long chain 5) was decreased. Overall, Liang's results suggest that topiramate regulates hepatic expression of genes involved in lipid metabolism, which in our opinion could potentially contribute to the mechanisms by which topiramate reduces plasma triglyceride levels in obese diabetic rodents.

In addition to its direct impact on gene expression, topiramate could also regulate the metabolism at posttranslation levels. This was hypothesized by Wilkes who reported that topiramate administration led to a three- to fourfold increase in circulating levels of total and high-molecular-weight adiponectin (Acrp30) and a twofold increase in phospho-AMPK in skeletal muscle in topiramate-treated rats (Wilkes et al. 2005b).

It was also hypothesized that topiramate exerts its effect on bipolar disorders by downregulation of enzymes involved in brain arachidonic acid (AA) release and cyclooxygenase (COX)-mediated metabolism, which seems to be a mechanism by which the anticonvulsants valproic acid, carbamazepine, and lithium exert their mood-stabilizing effects. However, topiramate treatment did not significantly modify expression of the enzymes involved in brain AA metabolism, thus suggesting that the AA cascade is not involved in the antiseizure properties of topiramate (Ghelardoni et al. 2005).

2.1.2 In Vivo Results: Topiramate as Single Agent

A review of the literature identified numerous studies of topiramate in animal models of either obesity, type 2 diabetes, or "diabesity." One of the first studies was performed in female Zucker rats (Picard et al. 2000). Obese Zucker rats (fa/fa) develop obesity as an autosomal recessive trait of the leptin receptor which is caused by a Gln269Pro mutation (Takaya et al. 1996), leading to a loss of function of this receptor. The group around Picard studied two cohorts of Zucker rats, namely lean (Fa/?) and obese (fa/fa) rats and applied two doses (15 mg/kg and 60 mg/kg) of topiramate controlled by one vehicle group. The animals had free access to high-carbohydrate diet (65% energy from CHO), and the drug was applied via gavage with one-third of dose in the morning and two-thirds 2 h before dark phase. The total treatment period was 4 weeks. Topiramate reduced body-weight gain in lean but much more so in obese rats. There was a significant dose-dependent reduction of food intake in obese but not in lean rats, but the food reduction effect subsided during the treatment period.

Topiramate dose independently reduced hyperinsulinemia of obese but failed to alter insulin levels of lean animals. Also, topiramate did not alter the digestibility of the food. Post hoc analysis showed that the low dose of topiramate decreased fat gain (with emphasis on subcutaneous fat) without affecting protein gain, whereas the high dose of the drug induced a reduction in both fat and protein gains. In lean but not in obese rats, apparent energy expenditure (as calculated by the difference between energy intake and energy gain) was higher in rats treated with topiramate than in animals administered with the vehicle. The calculated energetic efficiency (energy gain/energy intake) was decreased in both lean and obese rats after topiramate treatment. Picard concluded that topiramate is capable of reducing fat and energy gains through reducing energetic efficiency in both lean and obese Zucker rats. However, the mechanism remains to be elucidated.

York and his team studied Osborne–Mendel rats (York et al. 2000), a strain that readily becomes obese on a high-fat diet (HF) and that develops a high body weight within 10 weeks in males. In contrast to Picard, York did not use a high-CHO diet but rather allowed all animals free access to a high-fat diet (56% energy from fat, 20% from protein). Five groups of male Osborne–Mendel rats, namely a control (vehicle), two dose groups (10 mg/kg and 40 mg/kg of topiramate), a pair-fed group to the high dose, and a d-fenfluramine (1 mg/kg) group were studied in parallel. The drug was mixed into HF diet and supplied shortly before dark phase. The total treatment period was 80 days. Topiramate reduced body-weight gain in a manner similar to that of pair-fed rats. Topiramate initially reduced food intake in the high-dose group, but food consumption returned to control levels after about 4 days of treatment. According to a carcass analysis, the reduction of the body fat fully accounted for the weight reduction in treated groups but not in pair-fed or d-fen-treated rats. Increased energy expenditure (direct measurement) and a significant reduction of serum insulin were observed. The mechanisms underlying these observations are unclear.

A study performed in 2000 by Richard investigated female Sprague–Dawley rats subjected to an obesity-promoting diet (Richard et al. 2000). Topiramate administered by gavage at a daily dose of 30 mg/kg had no effect on the intake of chow or high-sucrose/high-fat diet (in grams per 100 g: sucrose, 45; maize oil, 10; lard, 10; casein, 22.5; dl-methionine, 0.3; vitamin mix, 1.2; mineral mix, 5.5; fiber, 5.5), but reduced 5-week energy gain through a slightly reduced energy efficiency with no significant effect on apparent energy expenditure. Apparent energy expenditure was calculated as the difference between energy intake and energy gain, and energetic efficiency, as the ratio of energy gain to digestible energy intake. It was hypothesized that topiramate increases lipoprotein lipase (LPL) activity in brown adipose tissue and enhances thermogenesis. In addition, topiramate stimulates LPL activity in skeletal muscles, further emphasizing its potential to promote substrate oxidation. The mechanisms whereby topiramate affects the regulation of energy balance have yet to be understood. Richard and coworkers (2000) also tried to deliver evidence that topiramate affects the neuropeptide Y (NPY), corticotrophin-releasing hormone (CRH), or pro-opiomelanocortinergic (POMC) systems, but in situ hybridization trials failed to demonstrate relevant changes in the respective mRNA expression levels.

The same group looked into gender-specific effects and evaluated the effects of topiramate and sex hormones on energy balance of male and female Wistar rats (Richard et al. 2002). For this purpose, intact or castrated rats with replacement therapies (testosterone administration in orchidectomized rats and estradiol or progesterone treatments in ovariectomized rats) were tested head-to-head. In orchidectomized male rats, energy and protein gains were decreased but could be blocked by treatment with testosterone. Female ovariectomized rats showed increases in energy, fat, and protein gains that were prevented by treatment with estradiol. A dose of 60 mg/kg topiramate (mixed into diet) was applied for 28 and 35 days in male and female rats, respectively. Topiramate reduced intake in males; energetic efficiency and fat content (energy deposition) were lowered in both male and female Wistars independently of hormone replacement therapies. In neither gender, the status of sex hormones interfered with the effects of topiramate on energy balance. Topiramate reduced LPL activity in white adipose tissue (WAT) and stimulated LPL activity in brown adipose tissue (BAT) in females. It reduced plasma glucose and plasma leptin levels in female rats as well as plasma insulin and liver triglycerides in male animals, suggesting that topiramate can enhance insulin sensitivity. The group concluded that the effects were due to a decrease in energetic efficiency, resulting from an effect exerted by the drug on both energy intake and thermogenesis.

Topiramate was also studied in female Zucker diabetic fatty (ZDF) rats (Wilkes et al. 2005a) and markedly lowered fasting glucose levels as well as glucose levels during an oral glucose tolerance test. Glucose clamp studies revealed a 30% increase in glucose disposal and a suppression of hepatic glucose output (HGO) from 30 to 60% as well as a suppression of plasma-free fatty acids from 40 to 75%. In summary, topiramate treatment led to a decrease in plasma glucose and an increase of in vivo insulin sensitivity, an insulin sensitization in adipocytes but not muscle (ex vivo; in vitro), and an enhanced insulin action in insulin-resistant adipose cells in vitro. The authors concluded that the effects of topiramate treatment appeared to be mediated by the adipose tissue.

2.1.3 In Vivo Results: Topiramate–Phentermine Combination

Phentermine is a sympathomimetic amine approved by the FDA as a short-term adjunct to a weight-loss regimen based on exercise, behavior modification, and caloric restriction. The anorectic effect is thought to be mediated via release of norepinephrine ($IC_{50} = 39.4$ nM) in the hypothalamus (Rothman et al. 2001). At clinically relevant doses, phentermine also stimulates the release of serotonin ($IC_{50} = 3{,}511$ nM) and dopamine ($IC_{50} = 262$ nM), but to a much lesser extent than that of norepinephrine (Rothman et al. 2001). It is also postulated that increased circulating catecholamines may cause appetite suppression by increasing blood leptin concentrations.

In vivo studies in DIO rats with phentermine and topiramate (Fig. 1) administered either alone or in combination indicate an additive (if not even synergistic) effect of combination therapy (Jackson et al. 2007).

Phentermine seems to have a faster onset of its weight-reducing action compared with topiramate. The progress of weight loss seen with topiramate alone is slower but sustained throughout the dosing period. The pharmacological actions of both drugs are distinct and complementary, which might be beneficial for the management of obesity with respect to weight loss and weight maintenance.

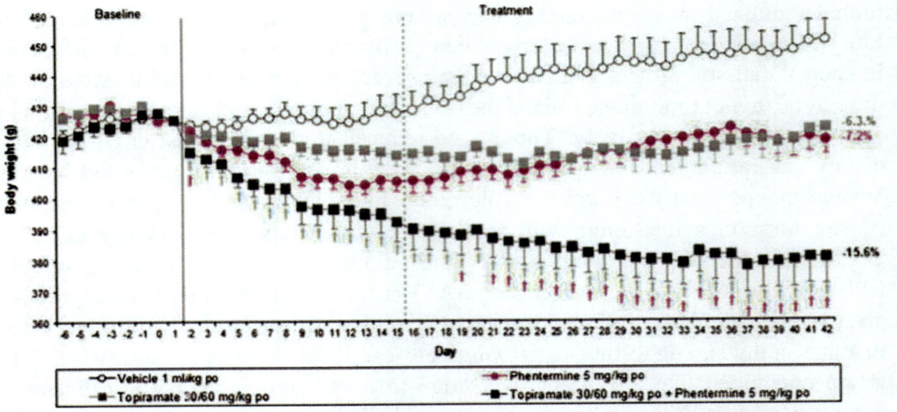

Fig. 1 Effect of chronic administration of phentermine and topiramate, alone and in combination, on body weight in dietary-induced obese, female Wistar rats. Results are means (adjusted for differences between the body weights of the different treatment groups at baseline (day 1)) ± SEM (calculated from the residuals of the statistical model), $n = 10$. The dose of topiramate was doubled on day 15. Numbers represent % reduction in body weight compared to the vehicle-treated control group on day 42 (after 41 days of drug administration). Comparisons between treatments (topiramate plus phentermine vs. topiramate or phentermine alone) and the interaction tests (topiramate plus phentermine vs. effect of topiramate alone plus the effect of phentermine alone) were by separate Sidak tests. Significant differences between treatments are denoted by $^{\dagger}p < 0.05$ (gray vs. topiramate, purple vs. phentermine, p values have only been given at one level for clarity). The interaction was significant on days −6 and −3 ($p < 0.05$, not indicated on figure) but not on any day of drug treatment. The significant values on days −6 and −3 were probably just false positives

2.1.4 In Vivo Results: Topiramate on Top of Olanzapine

Olanzapine (OLZ) is an atypical antipsychotic with clinically well-described side effects of weight gain and diabetes, hampering its clinical acceptance. Recent reports provide evidence that OLZ worsens insulin sensitivity independent of changes in body weight and composition. The underlying pharmacological mechanisms for these side effects of some atypical antipsychotics are unknown, and it is unclear whether the complications are mediated by peripheral or central actions. A complicating factor in studies of the mechanisms of antipsychotic-induced hyperglycemia or insulin resistance are the well-known confounders "weight gain" and "dyslipidemia," which show up in those studies and are known independent risk factors of diabetes. Therefore, Houseknecht and colleagues (2007) searched for acute metabolic effects that started prior to any weight gain. They studied the effects of four atypical antipsychotics on whole-body insulin resistance, using the hyperinsulinemic, euglycemic clamp technique in conscious rats. They discovered that olanzapine (OLAN) and clozapine (CLOZ) acutely impaired whole-body insulin sensitivity in a dose-dependent manner ($p < 0.001$ vs. vehicle), whereas ziprasidone and risperidone had no effect. CLOZ also induced a strong insulin resistance after dosing 10 mg/kg/day for 5 days ($p < 0.05$ vs. vehicle). From tracer studies, it was concluded that the acute changes may be mainly due to an increased hepatic glucose production (HGP), consistent with the lack of effect on glucose uptake. In conclusion, OLAN and CLOZ can rapidly induce insulin resistance independently of weight gain.

In another recent in vivo study (Martins et al. 2010), the effects of intravenous (IV) vs intracerebroventricular (ICV) infusion of OLZ or vehicle were compared in Sprague–Dawley rats. OLZ-iv caused a transient increase in endogenous glucose production rate (EGP) compared to vehicle-iv. Consistent with this effect, higher hepatic mRNA levels of the enzymes glucose-6-phosphatase and phosphoenolpyruvate carboxykinase were measured in the OLZ-iv group. The authors also observed an increased phosphorylation of hypothalamic AMPK in the OLZ-iv group compared to controls. Similarly, the icv infusion of OLZ resulted in a transient increase in glycemia as well as in a higher rate of glucose appearance in the basal period. OLZ-icv caused an increased EGP but no change in the rate of glucose utilization. In line with these findings, levels of hepatic gluconeogenic enzymes were elevated as well as hypothalamic NPY and agouti-related protein (AGRP) mRNA levels. Martins and coworkers concluded that acute CNS exposure to OLZ induces hypothalamic AMPK and hepatic insulin resistance. This would support a hypothalamic site of action for the metabolic dysregulation of atypical antipsychotics.

Antel and coworkers (Theisen et al. 2007) pursued the hypothesis that both olanzapine and clozapine lead to weight gain in predisposed individuals via binding to receptors in the CNS that are involved in body-weight regulation. The investigators compared radioligand binding affinities of clozapine, olanzapine, and haloperidol to both anorexigenic (bombesin receptor subtype 3, calcitonin gene-related peptide receptor, cholecystokinin receptor, melanocortin-4 receptor, neurotensin receptor 1) and orexigenic (cannabinoid receptor 1, galanin 1 receptor, melanin-concentrating hormone receptor, neuropeptide Y1 receptor) receptors. Clozapine, olanzapine, and haloperidol exhibited negligible affinities to all of these receptors except for the melanin-concentrating hormone receptor. With respect to other candidates from (neuro)transmitter systems suggested to be involved in antipsychotic-induced weight gain, the binding profile of olanzapine resembled that of clozapine, with high affinity for serotonin (5-HT) 5-HT2A, 5-HT2C, and 5-HT6; muscarinic M1; and histamine H1 receptors. In contrast, the binding profile of haloperidol was substantially different (high affinity only for the dopamine D1 receptor).

2.2 *Clinical Observations*

2.2.1 Clinical Trials of Topiramate in Obesity and Type 2 Diabetes

Topiramate is readily and completely absorbed from the gastrointestinal tract and negligibly bound to plasma proteins. Its bioavailability is 80%. When used as a monotherapy, topiramate is eliminated primarily in the urine in an unchanged form with a half-life of 20–30 h. Topiramate is usually titrated with an increasing dosing scheme from 2 × 25 mg/day up to about 2 × 200 mg/day or even higher.

One of the first reviews evaluating seven double-blind controlled studies of topiramate with a focus on safety and tolerability appeared in 1998 (Jones 1998). The review revealed no evidence of serious systemic side effects, but renal stones were reported in 1.5% of patients, possibly due to the inhibition of carbonic anhydrase by topiramate. The most common adverse events were somnolence, weight loss, mental slowing, fatigue, ataxia, and irritability. Most of these events were reversible. Jones stated that the unique observation of "abnormal thinking" seemed to be related to high doses and a rapid uptitration. Patients, rather than describing psychomotor slowing, described a phenomenon of slow thoughts, decreased cognition, and intermittent difficulty of, e.g., calculating. Last but not least, weight loss appeared in approximately 10–20% of patients and was hypothesized to be probably related to dulling of appetite. It is worth mentioning that in these studies, epileptic patients, but not obese patients were treated. Despite these side effects, it is remarkable that over 80% of the patients remained on the drug because of an improved quality of life. In general, doses beyond 600 mg/day were often not well tolerated.

In an uncontrolled prospective trial including nonobese and obese epileptic adults (40–110 kg) with partial-onset seizures, topiramate was added to the conventional anticonvulsants (Ben-Menachem et al. 2003). The observed weight reduction was dependent on the baseline BMI and was sustained for at least 1 year. Reduced caloric intake seemed to account for a part of the weight loss at the onset of treatment since caloric intake later returned to baseline levels. The more obese, the greater was the observed weight loss, and the strong correlation with baseline BMI ($p = 0.0007$) was considered a very promising pattern. In obese patients [body mass index (BMI) ≥ 30 kg/m^2] who completed 1 year of topiramate treatment, mean weight loss was 4.2 kg (4.3%) at 3 months and 10.9 kg (11.0%) at 1 year.

These observations of weight loss as secondary or tertiary endpoint in epilepsy or migraine studies led to the design of trials directly assessing the potential of topiramate as an antiobesity agent. Astrup and Toubro (2004) reviewed the first three clinical trials of topiramate for weight reduction in obese subjects, which included the 6-month dose-ranging (Bray et al. 2003), a 2-year, and a 44-week study of subjects who had previously lost weight on a low-calorie diet. All three studies found topiramate to be significantly more efficacious than placebo. Notably, weight loss continued for 1 year and, perhaps, could have continued for a longer period. A meta-analysis of several weight-loss trials including those with topiramate was published by Li and colleagues (2005). They used only data obtained for the higher dose and calculated that the overall pooled random-effects estimate of the percentage of weight loss in topiramate-treated patients compared with placebo recipients was 6.5% (CI, 4.8–8.3%).

All studies reported a statistically significant weight loss, but the individual magnitudes showed a substantial variability. It is important to note that at the time of Li's analysis, most studies were assessed on the basis of data in abstracts.

Adverse events that were reported more frequently in topiramate-treated patients than in placebo recipients included paresthesia and changes in taste. Other CNS and gastrointestinal effects were also reported more frequently in the topiramate groups and showed a dose dependency of the adverse effects.

Topiramate was also investigated in studies including patients with additional comorbidities such as essential hypertension, diabetes, and eating disorders. Tonstad and colleagues investigated topiramate (Tonstad et al. 2005) in obese subjects (body mass index 27–50 kg/m^2) with essential hypertension. They planned to study 531 obese subjects in three arms (placebo, topiramate 96, or 192 mg/day, respectively) in a randomized, placebo-controlled trial for 60 weeks. However, the study was ended earlier by the study sponsor (Johnson & Johnson). As a consequence, they could only asses the efficacy within a modified intent-to-treat population whose participants were on medication for 28 weeks. Weight loss was 1.9%, 5.9%, and 6.5% from baseline in the placebo, 96-mg, and 192-mg groups, respectively ($p < 0.001$ for each comparison with placebo). Diastolic and systolic blood pressure were reduced by 2.1, 5.5, and 6.3 mmHg ($p < 0.015$ vs. placebo) and 4.9, 8.6, and 9.7 mmHg ($p =$ NS) in the placebo, 96-mg, and 192-mg groups, respectively. Adverse events were paresthesia, fatigue, and difficulty with concentration and attention that might have led the manufacturer to stop the trial early and to considerations to develop a controlled-release formulation.

In addition, topiramate was evaluated in obese subjects with type 2 diabetes treated with metformin as standard of care. The study (Toplak et al. 2007) was a multicenter (68), double-blind, placebo-controlled trial defining mean percent change in weight and change in glycosylated hemoglobin (HbA1c) from baseline to end of study as joint primary efficacy parameters. All participants received an individualized diet (600 kcal/day less than the subject's individually calculated energy expenditure) and advice for exercise plus behavioral modification. After a 6-week single-blind placebo run-in, subjects were randomized to placebo or two arms with topiramate (96 mg/day or 192 mg/day, respectively). In order to mitigate CNS side effects, the trial started with an 8-week titration period, followed up by a treatment with the assigned dose for 52 weeks. However, this study was also terminated early by the sponsor Johnson & Johnson in order to develop a new controlled-release formulation with the potential to enhance tolerability via an improved pharmacokinetic. In order to make best use out of the study, a predetermined modified intent-to-treat (MITT) population of 307 subjects whose randomization date would have allowed them to complete 24 weeks of the study prior to the stop was used to assess the efficacy. In total, 646 obese men and women (age: 18–75 years; body mass index, 27–50 kg/m^2) with type 2 diabetes mellitus and metformin as the only background were randomized. After 24 weeks of treatment, subjects in the placebo, topiramate 96 mg/day, and topiramate 192 mg/day groups lost 1.7%, 4.5% ($p < 0.001$), and 6.5% ($p < 0.001$) of body weight, respectively. HbA1c decreased by 0.1%, 0.4% ($p < 0.001$), and 0.6% ($p < 0.001$) (MITT, LOCF). However, although topiramate had reduced weight and improved glycemic control in obese subjects with T2DM on top of metformin monotherapy, the side effects had led to an early stop of the trial.

Topiramate was also evaluated in five published controlled trials for eating disorders like bulimia nervosa and binge-eating disorders. A recent review by Arbaizar and coworkers (2008) concluded that topiramate is effective at least in short-term treatments of eating disorders associated with obesity. However, small sample sizes and a lot of dropouts ask for longer term and adequately powered studies prior to any generalization.

2.2.2 Clinical Trials of Topiramate–Phentermine (QNEXA®) Combination

QNEXA® is an investigational weight-loss therapy pursued by VIVUS Inc., USA, as a novel combination of low-dose immediate-release phentermine (1/8 to 1/2 of marketed dose) and controlled-release topiramate (1/16 to 1/4 of marketed dose).

It was hypothesized and finally proven that combinations of different therapeutic agents with complementary mechanisms of action may have greater efficacy at lower doses than each agent alone (Bays 2010). In addition, lower doses of the combined drugs may attenuate the intolerances and adverse effects associated with higher doses. A full package of clinical trials, mandatory for the submission of a New Drug Application (NDA) to the FDA, has been performed under the sponsorship of VIVUS. However, on July 15, 2010, the US FDA's advisory committee on endocrinologic and metabolic drugs voted ten to six that Vivus' weight-loss drug QNEXA® should not be allowed on the market.

The clinical development program for QNEXA® (VI-0521 (QNEXA®) Briefing Document 2010) consisted of three phase III studies, four phase II studies, and ten phase I studies. In total, more than 5,000 subjects were included, and about 3,000 subjects were treated for 6 months or at most 1 year duration. Three doses were tested: low (phentermine: IR 3.75 mg/topiramate, SR 23 mg), medium (phentermine: IR 7.5 mg/topiramate, SR 46 mg), and high (phentermine: IR 15 mg/topiramate, SR 92 mg).

The adult study population included a range of subjects, from overweight (BMI > 27 kg/m^2) to severely obese (BMI > 60 kg/m^2), with a range of obesity-related comorbidities, including type 2 diabetes, hypertension, and hypertriglyceridemia.

The three phase III studies of the QNEXA® clinical development program (OB-302 [EQUIP trial] and OB-303 [CONQUER trial], both 56-week pivotal trials; and OB-301 [EQUATE], a 28-week confirmatory study) evaluated the efficacy and safety of three fixed-dose combinations.

The EQUIP (OB-302) trial involved 1,267 obese patients with BMI > 35 kg/m^2. Subjects who had completed study OB-302 on QNEXA® Top lost 14.4% and on QNEXA® Low lost 6.7% of baseline body weight and showed by ITT analysis, reduction of 11.0%, 5.1%, and 1.6% for QNEXA® Top, QNEXA® Low, and placebo, respectively.

The CONQUER (OB-303) trial evaluated 2,487 patients with BMI between 27 and 45 kg/m^2 with two or more comorbidities. QNEXA® Top and placebo treatment groups comprised twice as many subjects as the QNEXA® Mid group. The coprimary endpoint was percent weight loss and percentage of subjects achieving

5% weight loss at study end (week 56). In total, 1,723 (69.3%) subjects completed all study visits, and 764 (30.7%) subjects discontinued the study. The percentages of subjects who completed the study were 62.0% in the placebo group, 75.1% in the QNEXA® Mid dose group, and 73.7% in the QNEXA® Top dose group (VI-0521 (QNEXA®) Briefing Document 2010). Subjects who had completed study OB-303 on QNEXA® Top lost 12.4% and on QNEXA® Mid lost 9.6% of baseline body weight. For the placebo group, mean percent weight loss was relatively stable from week 28 to the end of the study, with mean percent weight loss averaging 1.6% at the end of the study. By ITT analysis, the results were similar, with losses of 10.4%, 8.4%, and 1.8% for QNEXA® Top, QNEXA® Mid, and placebo, respectively.

There was a significant improvement of a number of cardiovascular risk markers in patients receiving the medication, such as waist circumference, systolic and diastolic blood pressure, and lipids. Significant improvements were also reported for fasting blood glucose, HbA1c, and HOMA index in diabetic patients on both doses. Overall, the QNEXA® treatment was very effective in a high proportion of obese subjects, and the greatest weight loss was observed in patients with the highest baseline BMI. The same applied also to benefits regarding cardiovascular, metabolic, and glycemic parameters.

The integrated safety analyses revealed that the commonly observed adverse events associated with QNEXA® treatment were, in order of frequency, dry mouth, paresthesias, constipation, upper respiratory tract infection, taste alteration, and insomnia. All these effects are known side effects of either topiramate or phentermine and do not represent novel side effects based on the combined pharmacology of the two drugs. Interestingly, some of the well-recognized side effects of topiramate, such as paresthesia, somnolence, psychomotor slowing, and difficulty with memory, occurred at a lower incidence than previously observed in monotherapy. This might be due to the lower doses of topiramate in the fixed-dose combinations, the opposing effects of the phentermine component (CNS side effects of this sympathomimetic amine are e.g., overstimulation, restlessness, dizziness, insomnia, euphoria, and dysphoria), or the altered pharmacokinetic of topiramate due to the modified-release formulation.

The dropout rate ranged from 31 to 43% for subjects on QNEXA® and 47% for the patients on placebo. However, 18% of patients taking QNEXA® high dose discontinued because of side effects compared to 9% in the placebo group (VI-0521 (QNEXA®) Briefing Document 2010).

2.2.3 Clinical Observations: Topiramate and Antipsychotics

Several case reports (Dursun and Devarajan 2000) and observational studies (Levý et al. 2002; Lévy et al. 2007; Nickel et al. 2005; Kim et al. 2006) reported the sometimes dramatic weight gain induced by clozapine and olanzapine, and the mitigating effect of coadministered topiramate. Apart from single weight-neutral antipsychotics, most increase appetite/caloric intake and/or modify the perception of satiety in a negative manner, leading to more or less significant weight gain

(Baptista et al. 2008; Maayan et al. 2010). A systematic study of healthy volunteers provided no evidence for a concomitant decrease in resting energy expenditure (Fountaine et al. 2010).

In 2000 Dursun and Devarajan (2000) described a weight gain of 45.5 kg in 25 months after treatment of a chronic paranoid schizophrenic male adult with clozapine. After initiation of additional treatment with topiramate, a significant weight loss of 21 kg in the 5 subsequent months was reported.

Kim and colleagues performed a 12-week, randomized, open-label, parallel-group trial of topiramate and evaluated the weight-gain trajectory during olanzapine treatment in 60 male outpatients meeting the DSM-IV criteria for schizophrenia (Kim et al. 2006). All eligible patients were started on olanzapine, 10 mg/day, with subsequent uptitration of the olanzapine dosage as clinically warranted. The subjects in the olanzapine + topiramate group were additionally started on topiramate, 25 mg b.i.d., increased to 50 mg b.i.d. on day 8, and remained on a fixed dosage throughout the rest of the 12-week study period. The latter group significantly differed from the olanzapine-only group with respect to the amount of weight gain at weeks 4, 8, and endpoint, respectively ($p = 0.042$, $p = 0.008$, $p = 0.038$). Mean changes from baseline to endpoint for the olanzapine + topiramate group and the olanzapine-only group were 2.66 ± 1.79 kg and 4.02 ± 2.52 kg.

Lévy and colleagues concluded from their retrospective case series study (Lévy et al. 2007) that topiramate may offer a potential adjunctive therapy to target weight loss in stable overweight schizophrenic patients, thus warranting further investigations.

Subsequently, topiramate was more systematically tested as a means to prevent drug-induced weight gain (Baptista et al. 2008). In a recent review and meta-analysis (Maayan et al. 2010), 15 different medications were evaluated: amantadine, dextroamphetamine, D-fenfluramine, famotidine, fluoxetine, fluvoxamine, metformin, nizatidine, orlistat, phenylpropanolamine, reboxetine, rosiglitazone, sibutramine, topiramate, and metformin + sibutramine. Compared with placebo, metformin led to the greatest weight loss ($N = 7$, $n = 334$, −2.94 kg) (confidence interval (CI, −4.89, −0.99)). Weight loss achieved with topiramate was significant and of similar magnitude ($N = 2$, $n = 133$, −2.52 kg) (CI, −4.87, −0.16). As stated by the authors (Maayan et al. 2010) themselves, the main limitation of this meta-analysis is the relative paucity of randomized controlled trials. As a result, there are too few studies for a number of medications, including orlistat and topiramate. The authors also concluded that none of the agents were able to entirely reverse weight gain induced by antipsychotics, and that no treatment can, at the time being, be recommended for broad clinical usage.

Interestingly, topiramate does not only mitigate clozapine-induced weight gain and improve metabolic parameters. The agent also appeared to augment the clinical benefits (e.g., improvement in total Brief Psychiatric Rating Scale scores) of clozapine in partial responders treated for refractory psychosis (Hahn et al. 2010). The mechanism(s) of this action/interaction is unknown.

More recently, in vitro, animal, and clinical studies of drug-induced weight alterations have also been used as an innovative translational research approach (TOP Institute Pharma; Project T2-105 2006–2011) to uncover the underlying mechanisms which should finally lead to the development of novel therapeutic strategies to treat human obesity and related metabolic disorders.

3 Zonisamide

Zonisamide (Zonegran™) can be chemically classified as a sulfonamide and is structurally unrelated to other antiepileptic agents except for topiramate which can be chemically classified as a sulfamate. Zonisamide was discovered by Uno in 1972 and subsequently launched by Dainippon Sumitomo Pharma as Excegran in Japan. Zonisamide has been on the market in Japan (Seino 2004; Yagi 2004; Ohtahara 2006) since 1989 and in the USA since 2000 (Jain 2000), for adjunctive therapy in the treatment of partial seizures in adults with epilepsy. More recently (March 2005), it was approved in Europe as adjunctive therapy for refractory partial seizures in adults.

3.1 Preclinical Observations

Zonisamide is structurally distinct from other AEDs and has multiple and complementary mechanisms of action, which likely contribute to its efficacy across a broad range of epilepsy types. Effective control of partial seizures is attained at doses of ≥300 mg/day.

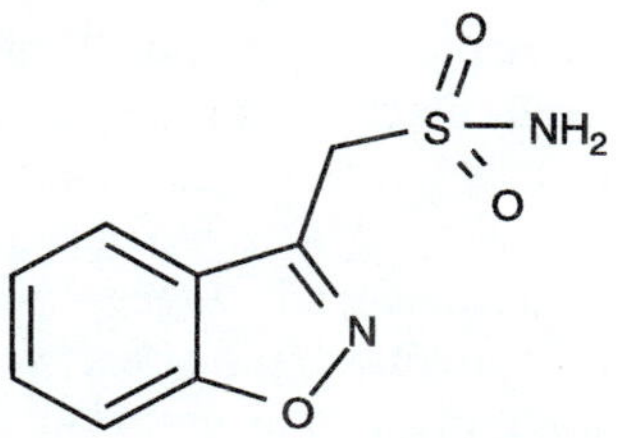

Anticonvulsant activity was demonstrated in several animal models. Its precise mechanism(s) of action (Biton 2004) is still unknown and requires further investigations. Zonisamide is like topiramate, a CA inhibitor (De Simone et al. 2005). However, CA inhibition is thought not to be the main mechanism of its antiepileptic effects (Thone et al. 2008), but may be relevant for its antiadipogenic effects (Supuran et al. 2008). Zonisamide blocks voltage-gated sodium channels and reduces T-type calcium channel currents (Leppik 2004). Furthermore, it binds allosterically to GABA receptors like the benzodiazepines (Mimaki et al. 1990a, b) and inhibits the uptake of the inhibitory neurotransmitter GABA, while enhancing the uptake of the excitatory neurotransmitter glutamate (Ueda et al. 2003).

Similar to other AEDs, most of the observed adverse events of zonisamide are CNS-related (e.g., somnolence, dizziness). In contrast to agents other than topiramate, zonisamide therapy is accompanied by a substantial weight loss, which has been demonstrated in animal studies (Walker et al. 1988) and investigated in several clinical trials (Li et al. 2005; McElroy et al. 2004; Kim 2003; Gadde et al. 2003).

It is interesting to note that in addition to the coinciding antiepileptic effects of topiramate and zonisamide, similarities can also be found for other interesting therapeutic areas. For example, both compounds are effective in animal models of neuropathic pain (Hord et al. 2003; Wieczorkiewicz-Plaza et al. 2004; Tanabe et al. 2005) and show some promise in clinical studies in patients with neuropathies (Chong and Libretto 2003; Vinik 2005; Hasegawa 2004; Atli and Dogra 2005). Other overlapping efficacies pertain to bipolar disorders (Chengappa et al. 2001; van Passel et al. 2006) and migraine prophylaxis (van Passel et al. 2006; Capuano et al. 2004; Drake et al. 2004; Wenzel et al. 2006; Bussone et al. 2006) which has become an approved indication for topiramate.

3.1.1 Potential Modes of Action

The precise mechanism(s) by which zonisamide exerts its antiseizure effect is still unknown. Zonisamide was tested in several experimental models for its anticonvulsant activity. It was effective against tonic extension seizures induced by maximal electroshock (MES), but ineffective against clonic seizures induced by subcutaneous pentylenetetrazol (PTZ). In the kindled rat model, zonisamide raised the threshold for generalized seizures. In cats, zonisamide reduced the duration of cortical focal seizures induced by electrical stimulation of the visual cortex. It was hypothesized that zonisamide may produce these effects through action at sodium and calcium channels (Leppik 2004). In vitro pharmacological studies suggest that zonisamide blocks sodium channels and reduces voltage-dependent, transient inward currents (T-type Ca^{2+} currents). This effect would lead to a stabilization of neuronal membranes and suppression of neuronal hypersynchronization. In vitro binding studies have demonstrated that zonisamide binds to the GABA/benzodiazepine receptor ionophore complex in an allosteric fashion (Mimaki et al. 1990a), however, without any effect on the chloride flux. Zonisamide interacts with this receptor ionophore complex (Mimaki et al. 1990b) in a manner similar to phenytoin at therapeutic serum levels (10^{-4} M). Since phenytoin is weight neutral, it is therefore unlikely that GABAergic signaling contributes to the antiadipogenic effects. Zonisamide does not appear to potentiate the synaptic activity of GABA, which was confirmed in in vitro studies where zonisamide at 10–30 mg/mL suppressed synaptically driven electrical activity without affecting postsynaptic GABA or glutamate responses (cultured mouse spinal cord neurons) or neuronal or glial uptake of [3H]-GABA (rat hippocampal slices) (Ueda et al. 2003). In vivo microdialysis studies demonstrated that zonisamide facilitates both dopaminergic and serotonergic neurotransmission. Like topiramate, zonisamide was known to have also carbonic anhydrase–inhibiting activity, (De Simone G et al. 2005) but this pharmacologic effect was not thought to be a major contributing factor in the antiseizure activity of zonisamide.

Whether this is a correct assumption still remains open, but from the observed profile of side effects (oligohidrosis, metabolic acidosis, and kidney stone formation (Knudsen et al. 2003; Fung and Nelson 2007)), it is obvious that the inhibition of some carbonic anhydrase isozymes may become clinically relevant in some patients treated with either zonisamide or topiramate. With regard to the rare cases of oligohidrosis, it is speculated that the inhibition of carbonic anhydrase in exocrine sweat glands leads to negative alterations in pH dynamics, hydrogen ion concentration, and available calcium transients (Knudsen et al. 2003).

3.1.2 In Vivo Results: Zonisamide as Weight-Loss Agent

Several clinical studies have evaluated either zonisamide alone or Zonisamide-Bupropion (Empatic®) as a combination for weight-loss effects. In contrast, so far, there is no publication about the efficacy of zonisamide or Empatic® in animal models for obesity or diabetes. However, there is one paper discussing study results of zonisamide explored in a model of drug-induced weight gain (see below).

3.1.3 In Vivo Results: Zonisamide and Antipsychotics

In a study performed by Wallingford, the olanzapine-associated hyperphagia, elevated blood glucose, and weight gain in female Sprague–Dawley rats were attenuated and even reversed by a concomitant application of zonisamide (Wallingford et al. 2008). Olanzapine was delivered via an osmotic minipump at 1.75 mg/day or control placebo (1.5% lactic acid: 100 ml of 10% lactic acid/10 mg of drug) and delivered 5 ml per h for 14 days. Zonisamide at 26 mg/kg was injected (IP) two times daily in a volume of 1 ± 0.2 ml on the basis of body weight. Eventually, zonisamide administration induced weight loss in vehicle-treated animals and inhibited chronic OLZ-associated hyperphagia, weight gain, and increased feed efficiency in OLZ-treated rats.

3.2 *Clinical Observations*

3.2.1 Clinical Trials of Zonisamide in Obesity and Type 2 Diabetes

Zonisamide has been shown to effectively induce weight loss in a 16-week randomized, double-blind, placebo-controlled trial with an optional single-blind extension of the same treatment for another 16 weeks (Gadde et al. 2003). A total of 60 subjects (92% female) with a mean (SE) BMI of 36.3 (0.5) kg/m^2 and a mean (SE) age of 37 (1.0) years participated at the single study center. All participants received a balanced hypocaloric diet (500 kcal/day deficit), and compliance was monitored with self-rated food diaries. Patients were randomly assigned to the zonisamide or the placebo arm ($n = 30$ each). All five participating men were randomized to zonisamide ($p = 0.08$) and baseline BMI was slightly lower ($p = 0.07$) in the zonisamide group. At the onset of the trial, 100 mg of zonisamide was given daily. Dosage was increased to 400 mg/day and further to 600 mg/day in those patients who had lost less than 5% of their body weight after 12 weeks; placebo dosage was adjusted accordingly. The primary outcome measure was change in absolute body weight; percent change in weight and the number of participants in each group who achieved weight losses of 5% or greater and 10% or greater were also examined. Secondary outcome measures included heart rate, blood pressure, frequency of adverse effects, fasting electrolytes and lipids, waist measurement, and impact of weight on quality of life (IWQOL) score, body composition, and bone mineral density (BMD) as determined at baseline and week 32 by dual-energy X-ray absorptiometry (DEXA). Fifty-one patients completed the 16-week acute phase.

In an intent-to-treat analysis with the last observation carried forward, the body-weight loss of the *verum* group significantly exceeded that of the placebo group (mean [SE], 5.9 [0.8] kg [6.0% loss] vs. 0.9 [0.4] kg) during the 16-week period (Gadde et al. 2003). Seventeen (57%) of 30 in the zonisamide arm and three (10%) of 30 in the placebo arm lost $\geq$5% of body weight by week 16. Of the 37 participants who entered the extension phase, 36 completed week 32. The zonisamide group ($n = 19$) had a mean weight loss of 9.2 kg (1.7 kg) (9.4% loss) at week 32 compared with 1.5 kg (0.7 kg) (1.8% loss) for the placebo group ($n = 17$). Weight loss for patients treated with zonisamide was significantly associated with a decrease in fat mass. At week 16, the IWQOL subscales health, work, mobility, and activities of daily living improved significantly in the *verum* group vs. placebo. Decrease in waist circumference was significantly greater with zonisamide therapy over the 16 weeks. Heart rate decreased by a mean of 2 beats per min in the total sample; no statistical difference between the groups was detected. Blood pressure did not change over time. Similarly, no significant changes in levels of lipids or fasting blood glucose with either treatment were detected. Total BMD showed a small but statistically significant increase in the overall sample although there was no difference between the groups. Fatigue was the only side effect reported more frequently by the patients treated with zonisamide both in the acute and extension phases. Mean serum creatinine concentrations increased significantly in the *verum* (from 0.78 mg/dL at baseline to 0.92 mg/dL) vs. the placebo group (0.75 to 0.77 mg/dL).

The efficacy of zonisamide to treat binge-eating disorder (BED) was initially addressed in an open-label, prospective, 12-week, flexible-dose (100–600 mg/day) study; 8 of the 15 subjects with BED completed the study (McElroy et al. 2004). Among these completers who received a mean (SD) zonisamide daily dose at endpoint evaluation of 513 (103) mg/day, binge-eating episode frequency, binge day frequency, and BMI decreased significantly. Four noncompleters suffered adverse events.

Zonisamide was subsequently assessed in a 16-week, single-center, randomized, double-blind, placebo-controlled, flexible-dose (100–600 mg/day) trial of 60 outpatients with BED and obesity (McElroy et al. 2007; NCT00221442). The primary outcome measure was weekly frequency of binge-eating episodes. The primary analysis of efficacy was a longitudinal analysis of the intent-to-treat sample, with treatment-by-time interaction as the effect measure. In comparison to placebo, zonisamide resulted in a significantly greater rate of reduction in binge-eating episode frequency, BMI, and scores on the Clinical Global Impressions-Severity scale, Yale-Brown Obsessive Compulsive Scale Modified for Binge Eating, and Three Factor Eating Questionnaire disinhibition scales. Interestingly, plasma ghrelin concentrations increased with zonisamide but decreased with placebo. The mean (SD) zonisamide daily dose at endpoint evaluation was 436 (159) mg/day. Twelve patients ($N = 8$ receiving zonisamide, $N = 4$ receiving placebo) discontinued because of adverse events. The most common reasons for discontinuing zonisamide were accidental injury with bone fracture ($N = 2$), psychological complaints ($N = 2$), and cognitive complaints ($N = 2$). The authors concluded that zonisamide was efficacious in the short-term treatment of BED but not well tolerated. According to a recent review of psychiatric side effects of centrally active antiobesity drugs, cognitive impairments have been most frequently associated with the antiepileptic drugs, topiramate and zonisamide (Nathan et al. 2011).

Zonisamide has also been administered to 25 recovered, overweight (mean BMI 34.2 ± 3.1 kg/m^2) euthymic adult outpatients with bipolar disorder in an open 6-month-long trial (Wang et al. 2008). Mean BMI loss was 1.2 ± 1.9 kg/m^2. The majority of patients ($n = 18$; 72%) discontinued study participation, 11 of these due to emergent mood and 5 due to adverse physical events. The investigators concluded that adjunctive zonisamide appeared effective; however, the high rates of mood adverse events warrant consideration.

3.2.2 Clinical Trials of Zonisamide–Bupropion (Empatic®) Combination

In a small 12-week-long randomized, open-label, parallel-group trial, 18 females with a mean BMI of 36.8 kg/m^2 were randomized to either zonisamide monotherapy or a combination therapy of zonisamide and bupropion. For all subjects, zonisamide was started at 100 mg/day and gradually increased to 400 mg/day. For the combination therapy arm, bupropion was started at 100 mg/day and increased to 200 mg/day after 2 weeks. A balanced hypocaloric diet (500 kcal/day deficit) was recommended to all subjects, and compliance was monitored with self-rated food diaries. Body weight in kilograms was the primary outcome measure. Six patients did not complete the study. In an intent-to-treat analysis, carrying the last observation forward for all randomly assigned participants with at least one postbaseline assessment, the combination group lost more body weight than the zonisamide group (mean [SE] = 7.2 [1.2] kg [7.5%] vs. 2.9 [0.7] kg [3.1%]; $F = 4.7$, df = 4,56; $p = 0.003$) during the 12-week period. For the subset of 12 patients (combination, $N = 7$; zonisamide, $N = 5$) who completed the 12-week treatment, the mean (SE) weight loss was 8.1 (1.4) kg (8.5%) for the combination group vs. 3.0 (0.9) kg (3.3%) for the zonisamide group ($F = 4.6$, df = 4.40; $p = 0.004$). Six subjects in the combination group and two in the zonisamide group lost at least 5% of body weight (Gadde et al. 2007).

Subsequent to the aforementioned small trial, Orexigen Therapeutics Inc. conducted the phase II trial "A Dose Parallel, Randomized, Placebo-Controlled, Multicenter Study of the Safety and Efficacy of Multiple Regimens of the Combination of Zonisamide CR Plus Bupropion SR in the Treatment of Subjects With Uncomplicated Obesity" (NCT00339014) between 2006 and 2008 in order to assess which of six combinations of zonisamide controlled release (CR; 120, 240, and 360 mg) and bupropion slow release (SR; 280 and 360 mg) gives the best weight loss in comparison to a placebo arm and is safe and well tolerated for the treatment of obesity not associated with the complications of obesity such as type 2 diabetes mellitus. Thirteen US treatment centers are involved; 611 subjects were enrolled. The primary outcome criterion was % change in total body weight as measured between baseline and week 24 (ITT-LOCF analysis). In addition to other weight-related outcomes such as blood pressure, triglyceride levels, and fasting glucose levels, respectively, quality of life, sleep quality and quantity, HAMD-17 Maier subscale scores, and change in "brief assessment of cognition" composite scores were the secondary outcomes. Relevant inclusion criteria for obese individuals with a BMI between 30 and 43 kg/m^2 and within the age range 18 to 60 included absence of clinically significant illness or disease as determined by medical history and physical examination. The study has been completed; results have not yet been published.

For the trial "A Phase IIB, Multi-Center, Dose-Parallel, Randomized, Double-Blind, Monotherapy and Placebo-Controlled Safety and Efficacy Study of Zonisamide SR Plus Bupropion SR Combination Therapy in Subjects With Uncomplicated Obesity" [NCT00709371] Orexigen Therapeutics Inc. included 20 treatment centers; the estimated enrollment is given at 600 subjects within the age range of 18–65 with uncomplicated obesity (BMI between 30 and 45 kg/m^2). The six treatment arms include placebo and zonisamide SR (120 and 360 mg/day) with bupropion (placebo, 360 mg/day). The primary outcome criterion was also %

change in total body weight as measured between baseline and week 24. The study has also been completed; results have not yet been published.

3.2.3 Clinical Observations: Zonisamide and Antipsychotics

Based on the suppression of typical side effects of olanzapine with ziprasidone in rodents (Wallingford et al. 2008), Orexigen Therapeutics Inc. conducted a small trial (NCT00734435) with enrollment of $n = 26$ patients to assess the effect of Ziprasidone SR (placebo, 90 and 360 mg/day) in patients with schizophrenia, schizoaffective disorder, or schizophreniform disorder who received either 10 or 20 mg of olanzapine (A Proof of Concept, Multi-Center, Randomized, Double-Blind, Parallel, Placebo-Controlled, Study of Zonisamide Sustained Release (SR) 360 mg Versus Placebo in the Prevention of Weight Gain Associated With Olanzapine Therapy for Psychosis). The study was terminated based on financial considerations.

Currently (August 2010), a single-center, 16-week, randomized, double-blind, placebo-controlled, parallel-group, flexible-dose study in 60 outpatients with schizophrenia and related psychotic or bipolar disorders and with a BMI > 22 treated with olanzapine (5–20 mg/day) is ongoing (ClinicalTrials.gov Identifier: NCT00363376). Aim of this study is to evaluate the efficacy, tolerability, and safety of zonisamide therapy in the prevention of weight gain associated with olanzapine treatment for psychotic or bipolar disorders.

The primary outcome will be change in weight. Secondary outcome measures will include BMI, waist circumference, and other metabolic variables (fasting lipids, glucose, insulin) as well as outcomes from psychiatric scales (e.g., YMRS, IDS, PANSS, CGI, BES). In this study, zonisamide dose will be increased from 100 mg/day (7 days) up to a potential maximum of 600 mg/day (depending on clinical signs and tolerability) by the end of the sixth week of treatment in 100 mg/day weekly installments. Olanzapine is administered open-label at 5-20 mg/day and titrated to minimize side effects and optimize response.

Another ongoing placebo-controlled trial is investigating zonisamide in a more general fashion for the treatment of weight gain associated with psychotropic medication use (NCT00203450). The primary objective of this study is to compare the efficacy of zonisamide (Zonegran; 100–400 mg/day) and placebo as an adjunctive agent on lowering weight in subjects (BMI > 25 kg/m^2) who are on atypical neuroleptics except aripiprazole or ziprasidone, all forms of valproate, all forms of lithium, or all forms of carbamazepine.

Another trial aiming to provide proof of concept for the effect of zonisamide (SR 360 mg) to prevent weight gain associated with olanzapine therapy was terminated for financial reasons (NCT00734435).

A fourth study (NCT00401973) involving zonisamide and olanzapine has just been completed but not yet published.

4 Conclusions and Outlook

It remains speculative whether the mechanism of action of topiramate and zonisamide as AEDs is different from that of their weight-modifying and antidiabetic effect. The jury remains still out since highly selective and potent tool compounds for each of the known mechanisms of topiramate and zonisamide are still missing. Mechanisms such as enhancing GABAergic and reducing glutamatergic neurotransmissions or interactions with calcium and brain sodium channels are shared with many other AEDs and may well be "main mechanisms" for the antiepileptic activity. It seems unlikely, however, that these mechanisms are the main contributors to the pronounced weight-loss effects of topiramate and zonisamide since these mechanisms are shared with weight-neutral or even weight-gain-inducing AEDs. Whether topiramate and zonisamide act on the same channel subtypes or on the same AMPA/kainate receptor isotypes is open to question. There is evidence that topiramate may selectively modulate the kainate type of receptors that contain GluR5 subunits (Gryder and Rogawski 2003). On the other hand, studies on the interaction of zonisamide with individual AMPA/kainate receptor subunits have not been conducted so far.

There is some evidence that carbonic anhydrase (CA) inhibition might be a contributing factor for the antiadipogenic effect of topiramate and zonisamide since CA inhibition is clearly a common mechanistic component of those two drugs which is not shared by the other weight-neutral or weight-gain-inducing AEDs. But also here, the final proof with selective and potent CA inhibitors is missing. The possibility cannot be excluded that a combination of different mechanisms is responsible for the observed antiadipogenic effects which warrants further studies.

Currently, there is no pharmacological treatment available that achieves an average weight loss of 10% or more without unacceptable side effects. Antiadipogenic antiepileptics like topiramate and zonisamide, especially when used in a combination, achieve the targeted efficacy. However, adverse side effects have so far led to controversial discussions at advisory board meetings and rejections of applications for marketing approval. It remains to be seen, whether modified dosing schemes, optimized risk plans, or outcome studies will be performed by the sponsors in order to eventually obtain marketing approval.

References

Angehagen M, Ben-Menachem E, Rönnbäck L, Hansson E (2003a) Topiramate protects against glutamate- and kainate-induced neurotoxicity in primary neuronal-astroglial cultures. Epilepsy Res 54(1):63–71

Angehagen M, Ben-Menachem E, Rönnbäck L, Hansson E (2003b) Novel mechanisms of action of three antiepileptic drugs, vigabatrin, tiagabine, and topiramate. Neurochem Res 28 (2):333–340

Angehagen M, Ben-Menachem E, Shank R, Rönnbäck L, Hansson E (2004) Topiramate modulation of kainate-induced calcium currents is inversely related to channel phosphorylation level. J Neurochem 88(2):320–325

Angehagen M, Rönnbäck L, Hansson E, Ben-Menachem E (2005) Topiramate reduces AMPA-induced Ca(2+) transients and inhibits GluR1 subunit phosphorylation in astrocytes from primary cultures. J Neurochem 94(4):1124–1130

Antel J, Weber A, Sotriffer CA, Klebe G (2004) Multiple binding modes observed in X-ray structures of carbonic anhydrase-inhibitor complexes and other systems: consequences for structure-based drug design. In: Supuran CT, Scozzafava A, Conway J (eds) Carbonic anhydrase: its inhibitors and activators, CRC enzyme inhibitors series. CRC Press, Boca Raton

Antel J, Gregory PC, Nordheim U (2006) CB1 cannabinoid receptor antagonists for treatment of obesity and prevention of comorbid metabolic disorders. J Med Chem 49(14):4008–4016

Arbaizar B, Gómez-Acebo I, Llorca J (2008) Efficacy of topiramate in bulimia nervosa and binge-eating disorder: a systematic review. Gen Hosp Psychiatry 30(5):471–475

Aribi AM, Stringer JL (2002) Effects of antiepileptic drugs on extracellular pH regulation in the hippocampal CA1 region in vivo. Epilepsy Res 49:143–151

Astrup A, Toubro S (2004) Topiramate: a new potential pharmacological treatment for obesity. Obes Res 12(Suppl):167S–173S

Atli A, Dogra S (2005) Zonisamide in the treatment of painful diabetic neuropathy: a randomized, double-blind, placebo-controlled pilot study. Pain Med 6:225–234

Baptista T, ElFakih Y, Uzcátegui E, Sandia I, Tálamo E, Araujo de Baptista E, Beaulieu S (2008) Pharmacological management of atypical antipsychotic-induced weight gain. CNS Drugs 22 (6):477–495

Bays H (2010) Phentermine, topiramate and their combination for the treatment of adiposopathy ('sick fat') and metabolic disease. Expert Rev Cardiovasc Ther 8(12):1777–1801

Ben-Menachem E (2007) Weight issues for people with epilepsy–a review. Epilepsia 48(Suppl 9):42–45

Ben-Menachem E, Axelsen M, Johanson EH, Stagge A, Smith U (2003) Predictors of weight loss in adults with topiramate-treated epilepsy. Obes Res 11(4):556–562

Berghöfer A, Pischon T, Reinhold T, Apovian CM, Sharma AM, Willich SN (2008) Obesity prevalence from a European perspective: a systematic review. BMC Public Health 8:200

Bischofs S, Zelenka M, Sommer C (2004) Evaluation of topiramate as an anti-hyperalgesic and neuroprotective agent in the peripheral nervous system. J Peripher Nerv Syst 9:70–78

Biton V (2004) Zonisamide: newer antiepileptic agent with multiple mechanisms of action. Expert Rev Neurother 4:935–943

Biton V (2007) Clinical pharmacology and mechanism of action of zonisamide. Clin Neuropharmacol 30:230–240

Bray GA, Hollander P, Klein S, Kushner R, Levy B, Fitchet M, Perry BH (2003) A 6-month randomized, placebo-controlled, dose-ranging trial of topiramate for weight loss in obesity. Obes Res 11:722–733

Bussone G, Usai S, D'Amico D (2006) Topiramate in migraine prophylaxis: data from a pooled analysis and open-label extension study. Neurol Sci 27(Suppl 2):S159–S163

Campistol J, Campos J, Casas C, Herranz JL (2005) Topiramate in the prophylactic treatment of migraine in children. J Child Neurol 20:251–253

Capuano A, Vollono C, Mei D, Pierguidi L, Ferraro D, Di Trapani G (2004) Antiepileptic drugs in migraine prophylaxis: state of the art. Clin Ter 155:79–87

Casini A, Antel J, Abbate F, Scozzafava A, David S, Waldeck H, Schafer S, Supuran CT (2003) Carbonic anhydrase inhibitors: SAR and X-ray crystallographic study for the interaction of sugar sulfamates/sulfamides with isozymes I, II and IV. Bioorg Med Chem Lett 13:841–845

Catterall WA (1999) Molecular properties of brain sodium channels: an important target for anticonvulsant drugs. Adv Neurol 79:441–456

Chengappa KN, Gershon S, Levine J (2001) The evolving role of topiramate among other mood stabilizers in the management of bipolar disorder. Bipolar Disord 3:215–232

Chong MS, Libretto SE (2003) The rationale and use of topiramate for treating neuropathic pain. Clin J Pain 19:59–68

Costa C, Tozzi A, Luchetti E, Siliquini S, Belcastro V, Tantucci M, Picconi B, Ientile R, Calabresi P, Pisani F (2010) Electrophysiological actions of zonisamide on striatal neurons: selective neuroprotection against complex I mitochondrial dysfunction. Exp Neurol 221:217–224

Davidson MH, Hauptman J, DiGirolamo M, Foreyt JP, Halsted CH, Heber D, Heimburger DC, Lucas CP, Robbins DC, Chung J, Heymsfield SB (1999) Weight control and risk factor reduction in obese subjects treated for 2 years with orlistat: a randomized controlled trial. JAMA 281:235–242

De Simone G, Di Fiore A, Menchise V, Pedone C, Antel J, Casini A, Scozzafava A, Wurl M, Supuran CT (2005) Carbonic anhydrase inhibitors. Zonisamide is an effective inhibitor of the cytosolic isozyme II and mitochondrial isozyme V: solution and X-ray crystallographic studies. Bioorg Med Chem Lett 15:2315–2320

Dhellemmes P, Defoort S, Vinchon M (2008) Benign intracranial hypertension: the role of medical treatment. Neurochirurgie 54:717–720

Di Marzo V, Bifulco M, De Petrocellis L (2004) The endocannabinoid system and its therapeutic exploitation. Nat Rev Drug Discov 3:771–784

Dodgson SJ, Shank RP, Maryanoff BE (2000) Topiramate as an inhibitor of carbonic anhydrase isoenzymes. Epilepsia 41(Suppl 1):S35–S39

Drake ME Jr, Greathouse NI, Renner JB, Armentbright AD (2004) Open-label zonisamide for refractory migraine. Clin Neuropharmacol 27:278–280

Dursun SM, Devarajan S (2000) Clozapine weight gain, plus topiramate weight loss. Can J Psychiatry 45(2):198

Elterman RD, Glauser TA, Wyllie E, Reife R, Wu SC, Pledger G (1999) A double-blind, randomized trial of topiramate as adjunctive therapy for partial-onset seizures in children. Topiramate YP Study Group. Neurology 52:1338–1344

Fakhoury T, Murray L, Seger D, McLean M, Abou-Khalil B (2002) Topiramate overdose: clinical and laboratory features. Epilepsy Behav 3:185–189

Flegal KM, Carroll MD, Ogden CL, Curtin LR (2010) Prevalence and trends in obesity among us adults, 1999–2008. JAMA 303(3):235–241

Fountaine RJ, Taylor AE, Mancuso JP, Greenway FL, Byerley LO, Smith SR, Most MM, Fryburg DA (2010) Increased food intake and energy expenditure following administration of olanzapine to healthy men. Obesity (Silver Spring) 18:1646–1651

Francis J, Burnham WM (1992) [3H]Phenytoin identifies a novel anticonvulsant-binding domain on voltage-dependent sodium channels. Mol Pharmacol 42(6):1097–1103

Fung EL, Nelson EA (2007) Oligohydrosis underestimated side effect with topiramate treatment. Indian J Pediatr 74(7):694

Gadde KM, Franciscy DM, Wagner HR 2nd, Krishnan KR (2003) Zonisamide for weight loss in obese adults: a randomized controlled trial. JAMA 289:1820–1825

Gadde KM, Yonish GM, Foust MS, Wagner HR (2007) Combination therapy of zonisamide and bupropion for weight reduction in obese women: a preliminary, randomized, open-label study. J Clin Psychiatry 68:1226–9

Gaspari CN, Guerreiro CA (2010) Modification in body weight associated with antiepileptic drugs. Arq Neuropsiquiatr 68(2):277–281

Gebhardt S, Theisen FM, Haberhausen M, Heinzel-Gutenbrunner M, Wehmeier PM, Krieg JC, Kühnau W, Schmidtke J, Remschmidt H, Hebebrand J (2010) Body weight gain induced by atypical antipsychotics: an extension of the monozygotic twin and sib pair study. J Clin Pharm Ther 35(2):207–211

Ghelardoni S, Bazinet RP, Rapoport SI, Bosetti F (2005) Topiramate does not alter expression in rat brain of enzymes of arachidonic acid metabolism. Psychopharmacology (Berl) 180:523–529

Gibbs JW III, Sombati S, DeLorenzo RJ, Coulter DA (2000) Cellular actions of topiramate: blockade of kainate-evoked inward currents in cultured hippocampal neurons. Epilepsia 41 (Suppl 1):S10–S16

Gryder DS, Rogawski MA (2003) Selective antagonism of GluR5 kainate-receptor-mediated synaptic currents by topiramate in rat basolateral amygdala neurons. J Neurosci 23 (18):7069–7074

Hahn MK, Remington G, Bois D, Cohn T (2010) Topiramate augmentation in clozapine-treated patients with schizophrenia: clinical and metabolic effects. J Clin Psychopharmacol 30 (6):706–710

Hasegawa H (2004) Utilization of zonisamide in patients with chronic pain or epilepsy refractory to other treatments: a retrospective, open label, uncontrolled study in a VA hospital. Curr Med Res Opin 20:577–580

Hebebrand J, Volckmar AL, Knoll N, Hinney A (2010) Chipping away the 'missing heritability': GIANT steps forward in the molecular elucidation of obesity – but still lots to go. Obes Facts 3:294–303

Herrero AI, Del Olmo N, Gonzalez-Escalada JR, Solis JM (2002) Two new actions of topiramate: inhibition of depolarizing GABA(A)-mediated responses and activation of a potassium conductance. Neuropharmacology 42:210–220

Hord AH, Denson DD, Chalfoun AG, Azevedo MI (2003) The effect of systemic zonisamide (Zonegran) on thermal hyperalgesia and mechanical allodynia in rats with an experimental mononeuropathy. Anesth Analg 96:1700–1706; table of contents

Houseknecht KL, Robertson AS, Zavadoski W, Gibbs EM, Johnson DE, Rollema H (2007) Acute effects of atypical antipsychotics on whole-body insulin resistance in rats: implications for adverse metabolic effects. Neuropsychopharmacology 32(2):289–297

TOP Institute Pharma; Project T2-105 (2006–2011) Investigation of drug-induced weight alterations to identify novel therapeutic strategies for the treatment of obesity, dyslipidemia and diabetes. Principal investigator: Elwin Verheij; partners: Abbott (formerly Solvay Pharmaceuticals), Danone Research (formerly Numico), Leiden University Medical Center, Netherlands Institute for Neurosciences, PRA International, TNO, University Medical Center Groningen http://www.tipharma.com/projects/cardiovascular-diseases/drug-induced-weight-alteration.html

Jackson HC, Cheetham SC, Gregory PC, Antel J (2007). Effect of chronic administration of topiramate and phentermine, alone and in combination, in an animal model of dietary-induced obesity. Neuroscience Program No 629.15, Society for Neuroscience: San Diego, CA

Jain KK (2000) An assessment of zonisamide as an anti-epileptic drug. Expert Opin Pharmacother 1:1245–1260

James WP, Astrup A, Finer N, Hilsted J, Kopelman P, Rössner S, Saris WH, Van Gaal LF (2000) Effect of sibutramine on weight maintenance after weight loss: a randomised trial. STORM study group. Sibutramine trial of obesity reduction and maintenance. Lancet 356 (9248):2119–2125

James WP, Caterson ID, Coutinho W, Finer N, Van Gaal LF, Maggioni AP, Torp-Pedersen C, Sharma AM, Shepherd GM, Rode RA, Renz CL, Investigators SCOUT (2010) Effect of sibutramine on cardiovascular outcomes in overweight and obese subjects. N Engl J Med 363(10):905–917

Johannessen Landmark C (2008) Antiepileptic drugs in non-epilepsy disorders: relations between mechanisms of action and clinical efficacy. CNS Drugs 22(1):27–47

Jones MW (1998) Topiramate – safety and tolerability. Can J Neurol Sci 25(3):S13–S15

Jones D (2008) End of the line for cannabinoid receptor 1 as an anti-obesity target? Nat Rev Drug Discov 7(12):961–962

Kelly KM (1998) Gabapentin. Antiepileptic mechanism of action. Neuropsychobiology 38 (3):139–144

Kennett GA, Clifton PG (2010) New approaches to the pharmacological treatment of obesity: can they break through the efficacy barrier? Pharmacol Biochem Behav 97(1):63–83

Kim CS (2003) Zonisamide effective for weight loss in women. J Fam Pract 52:600–601

Kim JH, Yim SJ, Nam JH (2006) A 12-week, randomized, open-label, parallel-group trial of topiramate in limiting weight gain during olanzapine treatment in patients with schizophrenia. Schizophr Res 82(1):115–117

Klein KM, Hamer HM, Reis J, Schmidtke J, Oertel WH, Theisen FM, Hebebrand J, Rosenow F (2005) Weight change in monozygotic twins treated with valproate. Obes Res 13 (8):1330–1334

Knudsen JF, Thambi LR, Kapcala LP, Racoosin JA (2003) Oligohydrosis and fever in pediatric patients treated with zonisamide. Pediatr Neurol 28(3):184–189

Kuo RL, Moran ME, Kim DH, Abrahams HM, White MD, Lingeman JE (2002) Topiramate-induced nephrolithiasis. J Endourol 16:229–231

Kushner SF, Khan A, Lane R, Olson WH (2006) Topiramate monotherapy in the management of acute mania: results of four double-blind placebo-controlled trials. Bipolar Disord 8:15–27

Kuzmiski JB, Barr W, Zamponi GW, MacVicar BA (2005) Topiramate inhibits the initiation of plateau potentials in CA1 neurons by depressing R-type calcium channels. Epilepsia 46 (4):481–489

Kuzniecky R, Ho S, Pan J, Martin R, Gilliam F, Faught E, Hetherington H (2002) Modulation of cerebral GABA by topiramate, lamotrigine, and gabapentin in healthy adults. Neurology 58:368–372

Leniger T, Wiemann M, Bingmann D, Widman G, Hufnagel A, Bonnet U (2002) Carbonic anhydrase inhibitor sulthiame reduces intracellular pH and epileptiform activity of hippocampal CA3 neurons. Epilepsia 43:469–74

Leniger T, Thöne J, Wiemann M (2004) Topiramate modulates pH of hippocampal CA3 neurons by combined effects on carbonic anhydrase and Cl-/HCO3-exchange. Br J Pharmacol 142 (5):831–842

Leppik IE (2004) Zonisamide: chemistry, mechanism of action, and pharmacokinetics. Seizure 13 (Suppl 1):S5–S9; discussion S10

Levy E, Margolese HC, Chouinard G (2002) Topiramate produced weight loss following olanzapine-induced weight gain in schizophrenia. J Clin Psychiatry 63:1045

Lévy E, Agbokou C, Ferreri F, Chouinard G, Margolese HC (2007) Topiramate-induced weight loss in schizophrenia: a retrospective case series study. Can J Clin Pharmacol 14(2):e234–e239

Li Z, Maglione M, Tu W, Mojica W, Arterburn D, Shugarman LR, Hilton L, Suttorp M, Solomon V, Shekelle PG, Morton SC (2005) Meta-analysis: pharmacologic treatment of obesity. Ann Intern Med 142:532–546

Liang Y, She P, Wang X, Demarest K (2006) The messenger RNA profiles in liver, hypothalamus, white adipose tissue, and skeletal muscle of female Zucker diabetic fatty rats after topiramate treatment. Metabolism 55:1411–1419

Linne Y, Hemmingsson E, Adolfsson B, Ramsten J, Rössner S (2002) Patient expectations of obesity treatment—the experience from a day-care unit. Int J Obes 26:739–741

Maayan L, Vakhrusheva J, Correll CU (2010) Effectiveness of medications used to attenuate antipsychotic-related weight gain and metabolic abnormalities: a systematic review and meta-analysis. Neuropsychopharmacology 35:1520–1530

Martins PJF, Haas M, Obici S (2010) Central nervous system delivery of the antipsychotic olanzapine induces hepatic insulin resistance. Diabetes 59(10):2418–2425

Maryanoff BE (2009) 2009 Edward E Smissman Award. Pharmaceutical "gold" from neurostabilizing agents: topiramate and successor molecules. J Med Chem 52:3431–3440

Maryanoff BE, Nortey SO, Gardocki JF, Shank RP, Dodgson SP (1987) Anticonvulsant O-alkyl sulfamates. 2,3:4,5-Bis-O-(1-methylethylidene)-beta-D-fructopyranose sulfamate and related compounds. J Med Chem 30:880–887

Maryanoff BE, Costanzo MJ, Nortey SO, Greco MN, Shank RP, Schupsky JJ, Ortegon MP, Vaught JL (1998) Structure-activity studies on anticonvulsant sugar sulfamates related to topiramate. Enhanced potency with cyclic sulfate derivatives. J Med Chem 41:1315–1343

Masereel B, Rolin S, Abbate F, Scozzafava A, Supuran CT (2002) Carbonic anhydrase inhibitors: anticonvulsant sulfonamides incorporating valproyl and other lipophilic moieties. J Med Chem 45:312–320

McAllister EJ, Dhurandhar NV, Keith SW, Aronne LJ, Barger J, Baskin M, Benca RM, Biggio J, Boggiano MM, Eisenmann JC, Elobeid M, Fontaine KR, Gluckman P, Hanlon EC, Katzmarzyk P, Pietrobelli A, Redden DT, Ruden DM, Wang C, Waterland RA, Wright SM, Allison DB (2009) Ten putative contributors to the obesity epidemic. Crit Rev Food Sci Nutr 49(10):868–913

McElroy SL, Kotwal R, Hudson JI, Nelson EB, Keck PE (2004) Zonisamide in the treatment of binge-eating disorder: an open-label, prospective trial. J Clin Psychiatry 65:50–56

McIntyre RS, Riccardelli R, Binder C, Kusumakar V (2005) Open-label adjunctive topiramate in the treatment of unstable bipolar disorder. Can J Psychiatry 50:415–422

McLean MJ, Bukhari AA, Wamil AW (2000) Effects of topiramate on sodium-dependent action-potential firing by mouse spinal cord neurons in cell culture. Epilepsia 41(Suppl 1):S21–S24

McNaughton NC, Davies CH, Randall A (2004) Inhibition of alpha(1E) Ca(2+) channels by carbonic anhydrase inhibitors. J Pharmacol Sci 95(2):240–247

Mimaki T, Suzuki Y, Tagawa T, Karasawa T, Yabuuchi H (1990a) Interaction of zonisamide with benzodiazepine and GABA receptors in rat brain. Med J Osaka Univ 39(1–4):13–17

Mimaki T, Suzuki Y, Tagawa T, Karasawa T, Yabuuchi H (1990b) [3H]Zonisamide binding in rat brain. Med J Osaka Univ 39(1–4):19–22

Nathan PJ, O'Neill BV, Napolitano A, Bullmore ET (2011) Neuropsychiatric Adverse Effects of Centrally Acting Antiobesity Drugs. CNS Neurosci Ther. 17:490–505

NCT00203450 (2003) Zonegran for the treatment of weight gain associated with psychotropic medication use: a placebo-controlled trial; Sponsor: Tuscaloosa Research & Education Advancement Corporation; Study Start Date: May 2003

NCT00363376 (2008) A double-blind, placebo-controlled study of zonisamide to prevent olanzapine-associated weight gain; Sponsor: Lindner Center of HOPE; Study Start Date: January 2008

NCT00401973 (2006) The assessment of the safety, efficacy, and practicality of an algorithm including amantadine, metformin and zonisamide for the prevention of olanzapine-associated weight gain in outpatients with schizophrenia; Sponsor: Eli Lilly and Company; Study Start Date: November 2006

NCT00734435 (2008) A proof of concept, multi-center, randomized, double-blind, parallel, placebo-controlled, study of zonisamide sustained release (SR) 360 mg versus placebo in the prevention of weight gain associated with olanzapine therapy for psychosis; Sponsor: Orexigen Therapeutics, Inc.; Study Start Date: September 2008

Nickel MK, Nickel C, Muehlbacher M, Leiberich PK, Kaplan P (2005) Influence of topiramate on olanzapine-related adiposity in women: a random, double-blind, placebo-controlled study. J Clin Psychopharmacol 25:211–217

Ohtahara S (2006) Zonisamide in the management of epilepsy – Japanese experience. Epilepsy Res 68(Suppl 2):S25–S33

Oommen KJ, Mathews S (1999) Zonisamide, a new anti-epileptic drug. Clin Neuropharmacol 22:192–200

Perucca E (1997) A pharmacological and clinical review on topiramate, a new antiepileptic drug. Pharmacol Res 35:241–256

Picard F, Deshaies Y, Lalonde J, Samson P, Richard D (2000) Topiramate reduces energy and fat gains in lean (Fa/?) and obese (fa/fa) Zucker rats. Obes Res 8:656–663

Post RM, Ketter TA, Uhde T, Ballenger JC (2007) Thirty years of clinical experience with carbamazepine in the treatment of bipolar illness: principles and practice. CNS Drugs 21(1):47–71

Raskin P, Donofrio PD, Rosenthal NR, Hewitt DJ, Jordan DM, Xiang J, Vinik AI (2004) Topiramate vs placebo in painful diabetic neuropathy: analgesic and metabolic effects. Neurology 63:865–873

Reis J, Tergau F, Hamer HM, Muller HH, Knake S, Fritsch B, Oertel WH, Rosenow F (2002) Topiramate selectively decreases intracortical excitability in human motor cortex. Epilepsia 43:1149–1156

Ribacoba Montero R, Salas Puig X (2002) Efficacy and tolerability of long term topiramate in drug resistant epilepsy in adults. Rev Neurol 34:101–105

Richard D, Ferland J, Lalonde J, Samson P, Deshaies Y (2000) Influence of topiramate in the regulation of energy balance. Nutrition 16:961–966

Richard D, Picard F, Lemieux C, Lalonde J, Samson P, Deshaies Y (2002) The effects of topiramate and sex hormones on energy balance of male and female rats. Int J Obes Relat Metab Disord 26:344–353

Rinaldi-Carmona M, Barth F, Congy C, Martinez S, Oustric D, Pério A, Poncelet M, Maruani J, Arnone M, Finance O, Soubrié P, Le Fur G (2004) SR147778 [5-(4-bromophenyl)-1-(2,4-dichlorophenyl)-4-ethyl-N-(1-piperidinyl)-1H-pyrazole-3-carboxamide], a new potent and selective antagonist of the CB1 cannabinoid receptor: biochemical and pharmacological characterization. J Pharmacol Exp Ther 310:905–914

Rogawski MA (2006) Diverse mechanisms of antiepileptic drugs in the development pipeline. Epilepsy Res 69(3):273–294

Rogawski MA, Löscher W (2004) The neurobiology of antiepileptic drugs. Nat Rev Neurosci 5 (7):553–564

Rosenfeld WE (1997) Topiramate: a review of preclinical, pharmacokinetic, and clinical data. Clin Ther 19:1294–1308

Rothman RB, Baumann MH, Dersch CM, Romero DV, Rice KC, Carroll FI, Partilla JS (2001) Amphetamine-type central nervous system stimulants release norepinephrine more potently than they release dopamine and serotonin. Synapse 39(1):32–41

Russo E, Constanti A, Ferreri G, Citraro R, De Sarro G (2004) Nifedipine affects the anticonvulsant activity of topiramate in various animal models of epilepsy. Neuropharmacology 46 (6):865–878

Scheen AJ (2010) Cardiovascular risk-benefit profile of sibutramine. Am J Cardiovasc Drugs 10 (5):321–334

Schubert S, Brandl U, Brodhun M, Ulrich C, Spaltmann J, Fiedler N, Bauer R (2005) Neuroprotective effects of topiramate after hypoxia-ischemia in newborn piglets. Brain Res 1058(1–2):129–136

Seino M (2004) Review of zonisamide development in Japan. Seizure 13(Suppl 1):S2–S4

Shank RP, Gardocki JF, Vaught JL, Davis CB, Schupsky JJ, Raffa RB, Dodgson SJ, Nortey SO, Maryanoff BE (1994) Topiramate: preclinical evaluation of structurally novel anticonvulsant. Epilepsia 35:450–460

Shank RP, Gardocki JF, Streeter AJ, Maryanoff BE (2000) An overview of the preclinical aspects of topiramate: pharmacology, pharmacokinetics, and mechanism of action. Epilepsia 41(Suppl 1):S3–S9

Sills GJ, Leach JP, Kilpatrick WS, Fraser CM, Thompson GG, Brodie MJ (2000) Concentration-effect studies with topiramate on selected enzymes and intermediates of the GABA shunt. Epilepsia 41(Suppl 1):S30–S34

Skradski S, White HS (2000) Topiramate blocks kainate-evoked cobalt influx into cultured neurons. Epilepsia 41(Suppl 1):S45–S47

Stringer JL (2000) A comparison of topiramate and acetazolamide on seizure duration and paired-pulse inhibition in the dentate gyrus of the rat. Epilepsy Res 40:147–153

Supuran CT (2008) Carbonic anhydrases: novel therapeutic applications for inhibitors and activators. Nat Rev Drug Discov 7:168–181

Supuran CT, Di Fiore A, De Simone G (2008) Carbonic anhydrase inhibitors as emerging drugs for the treatment of obesity. Expert Opin Emerg Drugs 13(2):383–392

Takaya K, Ogawa Y, Isse N, Okazaki T, Satoh N, Masuzaki H, Mori K, Tamura N, Hosoda K, Nakao K (1996) Molecular cloning of rat leptin receptor isoform complementary DNAs –

identification of a missense mutation in Zucker fatty (fa/fa) rats. Biochem Biophys Res Commun 225(1):75–83

Tanabe M, Sakaue A, Takasu K, Honda M, Ono H (2005) Centrally mediated antihyperalgesic and antiallodynic effects of zonisamide following partial nerve injury in the mouse. Naunyn Schmiedebergs Arch Pharmacol 372:107–114

Taverna S, Sancini G, Mantegazza M, Franceschetti S, Avanzini G (1999) Inhibition of transient and persistent Na^+ current fractions by the new anticonvulsant topiramate. J Pharmacol Exp Ther 288:960–968

Theisen FM, Cichon S, Linden A, Martin M, Remschmidt H, Hebebrand J (2001) Clozapine and weight gain. Am J Psychiatry 158(5):816

Theisen FM, Haberhausen M, Firnges MA, Gregory P, Reinders JH, Remschmidt H, Hebebrand J, Antel J (2007) No evidence for binding of clozapine, olanzapine and/or haloperidol to selected receptors involved in body weight regulation. Pharmacogenomics J 7:275–281

Thone J, Leniger T, Splettstosser F, Wiemann M (2008) Antiepileptic activity of zonisamide on hippocampal CA3 neurons does not depend on carbonic anhydrase inhibition. Epilepsy Res 79:105–111

Tonstad S, Tykarski A, Weissgarten J, Ivleva A, Levy B, Kumar A, Fitchet M (2005) Efficacy and safety of topiramate in the treatment of obese subjects with essential hypertension. Am J Cardiol 96:243–251

Toplak H, Hamann H, Moore R, Masson E, Gorska M, Vercruysse F, Sun X, Fitchet M (2007) Efficacy and safety of topiramate in combination with metformin in the treatment of obese subjects with type 2 diabetes: a randomized, double-blind, placebo-controlled study. Int J Obes 31(1):138–146

Ueda Y, Doi T, Tokumaru J, Willmore LJ (2003) Effect of zonisamide on molecular regulation of glutamate and GABA transporter proteins during epileptogenesis in rats with hippocampal seizures. Brain Res Mol Brain Res 116(1–2):1–6

Van Gaal L, Pi-Sunyer X, Després JP, McCarthy C, Scheen A (2008) Efficacy and safety of rimonabant for improvement of multiple cardiometabolic risk factors in overweight/obese patients: pooled 1-year data from the Rimonabant in Obesity (RIO) program. Diabetes Care 31(Suppl 2):S229–S240

Van Passel L, Arif H, Hirsch LJ (2006) Topiramate for the treatment of epilepsy and other nervous system disorders. Expert Rev Neurother 6:19–31

Vasudev K, Macritchie K, Geddes J, Watson S, Young A (2006) Topiramate for acute affective episodes in bipolar disorder. Cochrane Database Syst Rev CD003384

VI-0521(QNEXA®) Briefing document (2010) http://www.fda.gov/downloads/advisorycommittees/committeesmeetingmaterials/drugs/endocrinologicandmetabolicdrugsadvisorycommittee/ucm218821.pdf

Vinik A (2005) Clinical review: use of antiepileptic drugs in the treatment of chronic painful diabetic neuropathy. J Clin Endocrinol Metab 90:4936–4945

Vohora D, Saraogi P, Yazdani MA, Bhowmik M, Khanam R, Pillai KK (2010) Recent advances in adjunctive therapy for epilepsy: focus on sodium channel blockers as third-generation antiepileptic drugs. Drugs Today (Barc) 46(4):265–277

Walker RM, DiFonzo CJ, Barsoum NJ, Smith GS, Macallum GE (1988) Chronic toxicity of the anticonvulsant zonisamide in beagle dogs. Fundam Appl Toxicol 11:333–342

Wallingford NM, Sinnayah P, Bymaster FP, Gadde KM, Krishnan RK, McKinney AA, Landbloom RP, Tollefson GD, Cowley MA (2008) Zonisamide prevents olanzapine-associated hyperphagia, weight gain, and elevated blood glucose in rats. Neuropsychopharmacology 33(12):2922–2933

Wang PW, Yang YS, Chandler RA, Nowakowska C, Alarcon AM, Culver J, Ketter TA (2008) Adjunctive zonisamide for weight loss in euthymic bipolar disorder patients: a pilot study. J Psychiatr Res 42:451–7

Weiergräber M, Henry M, Krieger A, Kamp M, Radhakrishnan K, Hescheler J, Schneider T (2006) Altered seizure susceptibility in mice lacking the Ca(v)2.3 E-type Ca^{2+} channel. Epilepsia 47 (5):839–850

Wenzel RG, Schwarz K, Padiyara RS (2006) Topiramate for migraine prevention. Pharmacotherapy 26:375–387

Wieczorkiewicz-Plaza A, Plaza P, Maciejewski R, Czuczwar M, Przesmycki K (2004) Effect of topiramate on mechanical allodynia in neuropathic pain model in rats. Pol J Pharmacol 56:275–278

Wilding J, Van Gaal L, Rissanen A, Vercruysse F, Fitchet M (2004) A randomized double-blind placebo-controlled study of the long-term efficacy and safety of topiramate in the treatment of obese subjects. Int J Obes Relat Metab Disord 28:1399–1410

Wilkes JJ, Nelson E, Osborne M, Demarest KT, Olefsky JM (2005a) Topiramate is an insulin-sensitizing compound in vivo with direct effects on adipocytes in female ZDF rats. Am J Physiol Endocrinol Metab 288:E617–E624

Wilkes JJ, Nguyen MT, Bandyopadhyay GK, Nelson E, Olefsky JM (2005b) Topiramate treatment causes skeletal muscle insulin sensitization and increased Acrp30 secretion in high-fat-fed male Wistar rats. Am J Physiol Endocrinol Metab 289:E1015–E1022

Yagi K (2004) Overview of Japanese experience-controlled and uncontrolled trials. Seizure 13 (Suppl 1):S11–S15; discussion S16

Yamamura S, Hamaguchi T, Ohoyama K, Sugiura Y, Suzuki D, Kanehara S, Nakagawa M, Motomura E, Matsumoto T, Tanii H, Shiroyama T, Okada M (2009) Topiramate and zonisamide prevent paradoxical intoxication induced by carbamazepine and phenytoin. Epilepsy Res 84(2–3):172–186

York DA, Singer L, Thomas S, Bray GA (2000) Effect of topiramate on body weight and body composition of osborne-mendel rats fed a high-fat diet: alterations in hormones, neuropeptide, and uncoupling-protein mRNAs. Nutrition 16(10):967–975

Zullino DF, Krenz S, Besson J (2003) AMPA blockade may be the mechanism underlying the efficacy of topiramate in PTSD. J Clin Psychiatry 64:219–220

Index

H.-G. Joost (ed.), *Appetite Control*, Handbook of Experimental Pharmacology 209,
DOI 10.1007/978-3-642-24716-3, © Springer-Verlag Berlin Heidelberg 2012